Advances in Agricultural Machinery and Technologies

Advances in Agricultural Machinery and Technologies

Edited by
Guangnan Chen

CRC Press
Taylor & Francis Group
Boca Raton London New York

CRC Press is an imprint of the
Taylor & Francis Group, an **informa** business

CRC Press
Taylor & Francis Group
6000 Broken Sound Parkway NW, Suite 300
Boca Raton, FL 33487-2742

First issued in paperback 2021

© 2018 by Taylor & Francis Group, LLC
CRC Press is an imprint of Taylor & Francis Group, an Informa business

ISBN 13: 978-1-03-209567-7 (pbk)
ISBN 13: 978-1-4987-5412-5 (hbk)

Publisher's Note
The publisher has gone to great lengths to ensure the quality of this reprint but points out that some imperfections in the original copies may be apparent.

Visit the Taylor & Francis Web site at
http://www.taylorandfrancis.com

and the CRC Press Web site at
http://www.crcpress.com

Contents

SECTION I Farm Machinery and Technology

SECTION II Water and Irrigation Engineering

SECTION III Harvesting and Post-Harvest Technology

SECTION IV Computer Modeling

Preface

Machinery is an integral part of modern agriculture and farming systems. Worldwide, the agricultural sector is facing enormous challenges, including the sustainable uses of agricultural lands, energy, and water, and also a changing and more extreme climate. Furthermore, it is predicted that the world population will increase 30% from the present of about 7 billion to 9 billion by 2050, demanding a 50% increase in global food supply. The continuous improvement and innovation in agricultural machinery and technologies is thus essential to meet this challenge.

In the last decade, we have witnessed a huge advance in agricultural machinery and technologies, particularly through the development and applications of automation technologies and also the data and information gathering and analyzing capabilities of various machinery.

This book presents state-of-the-art information on the important innovations in the agricultural and horticultural industries. Different novel technologies and implementation of these technologies to optimize farming processes and food production are reviewed and presented.

This book is divided into four sections, each addressing a specific area of development. Section I describes and discusses the recent development of farm machinery and technology. Section II is focused on water and irrigation engineering. Section III deals with harvesting and post-harvest technology. Section IV is devoted to computer modeling and simulation. The current industry trend is also highlighted in these sections.

The chapters of this book are written by leading researchers who have extensive knowledge and practical experience in their respective fields. I wish to acknowledge their expert contributions here. I also hope that this book will assist all readers who are working in or are associated with the fields of agriculture, agri-food chain, and also technology development and promotion. After all, efficient mechanization and technology are essential and key factors underlying high agriculture productivity, future global food security, and ultimately, human survival and development.

Guangnan Chen (陈光南)
Editor

Acknowledgments

The editor of this book wishes to thank the following reviewers of the individual chapters for their valuable comments that significantly contributed to the quality of the book:

1. Ricardo Abadia Sanchez (Spain)
2. Sadegh Afzalinia (Canada)
3. Ricardo Aliod (Spain)
4. Megha P. Arakeri (India)
5. Avital Bechar (U.S.)
6. John Billingsley (Australia)
7. John Blackwell (Australia)
8. Juan Antonio Rodríguez Díaz (Spain)
9. Callum Eastwood (New Zealand)
10. Hamed Ebrahimian (Iran)
11. Sébastien Fournel (Canada)
12. Paolo Gay (Italy)
13. Fanis Gemtos (Greece)
14. César González Cebollada (Spain)
15. Rafael González Perea (Spain)
16. Saulo Guerra (Brazil)
17. Amir Haghverdi (U.S.)
18. Tamara Jackson (Australia)
19. Tapani Jokiniemi (Finland)
20. Manoj Karkee (U.S.)
21. Richard Koech (Australia)
22. Slawomir Kurpaska (Poland)
23. Aleksander Lisowski (Poland)
24. Gary Marek (U.S.)
25. Bernardo Martin-Gorriz (Spain)
26. John Reidar Mathiassen (Norway)
27. Guillermo Moreda (Spain)
28. Nuria Novas Castellano (Spain)
29. Roberto Oberti (Italy)
30. S. N. Omkar (India)
31. Yasin Osroosh (U.S.)
32. Candido Pomar (Canada)
33. Emma Prime (Australia)
34. Ar. Avinash Kumar Singh (India)
35. Claus Aage Grøn Sørensen (Denmark)
36. Konstantinos Soulis (Greece)
37. Trygve Utstumo (Norway)
38. Paul Van Liedekerke (Belgium)
39. Earl Vories (U.S.)
40. Dariush Zare (Iran)

The editor and authors of the book also wish to thank the staff of the Taylor & Francis Group for their assistance and the excellent typesetting of the manuscript.

About the Editor

Dr. Guangnan Chen is currently an Associate Professor in agricultural engineering at the University of Southern Queensland, Australia. He graduated from the University of Sydney with a PhD degree in 1994. Before joining the University of Southern Queensland in early 2002, he worked for two years as a postdoctoral fellow and more than five years in a private consulting company based in New Zealand.

Dr. Chen teaches and researches in the subjects of agricultural machinery, agricultural materials and post-harvest technologies, agricultural soil mechanics, and sustainable agriculture. He has so far published more than 100 papers in various international journals and conferences, including two edited books and ten invited book chapters. He is currently also the Secretary of the Board of Technical Section IV (Energy in Agriculture), CIGR (Commission Internationale du Genie Rural), which is one of the world's top professional bodies in agricultural and biosystems engineering.

Contributors

G. Adiletta
Faculty of Industrial Engineering
University of Salerno
Salerno, Italy

Akindele F. Alonge
Department of Agricultural and Food
  Engineering
Faculty of Engineering
University of Uyo
Uyo, Nigeria

C. Amiama
Department of Crop Production and
  Engineering Projects
Higher Polytechnic Engineering School
Universidade de Santiago de Compostela
Campus Universitario
Lugo, Spain

R. Ballesteros
University of Castile–La Mancha (UCLM)
Albacete, Spain

Thomas Banhazi
School of Civil Engineering and Surveying
National Centre for Engineering in Agriculture
Faculty of Health, Engineering and Science
University of Southern Queensland
Toowoomba, Australia

J. Bueno
Higher Polytechnic School Engineering
Universidade de Santiago de Compostela
Campus Universitario
Lugo, Spain

Guangnan Chen
School of Civil Engineering and Surveying
National Centre for Engineering in Agriculture
Faculty of Health, Engineering and Sciences
University of Southern Queensland
Toowoomba, Australia

J. I. Córcoles
University of Castile–La Mancha (UCLM)
Albacete, Spain

Ian Craig
National Centre for Engineering in Agriculture
University of Southern Queensland
Toowoomba, Australia

Josse De Baerdemaeker
KU Leuven Department of Biosystems MeBioS
Kasteelpark Arenberg 30
Leuven, Belgium

Jeff R. Esdaile
Agricultural Consultant
Scone, Australia

Joseph P. Foley
National Centre for Engineering in Agriculture
University of Southern Queensland
Toowoomba, Australia

Malcolm H. Gillies
National Centre for Engineering in Agriculture
University of Southern Queensland
Toowoomba, Australia

Chandima Gomes
Department of Electrical and Electronics
  Engineering
Faculty of Engineering
Universiti Putra Malaysia
Serdang, Malaysia

Felipe Gonzalez
Australian Research Centre for Aerospace
  Automation (ARCAA)
Queensland University of Technology
Brisbane, Australia

Y. T. Gu
Queensland University of Technology
Science and Engineering Faculty
School of Chemistry Physics and Mechanical
    Engineering
Brisbane, Australia

Lisa Guan
University of Technology Sydney
Faculty of Design Architecture and Building
Ultimo, New South Wales, Australia

D. M. C. C. Gunathilake
Fiji National University
Samabula, Fiji

Nigel Hancock
National Centre for Engineering in Agriculture
University of Southern Queensland
Toowoomba, Australia

Marcus Harmes
Open Access College
University of Southern Queensland
Toowoomba, Australia

Norhashila Hashim
Department of Biological and Agricultural
    Engineering
Faculty of Engineering
Universiti Putra, Malaysia
Serdang, Malaysia

Elijah Ikrang
Department of Agricultural and Food
    Engineering
Faculty of Engineering
University of Uyo
Uyo, Nigeria

H. C. P. Karunasena
Department of Chemical and Process
    Engineering
Faculty of Engineering
University of Moratuwa
Moratuwa, Sri Lanka

Alfadhl Y. Khaled
Department of Biological and Agricultural
    Engineering
Faculty of Engineering
Universiti Putra Malaysia
Serdang, Malaysia

Tom Leblicq
KU Leuven Department of Biosystems
    MeBioS
Kasteelpark Arenberg 30
Leuven, Belgium

A. Martínez-Romero
University of Castile–La Mancha (UCLM)
Albacete, Spain

Alison C. McCarthy
National Centre for Engineering in Agriculture
University of Southern Queensland
Toowoomba, Australia

Aaron Mcfadyen
Australian Research Centre for Aerospace
    Automation (ARCAA)
Queensland University of Technology
Brisbane, Australia

John McPhee
Tasmanian Institute of Agriculture
University of Tasmania
Burnie, Australia

Jeffrey P. Mitchell
University of California (Davis)
Kearney Agricultural Center
Parlier, California

M. A. Moreno
University of Castile–La Mancha (UCLM)
Albacete, Spain

Maciej Neugebauer
Faculty of Technical Sciences
University of Warmia and Mazury
Olsztyn, Poland

Kenny Nona
KU Leuven Department of Biosystems
    MeBioS
Kasteelpark Arenberg 30
Leuven, Belgium

Daniel I. Onwude
Department of Biological and Agricultural
    Engineering
Faculty of Engineering
Universiti Putra Malaysia
Serdang, Malaysia

and

Department of Agricultural and Food
    Engineering
Faculty of Engineering
University of Uyo
Uyo, Nigeria

Susan A. O'Shaughnessy
USDA-ARS
Conservation and Production Research
    Laboratory
Bushland, Texas

Hans Henrik Pedersen
Department of Engineering
Aarhus University
Aarhus, Denmark

J. M. Pereira
Higher Polytechnic Engineering School
Universidade de Santiago de Compostela
Campus Universitario
Lugo, Spain

Janusz Piechocki
Faculty of Technical Sciences
University of Warmia and Mazury
Olsztyn, Poland

Santosh K. Pitla
Advanced Machinery Systems (AMS)
    Laboratory
Department of Biological Systems Engineering
207 L.W. Chase Hall
University of Nebraska–Lincoln
Lincoln, Nebraska

Pam Pittaway
National Centre for Engineering in Agriculture
University of Southern Queensland
Toowoomba, Australia

Eduard Puig
Australian Research Centre for Aerospace
    Automation (ARCAA)
Queensland University of Technology
Brisbane, Australia

C. M. Rathnayaka
Queensland University of Technology
Science and Engineering Faculty
School of Chemistry Physics and Mechanical
    Engineering
Brisbane, Australia

and

Department of Chemical and Process
    Engineering
Faculty of Engineering
University of Moratuwa
Moratuwa, Sri Lanka

Wouter Saeys
KU Leuven Department of Biosystems
    MeBioS
Kasteelpark Arenberg 30
Leuven, Belgium

Michael Scobie
National Centre for Engineering in Agriculture
University of Southern Queensland
Toowoomba, Australia

Wiji Senadeera
School of Mechanical and Electrical
    Engineering
University of Southern Queensland
Toowoomba, Australia

D. P. Senanayaka
Institute of Post-Harvest Technology
Anuradhapura, Sri-Lanka

Piotr Sołowiej
Faculty of Technical Sciences
University of Warmia and Mazury
Olsztyn, Poland

Ruixiu Sui
USDA-ARS
Crop Production Systems Research Unit
Stoneville, Mississippi

J. M. Tarjuelo
University of Castile–La Mancha (UCLM)
Albacete, Spain

Matthew Tscharke
National Centre for Engineering
    in Agriculture
University of Southern Queensland
Toowoomba, Australia

Jeff N. Tullberg
Honorary Associate Professor
School of Agriculture and Food Sciences
University of Queensland
St Lucia, Australia

Kerry B. Walsh
Central Queensland University
Rockhampton, Australia

Section I

Farm Machinery and Technology

1 Mechanization of Agricultural Production in Developing Countries

Daniel I. Onwude, Guangnan Chen, Norhashila Hashim, Jeff R. Esdaile, Chandima Gomes, Alfadhl Y. Khaled, Akindele F. Alonge, and Elijah Ikrang

CONTENTS

1.1 INTRODUCTION

Agricultural production involves almost all aspects of cultivation, harvesting, processing, storage, and transportation of crops, animals, food, and fiber. Agricultural production depends heavily on the availability of agricultural inputs such as labor, water, arable land, and other resources (energy, fertilizer, etc.), which are significantly affected by the type and scale of farm practices.

In many developing countries, agricultural production offers significant employment opportunities, food security, and economic development to local people. However, for sustainable agricultural production, efforts must be made to introduce changes in order to increase crop yields, efficiency, and sustainability. According to Akinyemi (2007), this can only be effectively achieved by the application of adequate, mechanized agricultural practices.

The advancement in the mechanization of agricultural production in developing countries in Asia, Africa, and Latin America has been discussed and reported in the past few years (Diao *et al.*, 2014; Deininger and Byerless, 2012). Sims, Helmi, and Keinzle (2016) also gave an excellent summary of the context of agricultural mechanization in sub-Saharan Africa. They analyzed the current challenges being faced, including affordability, availability, lack of farmer skills, constraints within the private sector, and gender issues. Opportunities for improvement by sustainable crop production systems, agricultural mechanization development, and further investment were studied. Progress through more sustainable systems, business models, economic and social advantages, institutional involvement (both in the public and private sector), cooperation, and knowledge sharing were also examined. However, especially on the level of mechanization, many studies only focused on a limited aspect of the subject. Hence, a more in-depth review on this subject becomes indispensable.

This chapter focuses on an evaluation of relevant technology in the mechanization of agricultural production in developing countries, with a view to bridging the gap in the information as presented. First, the levels of agricultural mechanization technology are discussed. Then, mechanization in both large- and small-scale agricultural fields are examined. Opportunities concerning the adoption of specific advanced technology are also evaluated. Finally, case studies of agricultural mechanization are presented, and the challenges of mechanization are discussed. Overall, the main emphasis of this chapter is to focus on the key factors affecting mechanization, with a future outlook to improving agricultural productivity and reducing cost.

1.2   AGRICULTURAL MECHANIZATION

Agricultural mechanization is the application of equipment and machinery, as well as the implementation of farm tools to improve the productivity of farm labor and land, in order to maximize outputs and increase agricultural and food production. Ulusoy (2013) defined agricultural mechanization as the use of machines for agricultural production. In a similar manner, Ulger *et al.* (2011) viewed mechanization as the use of modern agricultural machines in place of traditional tools, equipment, machinery, and facilities.

In practice, agricultural mechanization involves the provision and use of all forms of power sources (manual, animal, and motorized) and engineering technologies to enhance agriculture production (Viegas, 2003; Clarke and Bishop, 2002). These engineering technologies include post-harvest handling methods, storage system, farm structures, erosion control, water management (water resources development, as well as irrigation and drainage), meteorological systems, and the techniques for optimally utilizing the above (Chisango and Ajuruchukwu, 2010; Asoegwu and Asoegwu, 2007). Furthermore, agricultural mechanization also encompasses the design, manufacture, distribution, maintenance, repair, and general utilization of farm tools and machines (FAO, 2013). According to Akdemir (2013), the most commonly used indicators of the level of agricultural mechanization are the instrument/machine weight per tractor (kg/tractor), tool/machine number per tractor, tractor power per cultivated area (kW/ha), number of tractors per cultivated 1000 hectares field (tractors/1000 ha), and cultivated area per tractor (ha/tractor).

In a nutshell, agricultural mechanization minimizes drudgery, which hitherto makes it difficult or rather impossible to achieve or practice effective food and agricultural production. Effective agricultural mechanization can help in maintaining improved competitiveness and low consumer price. This can go beyond the application of tools and power machinery, to the application of automation, control, and robotics (Reid, 2011). In fact, agricultural mechanization was identified as one of the top ten engineering achievements of the twentieth century. Figure 1.1 shows the demand for new and improved agricultural technology, for mechanization has continued to increase in tandem with demand for food and fiber increases, particularly in view of the rural–urban migration among the

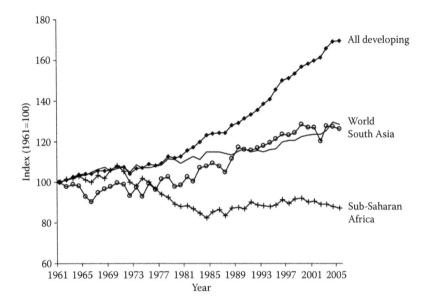

FIGURE 1.1 Index of total agricultural output per capita by region (index 1961–2005). (From Hazell, P. and Wood, S., 2008, *Philosophical Transactions of the Royal Society B: Biological Sciences* 363 (1491), 495–515.)

younger generation that reduces the available labor in many developing countries in Asia, Africa, and Latin America (Cohen, 2006).

1.2.1 LEVELS OF AGRICULTURAL MECHANIZATION TECHNOLOGY

Agricultural mechanization has been in progress for around 100 years. Farmers have previously farmed principally with high manual labor input on small areas using hand-hoe technology, and/or animal traction systems.

In the 1950s, as a result of government policy and various aid programs, schemes were devised by various agencies to introduce Western mechanized agriculture to many areas, particularly in Africa. This involved the import of four-wheel diesel tractors (often with government assistance) and associated tillage implements (mainly inversion plows). These units were then loaned or hired to farmers.

However, in practically all cases these schemes failed, often due to the poor technical knowledge and the educational skills of the local farming community. Tractors and implements were often not maintained correctly and predictable breakdowns occurred. There were also insufficient resources and mechanical skills available, as well as insufficient spare parts to keep these machines operational. These schemes have now been largely abandoned (Sims *et al.*, 2006).

Overall, mechanization in agriculture may be broadly grouped into three levels, i.e., low (manual), medium or fair (animal), and high (mechanical power), with various degrees of sophistication based on the capacity, cost, precision, and effectiveness (Henríquez *et al.*, 2014; Rijk, 1997).

Table 1.1 shows the achievements of three levels of mechanization in selected African, Asian, and Latin American countries in the year 2005. Overall, the number of human and animal power usage in Asian countries was mechanical 65.46%, animal 13.65%, and human 14.11%; while the case was different for countries in Africa, with a percentage average value of 21.38%, 33.79%, and 39.81% for mechanical, animal, and human mechanization levels, respectively. In sub-Saharan Africa, 65% of agriculture is currently carried out by manual labor, 25% by animal traction, and only 10% is mechanized (Esdaile, 2016). Also, the data illustrate 41.5% for mechanical, 14.55% for animal, and 19.95% for human power in Latin America.

TABLE 1.1

Percentage of Power Sources Usage for Farming in Some Asian, African, and Latin American Countries between the Years 1997 and 2005

Continent	Country	Tractors %	Animal %	Human %	Reference
Asia	China	52	–	22	Sims *et al.* (2006)
	India	87.6	10.4	2	Singh (2007)
	Iran	96.48	1.28	2.24	Sharabiani (2008)
	Nepal	23	41	36	Shrestha (2012)
	Turkey	58.7	23.2	6.4	Ozkan *et al.* (2004)
	Oman	75	6	19	Ampratwum *et al.* (2004)
Africa	Mali	0.98	81.89	17	Fonteh (2010)
	Nigeria	10	–	85	Sims *et al.* (2006)
	Zimbabwe	55	–	15	Sims *et al.* (2006)
	Tunisia	66.67	29.63	3.70	Fersi *et al.* (2012)
	Ethiopia	2	85	13	Gebresenbet (1997)
	Kenya	5	15	80	Gebresenbet (1997)
Latin America	Mexico	14.0	–	–	Clara (1997)
	Ecuador	59	32	9	Hetz (2007)
	Brazil	75	10.2	14.8	Jasinski *et al.* (2005)
	Latin America	28	16	56	Gebresenbet (1997)

Figure 1.2 further depicts the variations in the use of mechanical power for different regions in Asia, Africa, and Latin America in the year 2003 (FAO, 2015). From the plot, it can be seen that all Asian and South American regions have a significant number of tractors in use for agricultural production when compared to Africa; where, due to the abundance of manual labor, the use of human power (the first level of mechanization) is still dominating, thus limiting the application of present-day technology in agricultural production. For post-harvest, most countries in Africa also still rely largely on sun drying. The commodities are either spread on suitable surfaces, or hung on farm buildings (Alonge and Onwude, 2013). Significant losses may result from this practice.

Alternatively, within the historical and economic context, some authors have proposed that agricultural mechanization has six stages of evolution (Viegas, 2003; Clarke and Bishop, 2002), while others report seven stages of evolution (Speedman, 1992; Rijk, 1989). The seven stages of agricultural mechanization are:

- Stationary power replacement, where human power is substituted by mechanical power.
- Motive power replacement, where mechanical power replaces operation systems previously based on human power.
- Human control replacement, where operations previously controlled by human decision-making are replaced by mechanized operations.
- Adjusting cropping systems to the mechanization requirements.
- Adjusting farming systems to the mechanization requirements.
- Adjusting plant physics to the requirements of mechanization (plant adaptation).
- Automation, where automation, control, and robotics are applied to agricultural production operations.

The sequence of these stages becomes obvious at the farm level (Onwude *et al.*, 2016). However, this sequence could vary according to the type of agricultural production and farming system used.

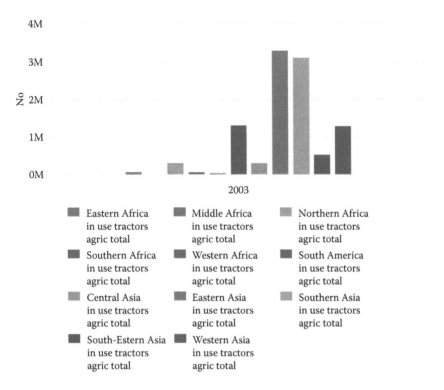

FIGURE 1.2 Variations in the use of tractor for different regions in Asia, Africa, and Latin America in the year 2003. (From FAO, 2015. Food and Agriculture Organization of the United Nations Statistics Division. Fao Stat: Agriculture Data. http://fenix.fao.org/faostat/beta/en/#compare [Accessed August 21, 2016].)

For example, in Asian and African countries, the adoption of labor-intensive rice production increases the demand for labor. This increases the hourly demand for machines which are largely used for land preparation and threshing, particularly in Thailand and the Philippines (Viegas, 2003). The mechanization of agricultural production in the above-mentioned countries has often been associated mainly with rice production (Viegas, 2003).

Similarly, countries that have crops other than rice as the main crops may promote a different degree of mechanization for their production operations, e.g., Malaysia (rubber, cotton, and palm oil), Pakistan (wheat), Sri Lanka (tea and spices), Ghana (Cocoa), Nigeria (oil palm and cassava), and Brazil (cotton) (Diao *et al.*, 2014; Clarke and Bishop, 2002). In addition, the different characteristics associated with the geographical locations and terrain of most countries in Africa, Asia, and Latin America may also necessitate different types of machinery and equipment advancement. These factors could also serve as a limitation to the application of present day technologies in the mechanization of agricultural production.

1.2.2 Large-Scale Agricultural Fields

Large-scale fields are associated with large-scale farming, which takes advantage of a proportionate saving in costs gained by the increased level of production, to produce quality and safe food in large quantities at a relatively low cost. A large-scale field can also be regarded as a large expanse of agricultural land used for the production of both livestock and plants (Bureau, 2012). According to the World Bank, the total large-scale mechanized farms are up to 20 million all over the world (World Bank, 2002).

Food security is the primary concern for all countries of the world (Ncube and Kang'ethe, 2015). This can be more effectively achieved through the applications of modern technology in the mechanization of its production, storage, and distribution process. Consequently, the mechanization of both the large-scale and small-scale agricultural farming system becomes vital (FFTC, 2005). These different agricultural farming systems are also practiced both in the developed and developing countries, although with different levels of sophistication.

According to Deininger and Byerlee, (2012), three main factors have recently contributed to the increase in the practice of large-scale agricultural farming systems in some developing countries, particularly in Southeast Asia, Latin America, and Southern Africa. They are

- New mechanized technologies that make it easier to supervise labor
- The limited availability of labor in some areas, perhaps exacerbated by high wage demands
- More emphasis on integrated supply chains

The recent development of advanced technology in the mechanization of plant breeding, tillage, and on-farm production has also made labor supervision easier, and it can reduce losses in the operations of large-scale agricultural production (Viegas, 2003). Nowadays, large-scale agricultural fields can be more easily managed and controlled due to improved plant and crop varieties that are pest-resistant and herbicide-tolerant, the adoption of conservation farming, the reduced number of steps in the planting process (Clarke and Bishop, 2002), the reduced labor demand, and the use of automated and mechanized machines for harvesting (Suprem et al., 2013; Viegas, 2003).

Furthermore, the remotely sensed information on climate and field conditions can also reduce the application of local and traditional knowledge (first level of mechanization). For example, the ability to use the Global Navigation Satellite System (GNSS) has assisted machinery operations, rather than relying on the driver's skills, which makes labor and its supervision dispensable (Shaw, 1987). However, the application of this type of technology in large-scale farms is generally inadequate in most African, Asian, and Latin American countries.

Figure 1.3a and b shows the available agricultural and arable land in selected African, Asian, and Latin American countries. From the plot, it can be seen that there are large areas of arable land for large-scale production, especially in Africa, where the available agricultural land area is estimated to be above 80% of the total land mass (Nigeria and South Africa). However, less than half of the available agricultural land is currently used for large-scale farming. Moreover, Figure 1.3b shows abundant arable lands in developing countries. From Figure 1.3b we can see that there is still much land available for large-scale farming. Thus, the challenge is the application of mechanization in these countries. Similar studies have been reported in the literature (Rotimi, 2010; Rasouli et al., 2009).

1.2.3 SMALL-SCALE AGRICULTURAL FIELDS

Globally, 70% of farmers are smallholders (<1 ha). Smallholders also manage over 80% of 500 million farms in the world. Small-scale agricultural farming promotes self-sufficiency in food. Normally, small-scale agricultural fields are traditionally associated with the type of farming that provides for the family's food needs. Sometimes it is a mixed farming system, combining crop production with livestock rearing in a way that promotes interdependency (FAO, 2008).

In developing countries, almost 2.8 billion people live with less than $2 per day, most of whom reside in rural areas (Boutayeb and Boutayeb, 2005). About 2.5 billion are practicing agriculture and 1.5 billion are on small-scale farms with an average of 2 ha or less in size (Altieri, 2008). Therefore, small-scale agricultural fields play a very important role in increasing global agricultural production and food availability.

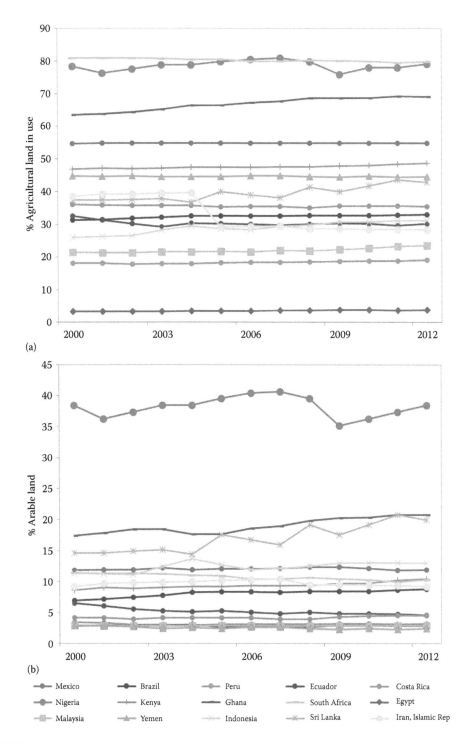

FIGURE 1.3 Available agricultural areas and arable land in selected African and Asian countries.

1.2.3.1 Mechanization in Small-Scale Agricultural Fields

Although this farming system can feed people in urban areas, it is largely considered for rural populations in developing countries (Reijntjes, 2009). In some countries, small-scale farms contribute drastically both to national and export food needs. Rukuni and Eicher (1994) stated that small-scale farming in Zimbabwe contributed almost 60% of national food needs and almost 20% of food exports. Similarly, Altieri (2008) reported that small farms in Latin America produced 77% of beans, 61% of potatoes, and 51% of maize consumed nationally.

Nonetheless, one of the major problems faced by small farmers is the adoption of present-day mechanized technologies to increase their productivity. Rukuni and Eicher (1994) stated that most small-scale farmers in the developing countries have yet to benefit from the research and advanced mechanization in agriculture. Additionally, many of the current education systems do not adequately support the improvement of family farms; rather, they promote industrial agriculture. In other cases, modern technology is not available, either because it is too expensive or because it is not appropriate for the system (Altieri, 2008).

The technological characteristics of small-scale agricultural fields in developing countries could be measured in terms of variables like economies of scale, the technological base, the technological disparity, learning, and labor intensity. Currently, the machinery for small-scale farming is becoming more readily available, which include: tractors, tillage, seeding implements, harvesters, and efficient human-propelled tools. Besides that, small-scale controlled environmental agriculture is also becoming viable. These can control the environmental factors such as light, heat, atmosphere, nutrients, and a much higher production can be obtained per unit of space (Fisseha, 1987).

In areas of the developing world where Western "commercial farmers" have introduced large farming enterprises, mechanization with advanced farm tractors and implements has achieved many successes. However, there have been little "trickle-down" effects for the small subsistence farmers, often due to poor financial resources, limited technical knowledge, and low educational standards.

1.2.3.2 Development and Applications of Two-Wheel Tractors

As a parallel to the development of large-scale agriculture by Western farmers, there has been the development of a range of small farm traction units. This was initially developed in the West, and was followed by development in the Far East, principally Japan after World War II. These have taken the form of what is commonly called the two-wheel tractor. They have also been variously described as "walking tractors," "garden tractors," "power tillers," or "iron buffaloes." These were first introduced in the 1920s and 1930s (Figure 1.4a–c).

These tractors were mainly purchased by small area farmers in Western countries. From the 1930s until the 1950s they were widely used in vegetable gardens and small horticultural enterprises. Often, they came equipped with a rotary tiller unit (rotavator) as the standard tillage implement (Figure 1.5).

During the 1950s and 1960s, the two-wheel tractor focus moved to Japan and various manufacturers (Kubota and Yanmar) moved into production for the local market, with some exports as well.

However, by the 1970s, the Chinese had taken over the technology. The local farm machinery manufacturers (e.g., Dong Feng and Sifang) have now become the major manufacturers of two-wheel tractors. Other major producers are Thailand (Siam Kubota), India (VST Shakti), and Indonesia (Quick).

There are now over 500,000 two-wheel tractors manufactured annually world-wide (Hossain *et al.*, 2009). They come in two main classes: The single-cylinder diesel motor 12–18 hp Asian made units, which have multi-speed gearboxes and are driven by a Vee belt from the motor. These are normally sold with a standard 60 to 80 cm wide rotavator. They have a tare weight of 400 to 500 kg. There are also many types of 6 to 10 hp units with various drive systems, made both in Asia, the United States, and Europe. These have either petrol or diesel motors and are essentially lighter in construction, having a tare weight of 80 to 150 kg.

(a)   (b)

(c)

FIGURE 1.4   1930s Planet Junior Garden tractor, 1940s David Bradley tractor, and Holder two-wheel tractor.

FIGURE 1.5   A 1950s Howard two-wheel tractor equipped with standard rotavator.

Traditionally, two-wheel tractors have often been used in paddy rice production. When fitted with steel cage wheels, they can be an ideal lightweight unit for land preparation, including initial ploughing, land leveling, and puddling before planting of a rice crop by the transplant method. In fact, one of the possible alternative strategies for small farm mechanization in sub-Saharan Africa is the appropriate use of two-wheel tractors as the first step down the mechanization path. A number of African countries like Ethiopia, Ghana, Kenya, and Nigeria are currently studying Bangladesh's experience in agricultural mechanization, which uses two-wheel single-cylinder diesel tractors to power well pumps, river boats, threshers, and mills, as well as producing crops.

Over the last 10 years, the two-wheel tractor technology has been further developed for upland crop production, in addition to the traditional paddy rice system. Several types of farm implements, mainly seeders and planters, are commercially available for these tractors. Farmers with access to the appropriate use of such smaller-horsepower tractors can also operate them with planters that deposit seeds directly into the soil with minimal disturbance, in line with zero tillage or conservation agriculture regimes (Esdaile, 2016). Two-wheel tractors are also a popular mode of transport and farm equipment in a number of developing countries like India and Bangladesh.

1.3 PRESENT-DAY TECHNOLOGY

It has been established that most developing countries currently practice the first and second levels of mechanization, compared to developed countries (Clarke and Bishop, 2002). However, in terms of agricultural mechanization, countries such as Japan, Brazil, Korea, and Egypt could be classified as developed countries because they currently practice a more advanced level of mechanization when compared to their counterparts (Diao *et al.*, 2014). In spite of this, most countries in Africa, Asia, and Latin America are still classified as developing countries (Anelich, 2014; Mondal and Basu, 2009).

Overall, some of the most important present-day technology for the mechanization of large- and small-scale agricultural fields includes precision agriculture, mobile and web applications for agriculture, digitalizing crop varieties and yield, forecasting farm weather and modeling, Geographical Positioning Systems and Geographic Information System (GNSS/GIS) applications, remote sensing, automated tractors and farm equipment, robotics, data mining and warehousing, and Internet of Things (Suprem *et al.*, 2013; Zhang *et al.*, 2002).

Many countries such as the United States, Canada, the Netherlands, England, and Germany have already applied some of these advanced technologies in agricultural production, and have been able to improve their agricultural technology, reducing the total cost of agricultural production, and increasing farm size successfully (Figure 1.6). For example, farmers in these countries can now routinely use portable mobile devices like PDAs to collect and share data and information to interested parties and stakeholders (Suprem *et al.*, 2013). Furthermore, electronic sensors and imaging tools are also used to characterize crop growth and development (Onwude *et al.*, 2016). Remote sensing has now been applied using Unmanned Aerial Vehicles (UAV) (popularly known as "drones") (Everaerts, 2008). Precision agriculture has been around since the 1990s, but it really took off when GPS technology became cheap and ubiquitous in the mid-2000s.

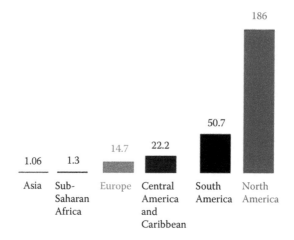

FIGURE 1.6 Agricultural "treadmill" based on present day technological boosts achieved through mechanization, plant breeding for high-yielding varieties, the use of agrochemicals and genetic engineering, etc. (From IFAD, 2010. The International Fund for Agricultural Development. Rural Poverty Report 2000/2001 Fact Sheet: Technology, Natural Resources and Rural Poverty Reduction. (2010). http://www.ifad.org/media/pack /rpr/4.htm.)

Wireless remote sensing is also being increasingly applied to various equipment (Wang *et al.*, 2006). In the future, robotics and automation can play a significant role in meeting the future agricultural production needs, and will revolutionize the way food is grown, tended, and harvested. Research development in this area, although still largely experimental, has received an enormous amount of attention from both the government and private sectors due to robots improving productivity, the scarcity of labor, and practicable design that is easier to handle (Suprem *et al.*, 2013).

Compared with developed countries, there are significant limitations to the application of these present-day technologies in the mechanization of agricultural production units in many parts of Africa, Asia, and Latin America (Clarke and Bishop, 2002; Kishida, 1984). This could be because of the high purchasing cost of advanced technology, amount of unskilled labor, education of farmers, government policies, and high cost of maintenance, among other reasons.

Nonetheless, modern day technology has also been progressively adopted and applied in certain activities in developing countries. For example, recent agricultural projects in South Africa, Egypt, Malaysia, Brazil, Mexico, Thailand, the Philippines, and India utilize satellite positioning systems and geographic information systems to aid in farming management. This technology also helps to select the appropriate type of fertilizer and application method to the soil (Devi *et al.*, 2011).

Furthermore, one of the important operations in agricultural production is harvesting. Currently, this operation is done manually in many parts of Asia, Africa, and Latin America. However, the situation is different for countries like China, Japan, Korea, India, Brazil, and South Africa, where most harvesting activities are now carried out with modern machines (Binswanger, 2014; Singh, 2006; Spoor *et al.*, 2000). Harvesting is actually one of the most labor-intensive types of work in crop production, and mechanization of this activity has greatly improved the agricultural productivity.

1.4   CASE STUDIES

The following case studies concern agricultural mechanization in some developing countries. These include rice production in the Philippines, palm oil production in Malaysia, and the implementation of an agricultural mechanization development program in China.

1.4.1   Rice Production in the Philippines

Rice production in the Philippines is currently carried out using advanced mechanization. The Philippines are considered as one of the world's largest rice exporters, with an average production of 18 million tons of rice in 2013 (Figure 1.7a). From Figure 1.7b, it can be seen that rice production in the Philippines occupies almost 4.8 million hectares out of the 9.5 million hectares of land used for agricultural production. In the Philippines, large-scale agricultural mechanization started in the middle of last century, whereby, tractors of both four-wheel and two-wheel types were imported and applied in large-scale agricultural production (Tuong and Bouman, 2003). Recently, agricultural mechanization has also been applied in small-scale rice production. This includes production processes ranging from land preparation, harvesting, threshing (shelling), drying, and milling of rice.

Ahammed and Herdt (1984) assessed the effects of agricultural mechanization on the production of rice in the Philippines. They reported that agricultural mechanization is a necessary part of the agricultural production process. They further stated that the agricultural production of rice, in order to meet increased demands, requires different levels of capital and labor investment, depending on the technologies used. The authors also studied the indirect and direct production effects, as well as the employment impact of alternative technologies for rice production in the Philippines. The direct impact of adapting new, mechanized rice production technology simply involves the change of production inputs, while the indirect effects can be seen based on the interaction between production and consumption processes.

Ahammed and Herdt (1984) also used the method of a social accounting matrix (SAM) to identify and measure the effects of a series of different technologies for rice production. The three techniques

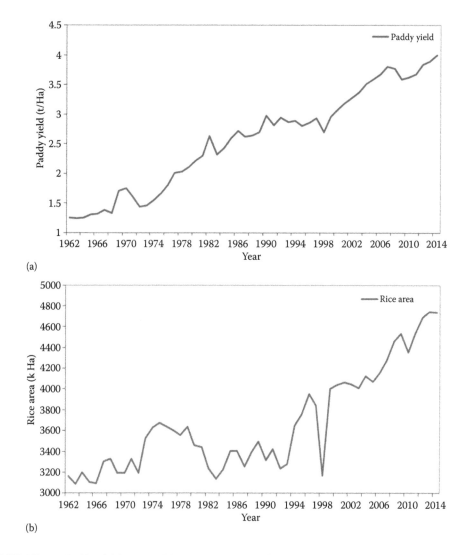

FIGURE 1.7 (a) Paddy yield measured in tonnes per hectare from 1961 to 2014; and (b) rice area measured in hectare from 1961 to 2014 in the Philippines. (Courtesy of Ricepedia in the Philippines.)

used in land preparation are specified in the rice production systems: water buffalo, two-wheel tractor, and four-wheel tractor. The three threshing techniques are manual, portable thresher, and large axial flow thresher.

It was found that the effects of agricultural mechanization resulted in increased total agricultural production with a reduction of labor requirements—with the two-wheel tractors alone this reduction was 15.42 thousand man-years. For the other two options of two-wheel tractors and small threshers, and four-wheel tractors and large threshers, the corresponding reductions were 16.54 thousand man-years, and 19.01 thousand man-years, respectively.

A major challenge in the value chain of rice production for the Philippines is in its postharvest preservation and handling processes. A total of 16.47% grain losses is incurred during the post-harvest activities. The estimated postharvest loss during rice production is between 9–37% (Mopera, 2016). Both drying and milling have the highest recorded losses with 36% and 34%, respectively (PHilMech, 2010; PhilRice, 2010). According to Mopera (2016), inadequate on-farm mechanized drying and processing facilities are some of the reasons for the huge losses during the postharvest value chain. Several researches have been conducted in the area of improving drying efficiency, quality of dried products, and reducing energy cost during drying. The application of a cost-effective advance novel thermal (Onwude *et al.*, 2016) and nonthermal (Onwude *et al.*, 2017) hybrid drying system have gained significant interest. These drying methods could be employed for on-farm drying of rice after harvest in order to reduce losses and increase market value.

1.4.2 Palm Oil Production in Malaysia

In Malaysia, the advancement of agricultural mechanization in palm oil production has led to increased revenue earnings, and relevancy in the global palm oil market. According to Abdullah *et al.* (2001), mechanization of palm oil production in Malaysia involves planting activities, harvesting, storage, processing, and transportation of palm fruits from the farm to the markets. Among these production processes, the most significant contribution of agricultural mechanization was in the harvesting of fresh fruit bunches (FFB). Figure 1.8 shows the comparison between the harvesting of FFB using advanced harvesting aids (Ckat^{TM}) and manual methods (chisel or sickle). From this figure, it can be seen that manual harvesting can only produce 110 FFB hr^{-1}, while Ckat^{TM} produced up to 160 FFB hr^{-1}. This result was further collaborated by Jelani *et al.* (2008). They reported that manual harvesting can only produce 50 to 60 FFB hr^{-1}. Thus, the productivity of using Ckat^{TM} was 45% higher than that of manual harvesting, with a daily productivity between 3.2 and 6.4 tonnes per man per day. Consequently, applying advanced machinery to the production of palm oil has shown to be more efficient than conventional means; increasing productivity and reducing the cost of human labor (Evans *et al.*, 2004).

Similarly, the Cantas^{TM} machine was also used in oil palm plantations for FFB. Table 1.2 shows the differences between Cantas^{TM} and the conventional method. As it can be seen, the application of Cantas^{TM} reduced the labor demand by almost half, while the labor to land ratio was doubled, productivity nearly tripled, and the harvesting costs were reduced by 75%. Trials in many estates produced encouraging results where the average productivity was 14 tons per day, 2.8 per man per day, or an average of 50 to 100 bunches per hectare (cut only). Depending on cropping level and land topography, a team of workers could cover 5 to 10 hectares per day (one cutter, one helper, one tractor driver, and two loaders) (Jelani *et al.*, 2008).

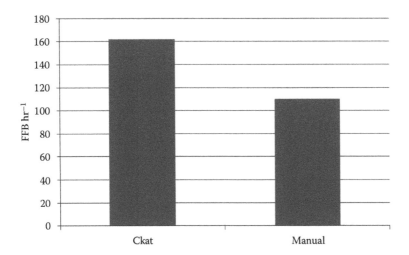

FIGURE 1.8 Performance comparison between the harvesting rates of fresh fruit bunches (FFB per hour) using advanced harvesting aids (Ckat™) and manual methods (chisel or sickle). (From Shuib, A. R. *et al.*, 2011, *Further Advances in Oil Palm Research*, 570–97.)

TABLE 1.2

Cantas™ Compared to the Use of Conventional Harvesting Pole at the Tereh Selatan Estate (Epa, Kluang, Johor)

	Cantas™	Conventional
Total workers (harvester + helper)	8	16
Land: labor (ha)	1:37	1:18
Average productivity (t/team)	11.60	4.19
Harvesting cost (RM t^{-1})	RM 20	RM 33

Source: Jelani, A. R. *et al.*, 2008, *Journal of Oil Palm Research* 20, 548–58.
Note: 626 ton FFB per month for 292 hectares.

In a similar manner, El Pebrian and Yahya (2013) studied the advancement of agricultural mechanization for oil palm FFB transportation in Malaysia. They designed a mini tractor-trailer with a grabber that has a single-chassis 50.5 kW universal prime mover, operated at 2600 rpm, four-wheel drive, and a collection-transportation attachment with a 1.5 payload storage bin. The machine system had an output of 2.526 tons per hour or 20.213 tons per day on sloping terrain and 2.620 tons per hour or 20.965 tons per day on gently undulating terrain. The machine was found to be more efficient, affordable, and easy to maintain when compared to the conventional means of in-field collection-transportation (van and Erreppi buffalo). Furthermore, this machine system presented a good technological solution for in-field collection-transportation of FFB for the oil palm plantation industry in Malaysia.

In addition, the method of collecting and evacuating oil palm bunches from the field to the collecting point affects the quality of the palm fruits (Jahis *et al.*, 2010). FFB can be harvested efficiently and with less damage to the fruits using advanced mechanization. A survey that was carried out in 2008 revealed that 83% of the in-field collection activity and mainline loading activity were mechanized as compared to 1995, which was only 62% mechanized (Halimah *et al.*, 2010).

1.4.3 China's Agricultural Mechanization Development Program

In China, agriculture has made tremendous progress since the start of implementing the reform and open-door policies in 1978. In order to further address the problems of limited resource availability and the widening gap between the urban and rural population, the Chinese government began to implement the "Agricultural Mechanization Promotion Law" in 2004 (Zhou *et al.*, 2009; Adekola *et al.*, 2014). The law established the status and role of agricultural mechanization in China. As a result, a significant amount of government funding has been invested to subsidize farmers to purchase various agricultural machinery, in particular, to promote the machine operations for major crops (such as wheat) and to enhance the role of agricultural engineering research and agricultural machinery application in agriculture and rural economic development. Overall, it was estimated that by the end of 2008, the total power of farm machinery in China has been increased to reach 822 million kW, and the overall mechanization level in agricultural operations of tillage, planting, and harvesting have reached an average of 46% (Gao, 2012). This has further increased to 59% in 2013 (Singh and Zhao, 2016). The machinery power density has also increased from 2.0 kW/ha in 1990 to 5.7 kW/ha in 2013.

Despite these great advancements, China has set a further target to reach the overall agricultural mechanization level of 70% by 2020 (Gao, 2012). To achieve this target, the structure of agricultural machinery industry would need to be optimized. Currently, there is an unbalanced crop mechanization level among different crops and different operations. For example, in 2008, the overall grain crop mechanization level was 87% for wheat, 61% for soybeans, 52% for maize, and 51% for rice; while for cash crops, they were respectively 21% for potatoes, 23% for rape seeds, and less than 10% for vegetables and fruit. The overall mechanization levels for different operations are tillage 76% and planting and harvesting 48% (Singh and Zhao, 2016). The standard of agricultural machinery in China would also need improvement (Yuan, 2005; Xu *et al.*, 2017).

1.5 CHALLENGES OF MECHANIZATION

There are a number of challenges to promote mechanization in developing countries. These could include: technical challenges, requirements of energy and fuels for machinery operations, government policies, and the adopted technology transfer mechanisms.

1.5.1 TECHNICAL CHALLENGES

Over the years, studies on the growth and challenges of mechanization in developing countries have been reported. In many parts of Africa, Asia, and Latin American, small-scale agricultural fields remain at the center of most farming practices. According to Viegas, (2003), this challenge persists largely due to the farmer's choice of alternative approaches to mechanization and technical change. Traditional farmers often make use of available resources according to their knowledge of technology, therefore tend to show skepticism about the adoption of new technologies (French and Schmidt, 1985). In fact, empirical evidence suggests that human capital investment through education and training is essential, and a very good incentive for the adoption of modern technologies in the mechanization of large-scale agricultural production.

For example, the design, development, and manufacturing of agricultural robotic technologies for harvesting crops in agricultural production involves several issues like engineering design, the employment of comprehensive architectural control and fixed technology for measurement and sensing, and the integration of data computing platform with communication systems (Suprem *et al.*, 2013; Blackmore and Griepentrog, 2006; Kyriakopoulos and Loizou, 2006). The integration of sensors into automated and controls system have also been reported as a major challenge (Suprem *et al.*, 2013), because of needed measurements and precision for controlling the rate of operation, and the utilization of an electronic system of operation in a tough agricultural field (Seelan *et al.*, 2003).

Land degradation has been reported as one of the challenges in the mechanization of agricultural fields (Fonteh, 2010). This problem can be found in the "hotspot" regions of Asia, such as the foothills of the Himalayas; sloping areas in Southern China; Southeast Asia; the Andes; forest margins in East Asia and the Amazon; rangelands in Africa and West and Central Asia; and in the Sahel. Such places have high concentrations of rural poor and ethnic minorities. Improved land management technologies can thus be applied to maintain the quality of the natural resources, which is often needed in mechanized agricultural production. Examples include: range management to reduce overgrazing; soil organic matter restoration through composts; animal-crop rotational grazing; crop rotation; agroforestry and fallowing systems; land reclamation; and earth or vegetative bunds against erosion.

1.5.2 REQUIREMENTS OF ENERGY AND FUELS FOR MACHINERY OPERATIONS

Due to problems in the availability of petroleum products in many developing countries, especially due to the recent fall in the price of crude oil which has led to reduced global production, there is a corresponding reduction in the use of fuel-dependent equipment for farming.

According to Jain and Sharma (2010), the increasing industrialization and modernization of many developing countries caused increased demand of petroleum products. They reported that economic development in these countries has led to a huge increase in the energy demand. In India, the energy demand is increasing at a rate of 6.5% per annum (Jain and Sharma, 2010). However, petroleum-based products are limited, due to a heavy dependence on the importation of 80% of the total required crude oil. This has led to the country's focus on alternative fuels, which can be produced from feedstocks available within the country. Biodiesel, an eco-friendly and renewable fuel substitute for diesel, has been getting the attention of researchers and scientists all over the world. This can be seen in various studies as reported in the scientific literature (Solaimuthu et al., 2015; Bietresato and Friso, 2014; Janaun and Ellis, 2010; Sarantopoulos et al., 2009; Barnwal and Sharma, 2005; Dorado et al., 2003).

Apart from the many vegetable oils that can be used to fuel diesel engines, which unfortunately induce some potential technical problems to the engines, recent studies have demonstrated the potentiality of using biodiesel derived from these oils (environmental friendliness, easiness of production, no need for adaptations to the existing engines, normal wear of metallic components). Overall, it is suggested that biodiesels can present a genuine opportunity as the future of renewable fuel for agricultural and other machinery, in terms of eliminating the further depletion of fossil fuels and to providing a significant reduction in greenhouse gas emissions (Chen et al., 2016).

1.5.3 IMPACT OF GOVERNMENT POLICIES

Another limiting factor to the adoption of the present-day technologies for agricultural production in developing countries is the government policies (Onwude et al., 2016). These policies include financial aid, importation, standard procedures to training programs, and a land tenure system. In Nigeria and Mali, for example, the government is the primary importer of farm machinery and it on-sells to farmers at subsidized prices (Fonteh, 2010; PrOpCom, 2011). Similarly, the Tanzanian government has sold more than 5000 sets of imported advanced agricultural machinery at subsidized prices since 2009 (Lyimo, 2011). Rijk (1989) reviewed the growth of mechanization in developing an Asian countries, and recommended the use of computer software (MECHMOD) for developing effective mechanization policy that will depend on data from economics of use of mechanization levels for different field operations.

However, despite the effort from governments to encourage mechanized agricultural production in most African countries, the distribution of the machinery to farmers there has been reported to be ineffective.

In Ghana, the approach to mechanization focuses on mechanized agricultural production of selected commodities as a business, based on their comparative and economical advantage (Diao et al., 2014). Agricultural mechanization in Ghana is presently carried out under the national development process plan, even without the adoption of any formal strategy (Fonteh, 2010). Some limitations in this approach have already been highlighted by Diao et al. (2014). These included that some principal stakeholders have been left out from the planning and implementation stages, especially large-scale farmers.

Furthermore, in Asia, various governments presently support the advancement of farm mechanization through policies on financing farm machinery advancement and research, as well as subsidies to farmers at half the purchase price (Mondal and Basu, 2009). In Thailand, India, Indonesia, Sri Lanka, Malaysia, Vietnam, and the Philippines, their governments have been responsible in supporting special projects for research, training, and programs for farm mechanization (Viegas, 2003; Deininger and Byerlee, 2012; Mondal and Basu, 2009). However, the overall level of mechanization is still medium-to-low because of factors such as inadequate resources, dilapidated infrastructure and institutional arrangements, dominance of manual operations, and lack of policies that encourage and help the general economic well-being of the different stakeholders in the agricultural mechanization and manufacturing industries. Thus, the government policies should be appropriately designed so that these policies serve their intended purposes. It was reported that, in some countries, a significant proportion of tractors actually go to transportation, because agricultural tractors in these countries have a reduced import tariff as opposed to trucks that can have tariffs over 100%. Furthermore, no corrupt government officer should be allowed to distort and make money out of these subsidy schemes.

1.5.4 TECHNOLOGY TRANSFER MECHANISMS

There are two broad modes of technology adoption in agricultural mechanization technologies. The first are the "drop-in" technologies which can be easily adopted or modified (Schueller, 2016). For this method, developing countries may either adopt or try to scale-up and out with small mechanization equipment used in developed countries, or they can scale-down existing equipment and concepts with some necessary modifications.

The second mode of technology adoption is for the "system" technologies, which may take longer to be adopted because many changes and supporting infrastructure may be required. One of the examples of this method may be some of the precision agriculture technologies being researched today.

FFTC (2005) classified the limitations of the application of present day technologies to the agricultural mechanization as technological constraints, sociocultural and behavioral limitations, financial and economic challenges, and environmental issues. Based on contemporary research findings, Viegas (2003) highlighted four key factors limiting the application of the present-day technologies in developing countries as follows: (1) technology compatibility with the environments, (2) availability of resources to expedite the adoption of technology, (3) suitability of the technology to deal with the needs and yearnings of the target population, and (4) appropriateness of the technology transfer mechanism.

In order for a technology to be adopted, the first three factors have to be met and channeled through an efficient transfer process (Francks, 1996). Thus, the level, adoption, and subsequent use of present day technology as mechanized agricultural inputs have a direct and significant effect on land productivity, cost reduction, production profitability, and eventually the quality of life.

Generally, for traditional farmers to adopt modern technology, they need to first improve or change the current farm practice. Second, the accessibility and affordability of the modern technology transfer mechanism must be within the reach of the farmer.

Detailed analysis of the results (Mottaleb *et al.*, 2016) showed that machinery ownership is positively associated with household assets, credit availability, electrification, and road density. Donors and policy makers should therefore not focus only on short-term projects to boost machinery adoption. Rather, sustained emphasis on improving the physical and civil infrastructure and services, as well as assuring credit availability, are also necessary to create an enabling environment in which the adoption of scale-appropriate farm machinery can occur.

1.6 FUTURE OUTLOOK

FAO (2014) predicts that there will be over nine billion people globally by 2050. In order to feed them, agricultural production should increase by 60 to 70% by 2050. The prospects for mechanization in the Asian, African, and Latin American regions is based on the projections of high economy growth rate on income per capital indices and given conditions of political stability (Clarke and Bishop, 2002). The process of mechanization can therefore be facilitated by the development of local manufacturing capacity in the region. For example, India is becoming the world's largest manufacturer of tractors, while China is a major source of affordable tractors and power tillers.

In Africa, the process of urbanization will also stimulate a switch to a higher mechanization level (Onwude *et al.*, 2016). The movement of rural populations to urban areas will see a switch to advanced mechanization of agricultural fields due to shortage in the available labor input in the rural areas (Viegas, 2003). This trend could adversely improve the rural economy.

Furthermore, the future design of advanced farm machinery should take into consideration the nature of operations and the need to utilize the use of effective control solution devices (sensor, actuator, drive, switch, etc.). Although literature has been widely published on the approach to designing individual control systems for simple, lab-based automation machines (Mondal and Basu, 2009; Blackmore and Griepentrog, 2006; Kyriakopoulos and Loizou, 2006; Cox, 2002), there has been little improvement in tackling problems associated with complex agricultural robotic machines. Overall, for wide adoptions, machines should be low-powered, multi-purpose, precise, compact, light, and affordable. Locally-available materials should be incorporated in fabricating machines to reduce the manufacturing costs. Lastly, the overall recommendations on factors to be considered for efficient and effective application of present technology to the mechanization of agricultural production of countries in Africa, Asia, and Latin America are highlighted as follows:

- The needs of farmers (both large- and small-scale) should be met by manufacturing automated tractors, power tillers, and other large farm equipment locally. More so, the operator's safety and comfort must also be considered.
- Energy efficient machines should be developed by harnessing unconventional sources of energy. This is because the cost of fossil fuel in many African, Asian, and Latin American countries is very high.
- Information and communication technology through multimedia, fairs, and exhibitions should be actively engaged in strategic locations where agricultural mechanization programs will be carried out.
- Training local craftsmen in manufacturing technology, machine operation, and repair and maintenance that would promote local agricultural machinery manufacturing should be encouraged.
- Small groups, organizations, or cooperatives for farmers can also be harnessed, particularly in setting up joint use of farm machinery and other modern farm facilities in agricultural production.
- Government agricultural policies should be private sector–driven and should have a direct impact on the farmers and stakeholders in agricultural production.
- Government–private sector partnership in advocating agricultural mechanization should be embarked upon. Service centers should be established in rural areas. Financial assistance and subsidies should also be provided to machinery owners and stake holders of large- and small-scale agricultural fields.

1.7 CONCLUSION

Agriculture is a main source of income, employment, and livelihood of a significant proportion of the populations of developing countries. Agricultural mechanization has now been in progress for several decades. However, it has been mainly confined to developed countries and a small number of developing countries. Some of the developing nations have had little or no progress in this area of agriculture production until recent times. Particularly, in sub-Saharan Africa, where 65% of agriculture is still carried out by manual labor, 25% by animal traction, and only 10% is mechanized (Esdaile, 2016). This is compared with the rapidly improved situation in countries like China, Sri Lanka, and Cambodia.

Mechanization of agricultural production requires the applications of modern technologies. These technologies, however, are generally associated with relatively well-developed economies or large-scale farms. The application of these technologies in many developing countries in Africa, Asia, and Latin America are limited by factors such as the technology's compatibility with the environment, availability of resources to facilitate the technology's adoption, cost of the technology, government policies, adequacy of the technology, and appropriateness in addressing the needs of the population. As a result, many of the available resources have been inadequately used by farmers. This has led to low productivity and high cost of production.

The chapter has emphasized that the success of agricultural mechanization will require a clear institutional framework, and also a coherent strategy based on the actual needs and priorities of the farmers. To increase the level of mechanization, all basic farm machinery requirements must be met, such as: suitability to farm type and scales; simple engineering design and technology; affordability of technology in terms of cost to farmers; versatility for use in different on-farm operations; and, significantly, the provision of support services including repairing facilities from the government, private sector, and manufacturers. With good implementation, the overall mechanization level in agricultural operations of tillage, planting, and harvesting in China has now reached over 59% (Singh and Zhao, 2016). The machinery power density has also increased from 2.0 kW/ha in 1990 to 5.7 kW/ha in 2013. Thus, the model of China's experience may be consulted and adapted to advance agricultural mechanization in other countries (Adekola *et al.*, 2014).

As another example, the use of mechanized agricultural equipment in Cambodia has doubled in the past five years and over 90% of farming land preparation in that country is now done by machinery instead of draft animals (Cheng Sokhorng, 2017). Overall, the use of agricultural machinery is rapidly increasing in Cambodia and most farming has transformed from manual labor or cattle-driven equipment to machinery. Almost every household has a two-wheeled tractor now for their daily activities in the field, while for larger jobs, member farmers can hire the cooperative's single tractor for use. As a result, mechanization in Cambodia plays an important role in furthering the productivity of farming.

It has been highlighted in this chapter that cost and also local recommendation are the two most important factors in any mechanization scheme. This is because not all farming activities can have acceptable cost and can be assisted by machinery. Therefore, before implementing a mechanization scheme, it is important to seek the input of local farmers and identify the agricultural tasks that are considered critical and strenuous to workers so that they can be prioritized to be assisted by machines (Nawi *et al.*, 2012). In many countries, low-cost, two-wheeled tractors could be the most basic step one in farm mechanization. While many Asian fields are often small, Africa has usually more space around the fields. Currently, smallholders manage over 80% of 500 million farms in the world.

This chapter has also reviewed the applications of some of advanced mechanization technologies. Precision agriculture which combines the use of information and technology will further promote agricultural productivity. Research and development in post-harvest technology and the mechanization of vegetable and fruit production is also very important (Bowman, 2015).

Overall, it can be concluded that proper mechanization in agriculture is one of the most important factors underlying high productivity. Proper mechanization and its appropriate use is necessary for economic, environmental, and social sustainability, and is an effective strategy to achieve food security. Translating and adapting technical knowledge to local applications should consider local and regional resources, both physical and human, as well as cultural acceptability. Appropriate technology, rather than advanced technology, should be promoted as a first priority. In some African, Asian, and Latin American countries, the first step may even be to have farmers progress from the hand hoe and bullock to a simple form of mechanization. Facilitating a local manufacturing, repairing, and adapting capacity is also a very important part of a successful mechanization strategy.

ACKNOWLEDGMENTS

The authors would like to thank the Department of Biological and Agricultural Engineering and the Department of Electrical and Electronics Engineering at Universiti Putra Malaysia, and the Department of Agricultural and Food Engineering at University of Uyo, Nigeria, for the contributions and facilities rendered in making this work a success.

REFERENCES

Abdullah, M. Z., Guan, L. C., and Azemi, B. M. (2001). Stepwise discriminant analysis for color grading of oil palm using machine vision system. *Food and Bioproducts Processing*, 79 (4), 223–31.

Adekola, K. A., Alabadan, B. A., and Akinyemi. T. A. (2014). China agricultural mechanization development experience for developing countries. *International Journal of Agricultural Innovations and Research*, 3 (2), 654–8.

Ahammed, C. S., and Herdt, R. W. (1984). Measuring the impact of consumption linkages on the employment effects of mechanisation in Philippine rice production. *The Journal of Development Studies*, 20 (2), 242–55.

Akdemir, B. (2013). Agricultural mechanisation in Turkey. *IERI Procedia* 5, 41–44.

Akinyemi, O. M. (2007). *Agricultural Production: Organic and Conventional Systems* (1st ed.). New Hampshire: Enfield, NH. Science Pub.

Alonge A. F., and Onwude D. I. (2013). Estimation of slar radiation for crop drying in Uyo, Nigeria using a mathematical model. *Advance Material Research* 824: 420–28.

Altieri, M. (2008). Small farms as a planetary ecological asset: Five key reasons why we should support the revitalization of small farming in the global South. Kuala Lumpur: Third World Network. Available from: http://twn.my/title/end/pdf/end07.pdf [Accessed 26 November 2011].

Ampratwum, D. B., Dorvlo, A. S. S., and Opara, L. U. (2004). Usage of tractors and field machinery in agriculture in Oman. *Agricultural Engineering International: CIGR Journal of Scientific Research and Development* 6: 1–9.

Anelich, L. E. (2014). African perspectives on the need for global harmonization of food safety regulations. *Journal of the Science of Food and Agriculture* 94 (10), 1919–21.

Asoegwu, S. N., and Asoegwu, A. O. (2007). An overview of agricultural mechanization and its environmental management in Nigeria. *Agricultural Engineering International: CIGR Journal* 9, 1–22.

Barnwal, B. K., and Sharma, M. P. (2005). Prospects of biodiesel production from vegetable oils in India. *Renewable and Sustainable Energy Reviews* 9 (4), 363–78.

Bietresato, M., and Friso, D. (2014). Durability test on an agricultural tractor engine fueled with pure biodiesel (B100). *Turkish Journal of Agriculture and Forestry* 38 (2), 214–23.

Binswanger, H. (2014). Agricultural mechanisation: A comparative historical perspective. *World Bank Res. Obs.* 1, 27–56.

Blackmore, B. S., and Griepentrog, H. W. (2006). Mechatronics and applications. In: *CIGR Handbook of Agricultural Engineering* (Ed. A. Munack), 204–15.

Boutayeb, A., and Boutayeb, S. (2005). The Burden of non-communicable diseases in developing countries. *International Journal for Equity in Health* 4 (1), 1–8.

Bowman, J. E. (2015). Role of postharvest loss reduction in USAID's Feed the Future Initiative. First International Congress on Postharvest Loss Prevention, Rome, Italy.

Bureau, N. (2012). Large scale farms. *National Sample Census of Agriculture* 4, 1–8. The National Bureau of Statistics and the Office of the Chief Government Statistician, Zanzibar.

Chen, G., Maraseni, T. N., Bundschuh, J., Banhazi, T., Antille, D. L. and Bowtell, L. (2016). Agriculture, Energy, and Global Food Security. Engineering and Technology Innovation for Global Food Security, An ASABE Global Initiative Conference, 24–7, Stellenbosch, South Africa.

Cheng, Sokhorng. (2017). Farmers weeding out drudgery with mechanized equipment. *Phnom Penh Post*. http://www.phnompenhpost.com/business/farmers-weeding-out-drudgery-mechanized-equipment.

Chisango, F. F. T., and Obi, A. (2010). Efficiency Effects Zimbabwe's Agricultural Mechanisation and Fast Track Land Reform Program: A Stochastic Frontier Approach. Presented at the Joint Third African Association of Agricultural Economists (AAAE) and 48th Agricultural Economists Association of South Africa (AEASA) Conference, Cape Town, South Africa.

Clara, I. N., Altieri, M. A. (1997). Conventional agricultural development models and the persistence of the pesticide treadmill in Latin America. *International Journal of Sustainable Development & World Ecology* 4, 93–111.

Clarke, L., and Bishop, C. (2002). Farm power–Present and future availability in developing countries. *Agricultural Engineering International: CIGR Journal* 4, 1–19.

Cohen, B. (2006). Urbanization in developing countries: Current trends, future projections, and key challenges for sustainability. *Technology in Society* 28 (1), 63–80.

Cox, S. (2002). Information technology: The global key to precision agriculture and sustainability. *Computers and Electronics in Agriculture* 36 (2), 93–111.

Deininger, K., and Byerlee, D. (2012). The rise of large farms in land abundant countries: Do they have a future? *World Development* 40 (4), 701–14.

Devi, D. A., Malakondaiah, K., and Babu, M. S. (2011). Measurement of potassium levels in the soil using embedded system based soil analyzer. *International Journal of Innovative Technology & Creative Engineering* 1 (1), 1–5.

Diao, X., Cossar, F., Houssou, N., and Kolavalli, S. (2014). Mechanization in Ghana: Emerging demand, and the search for alternative supply models. *Food Policy* 48, 168–81.

Dorado, M. P., Ballesteros, E., Arnal, J. M., Gomez, J., and Lopez, F. J. (2003). Exhaust emissions from a diesel engine fueled with transesterified waste olive oil. *Fuel* 82 (11), 1311–5.

El Pebrian, D., and Yahya, A. (2003). Design and development of a prototype trailed type oil palm seedling transplanter. *Journal of Oil Palm Research* 15 (1), 32–40.

El Pebrian, D., and Yahya, A. (2013). Mechanized system for in-field oil palm fresh fruit bunches collection-transportation. *AMA, Agricultural Mechanization in Asia, Africa and Latin America* 44 (2), 7–14.

Esdaile, R. (2016). Current and future ideas for small farm mechanisation in sub-Saharan Africa. *Two-Wheel Tractor Newsletter.*

Evans, P. J., Miniaci, A., and Hurtig, M. B. (2004). Manual punch versus power harvesting of osteochondral grafts. *Arthroscopy: The Journal of Arthroscopic & Related Surgery* 20 (3), 306–10.

Everaerts, J. (2008). The use of unmanned aerial vehicles (UAVS) for remote sensing and mapping. *The International Archives of the Photogrammetry, Remote Sensing and Spatial Information Sciences, 1VII (Part B1),* 1187–92.

FAO. (2008). Boosting Food Production in Africa's "Breadbasket Areas"—New Collaboration among Rome-Based UN Agencies and AGRA. [Online]. URL: http://www.fao.org/newsroom/en/news/2009/1000855/index.html. [Accessed May 10, 2009].

FAO. (2013). Food and Agriculture Organization of the United Nations Statistics Division. Fao Stat: Rome, Italy. <http://faostat3.fao.org/home/E> [Accessed May 10, 2013].

FAO. (2014) Walking the nexus talk: Assessing the water-energy-food nexus in the context of the Sustainable Energy for All Initiative. Food and Agriculture Organization of the United Nations, Rome, Italy. www.fao.org/3/a-i3959e.pdf [Accessed October 2016].

FAO. (2015). Food and Agriculture Organization of the United Nations Statistics Division. Fao Stat: Agriculture Data. <http://fenix.fao.org/faostat/beta/en/#compare > [Accessed August 21, 2016].

Fersi, S., Chtourou, N., and Bazin, D. (2012). Energy analysis and potentials of biodiesel production from Jatropha Curcas in Tunisia. *International Journal of Global Energy Issues* 35 (6), 441–55.

FFTC. (2005). Small farm mechanisation systems development, adoption, and utilization. FFTC Annual Report, Food and Fertilizer Technology Center, Taipei, Taiwan.

Fisseha, Y. (1987). Small-scale forest based processing enterprises. *Forestry Paper*, 79.

Fonteh, M. F. (2010). Agricultural mechanisation in Mali and Ghana: Strategies, experiences and lessons for sustained impacts. Agricultural and Food Engineering Working Document, FAO, Rome 2010, 1–47. http://www.fao.org/fileadmin/user_upload/ags/publications/K7325e.pdf [Accessed November 26, 2016].

Francks, P. (1996). Mechanizing small-scale rice cultivation in an industrializing economy: The development of the power-tiller in prewar Japan. *World Dev.* 24, 781–91.

French, E. C., and Schmidt, D. L. (1985). Appropriate technology: An important first step. *Sustainable Agriculture and Integrated Farming Systems* (Eds. T. Edens, C. Fridgen, and S. Battenfield) 262–67, East Lansing.

Gao, H. (2012). China Country Paper: Agricultural Mechanization Development in China.

Gebresenbet, G., and Kaumbutho, P. G. (1997). Comparative analysis of the field performances of a reversible animal-drawn prototype and conventional moldboard plows pulled by a single donkey. *Soil and Tillage Research* 40 (3), 169–83.

Ghazali, K. H., Razali, S., Mustafa, M. M., and Hussain, A. (2008). Machine Vision System for Automatic Weeding Strategy in Oil Palm Plantation Using Image Filtering Technique. Information and Communication Technologies: From Theory to Applications, 2008. ICTTA 2008. Third International Conference on 1–5. IEEE.

Halimah, M., Zulkifli, H., Vijaya, S., Tan, Y. A., Puah, C. W., Choo, Y. M. (2010). Life cycle assessment for oil palm fresh fruit bunch production from continued land use for oil palm planted on mineral soil (Part 2). *Journal of Oil Palm Research* 22, 887–94.

Hazell, P., and Wood, S. (2008). Drivers of change in global agriculture. *Philosophical Transactions of the Royal Society B: Biological Sciences* 363 (1491), 495–515.

Henríquez, C., Córdova, A., Almonacid, S., and Saavedra, J. (2014). Kinetic modeling of phenolic compound degradation during drum-drying of apple peel by-products. *Journal of Food Engineering* 143, 146–53.

Hetz, E. J. (2007). Evaluation of the agricultural tractor park of Ecuador. *Agricultural Mechanization in Asia, Africa and Latin America* 38 (3), 60–66.

Hossain, I., Esdaile, R. J., Bell, R., Holland, C., Haque, E., Sayre, K., and Alam, M. (2009). Actual challenges: Developing low-cost, no-till seeding technologies for heavy residues; small-scale no-till seeders for two-wheel tractors. *Proc Fourth World Congress on Conservation Agriculture, Delhi* 4–7, 171–77.

IFAD. (2010). The International Fund for Agricultural Development. Rural Poverty Report 2000/2001 Fact Sheet: Technology, Natural Resources and Rural Poverty Reduction. (2010). http://www.ifad.org/media/pack/rpr/4.htm.

Jahis, S., Deraman, M. S., and Jelani, A. R. (2010). Mechanical loader for in-field FFB evacuation–crabbie. In Palm Mech 2010: Geared for Full Throttle: Proceedings of the 4th National Seminar on Oil Palm Mechanization, 274. Malaysian Palm Oil Board.

Jain, S., and Sharma, M. P. (2010). Prospects of biodiesel from Jatropha in India: A review. *Renewable and Sustainable Energy Reviews* 14 (2), 763–71.

Janaun, J., and Ellis, N. (2010). Perspectives on biodiesel as a sustainable fuel. *Renewable and Sustainable Energy Reviews* 14 (4), 1312–20.

Jasinski, E., Morton, D., DeFries, R., Shimabukuro, Y., Anderson, L., and Hansen, M. (2005). Physical landscape correlates of the expansion of mechanized agriculture in Mato Grosso, Brazil. *Earth Interactions* 9 (16), 1–18.

Jelani, A. R., Hitam, A., Jamak, J., Noor, M., Gono, Y., and Ariffin, O. (2008). Cantas™—A tool for the efficient harvesting of oil palm fresh fruit bunches. *Journal of Oil Palm Research* 20, 548–58.

Kishida, Y. (1984). Farm tractors: A question of scale. *AMA (Agricultural Mechanization in Asia, Africa, and Latin America)* 15: 9.

Kyriakopoulos, K. J., and Loizou, S. G. (2006). Robotics: Fundamentals and prospects. *Mechatronics and Applications, in CIGR Handbook of Agricultural Engineering* (ed. A. Munack), 93–107.

Leng, T. (2002). Mechanisation in oil palm plantations: Achievement and challenges. *Malaysian Oil Science and Technology* 22, 70–7.

Lyimo, M. (2011). Country presentation on Agricultural Mechanisation in Tanzania. Workshop on boosting agricultural mechanisation in rice-based systems in sub-Saharan Africa. Saint Louis, Senegal.

Mondal, P., and Basu, M. (2009). Adoption of precision agriculture technologies in India and in some developing countries: Scope, present status and strategies. *Progress in Natural Science* 19 (6), 659–66.

Mopera, L. E. (2016). Food loss in the food value chain: The Philippine agriculture scenario. *Journal of Development in Sustainable Agriculture*, 16, 8–16.

Mottaleb, K. A., Krupnik, T. J., and Erenstein, O. (2016). Factors associated with small-scale agricultural machinery adoption in Bangladesh: Census findings. *Journal of Rural Studies* 46, 155–68.

Nawi, N. M., Yahya, A., Chen, G., Bockari-Geva, S. M., and Maraseni, T. N. (2012). Human energy expenditure in lowland rice cultivation in malaysia. *Journal of Agricultural Safety and Health* 18 (1), 45–56.

Ncube, N., and Kang'ethe, S. M. (2015). Pitting the state of food security against some millennium development goals in a few countries of the developing world. *Journal of Human Ecology* 49 (3), 293–300.

Nicholls, C. I., and Altieri, M. A. (1997). Conventional agricultural development models and the persistence of the pesticide treadmill in Latin America. *The International Journal of Sustainable Development & World Ecology* 4 (2), 93–111.

Onwude, D. I., Abdulstter, R., Gomes, C., and Hashim, N. (2016). Mechanisation of large-scale agricultural fields in developing countries—A review. *Journal of the Science of Food and Agriculture* 96 (12), 3969–76.

Onwude, D. I., Hashim, N., and Chen, G. (2016). Recent advances of novel thermal combined hot air drying of agricultural crops. *Trends in Food Science & Technology*, 57, 132–145.

Onwude, D. I., Hashim, N., Janius, R., Abdan, K., Chen, G., and Oladejo, A. O. (2017). Non-thermal hybrid drying of fruits and vegetables: A review of current technologies. *Innovative Food Science and Emerging Technologies*.

Ozkan, B., Akcaoz, H., and Fert, C. (2004). Energy input–output analysis in Turkish agriculture. *Renewable Energy* 29 (1), 39–51.

PHilMECH (Philippine Center for Postharvest Development and Mechanization), (2010). Postharvest Losses. http://www.philmech.gov.ph/?page=phlossinfo accessed August 15, 2016.

PhilRICE (Philippine Rice Research Institute). (2010). Annual Report—Philippine Rice Research Institute. http://www. philrice.gov.ph/?s=annual+report+2010 accessed August 13, 2016.

PrOpCom. (2011). *Making Tractor Markets Work for the Poor in Nigeria: A PrOpCom Case Study*.

Rasouli, F., Sadighi, H., and Minaei, S. (2009). Factors affecting agricultural mechanisation: A case study on sunflower seed farms in iran. *Journal of Agricultural Science and Technology* 11, 39–48.

Reid, J. (2011). Agriculture and information technology. *The Bridge: National Academy of Engineering*, 22–29.

Reijntjes, C. (2009). Small-scale farmers: The key to preserving diversity. *LEISA* 25:1.

Rijk, A. G. (1989). Agricultural mechanisation policy and strategy. The case of Thailand. *Asian Productivity Organization*.

Rijk, A. G. (1997). Agricultural mechanisation strategy. *unapcaem.org*.

Rotimi, A. O. (2010). Measurement of agricultural mechanisation index and analysis of agricultural productivity of some farm settlements in Southwest Nigeria. *Agricultural Engineering International: CIGR* 12 (1), 125–34.

Rukuni, M. and Eicher, K. (1994). Zimbabwe's agricultural revolution. Harare: University of Zimbabwe Publication Office (xii).

Salokhe, V., and Ramalingam, N. (1998). Agricultural mechanisation in South and Southeast Asia. *International Conference of the Philippines Society of Agricultural Engineers*.

Sarantopoulos, I., Che, F., Tsoutsos, T., Bakirtzoglou, V., Azangue, W., Bienvenue, D., and Ndipen, F. M. (2009). An evaluation of a small-scale biodiesel production technology: Case study of Mango'o village, Center province, Cameroon. *Physics and Chemistry of the Earth, Parts A/B/C* 34 (1), 55–8.

Schueller, J. K. (2000). In the service of abundance: Agricultural mechanization provided the nourishment for the 20th century's extraordinary growth. *Mechanical Engineering*.

Schueller, J. K. (2016). Role of mechanization and precision agriculture in food availability in the context of smallholder farms. Engineering and Technology Innovation for Global Food Security, An ASABE Global Initiative Conference, 24–27. October 2016, Stellenbosch, South Africa.

Seelan, S. K., Laguette, S., Casady, G. M., and Seielstad, G. A. (2003). Remote sensing applications for precision agriculture: A learning community approach. *Remote Sensing of Environment* 88 (1), 157–69.

Sharabiani, V. R. (2008). The situation of agricultural mechanization in Sarab City, Iran. *Agricultural Mechanization in Asia, Africa, and Latin America* 39 (2), 57–63.

Shaw, A. B. (1987). Approaches to agricultural technology adoption and consequences of adoption in the third world: A critical review. *Geoforum* 18 (1), 1–19.

Shrestha, S. (2012). Status of agricultural mechanization in Nepal. *United Nations Asian and Pacific Center for Agricultural Engineering and Machinery (UNAPCAEM)*.

Shuib, A. R., Khalid, M. R., and Deraman, M. S. (2011). Innovation and technologies for oil palm mechanization. *Further Advances in Oil Palm Research*, 570–97.

Sims, B. G., Josef, K., Roberto, C., and Wall, G. (2006). Addressing the challenges facing agricultural mechanisation input supply and farm product processing. Agricultural and Food Engineering Technical Report Proceedings: CIGR World Congress on Agricultural Engineering.

Sims, B. G., Kienzle, J., Hilmi, M. (2016). Agricultural mechanization: A key input for sub-Saharan African smallholders. Food and Agriculture Organization of the United Nations, Rome, Italy.

Singh, G. (2006). Estimation of a mechanisation index and its impact on production and economic factors— A case study in India. *Biosystems Engineering* 93 (1), 99–106.

Singh, G. and Zhao, B. (2016). Agricultural mechanization situation in Asia and the Pacific region, *Agricultural Mechanization in Asia, Africa, & Latin America* 47 (2), 15–25.

Singh, S. (2007). Hill agricultural mechanization in Himachal Pradesh: A case study in two selected districts. *Agricultural Mechanization in Asia, Africa & Latin America* 38 (4), 18–25.

Sistler, F. (2003). Robotics and intelligent machines in agriculture. *IEEE Robot. J. Autom. Soc.* 3, 3–6.

Solaimuthu, C., Ganesan, V., Senthilkumar, D., and Ramasamy, K. K. (2015). Emission reduction studies of a biodiesel engine using EGR and SCR for agriculture operations in developing countries. *Applied Energy* 138, 91–8.

Speedman, B. (1992). Changes in agriculture; challenges for education in agricultural engineering. Agricultural Engineering and Rural Development Conference 2, 12–14 (Pergamon-CNPIEC Joint Publication).

Spoor, G., Carillon, R., Bournas, L., and Brown, E. H. (2000). The impact of mechanisation. *Land Transformation in Agriculture* (Ed. M. G. Wolman and F. G. A. Fournier), 133–52. Wiley, Chichester.

Suprem, A., Mahalik, N., and Kim, K. (2013). A review on the application of technology systems, standards and interfaces for agriculture and the food sector. *Computer Standards & Interfaces* 35 (4), 355–64.

Tuong, T. P., and Bouman, B. A. M. (2003). Rice production in water-scarce environments. (Ed. J. W. Kijne, R. Barker, and D. J. Molden), *Water Productivity in Agriculture: Limits and Opportunities for Improvement* 1, 53–67, Wallingford: Cabi.

Ulger, P., Guzel., E., Kayisoglu, B., Eker B., Akdemir, B., Pinar, Y., Bayan Y. *et al.* (2011). *Principles of Agricultural Machines (Tarim Makinalari IIkeleri).* (Third edition).

Ulusoy, E. (2013). Agricultural mechanisation in Turkey. *IERI Procedia* 5, 41–4.

Viegas, E. (2003). Agricultural mechanization: Managing technology change. *Agriculture: New Directions for a New Nation East Timor (Timor-Leste)* 113, 32–44.

Wang, N., Zhang, N., and Wang, M. (2006). Wireless sensors in agriculture and the food industry—Recent developments and future perspectives. *Computers and Electronics in Agriculture* 50 (1), 1–14.

World Bank. (2002). Globalization, growth and poverty. Washington, DC: World Bank.

Xu, Y., Li, J., and Wan, J. (2017). Agriculture and crop science in China: Innovation and sustainability. *The Crop Journal* 5 (2): 95–99.

Yahya, Z., Mohammed, A. T., Harun, M. H., and Shuib, A. R. (2012). Oil palm adaptation to compacted alluvial soil (typic endoaquepts) in Malaysia. *Journal of Oil Palm Research* 24, 1533–41.

Yang, Z., Chen, G., Duan, J., Peng, T., and Wang, J. (2009). Development Strategy of Agricultural Machinery Based on Energy-Saving in China, Conference Proceedings, 2009 CIGR International Symposium of the Australian Society for Engineering in Agriculture—Agricultural Technologies in a Changing Climate, 13–16, Brisbane, Queensland.

Yuan, J. The Status of China's Agricultural Machinery Industry and the Prospects for International Cooperation. *Agricultural Engineering International: CIGR Journal of Scientific Research and Development.* Invited Overview Paper. Vol. VII. Presented at the Club of Bologna meeting, November 12, 2004. March, 2005.

Zhang, N., Wang, M., and Wang, N. (2002). Precision agriculture—A worldwide overview. *Computers and Electronics in Agriculture* 36 (2), 113–32.

2 Developments in Mechanization Technology
Controlled Traffic Farming

Jeff N. Tullberg

CONTENTS

2.1 INTRODUCTION

"Plants grow better in soft soil." To subsistence farmers or modern gardeners observing the effect of human and animal footprints on moist soil, this statement is obvious. It accounts for the history of traditional "bed" or "zone" farming designed to avoid surface and root zone compaction in manual cropping systems in China (Chi and Zou, 1988). Raised bed systems have been associated with the Aztec and Mayan civilizations of South America, and similar ideas were part of the rationale for the nineteenth century development of cable-drawn implements.

Concerns were first expressed about the production of "plow-pans" (compacted layers immediately below plowing depth) in animal-powered farming, but the mass and power of First World farm equipment has increased by a factor of more than 20 in the past century. Because most cropping systems use equipment with a variety of different operating widths, field traffic patterns are effectively random, even when guidance systems are used. Some growers have claimed that this random element is a positive contribution to achieving uniformity of soil conditions within fields, but they ignore the costs of what is effectively random and unmanaged compaction. Half a century ago, Arndt and Rose (1966) observed that excessive traffic necessitates excessive tillage, but their message was not heeded. At that time, tillage was still seen as an essential element of cropping, combining the functions of weed control and crop residue management, and compaction could be dealt with by deeper tillage. In many parts of Europe and North America, increased depth of tillage exposed new organic matter, released more nutrients and increased yields, so this process looked completely reasonable at a time when increasing tillage depth was made possible by increasing tractor power and decreasing power costs.

The economic and environmental problems of tillage were not widely understood until the later part of the twentieth century, with the development and large-scale adoption of no-till or minimum till in many regions of the world. No-till systems provide no mechanical disturbance to relieve compaction beneath seeding depth, but instead rely on natural amelioration and the structural resilience of undisturbed root channels and earthworm burrows to maintain soil porosity. These natural mechanisms might be effective in some soils and cropping systems using relatively small-scale equipment. They are demonstrably not adequate to maintain soil productivity in systems using tractors and seeding equipment with a mass of 10–20 Mg, and harvesters with a mass of between 20 Mg (unloaded) and 30–40 Mg (fully loaded).

Compaction does ameliorate naturally in some soil types—particularly in the "self-mulching" cracking clay vertosols, where—in the absence of traffic—the combination of moisture-driven shrinking and swelling, crop root activity, and soil biota largely restored soil structure to greater than 250 mm depth within three cropping cycles over less than two years (McHugh et al., 2007). On the other hand, amelioration processes driven solely by root activity and soil biota are slower, and—in some soils—extremely slow or absent. They are rarely capable of compensating for the impacts of annual cycles of uncontrolled heavy machinery traffic. Tillage at the right moisture content is still the only rapid way to loosen compacted soil, and subsoilers have sometimes been advertised (with no apparent irony) as the ideal "no tillage" implements. When increasing axle loads produce compaction at depths too great for conventional subsoiling (Bennett et al., 2015), the solution must lie in traffic management rather than tillage management.

2.1.1   Controlled Traffic Farming—The Evolution of CTF

Compaction is usually a product of machinery or animal management, the corollary of which is that compaction can be avoided, or confined to defined areas, by management. Controlled traffic farming achieves this by restricting all heavy field traffic to permanent traffic* lanes whose primary function is trafficability—a function which is clearly improved by compaction. The corollary of "plants grow better in soft soil" must surely be "and wheels work better on roads." This important aspect was recognized in the earliest discussions of controlled traffic systems by Cooper et al. (1969), and Taylor (1982; 1989) who recognized the value of minimizing the area occupied by permanent traffic lanes, and the importance of wider traffic lane spacing. Controlled traffic also requires modular track gauge widths and implement widths.

Recognition of these factors, and the obvious incompatibility between the tractor and grain harvesters' track gauges, resulted in an early focus on the advantages of the wide-span field "gantry tractor" as the base unit of farm mechanization (Figure 2.1). Much effort has since been expended on gantry tractor development in the United States and Europe (e.g., Carter et al., 1965; Chamen et al., 1992; Monroe and Burt, 1989). Research on traffic impact continued in the United Kingdom, Holland, Germany, and Australia, but the anticipated large-scale development of commercial gantry farming systems did not eventuate, and on-farm applications were rare.

This work nevertheless demonstrated the major soil condition and crop production benefits of controlling traffic (Dickson et al., 1992). One early outcome was the "tramlining" of post-planting operations in cereal crops, which, in Australia (Adem and Tisdall, 1984), demonstrated the value of the "permanent bed" cropping systems practiced by forward-looking growers. The energy effects of

---

* The terms "wheel," "wheeled," and "wheeling" are used here synonymously with "traffic," "trafficked," and "trafficking" and, unless specifically mentioned, assume no distinction between the effects of steel track or rubber belt traction systems.

FIGURE 2.1   "Dowler" 12 m span gantry tractor equipped with seeder tines at the University of Queensland Gatton campus in 1990.

controlled traffic were confirmed by Tullberg (1988), who noted a number of growers already controlling traffic and pointed to other advantages. This work also drew attention to the opportunity to provide economic controlled traffic systems using conventional tractors, modified to provide a track gauge of about 3 m to match that of grain harvesters. Large scale adoption started in central Queensland with a participatory research, development, and extension program in the 1990s (Yule *et al.*, 2000). This demonstrated substantial productivity improvements when controlled traffic was combined with no-till and increased cropping intensity. It also emphasized the soil conservation benefits of "downslope" field layouts designed for surface water management and suggested the label "Controlled Traffic Farming" or CTF for this set of practices. Parallel developments occurred in "raised bed" farming, aimed primarily at water management in the high rainfall areas of Victoria (Anderson *et al.*, 1999; Wightman *et al.*, 2005), Western Australia (Hamilton *et al.*, 2005), and the irrigation areas of Southern New South Wales. Growers and scientists who observed this quickly appreciated the value of maintaining these as permanent beds, and, where possible, confining traffic to the intervening furrows.

This work and controlled traffic research occurring elsewhere in Australia (Blackwell *et al.*, 1995; Sedaghatpour *et al.*, 1995) was brought together with the experience of innovative farmers and widely publicized by a first Controlled Traffic Conference (Yule and Tullberg, 1995). Rapid adoption in the grain industries was facilitated by the advent of "2 cm" precision RTK "GPS autosteer" for farm equipment in the 1990s, which overcame the guidance issues of CTF. The other major problem of incompatibility between tractors and harvesters was largely overcome with the development of 3 m tractor modifications by small manufacturers, and subsequently by production of tractors capable of working on a 3 m track gauge (Figure 2.2). Recent surveys (GRDC, 2015) indicate that less than 20% of Australian grain production is in CTF systems, and adoption is continuing to grow. The CTF approach used in Australia has been effective in Western Canada (Laroque, 2013; Gamache, 2013), but will not apply everywhere,* but the success in achieving large-scale adoption explains this chapter's focus on Australian research, development, and on-farm application.

---

* In many densely settled areas, traffic regulations severely restrict 3 m track gauge equipment on public roads.

FIGURE 2.2   Standard unequal-wheeled four-wheel-drive tractor modified to 3 m operation (John Deere).

2.1.2 CTF Organizations

A number of controlled traffic farming conferences were organized by ad hoc groups before the Australian Controlled Traffic Farming Association (ACTFA) was formed in 2006. ACTFA has been a CTF advocate and organizer of subsequent conferences, and has led some significant CTF projects. The ACTFA website provides access to the papers from all nine controlled traffic farming conferences held in Australia to date (http://actfa.net/actfa-conferences/). CTF Europe was formed in 2007 with similar objectives, and its website provides case studies of CTF systems operating in a number of countries. It has facilitated a number of research projects and provides an active program of visits to farms operating CTF systems. CTF Europe organized the second international CTF conference in Prague in 2013, papers from which can be found at Smart Agri-Systems (http://www.smartagriplatform.com/CTF2015Presentations). Similarly, Controlled Traffic Farming Alberta (http://www.controlledtrafficfarming.org/) has facilitated research and extension in Western Canada, and organized one conference.

2.2   MINIMIZING OR AVOIDING SOIL COMPACTION

Soil compaction is a major factor in land degradation and the topic of an extensive collection of literature (Hamza and Anderson, 2005; Batey, 2009). While compacted layers are a natural feature of some soils, compaction is increasingly seen as a product of current management leading to the structural disruption and increase in density that occurs when soil stress exceeds soil strength. Compaction reduces productivity by a number of mechanisms, largely associated with reduced permeability to air, water, and crop roots. The severity of compaction produced by any given soil stress is a complex function of soil bulk density, texture, and moisture content, and is also influenced by organic matter content and root density. These parameters vary down the soil profile, depending on cropping history and rainfall. The stresses causing compaction (such as farm machinery or livestock) are produced by loads applied at or near the surface, and are a function of the mass supported and stress distribution across the area supporting it. This is an extremely complex situation, but some generalizations can be made:

- Compaction caused by livestock is generally confined to a surface of 50–100 mm.
- Susceptibility to compaction is greater when soil is wet.
- Soil stress is usually greatest close to the surface, and diminishes with depth.
- Depth of compaction caused by vehicles is proportional to axle weight.
- Surface compaction is proportional to contact pressure.
- The first heavy trafficking traffic event does most of the compaction damage.
- The impact of subsequent similar trafficking is variable, but limited.
- Sometimes subsoil can be protected from compaction by shallower compacted layers.
- "Healthy" soils, with high levels of organic matter, are rather more resistant to compaction.

The vulnerability of soil to compaction damage is a complex subject, reviewed from a number of soil impact perspectives in Soane and van Ouwerkerk (1994). Clearly, the potential for serious compaction increases with moisture content, but crop production always requires crop establishment (seeding and planting) which must usually be carried out when the soil is moist and vulnerable to compaction. Similarly, spraying and fertilizing must often be completed shortly after rainfall, and dry soil is not even assured for grain harvesting. Growers are sometimes advised to minimize compaction damage by delaying operations until soil is dry, but most field operations must be completed within a restricted "window of operation" dictated by crop development, soil, and weather. Delayed operation incurs significant economic penalties, as recognized in machinery management publications (ASABE, 2011). Greater equipment capacity can of course reduce operating time and economic penalties due to delay, but it usually entails equipment of greater weight and soil compaction impact. If farmers judge that the cost of soil compaction damage will be less than the immediate crop loss costs of delay, they will usually proceed with operations despite the inevitable soil compaction.

Compaction has recently been considered in the context of other soil stresses by Dang and Moody (2013). These authors note that compaction affects approximately 30% of Australian cropping soils, but also point out that surface evidence of compaction is rare, so it is difficult to detect. They noted compaction-related production losses estimated at $144M in the Murray-Darling basin alone, but other soil problems (e.g., acidity and salinity) are greater constraints to production. Compaction is usually a product of machinery or animal management, the corollary of which is that compaction can be avoided, or confined to defined areas, by management. This chapter will focus largely on machinery management.

Crop production requires the methodical movement of equipment over production areas for seeding, harvesting, etc. With a few exceptions, this means the operative components of equipment must traverse the whole area (this process is known as "wheeling" or "trafficking"), with pneumatic wheels or flexible belts directly affecting a proportion of the field area determined by the ratio of effective working width to the total soil contact width of the traction system tires or belts. Soil impact will be determined by a range of factors including equipment weight, traction system, contact pressures, soil condition, and moisture content (Chamen *et al.*, 1990); but the productivity impact will be proportional to the area actually trafficked. In the literature, cropping system traffic effects have been quantified in terms of parameters such as Field Load Index, Traffic Intensity (Mg.km/ha), and Compaction Risk Factor (Kuipers and van de Zande, 1994). In this approach, consideration of the spatial distribution of traffic (Duttmann, 2013) is rare, and compaction damage is often assumed to be uniform and proportional to some function of load, distance, and soil susceptibility. These parameters might be a useful basis for comparison in random traffic systems, but are largely inappropriate for CTF, where all traffic is restricted to permanent lanes, so additional traffic has little or no effect.

If soil compaction is to be reduced, the obvious solution is to use smaller, lighter equipment, but equipment weight is closely related to capacity and productivity (Chamen *et al.*, 2003). Despite clear evidence of compaction problems in the 1960s (Arndt and Rose, 1966), the economic pressure to increase productivity has resulted in a steady increase in the weight of developed-world farm equipment. It has been suggested that robotics will reverse this trend, citing examples such as use of "swarms" of small robots for weed control. Removing the operator might well allow the substitution of a number of smaller units for large, high-capacity machines, but, within the scope of current technology there are limits to the potential weight reductions for operations where input or output quantities are significant. Without major change to the operating principles of the machines involved, these logistic issues will continue to apply to most harvesting, fertilizing, and seeding tasks, so it is important that all robotic equipment should operate on the one set of permanent traffic lanes.

The limited capacity to reduce equipment mass or avoid operation on vulnerable soil leads to the focus on traction system characteristics. Support of any given load can be looked at in terms of contact pressure and undercarriage design, and contact pressure can be reduced by spreading the equipment weight over a larger area. This is most easily achieved by increasing the effective tire width by fitting larger section widths with dual or triple tires. Increased width is usually a cheaper way to support greater load, and is sometimes the only option where other design constraints limit tire diameter. Greater width can certainly allow reduced pressure and reduced rut depth, but inevitably increases the proportion of land area affected, and can increase compaction at greater depth (Batey, 2009). Unfortunately, tire pressure can rarely be reduced to the point where it results in no loss of crop productivity (Smith et al., 2015), and dual tires at moderate to high pressure are often the standard fitting for the largest tractors and harvesters. This appears to be driven by cost and tire/axle load considerations, rather than issues of soil impact or tractive efficiency.

An increased tire diameter provides a greater load capacity for any given contact pressure, but there are practical constraints on the tire's diameter. A major advantage of greater diameter is the increase in contact patch length, which also provides greater tractive efficiency. Similar effects are achieved by tandem driven wheels (four-wheel drive) and flexible "rubber" belt traction systems. Both of these systems increase load carrying capacity at a given mean contact pressure by increasing effective contact patch length, but peak loads under drive or idler rollers are always greater than the mean ground pressure under the track systems. This effect can be exacerbated by poor weight distribution, but rubber belts commonly support greater loads on a narrower footprint.

Tandem tire or belt systems can rarely reduce ground pressures to levels that cause no soil damage, but studies generally show a small performance advantage to the belt system compared to four-wheel drive traction systems (Bashford and Kocher, 1999). They also restrict the impact to a smaller proportion of field area, which is another important factor affecting the productivity impact of field traffic, and is much easier to assess. In individual farm machine operations, this can be less than 3% of field area in the case of large, wide, self-propelled sprayers. In contrast, equipment units with relatively narrow working widths (e.g., cotton, sugar cane, root crop, or silage harvesters) can wheel more than 50% of the field area in a single operation. The same can be true for heavy tillage. The total area wheeled in the course of producing one crop varies with the number and characteristics of each operation and the extent to which individual wheel pathways overlap previous wheelings, so the total is almost always less than the sum of areas wheeled in each operation.

Several studies of the area wheeled by cropping systems are reported in Soane and Van Ouwerkerk (1994), and the topic has been considered more recently by Kroulik et al. (2009), in the Czech Republic. These studies show that total wheeled area always exceeded 55% and was often greater than 100% of field area in European systems. Despite the much greater width of equipment used in extensive dryland systems, such as those of Australia, unpublished surveys have demonstrated that (without controlled traffic) heavily wheeled areas usually represent about 50% of crop area. The best controlled traffic grain production systems can reduce wheeled area to less than 10%, but 12–15% is still more common. The Australian surveys have used a spreadsheet approach to the evaluation of total wheeled area, an online version of which is available at http://www.ctfcalculator.org.

In summary, it has been shown that:

- Economic considerations dictate the use of heavy farm equipment.
- Practicable low-pressure traffic systems still impose excessive soil loads.
- Delaying operation until soil conditions improve is not generally feasible.

- Unless traffic is controlled, a large proportion of field area is driven over in each cropping cycle, often when soil is moist and vulnerable to compaction.
- Some soils ameliorate naturally over two or three years under the influence of wetting and drying cycles, root activity and soil biota (McHugh *et al.*, 2009), but this is not helpful if they are re-compacted by traffic before full amelioration. In other soils, amelioration is slower or nonexistent, so the whole area of fields in long-term cropping is compacted. Impacts can sometimes be estimated from fence-line studies.

Growers are experimenting with agronomic measures to enhance amelioration (e.g., rotations, "tillage radish"), but in soils that do not ameliorate rapidly the only short-term solution is deep tillage. In systems where annual tillage (e.g., moldboard plow) is conventional practice, tillage-based solutions appear reasonable, but the steady increase in equipment weight ensures increasing compaction in the sub-tillage zone. In these circumstances, it is unsurprising that compaction is seen by many authors as an important, widespread soil degradation problem (Dang and Moody, 2013). A cynic might infer that current systems ensure a continuing market for farm equipment which both creates soil compaction and provides "practical" (short-term) solutions to the problems. The historical reality is that our systems have evolved gradually over many decades of economically effective farm mechanization, before the availability of effective field guidance systems. This approach has now reached its limits, and it is difficult to see any effective alternative to restricting heavy field traffic to permanent lanes—CTF. This can be achieved in permanent bed systems using standard tractors with all wheels operating in tire-width furrows, but current 3 m CTF systems compromise a smaller area of permanent traffic lanes. This area could be further reduced by the development of wide-span farming systems.

2.3 CTF: DEFINITION AND COROLLARIES

Controlled traffic systems restrict all load-bearing field traffic to permanent lanes, allowing most of the field area to be managed as crop beds for optimal plant production, uncompromised by traffic effects. The permanent lanes, occupying the smallest possible area, can then be managed for optimum trafficability. This is the simple rationale for Controlled Traffic Farming (CTF), but there are a number of practical corollaries of sufficient importance to be added to that fundamental idea—for instance, the requirement for precision. Precise guidance has been achieved manually, but in practice high-quality, large-scale CTF is usually associated with the increasingly affordable "precision GPS autosteer." These systems use signals from Global Navigation Satellite Systems (GNSS) with base stations or networks to provide real-time kinematic (RTK) correction and a nominal 25 mm precision for all operations.

Permanent traffic lanes will also be depressed below the level of the cropping beds, and so tend to channel surface runoff. This can be a major problem, such as when a wheel track running diagonally across a slope concentrates large volumes of runoff and initiates significant erosion. With careful layout on the other hand, the permanent traffic lanes of CTF can prevent runoff concentrating from more than one machine width and provide a safe and effective distributed surface drainage system. Optimum layouts also benefit from improved infiltration and surface protection under no-till CTF to reduce runoff volumes. They have been shown to perform well in extreme rainfall events in Central Queensland, and reduce the need for contour banks as erosion control measures. Designed layouts are an important element of effective CTF, and must take account of both topography and logistics (transport of crop inputs and harvest outputs). Discussion on different aspects of this can be found in Rohde and Yule (1998) and Isbister *et al.* (2013).

The fundamental elements of CTF have been defined by The Australian Controlled Traffic Farming Association (ACTFA) as:

1. All machinery has the same or modular working and track gauge width to allow the establishment of permanent traffic lanes.

2. All machinery is capable of precise guidance along the permanent traffic lanes.
3. Farm, paddock, and permanent traffic lane layouts are arranged to optimize drainage and logistics.

ACTFA also notes the following practices that are facilitated by and congruent with CTF:

1. Minimal soil disturbance, preferably zero-till, or (at most) strip-till
2. More intensive cropping frequency and cover crops to maximize biomass production and provide greater residue return to the system
3. Precise management, such as inter-row seeding, and accurate placement of chemicals and fertilizer
4. Spatial monitoring, mapping, and management (e.g., yield mapping and subsequent zonal management, if required) at progressively finer scale within a defined spatial framework (i.e., permanent traffic and crop zones)
5. Accurate and repeatable on-farm research trials based on a defined spatial framework and spatial technologies

Until widespan systems are available, CTF adoption must depend on existing equipment or its modifications. The mismatch between wheel track gauge widths of farm equipment (e.g., between tractors and harvesters) has been a major impediment to greater adoption of CTF. This has been largely overcome in Australian grain production by modification of tractors for operation on a 3 m track gauge width to match harvesters. Matching track gauges is a much more severe problem for grain production in regions and systems where frequent road travel is required, making the frequent use of 3 m track gauge machinery impractical. It should nevertheless have been noted, that most countries have no problems accepting other large, infrequently used equipment on public roads subject to restrictions on timing, lights, signage, and escorts.

Matching track gauges presents a different problem in horticultural operations where rotation demands a range of crops requiring harvest contractors with equipment of varying track gauges and operating widths (McPhee *et al.*, 2013). Sugar cane, forage, and some horticultural crops present a different issue in the need to transport large masses of material from the field. Conceptual solutions for these issues are straightforward in most cases, but practical, commercial solutions are more difficult. They usually involve working with different contractors to coordinate width modules, which in turn requires consultation with different equipment manufacturers. The result is that high-quality CTF is relatively rare in these industries, except where large-scale producers, with complete control over machinery investment decisions, have designed systems to service a defined range of crops.

2.4 TRAFFIC IMPACTS ON SOIL AND CROPS

Proper assessment of the impact of field traffic requires access to a "control" area of non-trafficked soil—a condition that does not exist in most current cropping systems. Comparisons of soil cropped for varying periods and previously non-cropped soil over a range of soil types (Connolly *et al.*, 1997) have demonstrated major differences, but these are an outcome of all cropping-system activities, usually including many years of tillage and depletion of soil organic matter, and nutrient extraction and fertilizer inputs, in addition to machinery traffic. Specific study of traffic impact requires cropping in a controlled traffic situation, after the removal of prior traffic effects to the greatest extent possible by deep tillage. Much of the work reported here used modified conventional farm machinery, and followed a procedure similar to that described by Morrison *et al.* (1990). The alternative, a purpose-designed wide-span gantry, was used by Chamen *et al.* (1992), in the United Kingdom, working with broadly similar objectives, but in a very different environment.

Traffic effects on soil and grain cropping were studied at Gatton, Queensland between 1993 and 2001 in plots on a self-mulching black vertosol, overlying a permeable sand/gravel at 0.6–1.0 m.

The plots were installed on a 6% slope, and bounded by 3 m spaced wheel track furrows enabling pit-mounted tipping bucket units to intercept runoff from each plot. An automatic weather station and logging of the runoff data allowed continuous monitoring of rainfall and runoff. Replicated pairs of wheel-trafficked and CTF (non-wheeled) plots were managed in no-till or with conventional stubble mulch tillage weed control. An annual wheeling treatment was imposed over the full surface area of the trafficked plots by repeated passes of a 5 Mg tractor, leaving overlapping wheel marks. This usually occurred following the first rainfall after winter crop harvest, coinciding with the first "conventional" post-harvest mulch tillage operation. Summer crops were planted on an opportunity basis, with occasional irrigation when necessary for crop survival.

The results of this work are summarized here in terms of wheel traffic effects on infiltration and runoff, crop yield, and soil biological activity but these effects were primarily the outcome of wheeling effects on soil porosity. The impact was clearly illustrated in soil profile images of adjacent wheeled and non-wheeled plots (Figure 2.3).

Over the five years between 1995 and 1999, the overall mean annual runoff from wheeled plots was 57% greater than from CTF (non-wheeled). The magnitude of this effect in wet years is illustrated in Figure 2.4 (reproduced from Li *et al.*, 2007). Large increases in the infiltration rates of CTF, and many indications of improvement in a range of soil parameters were also noted by McPhee *et al.* (2015), in a study of CTF impacts on a red ferrosol in Northern Tasmania. The work was, however, carried out over three years in an intensive vegetable production system which included tillage and when required to produce a range of crops. Soil condition improvements were evident at most times throughout this trial, but they were not always maintained from crop to crop or statistically significant.

Over the eight grain crops produced in the Gatton trial, CTF alone increased overall crop yields by 9.5% and reduced runoff by 36%. No-tillage alone increased yield by 8% and reduced runoff by 16%. CTF combined with no-tillage increased yield by 14.5% and reduced runoff by 47%, all compared with the "conventional" treatment: stubble mulch management with annual traffic.

The data illustrated in Figure 2.4 is from one year with above average rainfall (greater than 1100 mm), demonstrating large and highly significant differences between treatments. In two years of less rainfall, the plots produced no more than 50 mm of runoff, and treatment runoff differences were not significant. It illustrates the importance of year-to-year and spatial variability of high-intensity rainfall, and the impact of extreme events. These might have a ten-year recurrence interval, but they can produce severe soil erosion and deep rills that require significant rectification.

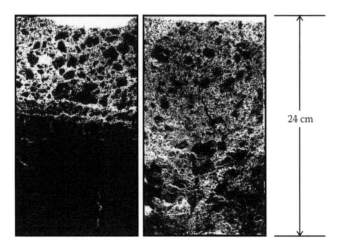

24 cm

FIGURE 2.3 The impact of three-year annual traffic (L) compared with three years of CTF®. Soil profiles from adjacent plots in a clay vertosol. (Courtesy of the late Dr. Des McGarry.)

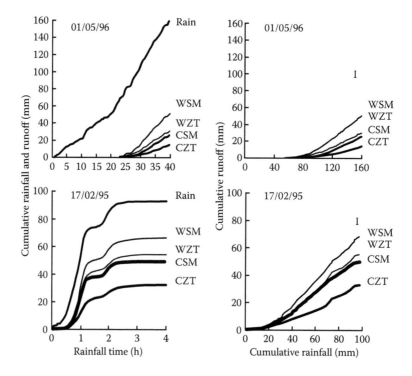

FIGURE 2.4 Examples of traffic and tillage effects on cumulative runoff from dry soil (May 1, 1996) and wet soil (February 17, 1995). Cumulative rainfall and runoff are plotted against rainfall time (left plots), and cumulative runoff against cumulative rainfall (right plots). Vertical bars indicate average standard error of means cumulative runoff (P is less than 0.05). Treatments are wheeled stubble mulch (WSM), wheeled zero tillage (WZT), controlled traffic zero tillage (CZT), and controlled traffic stubble mulch (CSM). (From Li, Y. X. *et al.*, 2007, *Soil & Tillage Research* 97, 282–92.)

Similarly, high levels of variability in traffic compaction effects on yield and runoff can be seen in the results of Rhode and Yule (1998), who also measured soil loss from a vertosol over two years and three crops, in central Queensland and noted the dramatic impact of compaction, and benefits of "down slope" (rather than contour) layouts when controlled traffic was combined with no till.

Soil loss was also monitored in a four-year collaborative study using similar methodology and equipment in the Shanxi province of the People's Republic of China (Wang *et al.*, 2008). This work was not replicated, but consistent results were found over four years of monitoring in a soil typical of the loess plateau (a non-self-mulching Chromic Cambisol described as porous and homogenous to depth), which demonstrated that minimum runoff and soil loss occurred under CTF no-till management in very different conditions. Interestingly, wheeled no-till plots without residue cover produced the greatest runoff and soil loss found in this work, which is consistent with anecdotal reports of similar outcomes in Australia when no-till fallows have been heavily grazed.

Over four years of the Gatton study, the soil moisture profile was monitored in 10 cm increments on a regular basis using a neutron probe unit using semi-permanent access tubes. There were no treatment differences between the moisture contents of wheeled and CTF soil at the drained upper limit (field capacity), but moisture content was significantly lower in CTF (non-wheeled) soil at the lower limit (wilting point) measured at times of severe crop water stress, indicating greater available water capacity in CTF soil. This was confirmed in a concurrent study, on a nearby area of slightly heavier soil that had been degraded by frequent tillage and traffic. The progress of natural amelioration was followed at this site after management changed to CTF-no-tillage during the production of three crops over a period of 22 months. The change in available water capacity at three

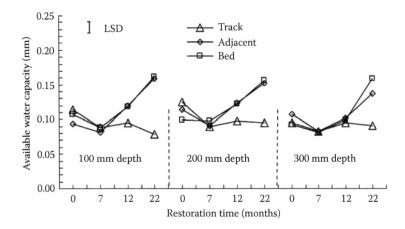

FIGURE 2.5 Available water capacity (mm) per 10 cm of soil profile at various times since the start of CTF management (restoration time), for depths of 100, 200, and 300 mm, in the tracks, adjacent to the tracks and in the beds. The difference between any two seasons is statistically significant when greater than 0.0178 g/cm^3 (LSD; $P < 0.05$).

depths in the crop beds, the zone adjacent to the wheel track, and the wheel track itself is summarized in Figure 2.5 (McHugh, 2009).

A number of different parameters were used in this study to demonstrate the amelioration of this self-mulching vertosol, but the large change in available water capacity (AWC) was the clearest evidence of wheeling effects on dryland cropping potential. Compaction is known to reduce AWC (Lipec *et al.*, 1995), and, in this case, its amelioration was driven largely by crop root activity and the shrink/swell properties of this soil with wetting and drying. For 18 months after the completion of this study, the area was closed off and grew weeds. This was fortuitous, but presented the opportunity to make another set of the same measurements. Interestingly, the available water capacity of the track and unwheeled bed profile was almost unchanged, but a single tractor wheeling on the bed reduced available water capacity to the same level as the track. This illustrated the major compaction damage resulting from a single wheeling of soil in a substantially-ameliorated condition. Schjonning and Rasmussen (1994) have demonstrated that multiple passes on the same path might increase some parameters of compaction, but a single pass can have a major impact on soil productivity.

During amelioration sampling activity at Gatton, McHugh (2009) noted an obvious increase in visible biological activity of all descriptions in the soil surface and subsurface. Detailed monitoring of soil biota was subsequently carried out in the runoff plots by Pangnakorn *et al.* (2003). In this work populations of macrofauna (earthworms), mesofauna (Acarina, Collembola), and microfauna (nematodes) were measured at intervals of approximately six weeks from August 1999 to June 2001. The microflora were monitored over a shorter period. Tillage and traffic effects on earthworm numbers found in this work are illustrated in Figure 2.6. Similar trends were found with mesofauna, but the impact on microfauna and flora was more complex. There is little doubt that better-structured, unwheeled soils are more "healthy" in many respects, as illustrated by authors such as Chamen (2009). Stirling (2008), for instance, has demonstrated the greater nematode suppressive capacity of unwheeled, no-till soils used for cane and vegetable production.

The unwheeled soil of controlled traffic beds emits substantially less nitrous oxide compared to wheeled soil and also absorbs small quantities of methane (Tullberg *et al.*, 2011). Nitrous oxide and methane are powerful greenhouse gasses, and continuing work on this topic has demonstrated the same effect in a wide range of grain production environments. This data shows that that emissions from wheeled soil exceed those from CTF soil by a factor of 2.3 (Tullberg, 2018), and indicates that CTF might be expected to reduce soil emissions by 30 to 50%. Preliminary calculations suggest this

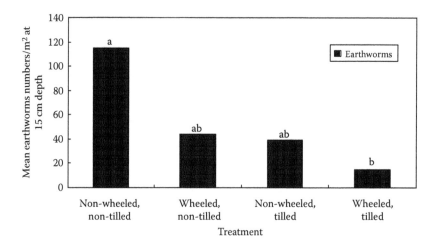

FIGURE 2.6 Tillage and traffic impacts on soil biology: mean earthworm numbers per square metre at 0–15 cm depth of soil under grain cropping subject to varying tillage and traffic treatments for greater than three years. (From Pangnakorn, U. *et al.*, 2003, Effect of Tillage and Traffic on Earthworm Populations in a Vertosol in Southeast Queensland. Proc. ISTRO CD, University of Queensland.)

might reduce annual soil emissions from Australian grain production in the order of 100 kg/ha CO_2-e and denitrification N loss of 2–11 kg/ha (depending on the $N_2/(N_2O\text{-}N)$ loss ratio). Losses were greater in regions of greater growing season rainfall, and much greater losses would be expected in intensive irrigated cropping where soil moisture levels and nitrogen inputs are greater. Antille *et al.* (2015) has explored the literature on this topic in detail, and particularly the potential CTF impact on soil carbon sequestration.

2.5 CTF AND CROPPING ENERGY REQUIREMENTS

Compacted layers can be produced naturally, but this is a slow process, occurring in a relatively small number of generally sandy soils. Comparison with nearby non-cropped soil has demonstrated that soil in cropping fields is more compact than adjacent undisturbed, non-cropped soil, and differences extend well below normal tillage depth (Connolly, 1997). A recent study in the southern Australian Mallee region, for instance, demonstrated mean penetrometer cone index readings at 100–150 mm depth of 0.9 MPa and 2.2 MPa, respectively, in moist sandy loam soils under uncleared native vegetation and adjacent no-till fields. At a 500 mm depth, the same mean readings were 1.25 MPa and 2.3 MPa, respectively (P.D. Fisher, pers. com.). Many factors might contribute to the degradation of cropping soils, but in this situation historical (disc or scarifier) tillage will never have exceeded 100 mm depth, so many years of cumulative traffic energy input to the sub-tillage profile is the most likely explanation of a severely compacted profile from 100 mm downward. The yield cost of this is sometimes apparent when excavation reveals moisture in soil 300 mm beneath a crop that has failed (or failed to yield) from lack of moisture.

The conventional, but energy-intensive solution of deep tillage—when effective—provides only a short-term solution unless traffic is controlled. In some environments, however, deep tillage is ineffective when soil settles again very rapidly, or has negative effects where an unfriendly subsoil layer is mixed with the topsoil. These conditions are not uncommon in the deep sands of Western Australia, where recent research and on-farm demonstration of deep (0.4–0.5 m) organic matter slotting, or even topsoil inclusion in deep tillage slots has produced very encouraging results (Blackwell *et al.*, 2016). In either case, there is little point in such expensive, energy-intensive tillage and soil amendment operations unless they are subsequently managed in CTF.

Williford (1980) has suggested that controlled traffic could be expected to reduce energy requirements, and this effect that has subsequently been quantified in different ways by a number of researchers (Chamen et al., 1996). It is common knowledge that the tines behind tractor wheels need replacing more frequently than others, and Tullberg (2000), working in the heavy clay soils of Queensland, demonstrated an increase in the draft of tractor wheel track tines by a factor of almost two, at tillage depths in the range of 100–150 mm. This work also showed that:

- If 50% of soil area is wheeled in conventional traffic systems, and CTF reduces the wheeled area to less than 15% of uncropped traffic lanes, the draft requirement of soil engaging tasks could be reduced by up to 40%.
- If the improved tractive efficiency of operation on permanent traffic lanes is taken into account, the overall power requirement of such tillage operations could be reduced by about 50%.

The energy aspect of the cost of soil compaction is probably of decreasing relevance in most dryland cropping environments, where tillage-based systems are rapidly declining. Research evidence and practical experience increasingly demonstrates the negative impact of routine tillage, which is expensive in terms of energy and time. Complete elimination of tillage is currently impractical in a number of environments, particularly horticulture, where it is difficult to envisage—for instance—a potato production system that does not involve substantial soil disturbance. Similarly, precise placement of small seeds requires some physical soil preparation, although this could be confined to strip tillage. The advantages of minimizing tillage are nevertheless clear, so, even where some tillage is still necessary, farmers are more likely to question the objectives and consider alternatives. While the development of herbicide-resistant weeds has stimulated some more surface-tillage operations in some dryland broadacre environments, innovative farmers are having considerable success with alternatives such as weed seed windrowing or collection and subsequent destruction, and weed-sensing spray technology.

Where seeding is the only regular soil disturbance, no-till seeder opener design usually attempts to minimize disturbance, reducing weed seed germination and moisture loss. Seeders with near-flat disc openers, or "bio blade" type units, cause very little soil movement. Energy input to disturbance—"tillage"—is therefore small (Ashworth et al., 2010), but this benefit is often compromised by the weight required to ensure disc penetration in hard soils. This weight is usually provided by heavy construction, so when carried on small rigid depth wheels in soft soil conditions, rolling resistance is substantial. Soil disturbance might thus be a relatively small proportion of the power require-ments of disc-type seeders, particularly with commonly used air seeders, where the "aircart" seed and fertilizer distribution system also transports large quantities of seed and fertilizer.

Popular images of agriculture still tend to focus on soil and cultivation, but the only soil dis-turbance involved in most extensive grain production systems in Australia now is only that of seeding, where soil disturbance often represents only a small proportion of the total mechanical energy input. The same is even more true of spraying, fertilizer spreading, and grain handling. Little published information is available on the power requirements of the different operations involved in grain harvesting, but simple calculations based on common values of rolling resistance coefficient and harvester mass suggest that self-propulsion is likely to account for at least 20% of total power requirements, and will often be greater. This means that the cumulative energy input to self-propulsion might well be the largest single component of the mechanical energy inputs to no-till grain cropping systems. Stated differently, the largest single energy input to modern extensive grain production in Australia is likely to be the energy requirement of overcoming rolling resistance—i.e., creating soil compaction.

An analysis of the energy requirements of each component of Australian dryland grain production systems by Tullberg (2013) suggested that the conventional minimum (conservation, non-inverting "stubble mulch") tillage-based Australian grain production system required 52 L/ha for field operations, while the most efficient non-controlled traffic, no-till system required a minimum fuel input of about 26 L/ha. When rolling resistance losses were minimized by controlling traffic, fuel input was reduced to 13–14 L/ha.

If these estimates are reasonable, at least 10 L/ha fuel is dissipated in unnecessary soil compaction where traffic is uncontrolled. Assuming an overall engine thermal efficiency of 40%, this is equivalent to a mechanical energy input of over 150 mJ/ha, or 15 kJ/m^2. These are estimates, based on a number of assumptions; nevertheless this compactive energy input is large compared with the energy input of an Australian chisel plough (4.5–5.5 kJ/m^2; Ward and Rickman, 1988) or with the mechanical energy impact from intense rainfall on unprotected soil (<5 kJ/m^2). There are large differences in uniformity and temporal impact between traffic and rainfall, but both represent compactive energy inputs to the soil, whereas the chisel plough is largely a loosening energy input, but it is interesting to consider them together. This was done by Li *et al.* (2009), who demonstrated a relationship between steady infiltration rate and the energy dissipated by farm equipment wheels, where "wheeling energy" (kilojoules per meter squared) is a function of motion resistance and wheel slip.

Li *et al.* (2009) also showed that when total energy input to the soil (wheeling and rainfall energy) was plotted against infiltration rate, all points appeared to lie within the same relationship ($R^2 > 0.94$), regardless of energy source. R, D, and E effort is often directed to the cost of compaction in terms of yield loss, and sometimes to the cost of deep tillage to rectify compaction. The cost of creating compaction is rarely discussed, but the unnecessary fuel energy input to non-controlled traffic farming systems clearly represents an ongoing and substantial penalty, which is also a major input to soil degradation The additional power requirement to overcome unnecessarily high rolling resistance also requires increased investment in tractor power. Controlled traffic farming substantially reduces fuel costs, and confines compaction to designated areas. Anecdotal reports by Australian CTF farmers indicate an overall reduction in fuel costs by 20 to 40%, and some have reported the replacement of large equal-wheeled four-wheel drives with smaller unequal-wheel (FWA) four-wheel drive tractors.

In summary, controlled traffic farming has been shown to provide direct benefits of:

- Increased rainfall infiltration rates and available water capacity
- Reduced runoff and erosion
- Increased soil biological activity and "soil health"
- Reduced fuel and power requirements
- Improved trafficability and field access
- Improved crop yields
- Reduced emissions of nitrous oxide and smaller N loss

Much of the data used here to demonstrate the benefits of CTF has been drawn from studies in the heavy clay vertosols of Queensland, because this is where large-scale CTF adoption first occurred in the 1990s. There is also substantial research evidence of similar benefits in a wide range of situations in different regions both in Australia and globally.

2.6   SYSTEM EFFECTS OF CTF

Most physical effects of controlling field traffic can be associated with reduced soil compaction and related mechanisms influencing crop yield, but yields do not always automatically increase in the absence of compaction. There are circumstances where no yield benefit could be expected, for instance, when rainfall and fertilizer input corresponds to crop requirements, when historical compaction fails to ameliorate in rigid soil, or where soil performance is constrained by naturally occurring compacted layers. Growers normally expect to see these direct advantages of CTF, but Australian growers often talk of a significant "system effect" producing outcomes "greater than the sum of the parts." Important contributing factors are those associated with trafficability and field access, and those associated with greater precision and uniformity.

Trafficability and Field Access. When permanent traffic lane layouts are well designed, they can provide an excellent system for rapid and safe disposal of surface water. This will overcome waterlogging risks (see raised beds, below), and also ensure that the permanent compacted lanes are drained, and become trafficable more rapidly after rain than would otherwise be the case. This can be extremely important when timeliness is so important to the outcome of most crop operations, and optimal timing is usually influenced by rainfall events.

Obviously, there are occasions when this makes little difference, and other occasions where it is critically important. In general, it is more important in what growers regard as "difficult" seasons. McPhee (1995), for instance, noted a case where crop establishment was possible only in the CTF treatment of a trial intended to compare crop yields under CTF and non-CTF. This is a case where a very large economic benefit might be calculated for that year, but in the absence of a recurrence interval, it is difficult to generalize.

Another example relates to a CTF grower's capacity to take advantage of an opportunity that occurred in a "difficult" harvesting season. In that case, permanent traffic lanes allowed a Queensland grain grower to harvest and plant the next crop, while neighbors were still waiting to start harvesting. The rainfall that delayed the neighbor's harvest thus contributed to the yield of the CTF grower's second crop. In this case, economic benefits would accrue from both the timely harvest (yield/quality losses avoided), and from increased probability of a second crop.

The examples above relate to events that might have recurrence intervals of perhaps 3 to 10 years, but timeliness impacts on spraying operations occur much more frequently. Effective weed management can be a major challenge in no-tillage situations, where the growth stage of weeds and crops can interact with weather to produce narrow windows of fully effective operation. Greater uniformity of crop and weed development (discussed below) also reduces the variables influencing decisions on spray and harvest operation timing.

Greater Uniformity and Precision. CTF also facilitates greater precision in the relationship between crop and machine. RTK GPS autosteer systems have been shown to achieve precision in the order of ±25 mm 68% of the time, but this precision can be substantially degraded by a number of factors, including the transverse movement or "yaw" allowed by tractor tires. Transverse movement of implements relative to tractors can be even greater. Controlled traffic does not eliminate these effects, but they are reduced, and their effects largely negated by the use of consistent field patterns.

The matched implement widths and consistent traffic patterns also provide a more consistent machine/crop relationship. This effect has been demonstrated by Bromet (2006) who found that RTK GPS auto steer could provide ±20 mm precision 96% of the time when following a consistent field pattern with a three-point linkage implement. Its potential value was illustrated when 100% weed control was achieved with small sweep tines in wheat at 20 cm row spacing with less than 15% crop damage. Physical control might be a valuable strategy in dealing with herbicide-tolerant weeds in the future, but a more important current application of precision is in its application to herbicide weed control with interrow (shield) spraying and band spraying. Interrow seeding is probably the most common application of high levels of precision. This involves seed placement for a following crop in a particular relationship to the standing residue of a previous crop. This is often and most easily in the center of the interrow space, but in some cases greater precision is needed to take advantage of reduced water repellence and improved moisture-holding capacity of soil close to previous crop rows.

These effects all contribute to the idea encapsulated in the oft-repeated comments of controlled traffic farmers: "CTF adoption involves thought and effort, but once it's done, it's just an easier way to farm."

Interface with Precision Agriculture. These factors, together with well-designed layouts all contribute to greater uniformity in CTF crops, a characteristic which is usually apparent when comparing yield maps of adjoining CTF and non-CTF fields. Yule (pers. com.) attributed the improvement under CTF to more uniform water availability and the absence of short-term water-logging when layouts were designed for efficient surface drainage. He also noted that the small-scale variability lost from yield maps by kriging* is visible—along with field scale variability—in high-quality satellite imagery, as further evidence of the proposition that much field scale variability was a direct product of management activity, rather than a reflection of soil type.

The input optimization processes of precision agriculture rarely acknowledges that the basic "pixel" of yield maps is an approximate square with one side equal to the harvester cutting width. Current yield maps are unable to recognize any smaller scale soil state variability, such as that produced by wheel tracks. Similarly, it cannot see yield variability resulting from the non-uniformity of air seeder and fertilizer spreader distribution patterns. For these units, a coefficient of distribution variation of 15% is regarded as good, although this implies that the best area deciles receive 67% more inputs than the worst. A kriged yield map will almost certainly hide repetitive, small-spatial-scale crop performance variability of similar or greater magnitude to the field-scale variability addressed by variable rate technology. This small-scale variability is often quite clear in high-quality (less than 1 m^2 pixel) satellite or UAV ("drone") images and is largely a function of management of inputs, subject to immediate and relatively inexpensive management control.

The combination of precise positioning, connectivity with built-in farm equipment monitors, and wireless data networks already make it possible to harvest and analyze large volumes of soil, crop, and equipment-related data. This presents exciting technical challenges and opportunities, but it should not obscure the simpler and cheaper methods of avoiding unwanted variability. "Precision agriculture" is sometimes seen as the silver bullet of improved productivity but it ignores variability at sub-machine width spatial scales. In the case of traffic effects this is of known economic significance, and the same is very likely to apply to variability imposed by non-uniform machine performance. These can be addressed only by CTF and more thorough equipment calibration.

2.7 ON-FARM CTF IN DIFFERENT REGIONS

2.7.1 Sparsely Populated Sub-humid and Arid Regions

Much of the work reported above was carried out in the relatively arid grain production regions of Australia, which probably accounts for the greater extent of CTF adoption in these areas. Survey data, for instance, shows CTF adoption rates of 45% in Northern Australia, compared with a national adoption rate of 22%. Both examples appeared to be substantial overestimates of the area under full, high-quality CTF (i.e., with *all* heavy wheels on permanent traffic lanes in designed layouts), but it still indicates that a large proportion of Australia grain growers are moving toward controlled traffic.

The beneficial effects of controlled traffic are clearly not peculiar to Northern Australia, but they are most immediately beneficial in relatively arid, non-irrigated grain production. These are the conditions under which crop yields are often water-limited, and CTF-driven improvements in soil structure are likely to be reflected in increased yields. Thus, improvements similar to those in the black clays of Northern Australia were observed in a six-year, large-scale comparison of conventional versus controlled traffic cropping in a red chromosol in Southern Australia (Sedaghatpour *et al.*, 1995). This work demonstrated yield improvements of 12 to 17%, associated with beneficial CTF effects on porosity, soil bulk density, cone index, root extension, and infiltration rates.

Many years work in Western Australia have also shown similar outcomes in dryland grain cropping of (usually) sandy soils. The outcomes of this work have been summarized by Blackwell *et al.* (2003; 2013), and supported by excellent extension material (Isbister *et al.*, 2013), citing the outcomes of CTF systems applied by a number of farmers.

---

* The data smoothing process used to hide minor variability and emphasize field scale variability in yield maps.

Research with similar CTF systems in the winter-freeze environment of Western Canada has demonstrated benefits in terms of soil state and soil biology. Yield effects appeared to be limited and nonsignificant (Andreiuk, 2017), but economic and farming system outcomes were very good in some cases. Positive results were also demonstrated in systems using much smaller and lighter farm equipment on the loess soils of Shanxi province, Northwest China by Bai (2008). Working in the same plots, Chen (2008) found that controlled traffic was associated with increased levels of soil organic matter and microbial biomass. More recently, Lu *et al.* (2016) have also demonstrated soil fertility and yield benefits of no-till controlled traffic systems in double-cropping on the north China plain, despite a high wheel track/bed area ratio using small-scale equipment.

Most of the work referred to above has been carried out in relatively arid environments (occasionally with strategic irrigation), where yield outcomes have been driven by improved water availability in high-quality controlled traffic systems. These have a fixed wheel track gauge for all heavy equipment, modular working widths, and precise guidance.

2.7.2   MORE POPULOUS, HUMID REGIONS

In Australia and Canada, the common wheel track gauge of most grain harvesters is 3 m, a wheel setting that can be achieved with larger tractors. This wheel track gauge presents no problem where road travel is not required, or road transport regulations are not unduly restrictive of 3.3–3.6 m wide equipment. This is generally not the case in areas of greater population density, such as Europe, where much of the early work on wheel track impact was carried out. Similar constraints apply in the Eastern United States, and all areas of high population density, where traffic regulations inhibit the movement of wide agricultural equipment. This is why researchers focused on long-term solutions where controlled traffic could be achieved by wide-span, "gantry" equipment.

Gantry units of 4–12 m width have been built by research organizations (Carter, 1965) and produced commercially in limited numbers by innovative small companies in the United Kingdom and Israel. These overcame the road transport problem by providing a rapid switch from field to "road" mode allowing longitudinal travel on public roads. Unfortunately, the development of complete gantry systems—including harvesting—is beyond the capacity of small companies. Few farmers will invest in expensive new technology unless the system is complete, and without assurance of long-term dealer/service backup. More recently an established Danish manufacturer of root crop harvesting equipment developed a wide-span unit for horticultural crops (http://ctfeurope .com/2013/ws/). Design of this unit was clearly intended to facilitate the transition from conventional to wide-span equipment, but progress has been very slow.

Adoption of CTF continues to be encouraged by organizations such as CTF Europe Ltd, a product of the enthusiasm of researchers and interested farmers (Vermeulen *et al.*, 2010). (See also http:// ctfeurope.com/ and http://www.controlledtrafficfarming.com/Home/Default.aspx). While there are some instances of large growers adopting 3 m systems, particularly in Eastern Europe, these are not feasible for most Western European systems, where road travel is essential. In contrast to Australia or Canada, rainfall in Europe is generally not limiting, so crop inputs and grain yields are greater and machinery widths are smaller. The moldboard plow is also still the common basis of many cropping systems, with important residue management and weed control functions, in addition to other forms of tillage.

Transverse soil movement ensures that the moldboard plow is incompatible with controlled traffic, so CTF represents a bigger step change in European cropping systems, and farmers moving to CTF are commonly also in the process of adopting direct drilling systems. European CTF systems generally match the harvester and seeder operating widths, with sprayers and spreaders operating on a multiple of this module. Road travel capability is often maintained by using the "OutTrac" CTF systems (Chamen, 2009) where the tractor wheels—at standard wide wheel settings—run within the much wider wheel marks of the harvester at its normal wheel settings. With seeder and harvester operating widths in the range 6–7.5 m, about 15% of field area is heavily wheeled, with an additional 5 to 10% trafficked only at harvest. This arrangement is probably the best that can be achieved given current track gauge width restrictions imposed by road transport. Proposals have previously been advanced for tractor designs that permit rapid (from within the cab) track gauge width adjustment. These would provide a major advance in flexibility not just for CTF in Europe.

2.8 CONCLUSION

Over the past 60 years, researchers in Europe, the United States, and Australia have demonstrated the substantial productivity and sustainability benefits of controlling field traffic. In many countries, a very small number of farmers have found ways to control traffic, demonstrated that these benefits apply in practice, can be achieved with modest investments, and provide significant economic benefits. Despite all these positives, large-scale adoption has occurred only in Australia, and even over there the adoption is only 20 to 25% of cropping area in grain production, and minuscule in most other industries. The question must be asked: why haven't we seen rapid adoption of such an obviously beneficial practice?

Similar questions were being asked in the 1990s about no-till. Despite initial opposition from the farm machinery industry and grower caution about a new practice, adoption gradually increased with support from the agricultural chemical industry backed by soil conservation extension services (Tullberg *et al.*, 2007). Unfortunately, government extension services have been drastically reduced, and commercial agronomists are probably now the major source of grower advice. CTF might provide benefits for farmers and the environment, but discussion of reduced tractor power, fertilizer, and agricultural chemicals has little appeal to the organizations providing those inputs. No-till also had the advantage of being a practice where adoption risk could be reduced by experimenting in one paddock only, but CTF involves machinery modifications. In practice, this means that CTF will affect the whole farm. It also involves tractor modifications that might void a tractor warranty, and other machinery system changes which can appear to be a risky proposition.

Every study of CTF economics to date has shown very significant positive results, even when some important benefits are difficult to evaluate—particularly timeliness effects (Kingwell and Fuchsbichler, 2011; Bowman, 2009; Strahan and Hoffman, 2009). With continued support from organizations such as the Australian Controlled Traffic Farming Association and farmer groups, CTF adoption can be expected to increase, particularly as the message of "an easier way to farm" becomes louder than that about adoption costs and risks. Many farmers have also come to appreciate that CTF adoption does not involve large capital expenditure when forward planning integrates the change into the normal machinery replacement program. Further acceleration will also occur as the machinery industry adjusts to accommodate the requirements of an increasingly profitable market segment.

REFERENCES

Adem, H. H. and Tisdall, J. M. (1984). Management of tillage and crop residues for double-cropping in a fragile soil. *Soil & Tillage Research* 4, 577–89.

Anderson, G., McKenzie, M., and Wightman, B. (1999). *Best Practice for Raised Bed–Controlled Traffic Cropping.* Department of Primary Industries, Victoria.

Andreiuk, R. (2017). Controlled Traffic Farming: Infiltration, Yields. Presentation at Alberta CTF conference. http://www.controlledtrafficfarming.org/index.php/links.

Antille, D. L., Chamen, W. C. T., Tullberg, J. N., and Lal, R. (2015). The potential of controlled traffic farming to mitigate greenhouse gas emissions and enhance carbon sequestration in arable land: A critical review. *Transactions of the ASABE* 58 (3), 707–31.

ASABE. (2011). *Agricultural Machinery Management Data.* American Society of Agricultural and Biological Engineers, St Joseph (Standards Manual).

Arndt, W. and Rose, C.W. (1966). Traffic compaction of soil and tillage requirements. *J Agr Eng Res* 11, 170–87.

Ashworth, M. B., Desboilles, J. M. A., and Tola, E. K. H. (2010). *Disc seeding in zero-till farming systems: A review of technology and paddock issues*. Western Australian No-Tillage Farmers Association, Perth, Western Australia.

Bai, Y., Chen, F., Li, H., Chen, H., Jin H., Wang, Q, Tullberg, J., and Gong, Y. (2008). Traffic and tillage effects on wheat production on the Loess Plateau of China: II. Soil physical properties. *Australian Journal of Soil Research* 46, 652–8.

Bashford, L. L. and Kocher, M. F. (1999). Belts vs. Tires, Belts vs. Belts, Tires vs. Tires. *Applied Engineering in Agriculture* 15 (3), 175–81.

Batey, T. (2009). Soil compaction and soil management—A review. *Soil Use and Management* 25, 335–45.

Bennett, J. M., Woodhouse, N. P., Keller, T., Jensen, T. A. and Antille, D. L. (2015). Advances in cotton harvesting technology: A review and implications for the john deere round baler cotton picker. *The Journal of Cotton Science* 19, 225–49.

Blackwell, P. S., Vlahov, V., and Malcolm, G. (1995). On-farm controlled traffic systems for improving benefits of deep tillage with broad acre cropping. WA. Proc. Controlled Traffic Conference, Rockhampton, 153–8.

Blackwell, P., Hagan, S., Davies, J., Riethmuller, S., Bakker, G., Hall, D., Knight, D., Lemon, Q., Yokwe, J. S., and Isbister, B. (2013). Pathways to more grain farming profit by Controlled Traffic Farming in WA. Grains Research and Development Corporation Update Papers. https://grdc.com.au/research-and -development/grdc-update-papers/2013/04/pathways-to-more-grain-farming-profit-by-ctf-in-wa.

Blackwell, P. S., Webb, B., Lemon, J. L., and Riethmuller, G. (2003). Tramline farming; pushing controlled traffic further for Mediterranean farming systems in Australia. Proc. 16th Triennial Conference of the International Soil Tillage Research Organization. The University of Queensland, Brisbane.

Blackwell, P. S, Isbister, B., Riethmuller, G., Barrett-Lennard, E., Hall, D., Lemon, J., Hagan, J., and Ward, P. (2016). Deeper ripping and topsoil slotting to overcome subsoil compaction and other constraints more economically. Report on GRDC Project Number: DAW000243, Dept. of Agriculture and Fisheries Western Australia, Perth, WA.

Bowman, K. (2009). Economic and environmental analysis of converting to controlled traffic farming. 7th Australian CTF Conference, Canberra, ACT, 61–8.

Bromet, N., (2006). Weed Management Using GPS Guidance. PhD thesis, School of Agronomy in Horticulture, University of Queensland.

Carter, L. M., Stockton J. R., Tavernetti, J. R., and Colwick, W. F. (1965). Precision tillage for cotton production. *Transactions of the ASAE* 8: 177–9.

Chamen, W. C. T., Watts, C. W., Leede, P. R., and Longstaff, D. J. (1992). Assessment of a wide span vehicle (gantry), and soil and crop responses to its use in a zero traffic regime. *Soil & Tillage Research* 24, 359–80.

Chamen, W. C. T., Chittey, E. T., Leede, P. R., Goss, M. J., and Howse, K. R. (1990). The effect of tire/soil contact pressure and zero traffic on soil and crop responses when growing winter wheat. *J Agr Eng Res* 47, 1–21.

Chamen, W. C. T., Cope, R. E., Longstaff, D. J., Patterson, D. E., and Richardson, C. D. (1996). The energy efficiency of seedbed preparation following moldboard plowing. *Soil & Tillage Research* 39, 13–30.

Chamen, T., Alakukku, L., Pires, S., Sommer, C., Spoor, G., Tijink, F., and Weisskopf, P. (2003). Prevention strategies for field traffic-induced subsoil compaction: A review. *Soil & Tillage Research* 73, 161–74.

Chamen, W. C. T. (2009). Controlled Traffic Farming—An essential part of reducing in-field variability. Home Grown Grains Authority (UK). "Precision in arable farming—Current practice and future potential."

Chen, H., Bai, Y., Wang, Q., Chen, F., Li, H., Tullberg, J. N., Murray, J. R., Gao H., and Gong Y. (2008). Traffic and tillage effects on wheat production on the Loess Plateau of China: I. Crop yield and SOM. *Australian Journal of Soil Research* 46: 645–51.

Chi, R. and Zuo, S. (1988). Development and evolution of the zonal tillage concept in China; a historical review. Proc. 11th ISTRO Conference Edinburgh 2, 601–6.

Connolly, R. D., Freebairn, D. M., and Bridge, B. J. (1997). Change in infiltration characteristics associated with cultivation history of soils in SE Queensland. *Aust. J. Soil Research* 35, 1341–58.

Cooper, A.W., Trouse, A. C., and Dumas, W. T. (1969). Controlled traffic in rowcrop production. Proc. Seventh International Congress of Agricultural Engineering, Baden-Baden, Germany, 1–6.

Dang, Y. and Moody, P. (2013). Costs of soil-induced stress to the Australian grains industry. Queensland Department of Science, Information Technology, Innovation and the Arts, Tor Street, Toowoomba, Queensland, Australia.

Dickson, J. W., Campbell, D. J., and Ritchie, R. M. (1992). Zero and conventional traffic systems for potatoes in Scotland, 1987–1989. *Soil Tillage Res.* 24, 397–419.

Duttmann, R., Brunotte, J., and Bach, M. (2013). Spatial analyses of field traffic intensity and modeling of changes in wheel load and ground contact pressure in individual fields during a silage maize harvest. *Soil & Tillage Research* 126, 100–11.

Gamache, P. (2013). A Path to CTF in Canada. Proc. 1st international CTF Conference, Toowoomba. http://actfa.net/actfa-conferences/international-ctf-conference-2013/.

GRDC. (2015). Farm Practices Survey Report 2015. https://grdc.com.au/Resources/Publications/2015/10/GRDC-Farm-Practices-Survey-2015.

Hamilton, G., Bakker, D., Houlbrooke, D., and Spann, C. (2005). Raised Bed Farming in Western Australia. Bulletin 4646, Department of Agriculture, Western Australia.

Hamza, M. A. and Anderson, W. K. (2005). Soil compaction in cropping systems: A review of the nature, causes, and possible solutions. *Soil and Tillage Research* 82, 121–45.

Isbister, B., Blackwell, P., Riethmuller, G., Davies, S., Whitlock, A., and Neale, T. (2013). Controlled traffic farming technical manual. NACC, Department of Agriculture and Food, Western Australia.

Kingwell, R. and Fuchsbichler, A. (2011). The whole-farm benefits of controlled traffic farming: An Australian appraisal. *Agricultural Systems* 104, 513–21.

Kroulik, M. Soil Protection and Sustainable Utilization of Soil through Modern Technologies. Proc. 7th International Soil Conference ISTRO Czech Branch, Křtiny, 77–88. Research Institute for Fodder Crops, Ltd., Troubsko, Cz.

Kuipers, K. and van De Zande, J. C. (1994). Quantification of traffic systems in crop production (Eds. B. D. Soane, and C. Van Ouwerkerk). *Soil Compaction in Crop Production*. Elsevier.

Larocque, S. (2013). CTF in Western Canada–Issues and Impacts. Proc. 1st international CTF Conference, Toowoomba. http://actfa.net/actfa-conferences/international-ctf-conference-2013/.

Li, Y. X., Tullberg, J. N., and Freebairn, D. M. (2007). Wheel traffic and tillage effects on runoff and crop yield. *Soil & Tillage Research* 97, 282–92.

Li, Y. X., Tullberg, J. N., Freebairn, D. M., and Li, H. W. (2009). Functional relationships between soil water infiltration and wheeling and rainfall energy. *Soil & Tillage Research* 104, 156–63.

Lipec, J., Hakansson, I., Tarkiewicz, S., and Kossowski, J. (1995). Soil physical properties and growth of spring barley as related to the degree of compactness of two soils. *Soil Till. Res.* 19, 307–17.

Lu, C., Li, H., He, J, Wang, Q., Sarker, K. K., Li, W., Lu, Z., Rasaily, R. R., Li, H., and Chen, G. (2016). Effects of controlled traffic no-till system on soil chemical properties and crop yield in annual double-cropping area of the North China Plain. *Soil Research* 54 (6), 760–6.

McHugh, A. D., Tullberg, J. N., and Freebairn, D. M. (2009). Controlled traffic farming restores soil structure. *Soil and Tillage Research* 104 (1), 164–72.

McPhee, J. E., Braunack, M. V., Garside, A. L., Reid, D. J., and Hilton, J. (1995). Controlled traffic for irrigated double cropping in a semi-arid tropical environment: Part III, timeliness and trafficability. *J. Agric. Engng Res.* 60, 191–99.

McPhee, J., Neale, T., and Aird, P. (2013). Controlled traffic for vegetable production: Part II. Layout considerations in a complex topography. *Biosystems Engineering* 2, 116.

McPhee, J. E., Aird, P. L., Hardie, M. A., and Corkrey, S. R. (2015). The effect of controlled traffic on soil physical properties and tillage requirements for vegetable production. *Soil & Tillage Research* 149, 33–45.

Monroe, G. E. and Burt, E. C. (1989). Wide frame tractive vehicle for controlled-traffic research. *Applied Engineering in Agriculture* 5 (1), 40–43.

Morrison, J. E. Jr., Gerik, T. J., Chichester, F. W., Martin, J. R., and Chandler, J. M. (1990). A no-tillage farming system for clay soils 1. *J. Prod. Agric.* 3, 219–27.

Pangnakorn, U., George, D. L., Tullberg, J. N., and Gupta, M. L. (2003). Effect of tillage and traffic on earthworm populations in a vertosol in Southeast Queensland. Proc. ISTRO CD, University of Queensland.

Rohde, K. W. and Yule, D. F. (1998). Controlling run-off, soil loss on soil degradation with controlled traffic and crop rotations. Proc. Controlled Traffic Conference, 108–103. University of Queensland, Gatton. http://actfa.net/actfa-conferences/1995-ctf-conference/.

Schjønning, P. and Rasmussen, K. J. (1994). Danish experiments on subsoil compaction by vehicles with high axle load. *Soil and Tillage Research* 29, 215–27.

Sedeghatpour, S., Ellis, T., Hignett, C., and Bellotti, W. (1995). Six Years of Controlled Traffic Cropping Research on a Red-Brown Earth at Roseworthy in South Australia. Proc.

Smith, E. K., Misiewicz, P. A., Chaney, K., White, D. R., and Godwin, R. J. (2014). Effects of Tracks and Tires on Soil Physical Properties in a Sandy Loam Soil. ASABE Paper No. 19122659. ASABE, St Joseph.

Soane, B. D. and van Ouwerkerk, C. (Eds). (1994). *Soil compaction in crop production*. Elsevier.

Stirling, G. H. (2008). The impact of farming systems on soil biology and soilborne diseases: Examples from the Australian sugar and vegetable industries—The case for better integration of sugarcane and vegetable production and implications for future research. *Australasian Plant Pathology* 37, 1–18.

Strahan, R. and Hoffman A. (2009). Estimating the economic implications for broadacre cropping farms in the Fitzroy Basin catchments of adoption of best management practices. http://www.fba.org.au/publication /downloads/Report-FINAL-Fitzroy-Basin-BMP-24-August-2009_RC.pdf.

Taylor, J. H. (1982). Benefits of permanent traffic lanes in a controlled traffic crop production system. *Soil Tillage Res.* 3, 385–95.

Taylor, J. H. (1989). Controlled traffic research: An international report. *Agricultural Engineering.* Proc. of the 11th International Congress on Agricultural Engineering, Dublin, Ireland, 1787–94.

Tullberg, J. N. (1988). Controlled traffic in subtropical grain production. *Proc. ISTRO* 11 1, 323–26.

Tullberg, J.N. (2000). Traffic effects on tillage energy. *Journal of Agricultural Engineering Research* 75 (4), 375–82.

Tullberg, J. N., Yule, D. F., and McGarry, D. (2007). Controlled traffic farming—From research to adoption in Australia. *Soil and Tillage Research* 97, 272–81.

Tullberg, J. N., McHugh A., Ghareel Khabbaz, Scheer C., and Grace P. (2011). Controlled traffic/permanent bed farming reduces GHG emissions. 5th World Congress of Conservation Agriculture incorporating Third Farming Systems Design Conference, Brisbane, Australia. www.wcca2011.org.

Tullberg, J. N. (2013). Energy in crop production systems (Eds. J. Bundschuh and G. Chen). *Sustainable Energy Solutions in Agriculture.* CRC Press, Boca Raton.

Tullberg, J. N., Antille, D. L., Bluett, C., Eberhard, J. and Scheer, C. (2018). Controlled traffic farming effects on soil emissions of nitrous oxide and methane. *Soil & Tillage Research*, 176, 18–25.

Vermeulen, G. D., Tullberg, J. N. and Chamen, T. T. C. (2010). Controlled traffic farming (Eds. A. P. Dedousis and T. Bartzanas). *Soil Engineering, Soil Biology*, Springer-Verlag. Berlin 20, 101–20.

Wang, Xiaoyan, Gao, Huanwen, Tullberg, J.N., Li, Hongwen, Kuhn, Nikolaus, McHugh A. D., and Li, Yuxia. (2008). Traffic and tillage effects on runoff and soil loss on the Loess Plateau of northern China. *Australian Journal of Soil Research* 46, 1–9.

Ward, L. and Rickman, J. (1988). *Tillage equipment selection and use.* Darling Downs Institute (now University of Southern Queensland) Press, Toowoomba, Queensland.

Wightman, B., Peries, R., Bluett, C., and Johnston, T. (2005). Permanent raised bed cropping in southern Australia: Practical guidelines for implementation. In Proceedings of the Workshop on Evaluation and Performance of Permanent Raised Bed Cropping Systems in Asia, Australia, and Mexico. *Aust. Centre for Int. Agr. Research.* ACIAR, Canberra (Eds: C. H. Roth, R. A. Fischer, and C. A. Meisner Trent Potter) SARDI, 618.

Williford J. R. (1980). A controlled-traffic system for cotton production. *Trans. ASAE.* 23 (1), 0065–70.

Yule, D. F. and Rohde., K. W. (1995). Land management systems including controlled traffic, Erosion Control for Dryland Cotton. Proc. Controlled Traffic Conference, Rockhampton. http://actfa.net/actfa-confer ences/1995-ctf-conference/.

Yule, D. F., Cannon, R. S., and Chapman, W. P. (2000). Controlled traffic farming—Technology for sustainability. Proc. CD, ISTRO 15, Fort Worth. http://iworx5.webxtra.net/~istroorg/p_frame.htm.

Yule, D. F. and Tullberg, J. N. (Eds.) (1995). Proc. Controlled Traffic Conference, Rockhampton. http://actfa .net/actfa-conferences/1995-ctf-conference/.

3 Mechanization of Vegetable Production

John McPhee, Hans Henrik Pedersen, and Jeffrey P. Mitchell

CONTENTS

3.1 INTRODUCTION

Diversity is a key feature of vegetable production, possibly more so than any other crop production system. Vegetable seeds range in size and shape from small and spherical (e.g., onions), to long and thin (e.g., carrots), or large and flat (e.g., pumpkins). Plant architecture comes in many different forms, such as leafy (e.g., lettuce), bunched (e.g., celery), bushes (e.g., beans), or vines (e.g., cucumber). The harvested part varies from individual leaves (e.g., spinach) to heads (e.g., cabbage), pods (e.g., peas), tubers (e.g., potatoes), roots (e.g., carrots), bulbs (e.g., onions), and large fruit (e.g., melons). This diversity of seed size, plant architecture, and harvested parts leads to a large diversity of mechanization systems in the vegetable industry. Crop type influences seedbed preparation requirements, selection of crop establishment system (seed or transplant), spatial distribution, in-crop management, tolerance of weeds or other foreign matter at harvest, and harvest technology. These factors interact to dictate mechanization systems, or sub-systems, which are often crop-specific, and which can be difficult to integrate into a holistic system that caters to the needs of each crop.

To add further complexity, most vegetable producers grow a range of crops. Sometimes these have similar characteristics (e.g., all transplants or all leafy vegetables), but it is common to find growers who produce a range of crop types. This requires either a large inventory of different types of machines, or the extensive use of contractors. Either approach influences the success or otherwise of integrating different machinery into a cohesive system that not only meets the needs of the different crops, but also supports the long-term sustainability of the production system. Although change has been slow, large-scale grain cropping systems in Austalia, Canada, and other regions (Coughenour and Chamala, 2000; Duiker and Thomason, 2014; Awada *et al.*, 2014) have been able to build on some common characteristics of crops and machinery to develop systems such as zero-till and Controlled Traffic Farming (CTF), which have significantly improved industry productivity and environmental sustainability. The diversity of crops and mechanization is a significant barrier to the development and adoption of similar production systems in the vegetable industry.

This chapter will cover the key features of vegetable production systems, demonstrate how crop diversity influences mechanization, consider the impacts of diverse mechanization on system sustainability, and highlight developments which can dramatically improve both the productivity and environmental sustainability of mechanized vegetable production.

3.2 GLOSSARY OF TERMS

Terms related to tillage systems listed below have been adapted from Mitchell *et al.* (2016).

- Standard or conventional tillage: Sequence of operations most commonly or historically used in a given field to prepare a seedbed and produce a given crop.
- Minimum tillage: Systems that reduce tillage passes and thereby conserve fuel use and do not fully incorporate crop residues.
- Pass-combining tillage: A subset of minimum tillage that combines two or more tillage elements into one machine, thereby reducing the number of passes, although not necessarily the extent of soil disturbance, to create a seedbed. Single pass seedbed preparation is often an intended goal of pass-combining tillage.
- No-tillage or zero-tillage: The soil is left undisturbed from harvest to planting except for the disturbance required for placement of seed and fertilizers, which is kept to the minimum possible within the constraints of the machinery used. For example, zero-till sowing of potatoes will inevitably disturb more soil than zero-till sowing of corn or crops with similar seeding requirements.

- Strip-tillage: A system in which the seed row is tilled prior to planting to allow residue removal, soil drying and warming, and, in some cases, subsoiling to a depth of 20–30 cm. Typically, less than one-third of the soil surface is disturbed. Ridge tillage is a special case of strip-tillage in which ridges for crop planting are retained from the previous season, and only the ridge is refurbished during tillage, leaving the furrows undisturbed.
- Controlled Traffic Farming (CTF): Maintains the same machinery wheel tracks in cropping fields year after year by integrating machinery so all field traffic travels on the smallest number and area of permanent traffic lanes. Machinery is used in an organized and precise way to increase productivity by minimizing the area of soil damaged by compaction.
- GNSS: Global Navigation Satellite System, being the constellations of satellites (e.g., GPS [Navstar], GLONASS, Galileo, Beidou) which provide signals and communication for geolocation purposes, with tractor and implement guidance being the most well-established uses in agriculture.

3.3 KEY MECHANIZATION FEATURES OF VEGETABLE PRODUCTION SYSTEMS

3.3.1 TILLAGE

High-disturbance tillage is a common feature of many vegetable production systems. Harvest operations commonly cause significant soil compaction, requiring intensive tillage to remediate soil to a condition suitable for seedbed preparation for the next crop. For some vegetable crops (e.g., baby leaf salad species), the presence of surface residue, whether from a preceding harvest or a cover crop, can present both an in-crop disease risk and a contamination risk in the harvested product. The implementation of management systems that retain surface residue, such as zero-till, present challenges for small seeded crops (e.g., carrots). Consequently, a common aim of vegetable seedbed preparation is to provide uniformly tilled soil, free of surface residue. Although vegetable seeding and transplanting machinery designed to work through surface residues is available, their use is not widespread, despite research and development into reduced and zero-till vegetable production systems dating back several decades (Hocking and Murison, 1989; Hoyt *et al.*, 1994; Loy *et al.*, 1987; Morse, 1989; Morse, 1999; Phatak *et al.*, 2002; Roberts and Cartwright, 1991; Stirzaker *et al.*, 1993; Stirzaker *et al.*, 1995; Mitchell *et al.*, 2004). Consequently, tillage equipment is still widely used in the vegetable industry.

Tillage equipment used for the preparation of seedbeds falls into three broad categories:

- Inversion (the moldboard plow is the most common implement of this type)
- Non-inversion (tined implements of various types, ranging from subsoilers to surface cultivators)
- Powered (the rotary hoe and power harrow are two common powered implements)

3.3.2 SEEDING AND TRANSPLANTING

Seeders used for vegetables can be broadly categorized as either precision or mass flow machines, depending on the type of metering system used. The purpose of precision seeders is to achieve single seed selection in the metering system in an effort to provide uniform, in-row placement of seeds at the time of sowing. Crops commonly sown with precision seeders include beans, onions, sweet corn, and carrots. A range of different metering systems has been designed for this purpose. Pneumatic seed metering systems are by far the most common, and come in two different forms—positive and negative (vacuum) pressure. Vacuum plate technology is used in a wide range of precision seeders from many different manufacturers. Vacuum pneumatic seeders rely on individual seeds being held against holes in a plate or cylinder until the air supply is cut off to release the seed, or the seed is mechanically removed into the seed drop tube. Plates and cylinders with different sizes and numbers of holes are used for different size and shape seeds, and to achieve different intra-row spacing of plants. Belt and plate metering systems are the next most common types used in vegetable precision seeders. Belt and plate metering systems rely on holes or notches of different size, shape, and spacing to accommodate different seed types. While precision seeders are designed to singulate individual seeds to improve uniformity of seed placement, their actual performance in the field is often considerably lower than their design (and name) would suggest (Bracy and Parish, 1998; Parish and McCoy, 1998). Precision is significantly influenced by seed shape and size, seeding rate, and operating speed. Seed shape and size is allowed for in the design of the singulation system. Seeding rate and field operating speed both influence the rotational speed of the metering system, and hence the time available for the single seed to be selected, and multiple seeds to be removed, prior to entering the seeding tube. Precision seeders often require large operating widths in order to meet seeding timeliness requirements, while also traveling slowly enough to maintain precision. More often than not, the cheaper solution is to travel faster with a smaller seeder, at the cost of lower precision. The precision placement of seeds for crops such as onions and carrots commonly suffers as a result of operating speeds that are faster than the design limits of the seeder.

The most common form of mass flow seeder used in vegetable production is the air seeder, in which the metering system design shares many features with air seeders used in cereal production. Seed is distributed at a predefined rate per hectare, or average number of seeds per linear meter. No attempt is made to singulate seed, and the target intra-seed spacing is an average defined by the inter-row spacing and the target number of seeds per hectare. Mass flow seeders are often used for sowing crops like peas.

Other specialist machines are used for crops such as potatoes. The potato is not planted as a seed, but as a tuber (or seed piece) that is much larger than most vegetable seeds, and therefore requires a different type of metering system. The three main types of potato planter metering systems are:

- Clamp: Seed pieces are singulated from a seed reservoir by a series of cam-operated spring-loaded clamps mounted to a vertical rotating disc.
- Cup: Seed pieces are singulated by a series of cups on a chain or belt which is dragged vertically through the seed reservoir, and lightly agitated to remove excess seed pieces when more than one is retrieved.
- Spike or needle: Seed pieces are singulated from a seed reservoir by needles mounted on a series arms mounted to a vertical rotating disc, and the seed piece is dropped to the furrow when the needle is retracted into the arm at a different point in the rotation of the mechanism.

Many vegetable crops are grown from transplants. Examples include lettuce, tomatoes, brassicas, celery, and leeks. Onions are sometimes transplanted to enable faster establishment. This improves competition against weeds, although transplanting may also be used to shorten the growing season to allow access to specific markets. Transplants have the advantage that crop establishment tends to be more uniform, as low-vigor plants are culled before planting, and there is no variation in emergence, as is the case with seeds. Transplants are often preferred for crops which are hand-harvested, as more uniform maturity improves harvest efficiency. Advanced transplanters feature automated systems for the supply of seedling trays to the planting system, although many technologies still rely on human assistance or intervention to ensure continued operation (Figure 3.1a and b).

A wide range of spatial arrangements is found in vegetable production. These arrangements are often dictated by seeder, transplanter, or harvester design. There is little research reported in the literature on the impact of rectangularity (the ratio of intra-row to inter-row spacing) on crop production. What does exist is decades old (Frappell, 1973; Hatridgeesh and Bennett, 1980; Sutherland *et al.*, 1989; Westcott and Callan, 1990). Nevertheless, it is anecdotally accepted that a uniform

(a)

(b)

FIGURE 3.1   (a) Semiautomatic broccoli transplanter. (b) Tray feed on semi-automatic broccoli transplanter.

distribution in which intra-row and inter-row dimensions are similar is desirable for optimum production. This influences the widespread use of precision seeders and transplanters in vegetable production. Transplanted crops like lettuce and broccoli are examples in which the intra-row and inter-row dimensions differ by no more than a factor of two. At the other end of the scale are crops like carrots, which are often sown in bands of paired or triple rows (approximately 100 mm inter-row) at close intra-row spacings (less than 50 mm), but with large spacings (greater than 750 mm) between each band. In such crops, the ratio of intra-row to inter-row dimension is in the order of two within the band, but may be up to 15 between bands. Precision seeders are widely used for carrots, but uniformity along the row is often poor. Design constraints in seeders and harvesters preclude the use of closer bands which would improve rectangularity. In many cases it appears that plant spatial arrangements have evolved to suit mechanization design. This may have advantages—for example, crops with close intra-row spacing may benefit from wider inter-row spacing in order to improve ventilation of foliage, and wider rows make operations like inter-row fertilizer banding and strip-till more achievable. Equally, it may be that agronomic performance could be improved with more closely matched intra- and inter-row dimensions, but there is little reported literature that would prompt alternative machine designs to achieve such an outcome.

3.3.3 In-Crop Management

Irrigation is widely used for vegetable production around the world, and vegetable crops tend to be more intensively managed than rain-fed crops. In-crop management practices include irrigation, inter-row, and intra-row tillage for weed control; addition of fertilizer as top-dressed or side-banded applications; hand or mechanical thinning; and chemical applications for control of weeds, insect pests, and diseases. As a result, most vegetable crops tend to be subject to more machinery passes during the growing season than other crop production industries. Further, many vegetables have short growing seasons, and hence machinery movements in the growing crop are much more frequent. Some machinery (e.g., sprayers and fertilizer applicators) has increased in operating width as production areas have increased, while inter-row and intra-row tillage equipment still tends to have relatively small working widths. This is gradually changing with the increased uptake of GNSS guidance technology to provide accurate steering control for tractors and implements (Figure 3.2), and the use of semi-permanent beds, which are now commonly used in subsurface drip-irrigated vegetable fields (Mitchell *et al.*, 2012).

The combination of frequent field traffic, the narrow working widths of some operations, and soil which is often moist from irrigation means that in-crop management operations can have a significant impact on soil compaction during the growing season. Further, the relatively short growing seasons of many vegetables, and withholding periods required between application of pesticides and harvest, often constrain remedial management activities required to address shortcomings in crop nutrition, pest control, or disease control. This feature of vegetable production can heighten the importance of timeliness considerations in the selection and management of machinery. Machinery width, operating speeds, and operating windows are all important considerations in relation to mechanization choices. Further, soil moisture conditions, as a result of irrigation or rainfall, determine trafficability during in-crop management operations.

FIGURE 3.2   Implement guidance for inter-row weeding.

3.3.4 Harvest

A number of features of vegetable production dictate the nature and timing of harvest, and these factors interact with harvest mechanization. Whether crops are destined for fresh consumption or processing, timeliness of harvest is of utmost importance. The harvest window to achieve optimum yield and quality can be as short as one day for some crops. Fresh market buyers often require newly harvested crops for everyday delivery, so crops are sometimes harvested in very wet conditions (Figure 3.3a and b). Consequently, harvest systems require a reliable, high throughput capacity. Many vegetable crops have high yields per hectare, which makes materials' handling capacity a key mechanization design requirement. It also means that if harvesters are required to carry harvested product on-board, the combined machine and product weight can be very high. For example, wheel loads on a fully loaded two axle sugar beet harvester can exceed 12 Mg, while on loaded potato harvesters they are typically 4 Mg for single row, and 7 Mg for two row pulled machines. Four row self-propelled potato harvesters can weigh up to 50 Mg (Moitzi and Boxberger, 2007). A common design approach to accommodating this requirement is to provide multiple and/or high flotation tires (Figure 3.4).

Many vegetable crops are grown on beds or relatively narrow row spacings, and even if larger harvesters are capable of recovering multiple rows in a pass, the operating width of vegetable harvesters tends to be narrow compared to grain harvest machinery. Therefore, the intensity of traffic coverage is very high, with the use of road vehicles as haul out bins contributing to the problem of soil compaction (Figure 3.5). Seasonal traffic loads are estimated to be in excess of 300 tkm ha^{-1} for some crops, with seasonal tracked area of 3–5 ha ha^{-1}, with a significant portion coming from harvest operations (McPhee et al., 2015). Significant soil disturbance is caused during harvest of some crops, such as root and tuber crops. For example, it is estimated that in excess of 1500 m^3 of soil per hectare is lifted and sieved as part of potato harvesting operations. Most vegetable crops are harvested "green" (i.e., before senescence). Therefore, it is common to irrigate crops to within a day or two of harvest, which means that the soil is moist at the time of harvest. All of these factors—machine and produce weight, tire and axle configuration, intensity of traffic coverage, soil disturbance and soil moisture—combine to make harvest traffic one of the most severe negative impacts on soil quality in the vegetable production cycle.

3.4 IMPACT OF CROP TYPE, DIVERSITY, AND PRODUCTION SYSTEMS ON MECHANIZATION DESIGN

3.4.1 Tillage

As noted earlier, there are three key factors in vegetable production systems that largely dictate the tillage needs of various crops:

- The need to remediate soil compaction caused by harvest operations
- Management of surface residues to overcome disease and contamination risks associated with residues
- Management of surface tilth and residues to allow operation of seeders that are designed to work well only in a finely structured seed bed

In some situations, cultivation of weeds and mechanical incorporation of herbicides or fungicides are additional reasons for tillage.

3.4.1.1 Subsoil Loosening

Usually the operation of first choice in dealing with soil compaction is subsoil loosening. Tined implements used for subsoil loosening usually operate to 30–40 cm in depth, or deeper in rare circumstances (Figure 3.6). The number of tines will vary depending on the tractor power available

(a)

(b)

FIGURE 3.3 (a) Market supply schedules dictate that the harvest of fresh vegetables (in this case, leeks) often occurs in very wet conditions. (b) Market supply schedules dictate that the harvest of fresh vegetables (in this case, baby leaf lettuce) often occurs in very wet conditions.

and the depth of operation. Tine spacing is often about 30 cm. Implements used for subsoil loosening usually feature narrow ground-engaging tools, although research from the 1970s showed that considerable improvements could be made in the effectiveness of the operation through the use of winged-tines and shallow leading tines (Spoor and Godwin, 1978) (Figure 3.7). Unfortunately, such design features appear only rarely in the manufacture of subsoil loosening equipment. Subsoil

FIGURE 3.4 Multiple tires used in an attempt to reduce soil compaction caused by heavy harvest equipment.

FIGURE 3.5 Road transport used as haul out vehicles contribute to soil compaction through high loads and random traffic.

loosening is an expensive operation, reported by Miyao *et al.* (2008) as $US150/ha in Californian tomato production systems. Particularly in clay soils, subsoil loosening needs to be undertaken when soils are dry so shattering will occur. Operating sub-soiling implements in wet clay soils will cause smearing in the rip line, with negative impacts on water movement and root penetration (Batey, 2009).

FIGURE 3.6 Subsoiler with parabolic shanks.

FIGURE 3.7 Subsoiler with winged tines.

Some crops have a greater requirement for loosening than others. For example, root and tuber crops require a considerable depth of friable soil to allow unimpeded growth (e.g., for carrots) or to allow the formation of beds or ridges to cover the growing crop to prevent greening (e.g., for potatoes). However, subsoil loosening is not the only means of achieving the required depth of tilth for such crops. Integrated changes to machinery design and traffic management that allow the use of controlled traffic systems could provide sufficient depth of tilth in many soil types without deep tillage, as will be outlined later in this chapter.

3.4.1.2 Inversion Tillage

The primary purpose of inversion tillage is to bury residue and leave a clean soil surface for the next tillage operation in the seedbed preparation process. The moldboard plow is unquestionably the best implement for burying residue, although as a tillage operation, moldboard plowing is highly disruptive to the soil environment, and is prone to leave the soil in a highly erodible state. The moldboard plow has been used extensively in Europe, where vegetable production systems are often based on deep topsoils in regions with relatively moderate rainfall intensities, and hence less risk of soil erosion when there is no surface protection.

An alternative to the moldboard plow is the offset disc, which consists of gangs of discs operating at angles to each other such that the soil and residue is first thrown to one side and then the other. Offset discs are not as effective as moldboard plows for complete inversion of residue, but are capable of leaving some residue on the surface to reduce erosion risk, while burying enough residue to enable operation of equipment which is not designed for no-till situations.

The need for residue incorporation is determined by factors such as the capacity of seeding or transplanting equipment to operate in the presence of residue, and the potential of surface residues to contribute to disease risks during crop growth, or to contaminate fresh product at the time of harvest. The availability of vegetable seeders suited to no-till operations for small-seeded crops or crops that are sown at narrow inter-row spacings is quite limited. Larger seeded crops (e.g., beans, pumpkins) are much easier to accommodate in no-till situations as precision seeders designed to meter such large seeds are more readily available in no-till configurations. These crops are also grown on inter-row spacings that are large enough to allow trash flow between seeding units. No-till transplanters fitted with residue-cutting coulters ahead of each transplanter shoe have also been used to a limited extent (Mitchell *et al.*, 2012).

The issue of residue contamination in fresh harvested product is significant for some crops. For example, baby-leaf lettuce is usually grown on raised beds and harvested with a band-saw harvester which operates very close to the bed surface (Figure 3.8). Residue from previous crops, or cover crops, is a significant risk for producers of this product, and would not be tolerated by consumers. Similarly, green bean harvesters, which operate a rotating brush system to strip beans from the bush, would easily recover residue from the soil surface if it were present. While there are cleaning systems on the harvester and in post-harvest processing, additional residue reduces the efficiency of harvest and processing operations.

3.4.1.3 Seedbed Preparation

While subsoilers, moldboard plows, and offset discs are considered to be primary tillage implements, secondary tillage implements used in final seedbed preparation include lighter tined implements and

FIGURE 3.8   Baby leaf salad harvester.

powered implements such as rotary hoes and power harrows. Powered implements are often used to break up clods and residue, or to distribute residue more evenly vertically (rotary hoe) or horizontally (power harrow) in the upper layers of the soil profile (Figure 3.9). Powered implements are very disruptive to the soil environment; every implement has a range of soil conditions which is ideal for its intended purpose. It is an unfortunate consequence of the ever-increasing power of tractors that powered implements are often used in conditions which are suboptimal for the preparation of seedbeds. Powered implements are often the basis for bed-formers, used when crops are to be grown on raised beds. In these implements, the power harrow or rotary hoe usually has attached to it a mechanism for shaping the tilled soil into the desired profile and dimensions of the bed (Figure 3.10).

FIGURE 3.9   Rotary hoe.

FIGURE 3.10   Power harrow bed-former.

Powered implements are also used in single-pass tillage configurations. In these implements, the rotary hoe or power harrow may be attached to the back of a single bar subsoiler, so that subsoil loosening, residue incorporation, and clod reduction are accomplished in a single pass. With optimal soil moisture conditions, such an implement can create a seed bed in one pass.

Lighter tined implements are also used in secondary tillage operations to aid in forming seedbeds. These are most often used to disturb the surface soil to depths of approximately 10 cm, and to kill weeds prior to seeding.

3.4.1.4 In-Crop Mechanical Weeding

Despite the widespread availability and use of herbicides, mechanical inter-row and intra-row weed control is still widely used in some vegetable crops. A wide range of implements and specific ground-engaging tools is used for this purpose, some of which are designed to throw soil from the inter-row into the intra-row as a means of suppressing weed establishment. More advanced implements now use vision systems to control the action of intra-row tillage tools which move in and out between the plants to remove intra-row weeds (Figure 3.11). Vision activated intra-row weeders are expensive and still only used in small numbers, and used mainly for transplanted crops with wide intra-row spacing (e.g., cabbages, lettuce). Inter-row weeding systems find their most widespread use in crops with relatively wide inter-row spaces (e.g., beans, broccoli, corn) (Figure 3.12).

3.4.2 SEEDING AND TRANSPLANTING

A feature of many vegetable seeders, particularly precision seeders, is a very small, vertical drop from the metering mechanism to the soil. In some cases, the seed outlet from the metering system is almost at ground level. This is desirable from the perspective of minimizing the loss of accuracy from the metering system to the placement of the seed in the soil, as long drops through seed tubes inevitably decrease precision in the separation of seeds. This design feature can impact the capacity of vegetable seeders to provide adequate trash clearance in no-till situations, particularly when

FIGURE 3.11 Garford RoboCrop® intra-row weeder.

FIGURE 3.12   Inter-row weeding.

seeding units are placed close together to accommodate a narrow inter-row spacing requirement (Figure 3.13). Further, many of the furrow opener designs have not been developed to allow trash flow. While no-till precision vegetable seeders are available, they tend to be primarily used for large seeded crops with relatively large inter-row spacing (e.g., pumpkins, beans) (Figure 3.14) rather than small seeded crops that require narrow inter-row spacings (e.g., onions, carrots). It is common when sowing some crops, particularly small-seeded crops, to firm the surface of a bed by rolling to improve seed-soil contact, although this requirement could be met by the use of appropriate press wheels on the seeder, rather than full width rollers.

FIGURE 3.13   Precision seeding carrots with a seeder that has no trash handling capability.

FIGURE 3.14 Precision seeder with trash handling capability.

Similar to the situation with seeders, most transplanters are designed to operate in intensively tilled seedbeds exhibiting a fine tilth. There are exceptions, with transplanters capable of operating in surface residue conditions being available, but they are not common, and high-residue, ground-engaging systems seem not to be coupled with advanced automatic seedling supply and feed systems.

The diversity of crops grown by many vegetable farmers also impacts on the choice of implements used and the characteristics of the production system. For example, if crops which require a clean seedbed are grown in conjunction with other crops that might be suited to no-till production, it would be most common for the grower to make mechanization choices that accommodate the clean seedbed requirement, rather than own different suites of machinery to cater to both circumstances.

3.4.3 HARVEST

As has been mentioned previously, the diversity of vegetable crops and their production systems influences the design and choice of mechanization. Nowhere is this more apparent than in harvest. There are many contrasting features of vegetable crop harvest which help explain the diversity of harvest machinery design and function. Some examples include:

- Narrow and wide row crops (e.g., onions) are often sown at approximately 15–25 cm row spacing on a bed, requiring the entire bed to be lifted and windrowed in a single pass. An onion harvester may recover one or two beds in a pass, with an overall effective operating width of up to 4 m. In comparison, potatoes are usually grown in rows 75–80 cm apart, and harvester designs range from one to four row capacity, with an overall effective operating width ranging from 0.75–3.2 m. Windrowing systems for potatoes, which lift and convey tubers from adjacent rows to the furrows between undug rows of potatoes, range from two to six rows wide, and can increase the effective harvest width up to 12.8 m without change to the base harvester unit.
- Hard and soft vegetables (e.g., pumpkins) are mostly hand-harvested after the vines have started to senesce. Depending on variety, they may be quite resilient to materials handling processes. Green peas for processing, which are very soft and damage prone, are mechanically harvested from an actively growing plant and de-podded in the field before rapid transport to processing facilities for cleaning and freezing. The materials handling requirements of the harvest systems for these two crops are very different.
- Above and below ground crops—for example, sweet corn harvest—rely on a harvester that can pick a fresh whole cob from the stalk, remove unwanted leaf and trash, and place it into a

holding bin. In contrast, potato harvests require the recovery of about 100 m^3/ha of tubers of varying size from below the soil surface, in the process sieving about 1500 m^3/ha of soil.

- High and low yield crops, such as green beans, require recovery of some 10 t/ha of product, while onions, carrots, and potato harvest processes recover and transport product ranging from 50 to 120 t/ha.

It can be seen from the brief examples given above that the differences in crop and product requirements inevitably lead to a diversity of designs and functions in vegetable harvest machinery. Harvesters with high materials handling requirements often have high axle loads, large tire footprints, and relatively narrow working widths. At the other end of the design spectrum, hand-harvested crops often rely on the use of harvest aids, usually in the form of a tractor-mounted boom, which can provide a wide, effective working width.

3.5 KEY SUSTAINABILITY ISSUES FACING VEGETABLE PRODUCTION

3.5.1 SOIL

Compared to production systems for dry-land grain crops, the impacts of vegetable production on soil quality and resilience are substantial and largely negative. Low or no residue seedbeds, frequent, high disturbance tillage to considerable depths, frequent application of water through irrigation, high frequency traffic during the growing season, and harvest operations characterized by high load traffic at narrow swath widths on moist soil combine to make soil sustainability one of the most demanding aspects of vegetable production. Poor traffic conditions rarely stop a vegetable grower from undertaking essential field operations, such as spraying or harvest, but they can lead to significant soil damage. This can have negative impacts on future crop performance, and requires significant remedial tillage in an effort to restore soil tilth prior to the next crop.

3.5.1.1 Erosion

The widespread requirement for a clean, residue-free seedbed leaves soil exposed to the risks of erosion when not in crop, and the growth habit and production methods of many vegetable crops means that erosion can remain a risk even during the growing season (Basher and Ross, 2001; Basher and Ross, 2002). Cover crops are becoming more widely used for protection against erosion between cash crops, although in the absence of zero-till technologies, or driven by other demands for a clean seedbed, they tend to be incorporated prior to establishment of the next cash crop (Phatak *et al.*, 2002). Retention of the mulch cover from cover crops provides additional protection against erosion during the cropping season (Lounsbury and Weil, 2015; Morse, 1999; Stirzaker *et al.*, 1995) (Figure 3.15a and b). For production systems requiring a clean seed bed, other techniques, such as sowing cover crop species in the furrows between beds (Figure 3.16) and rip mulching (Cotching, 2002), can be effective at reducing erosion risk. Rip mulching is an alternative to contour banks as a means of reducing soil erosion caused by downslope runoff. The machine uses a subsoiling tine to loosen a slot along the contour, into which straw is placed to slow runoff water velocity and trap sediment (Figure 3.17a and b). As can be seen by Figure 3.17b, the rip-mulch line acts like a small contour bank, in that it slows water flow and traps sediment that has moved downslope. However, it is also at risk of over-topping, and therefore concentrating water flow downslope of the rip line. This emphasizes the importance of residue retention over the whole surface to slow overland flow and reduce erosion, which is a challenge for many horticultural crops, and optimal layout to avoid concentration of overland flow (McPhee and Aird, 2013).

3.5.1.2 Aggregate Stability

Aggregate stability refers to the ability of soil aggregates to maintain their integrity and structure when subjected to disruptive forces, such as those that might occur under the influence of droplet

(a)

(b)

FIGURE 3.15 (a) Cover crop residue retained between rows of tomato plants. (b) Cover crop being crimp rolled prior to zero-till seeding operation.

impact (rainfall or irrigation), wind blow, or tillage. Wet aggregate stability measures resistance to raindrop impact and water erosion, while the size distribution of dry aggregates provides a predictor of resistance to wind erosion. Declining aggregate stability has serious consequences for agricultural production (Gregory *et al.*, 2015; Munkholm, 2011). The ability of soil aggregates to resist disintegration when exposed to water or tillage forces is important for maintaining soil structure, the capacity to infiltrate rainfall or irrigation, and to aid internal drainage. Conversely, unstable

FIGURE 3.16   Cover crop species grown in furrows between beds to reduce erosion on slopes.

aggregates predispose a soil to slaking and soil erosion, while limiting the ability to absorb and store water (Paul *et al.*, 2013). Aggregate stability can be negatively influenced by tillage (Du *et al.*, 2013; Kasper *et al.*, 2009) and be improved with the use of techniques which increase the accumulation of soil carbon, such as no-till and cover cropping (Pikul *et al.*, 2009; Adem and Tisdall, 1984). Traffic has been shown to increase aggregate strength (Blanco-Canqui *et al.*, 2010; Voorhees *et al.*, 1978) although this generally has negative consequences for the formation of seedbeds using tillage. The negative impacts of tillage and traffic on aggregate stability further support the argument for vegetable production systems to embrace technologies which reduce the need for tillage and increase the capacity to retain organic matter in the form of crop residues or cover crops.

3.5.1.3   Compaction

The use of increasingly larger machinery, the frequency of field operations, and the moisture status of soil during field traffic all point to soil compaction as being an increasing problem in vegetable production. The occurrence of soil compaction and the consequent impacts on soil physical parameters and crop production are well documented (Akker *et al.*, 2004; Batey, 2009; Boone and Veen, 1994; Håkansson, 2005; Hamza and Anderson, 2005; Stalham *et al.*, 2007; Servadio, 2013). Soil compaction is known to reduce aeration and gas exchange, limit soil water storage, increase the risk of water-logging, increase the risk of soil borne disease, reduce water and nutrient uptake by plants, reduce quality in root crops, reduce crop growth and yield, and overall increase the costs of production as a result of more expensive tillage requirements (Bell *et al.*, 2007; Moots *et al.*, 1988; Raghavan *et al.*, 1982; Stirling, 2008; Tu and Tan, 1988; Alvarez and Steinbach, 2009; Bakker *et al.*, 2007; Blanco-Canqui *et al.*, 2010; Unger, 1996).

3.5.2   WATER

Most vegetable production worldwide is dependent on irrigation. Irrigation systems range from simple furrow application, to fixed sprinklers, big guns, lateral move, pivots, and drip. Irrigation

(a)

(b)

FIGURE 3.17 (a) Rip mulcher laying down straw lines to slow overland flow. (b) Eroded soil caught in rip mulch line.

water management is a topic of increasing interest in vegetable production due to water and power costs and water shortages in many regions. The use of variable-rate irrigators is increasing. Control systems are informed by digital spatial data and maps reflecting varying water holding capacities, or the varying needs for water as the crop grows. Automation and the remote control of irrigation systems is becoming popular as a means of reducing labor input to irrigation management.

Nevertheless, irrigation is not without its environmental sustainability issues. Runoff and deep drainage are negative consequences arising from over-irrigation, which should reduce in line with the uptake of more effective soil moisture monitoring and irrigation scheduling, management, and control systems. Soil management approaches can also have a significant effect on irrigation water management and the consequent environmental factors associated with irrigation. Surface cover, as either cover crops or retained residue after harvest, aids in reducing runoff and evaporative losses. Improved soil structure, through improved traffic, residue, and tillage management, improves infiltration and soil water storage, both of which help to reduce runoff and deep drainage (Mateos and Gomez-Macpherson, 2011; Reicosky *et al.*, 1999; Reyes *et al.*, 2005; Titmarsh *et al.*, 2003; Wang *et al.*, 2008).

3.6 MECHANIZATION AS A FACTOR THAT NEGATIVELY IMPACTS ON VEGETABLE PRODUCTION SUSTAINABILITY

3.6.1 TILLAGE

It has been noted that the main reasons for tillage in vegetable production are compaction removal after harvest of the previous crop, the management of crop residues, seedbed preparation, and, in some cases, weed control during the growing season. While the use of tillage to remove the effects of soil compaction is a prerequisite to seedbed establishment, it is increasingly recognized that repetitive tillage has negative consequences for soil structure and biology. Every time soil is disturbed with tillage, natural pore structures, created by decaying plant roots and a wide range of soil biological processes, are destroyed. These internal soil structures are important to assist infiltration, drainage, and the storage of plant-available water. Their repeated destruction leads to suboptimal performance of the soil as a substrate in which to grow plants (Heisler and Kaiser, 1995; Petersen, 2002; Schrader and Lingnau, 1997).

Vegetable seedbed preparation often relies on the use of powered implements which cause significant soil disturbance, resulting in further destruction of soil pore structures and accelerated breakdown of organic matter, which is then not available to assist in the recovery of soil biology and soil structure. The use of tillage for in-crop weed control also disturbs and rearranges soil structure, although not to the same depths as those experienced in subsoil loosening and seedbed formation. Extensive and repeated tillage makes it difficult to build and maintain soil structure and biology, which ultimately leads to negative consequences for plant growth.

3.6.2 SEEDING

As outlined earlier, many vegetable seeders and transplanters are designed to operate in intensively tilled seedbeds with fine tilth. This can be important for small seeded crops, but transplants and large seeded crops can be successfully established and grown in no-till or minimum tillage situations. Expanding the number of vegetable crops that can be successfully established and produced with no-tillage and minimum-tillage is an attractive, albeit difficult, goal of many farmers. This requirement for fine tilth is a major driver for excessive tillage in the vegetable production process, and dictates that maintenance of surface cover by crop residue or cover crops is uncommon in vegetable production systems. This, in turn, means that the only way to add significant amounts of organic matter to the soil is through the production and incorporation of cover or green manure crops. For crops where the seeding operation itself is one of high soil disturbance (e.g., potatoes) it is difficult to maintain soil structural integrity or surface cover, although there are examples of innovative approaches using cover crops in sandy soils in which potatoes are planted directly into the standing cover crop.

3.6.3 HARVEST AND MATERIALS HANDLING

Harvest is unquestionably the operation with the most soil damage potential in the vegetable cropping cycle, and hence a major determinant of the type of tillage operations required to prepare for subsequent crops. There are a number of interrelated reasons for this, some of which are due to mechanization choices and design.

Timeliness is important in the harvest of many vegetables. This may be because of the need to meet market supply contracts or short time windows for optimum maturity, quality, and yield. Regardless of the reason, high demands on timeliness often result in the selection of large-capacity machinery, which inevitably tends to be heavy, both in the machine itself, and the loads it carries. Many hand harvested crops are also processed and packed in the field on the harvester, which further adds to the size and weight of machinery which is traversing the field at harvest time (Figure 3.18a and b). Machine size and weight has direct impact on the level of damage caused to the soil, with resultant negative consequences for sustainability.

Harvest schedules for vegetables are most often dictated by market supply contracts in the case of fresh vegetables, or factory throughput requirements in the case of processed vegetables. Unlike grain crops, for which harvest occurs after senescence, most vegetable crops are harvested in the fresh state. This usually means they are irrigated up to the point of harvest, and there is little opportunity for the crop to deplete soil moisture reserves before harvest operations begin. Consequently, the soil is usually in a highly compactible state at the time of harvest. Coupled with machine weight, this increases the potential for soil damage due to harvest traffic.

3.6.4 SOIL IMPACTS OF INTENSIVE TILLAGE AND TRAFFIC REGIMES

The combined use of zero-till and controlled traffic in some grain production systems has demonstrated the value of avoiding soil compaction and soil disturbance as a means of improving soil structure and resilience. The characteristics of vegetable production systems which have been outlined in this chapter demonstrate that it is very difficult under current mechanization systems to avoid intensive traffic and tillage regimes, and the resultant negative impacts on the soil.

The impacts of traffic and tillage practices encompass detrimental changes to infiltration, drainage, soil water storage, structural stability, runoff, erosion, crop productivity, nutrient use efficiency, and operating costs, in terms of fuel use and labor hours to perform tasks (Adem and Tisdall, 1984; Alvarez and Steinbach, 2009; Boulal *et al.*, 2011; Chamen and Longstaff, 1995; Gerik *et al.*, 1987; McPhee *et al.*, 2015). The widespread use of commonly available machinery dictates many of the soil sustainability challenges of vegetable production. However, changes are occurring, some of which will be outlined in the next section of this chapter.

3.7 MECHANIZATION DEVELOPMENTS THAT COULD POSITIVELY IMPACT ON VEGETABLE PRODUCTION SUSTAINABILITY

Many changes could be made to improve the sustainability of vegetable production systems, although this requires some major changes to the design of individual pieces of machinery and the operation of the production system as a whole. The most important changes revolve around the need to better manage traffic and tillage through the adoption of controlled traffic and zero-till or reduced tillage options, and, in the process, improve surface cover and organic matter retention for improved soil health.

(a)

(b)

FIGURE 3.18   (a) Lettuce harvest aid with trailed packing facility. (b) In-field lettuce packing facility trailed behind harvest aid.

3.7.1 Tillage

One requirement for reducing the intensity of tillage in vegetable production is to adopt different approaches, such as minimum tillage, strip-tillage, and zero-till. All of these require changes to machinery design and operation.

3.7.1.1 Pass-Combining Tillage

Pass-combining tillage does not necessarily reduce the degree of soil disturbance, but rather it aims to reduce the number of passes required to prepare a seedbed. Pass-combining tillage uses implements that combine a variety of tillage tools on the one machine. Operations combined on the one machine may include loosening to a depth of 20–30 cm, disking to incorporate residue, and rolling to create a smooth seedbed (Figure 3.19; Mitchell *et al.*, 2004). Pass-combining tillage will usually aim to produce a seedbed in a single pass. Although less deep or vertical tillage is usually performed in pass-combining tillage, the extent of horizontal or shallow surface tillage is generally similar to conventional tillage systems, with the same negative consequences for soil structure and biology.

Pass-combining tillage implements are available for both bed production systems and crops grown on flat land. In the case of bed machines, the use of pass-combining implements generally leads to a permanent or semi-permanent bed production system. The pass-combining implements are designed to cultivate and reform the bed *in situ*, and are generally used with GNSS–guidance steering systems, so that beds and furrows remain in the same place (Figure 3.20a and b). Pass-combining bed tillage systems are widely used with subsurface drip irrigation beds, although the approach is applicable to beds regardless of the irrigation system used. The key requirement for successful use of these implements in a single pass is that harvest operations do not destroy the beds to the extent that complete primary tillage is required to re-establish the bed cropping system.

The second type of pass-combining tillage approach is used in production systems that do not use beds. The aim here is to incorporate residues and prepare a seedbed tilth in a single pass. Having provided a suitable tilth, beds can easily be reformed with simple bedding equipment, if required.

FIGURE 3.19 Pass-combining tillage equipment with rippers, discs, and roller.

(a)

(b)

FIGURE 3.20 (a) Bed reformer incorporating lettuce residue into existing beds. (b) Some of the tillage elements of bed reformers, showing the trash cutting discs and ripping tines used for single-pass bed renovation.

In some cases, the seeding operation may be included in the same pass, although this would be rare for vegetables.

In the Australian potato industry, a single pass implement, consisting of subsoiling tines and a power harrow or rotary hoe, is often used to form a seedbed in one pass. Because these approaches reduce the total number of tillage operations, fuel use is also reduced. Average fuel savings of 50%, and time savings of 72%, have been reported with one-pass tillage equipment compared with the standard tillage program of disking and landplaning in the Sacramento Valley, California (Upadhyaya *et al.*, 2001). Although not widely used in a global sense, progress is being made with a variety of reduced-pass minimum tillage and strip-tillage systems.

3.7.2 STRIP-TILL

Strip-till, sometimes referred to as zone tillage, is a targeted approach to tillage in which only the soil required to establish the crop is disturbed prior to the seeding operation (Luna *et al.*, 2012; Morse, 1989; Muresan *et al.*, 2011). This approach has generally been used for crops with a wide inter-row spacing for which zero-till seeding equipment is not available, and for which inter-row surface residues do not present an added complication to crop performance or at harvest. For example, for a crop such as broccoli, a strip-till approach to seedbed preparation could reduce soil disturbance by 60% while still providing a residue-free strip of soil for crop establishment. The closeness of crop rows that are suited to strip-till depends on the design of the strip tiller, with most being suited for wider row spacings (Figure 3.21a through c). The strip-till seeding of onions into short-stature cover crops (Figure 3.22a and b) has been used as a means of protecting seedlings from wind damage, while the strip-till establishment of tomato crops is widespread in California (Figure 3.23). In addition to the soil advantages, the operating costs of seedbed preparation are substantially reduced, both in terms of fuel and time, leading to improved timeliness in crop establishment operations.

3.7.3 ZERO-TILL

While steady progress to develop zero-till and cover crop systems for vegetable production has been made over many years (Abdul-Baki *et al.*, 1999; Morse, 1999; Mitchell *et al.*, 2016; Ciaccia *et al.*, 2016; Lounsbury and Weil, 2015), successful commercialization is still relatively rare due to the wide-ranging challenges identified previously. However, drawing heavily from no-till experience developed for crops such as corn, soybeans, and cotton, and because much of the crop is hand-harvested, certain regions of Brazil produce a major proportion of the country's processing tomatoes using no-tillage. Progress is also being made with zero-till systems for onions (Figure 3.24a through c). Commercial no-till vegetable production is used for pumpkins in the southeastern United States (Harrelson *et al.*, 2007; Harrelson *et al.*, 2008), and similar approaches have been applied on a limited scale for melon production in tropical Australia (Rogers *et al.*, 2004). Despite the extensive soil disturbance requirements, even zero-till potato planting has received research and on-farm experimentation attention in various parts of the world (Figure 3.25; Carrera *et al.*, 2005; Mundy *et al.*, 1999). As happened with the development of zero-till techniques for grain crops throughout the world (Coughenour and Chamaba, 2000; Kassam, 2014; Awada, 2014), the increasing interest in zero-till vegetable production, coupled with the efforts of pioneering farmers and the creation of local cultures that facilitate cropping system improvements, will accelerate progress toward no-till vegetable production systems in the near future.

3.7.4 CONTROLLED TRAFFIC

Controlled traffic farming (CTF) systems are based on the concept that all field machinery travels on permanently located traffic lanes, while crop production occurs in the untrafficked soil between the wheel tracks. Many vegetable production systems use beds (raised or flat) as the production unit, with wheel tracks on either side. If these beds are kept permanent, they are often referred to as permanent bed, or controlled traffic, systems. However, given the diversity of crop types and machinery, as outlined earlier in the chapter, it can be very difficult to maintain the beds in a permanent arrangement. In order for the system to become permanent, and hence meet the requirements

(a)

(b)

FIGURE 3.21   (a) Four row strip-till machine. (b) Powered strip-till unit, with a narrow rotary hoe under each shroud.   (*Continued*)

of a controlled traffic system, it is necessary for all machinery to have a common track gauge and working width, or multiples of it, so that all wheels can be confined to the defined traffic lanes (Baker, 2007). Elimination of traffic-induced compaction from the crop growth zone (Chen *et al.*, 2008b; Li *et al.*, 2007), allows the two zones (traffic and crop production) to be managed separately for the optimal performance of both.

(c)

FIGURE 3.21 (CONTINUED)   (c) Strip tillage machine operating with one strip per bed.

The productivity and sustainability benefits of controlled traffic in crop production are substantial, and cover a range of aspects:

- Machinery and energy use: Reduced tillage or no-till; reduced fuel use and tractor time; lower capital investment in tractors and tillage equipment (McPhee *et al.*, 1995a); reductions in fuel use of over 50%, and reductions in tractor power requirements of 30 to 50%, are common.
- Soil and water: Improved soil structure, biology, infiltration, water holding capacity and drainage, and reduced runoff and erosion (Bai *et al.*, 2009; Braunack and McGarry, 2006; McHugh *et al.*, 2009; Boulal *et al.*, 2011; Radford *et al.*, 2007; Tullberg *et al.*, 2001); increases in infiltration of 70%, water holding capacity of 30%, and reductions in erosion of 40% have been reported.
- Crop: Higher, more uniform yield; improved crop quality and more even maturity (Chen *et al.*, 2008a; Dickson *et al.*, 1992); yield increases of greater than 10% have been reported.
- Economics: Improved farm gross margins and return on investment (Kingwell and Fuchsbichler, 2011; Halpin *et al.*, 2008; McPhee *et al.*, 2016); improvements of over 10% in farm gross margin have been reported for cane production, and 50% increases in farm profitability for dryland grain cropping. Modeling indicates improvements in farm returns in the order of 30% for vegetable production.
- Farming system: Improved timeliness and more effective application of precision farming techniques such as yield mapping and variable application of inputs (McPhee *et al.*, 1995b; Bramley, 2009); greater cropping reliability and frequency has been reported, leading to significant improvements in economic performance.

The commercial uptake of controlled traffic has been most successful in the dry land grain industry in Australia (Tullberg, 1994; Tullberg *et al.*, 2007). Uptake in the vegetable industry is constrained by many of the mechanization issues outlined previously and reported by McPhee and Aird, (2013), although some progress has been made with a 3 m track gauge system for a limited

(a)

(b)

FIGURE 3.22 (a) Strip-till seeding onions into cover crop. (b) Sprayed-off short-statured cover crop with strip-tilled onions before emergence.

number of crops such as sweet corn and green beans. Successful uptake of controlled traffic in vegetable production requires a situation in which the machinery suite can be selected (or modified) to achieve commonality of working and track widths to provide dimensional integration of equipment and operations. This requires one of three circumstances to be successful:

- Collaboration between machinery manufacturers to design equipment to an agreed upon dimensional base; given the diverse requirements of different vegetable crops, particularly in relation to harvest mechanization, this appears very unlikely.
- An operator of sufficient scale to make their own decisions on machinery choice, with the capacity to request designs from manufacturers that meet their needs; this is more achievable

FIGURE 3.23 Strip-till transplanting tomatoes.

if the range of crops grown is limited, and there are similarities between the crops which reduce the design conflicts that arise through diversity.

- The development of a new approach to vegetable production mechanization, in which a wide variety of mechanisms, designed to cater to the inherent diversity in the industry, can be attached to a common multi-purpose tool carrier. Such an approach is represented by the prototype Wide Span tractor developed in Denmark (Pedersen *et al.*, 2013).

3.7.4.1 Tractor-Based Controlled Traffic

Early controlled traffic research in vegetable production dates from the 1980s (Lamers *et al.*, 1986; Perdok and Lamers, 1985). The system studied was based on conventional tractors and vegetable machinery modified to operate on a 3 m track width. Key findings of the research were a 50% reduction energy use and up to 10% increase in crop yield. Tractor-based controlled traffic systems are used in some parts of the vegetable industry, using permanent beds and harvest aids for hand-harvested crops. These systems are able to avoid the integration problems of harvest machinery that are present in other sectors of the industry. Because vegetable harvest machinery (e.g., potato, onion, bean, pea, etc.) is difficult to integrate (McPhee and Aird, 2013), some growers have adopted what is known as seasonal controlled traffic, in which tractor-based operations are integrated, but harvest is not (Vermeulen and Mosquera, 2009).

3.7.4.2 Wide Span (WS)–Based Controlled Traffic

The WS, or gantry tractor, concept is at least 155 years old, with a 9.1-m span steam driven tractor on iron rails being described as the basis of an efficient farming system (Halkett, 1858). Gantries were used for research in the 1980s (Taylor, 1994), primarily for the purpose of isolating traffic impacts from crop production soil in some of the early work demonstrating the benefits of controlled traffic. Early attempts at commercialization occurred in the late 1980s and early 1990s, with machines such as the Israeli-built 6 m wide Field Power Unit (FPU) and the 12 m U.K.-built Dowler gantry (Figure 3.26), of which at least five were built (Ball and Ritchie, 1999). Research conducted using WS tractors for cereal crop production operations, including harvest, showed improvements in soil

(a)

(b)

FIGURE 3.24   (a) No-till onion seeding. (b) Emerging no-till onions.   (*Continued*)

structure and reductions in fuel use (Ball and Ritchie, 1999). Some WS prototypes were built for specialized horticultural production tasks (Ball *et al.*, 2008; Tillett and Holt, 1987; Holt and Tillett, 1989).

The most recent efforts to bring the WS concept to commercial reality have been undertaken in Denmark, with the design and construction of a 9.6-m prototype WS tractor, which has been used in CTF production of onions for all operations from primary non-inversion tillage to harvest. The WS tractor is fitted with three standard, three-point linkage hitches to accommodate common tillage equipment, including PTO-driven machines, and by way of an interchangeable center beam, it can convert to a 15 t capacity bunker-style onion harvester (Figure 3.27a and b). Fields up to 550 m long can be harvested, with the bunker unloaded at the field ends only, thereby excluding transport trailers from the field (Pedersen, 2011).

(c)

FIGURE 3.24 (CONTINUED)   (c) Established no-till onions.

FIGURE 3.25   Experimental zero-till seeding of potatoes into sprayed off cover crop under controlled traffic.

FIGURE 3.26   12 m Dowler gantry.

3.7.4.3   Mixed Tractor and Wide Span Controlled Traffic

The introduction of WS mechanization to vegetable production would be a significant and trans-
formational change to both the mechanization and operational requirements of the industry. It would
require all equipment to be WS compatible, and while this would be a relatively minor change for
existing three-point linkage equipment, it would represent a major change to provide harvest capability for a wide range of crops. Onion harvesting capability in the current prototype was achieved
with existing technologies adjusted to fit the WS. Although at a conceptual level this is all that is
needed for all crops to use a WS harvest system, the practicalities of achieving such a change across a
range of harvester styles requires a great deal of redesign of existing mechanization elements.

Many operations take place at the same time on vegetable farms that grow a range of crops, so a
transition to full WS CTF production would require all field-working tractors to be replaced with a
number of WS tractors. In most cases, such a transition would take some years. In the meantime,
many benefits of CTF production would still be achievable through the use of a system which mixes
the use of conventional and WS tractors, allowing WS tractors to be integrated gradually into the
production system (Figure 3.28). The key requirement would be for the WS track gauge to be a direct
multiple of existing conventional tractor dimensions (Pedersen *et al.*, 2016). The 9.6 m prototype WS
tractor is used on a farm growing in 3.2 m beds (from center to center of tracks), allowing it to
straddle three beds that can be worked by conventional tractors on 3.2 m track gauge. Other systems
could be considered, such as 8.0/2.0 m. A larger portion of the area would be tracked in a combined
system, somewhat reducing the benefits of the WS, although CTF benefits would be gained at the
individual bed level with harvest performed by the WS.

3.8   CONCLUSION

There is a direct link between the diversity of vegetable crops, their production requirements, and the
diversity of machinery used. This diversity leads to a number of issues in vegetable production, not
least of which is widespread soil compaction, which requires extensive use of tillage equipment in an

(a)

(b)

FIGURE 3.27 (a) ASA-Lift WS9600 wide span tractor in light tillage configuration. (b) ASA-Lift WS9600 wide span tractor in onion harvest configuration.

effort to achieve remediation. The difficulty of achieving dimensional matching of such a diverse suite of machinery is a barrier to the widespread adoption of production systems based on controlled traffic. There is no question that the sustainability and productivity of vegetable production systems would be greatly enhanced if it were possible to adopt controlled traffic relatively easily. Such a development would be further enhanced with the adoption of zero-till systems, and, in turn, zero-till or other low intensity tilling systems such as strip-till would be more easily adopted in controlled traffic systems, where compaction is avoided in the crop growing zones. The development of a

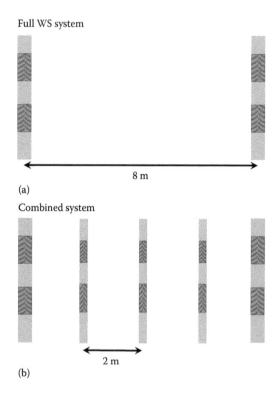

FIGURE 3.28   Bed and track layout for mixed WS and conventional tractor controlled traffic system based on 8 m WS (a), and 2 m tractor (b), track gauges.

prototype wide-span tractor points to one possible direction for overcoming the issue of mismatched equipment, and could provide the basis for the use of fully integrated controlled traffic in vegetable production.

In the meantime, the use of techniques such as strip-till provides opportunities to substantially reduce the amount of soil disturbance and energy use required for crop establishment. The relative lack of seeding and transplanting equipment suited to zero-till makes strip-till an attractive option for retaining surface residue and reducing soil disturbance, without having to make substantial changes to seeding technology in order to service a zero-till approach. Nevertheless, zero-till should be a goal of vegetable production systems wherever possible. While the challenges are substantial, the long-term sustainability of vegetable production requires a significant move away from random traffic and intensive tillage systems. This requires that future research, development, and design related to vegetable production mechanization should be built on a foundation of controlled traffic and zero-till, or at least strip-till.

REFERENCES

Abdul-Baki, A. A., Morse, R. D., and Teasdale, J. R. (1999). Tillage and mulch effects on yield and fruit fresh mass of bell pepper (*Capsicum Annum L.*). *Journal of Vegetable Crop Production* 5 (1), 43–58.

Adem, H. H., and Tisdall, J. M. (1984). Management of tillage and crop residues for double-cropping in fragile soils of Southeastern Australia. *Soil & Tillage Research* 4 (6), 577–89.

Akker, J. J., van Den, H., and Schjnning, P. (2004). Subsoil compaction and ways to prevent it. *Managing Soil Quality: Challenges in Modern Agriculture*. Wallingford UK: CABI Publishing, 163–84.

Alvarez, R., and Steinbach, H. S. (2009). A review of the effects of tillage systems on some soil physical properties, water content, nitrate availability, and crops yield in the Argentine Pampas. *Soil & Tillage Research* 104 (1), 1–15.

Awada, L., Lindwall, C. W., and Sonntag, B. (2014). The development and adoption of conservation tillage systems on the Canadian Prairies. *International Journal of Soil and Water Conservation Reserach* 2 (1), 47–65.

Bai, Y., He, J., Li, H. *et al.* (2009). Soil structure and crop performance after 10 years of controlled traffic and traditional tillage cropping in the Dryland Loess Plateau in China. [In English]. *Soil Science* 174 (2), 113–19.

Bakker, D. M., Hamilton, G. J., Houlbrooke, D. J., Spann, C., and van Burgel, A. (2007). Productivity of crops grown on raised beds on duplex soils prone to waterlogging in Western Australia. *Australian Journal of Experimental Agriculture* 47 (11), 1368–76.

Ball, B. C., Crichton, I., and Horgan, G. W. (2008). Dynamics of upward and downward N_2O and CO_2 fluxes in ploughed or no-tilled soils in relation to water-filled pore space, compaction, and crop presence. [In English]. *Soil & Tillage Research* 101 (1–2), 20–30.

Ball, B. C., and Ritchie, R. M. (1999). Soil and residue management effects on arable cropping conditions and nitrous oxide fluxes under controlled traffic in Scotland. I. Soil and Crop Responses. *Soil and Tillage Research* 52 (3–4), 177–89.

Basher, L. R., and Ross, C. W. (2001). Role of wheel tracks in runoff generation and erosion under vegetable production on a clay loam soil at pukekohe, New Zealand. *Soil & Tillage Research* 62 (3–4), 117–30.

Basher, L. R., and Ross, C. W. (2002). Soil erosion rates under intensive vegetable production on clay loam, strongly structured soils at pukekohe, New Zealand. *Australian Journal of Soil Research* 40 (6), 947–61.

Batey, T. (2009). Soil compaction and soil management: A review. *Soil Use and Management* 25 (4), 335–45.

Bell, M. J., Stirling, G. R., and Pankhurst, C. E. (2007). Management Impacts on Health of Soils Supporting Australian Grain and Sugarcane Industries. Paper presented at the 16th International Soil Tillage Research Organisation (ISTRO) conference, Queensland, Australia, 13–18.

Blanco-Canqui, H., Claassen, M. M., and Stone, L. R. (2010). Controlled traffic impacts on physical and hydraulic properties in an intensively cropped no-till soil. [In English]. *Soil Science Society of America Journal* 74 (6), 2142–50.

Boone, F. R., and Veen, B. W. (1994). Mechanisms of crop responses to soil compaction. *Soil Compaction in Crop Production.* (Eds. B. D. Soane and C. van Ouwerkerk) xvii, 662. Elsevier.

Boulal, H., Gomez-Macpherson, H., Gomez, J. A., and Mateos, L. (2011). Effect of soil management and traffic on soil erosion in irrigated annual crops. [In English]. *Soil & Tillage Research* 115–116, 62–70.

Bracy, R. P., and Parish, R. L. (1998). Seeding uniformity of precision seeders. *HortTechnology* 8 (2), 182–5.

Bramley, R. G. V. (2009). Lessons from nearly 20 years of precision agriculture research, development, and adoption as a guide to its appropriate application. [In English]. *Crop & Pasture Science* 60 (3), 197–217.

Braunack, M. V. and McGarry, D. (2006). Traffic control and tillage strategies for harvesting and planting of sugarcane (*saccharum officinarum*) in Australia. *Soil & Tillage Research* 89 (1), 86–102.

Carrera, L. M., Morse, R. D., Hima, B. L. *et al.* (2005). A conservation-tillage, cover-cropping strategy and economic analysis for creamer potato production. *American Journal of Potato Research* 82 (6), 471–9.

Chamen, W. C. T. and Longstaff, D. J. (1995). Traffic and tillage effects on soil conditions and crop growth on a swelling clay soil. *Soil Use and Management* 11 (4), 168–76.

Chen, H., Bai, Y. H., Wang, Q. J. *et al.* (2008a). Traffic and tillage effects on wheat production on the loess plateau of China: I. Crop Yield and Som. *Australian Journal of Soil Research* 46 (8), 645–51.

Chen, H., Li, H. W., Gao, H. W. *et al.* (2008b). Effect of long-term controlled traffic conservation tillage on soil structure. [In Chinese, English]. *Transactions of the Chinese Society of Agricultural Engineering* 24 (11), 122–5.

Ciaccia, C., Canali, S., Campanelli, G. *et al.* (2016). Effect of roller-crimper technology on weed management in organic zucchini production in a Mediterranean climate zone. *Renewable Agriculture and Food Systems* 31 (2),111–21.

Cotching, B. (2002). Stopping soil erosion: Mulched rip lines under annual cropping on steep slopes in Northwest Tasmania. *Natural Resource Management* 5 (2), 28–30.

Coughenour, C. M., and Chamala, S. (Eds.). (2000). *Conservation Tillage and Cropping Innovation: Constructing the New Culture of Agriculture.*

Dickson, J. W., Campbell, D. J., and Ritchie, R. M. (1992). Zero and conventional traffic systems for potatoes in Scotland, 1987–1989. *Soil & Tillage Research* 24 (4), 397–419.

Du, Z. L., Ren, T. S., Hu, C. S., Zhang, Q. Z., and Blanco-Canqui, H. (2013). Soil aggregate stability and aggregate-associated carbon under different tillage systems in the North China Plain. *Journal of Integrative Agriculture* 12 (11), 2114–23.

Duiker, S. W. and Thomason, W. (2014). *Conservation Agriculture in the USA*. Conservation Agriculture: Global Prospects and Challenges. (Eds. R. A. Jat, K. L. Sahrawat, and A. H. Kassam) doi:10.1079 /9781780642598.0026.

Frappell, B. D. (1973). Plant spacing of onions. *Journal of Horticultural Science & Biotechnology* 48 (1), 19–28.

Gerik, T. J., Morrison, J. E. Jr., and Chichester, F. W. (1987). Effects of controlled traffic on soil physical properties and crop rooting. *Agronomy Journal* 79 (3), 434–8.

Gregory, A. S., Ritz, K., McGrath, S. P. *et al.* (2015). A review of the impacts of degradation threats on soil properties in the United Kingdom. *Soil Use and Management* 31, 1–15.

Håkansson, Inge. (2005). *Machinery-Induced Compaction of Arable Soils*. Swedish University of Agricultural Sciences, Department of Soil Sciences, Reports from the Division of Soil Management. 109, Uppsala. Technical Report.

Halkett, P. A. (1858). On guideway agriculture: Being a system enabling all the operations of the farm to be performed by steam-power. *The Journal of the Society of Arts* 7 (316), 41–58.

Halpin, N. V., Cameron, T., and Russo, P. F. (2008). Economic Evaluation of Precision Controlled Traffic Farming in the Australian Sugar Industry: A Case Study of an Early Adopter. Proc. of the 2008 Conference of the Australian Society of Sugar Cane Technologists held at Townsville, Queensland, Australia, 34–42.

Hamza, M. A., and Anderson, W. K. (2005). Soil compaction in cropping systems: A review of the nature, causes and possible solutions. [In English]. *Soil & Tillage Research* 82 (2), 121–45.

Harrelson, E. R., Hoyt, G. A., Havlin, J. L., and Monks, D. W. (2007) Effect of winter cover crop residue on no-till pumpkin yield. *Hortscience* 42 (7), 1568–74.

Harrelson, E. R., Hoyt, G. D., Havlin, J. L., and Monks, D. W. (2008). Effect of planting date and nitrogen fertilization rates on no-till pumpkins. *Hortscience* 43 (3), 857–61.

Hatridgeesh, K. A. and Bennett. J. P. (1980). Effects of seed weight, plant-density, and spacing on yield responses of onion. *Journal of Horticultural Science* 55 (3), 247–52.

Heisler, C., and Kaiser, E. A. (1995). Influence of agricultural traffic and crop management on collembola and microbial biomass in arable soil. [In English]. *Biology and Fertility of Soils* 19 (2–3), 159–65.

Hocking, D. F., and Murison, J. A. (1989). Minimum tillage of vegetable crops. *Acta Horticulturae* (247), 263–6.

Holt, J. B., and Tillett, N. D. (1989). The development of a nine-meter span gantry for the mechanised production and harvesting of cauliflowers and other field vegetables. *Journal of Agricultural Engineering Research* 43 (2), 125–35.

Hoyt, G. D., Monks, D. W., and Monaco, T. J. (1994). Conservation tillage for vegetable production. *HortTechnology* 4 (2), 129–35.

Kasper, M., Buchan, G. D., Mentler, A., and Blum, W. E. H. (2009). Influence of soil tillage systems on aggregate stability and the distribution of C and N in different aggregate fractions. *Soil & Tillage Research* 105 (2), 192–9.

Kingwell, R., and Fuchsbichler, A. (2011). The whole-farm benefits of controlled traffic farming: An Australian appraisal. *Agricultural Systems* 104 (7), 513–21.

Lamers, J. G., Perdok, U. D., Lumkes, L. M., and Klooster, J. J. (1986). Controlled traffic farming systems in the Netherlands. *Soil & Tillage Research* 8, 65–76.

Li, Y. X., Tullberg, J. N., and Freebairn, D. M. (2007). Wheel traffic and tillage effects on runoff and crop yield. *Soil & Tillage Research* 97 (2), 282–92.

Lounsbury, N. P., and Weil, Ray R. (2015). No-till seeded spinach after winterkilled cover crops in an organic production system. *Renewable Agriculture and Food Systems* 30 (5), 473–85.

Loy, S. J. W., Peirce, L. C., Estes, G. O., and Wells, O. S. (1987). Productivity in a strip tillage vegetable production system. *Hortscience* 22 (3), 415–7.

Luna, J. M., Mitchell, J. P., and Shrestha, A. (2012). Conservation tillage for organic agriculture: Evolution toward hybrid systems in the Western United States. *Renewable Agriculture and Food Systems* 27 (1), 21–30.

Mateos, L., and Gomez-Macpherson, H. (2011). Soil management and traffic effects on infiltration of irrigation water applied using sprinklers. *Irrigation Science* 29 (5), 403–12.

McHugh, A. D., Tullberg, J. N., and Freebairn, D. M. (2009). Controlled traffic farming restores soil structure. *Soil and Tillage Research* 104 (1), 164–72.

McPhee, J. E., and Aird, P. L. (2013). Controlled traffic for vegetable production: Part I. Machinery challenges and options in a diversified vegetable industry. *Biosystems Engineering* 116 (2), 144–54.

McPhee, J. E., Aird, P. L., Hardie, M. A., and Corkrey, S. R. (2015). The effect of controlled traffic on soil physical properties and tillage requirements for vegetable production. *Soil & Tillage Research* 149, 33–45.

McPhee, J. E., Braunack, M. V., Garside, A. L., Reid, D. J., and Hilton, D. J. (1995a). Controlled traffic for irrigated double cropping in a semi-arid tropical environment: Part II., Tillage operations and energy use. *Journal of Agricultural Engineering Research* 60 (3), 183–9.

McPhee, J. E., Braunack, M. V., Garside, A. L., Reid, D. J., and Hilton, D. J. (1995b). Controlled traffic for irrigated double cropping in a semi-arid tropical environment: Part III., Timeliness and trafficability. [In English]. *Journal of Agricultural Engineering Research* 60 (3), 191–9.

McPhee, J. E., Neale, T., and Aird, P. L. (2013). Controlled traffic for vegetable production: Part II. Layout considerations in a complex topography. *Biosystems Engineering* 116 (2), 171–8.

McPhee, J. E., Maynard, J. R., Aird, P. L., Pedersen, H. H., and Tullberg, J. N. (2016). Economic modelling of controlled traffic for vegetable production. *Australian Farm Business Management Journal* 13, 1–17.

Mitchell, J. P., Carter, L. M., Reicosky, D. C. *et al.* (2016). A history of tillage in California's central valley. *Soil & Tillage Research* 157, 52–64.

Mitchell, J. P., Klonsky, K. M., Miyao, E. M. *et al.* (2012). Evolution of conservation tillage systems for processing tomatoes in California's central valley. *HortTechnology* 22 (5), 617–26.

Mitchell, J. P., Miyao, G. M., Jackson, J. J. *et al.* (2004). Reduced tillage tomato/wheat rotations in California's central valley. *Hortscience* 39 (4), 749–50.

Mitchell, J. P., Jackson, L., and Miyao, G. M. (2004). *Minimum Tillage Vegetable Crop Production in California*. University of California, Division of Agriculture and Natural Resources.

Miyao, G., Klonsky, K. M., and Livingston, P. (2008). *Sample Costs to Produce Processing Tomatoes Transplanted in the Sacramento Valley*. 18: University of Califonia Cooperative Extension.

Moitzi, G., and Boxberger, J. (2007). *Vermeidung Von Bodenschadverdichtungen Beim Einsatz Von Schweren Landmaschinen–Eine Aktuelle Herausforderung*. Ländlicher Raum, Online-Fachzeitschrift des Bundesministeriums für Land- und Forstwirtschaft, Umwelt und Wasserwirtschaft.

Moots, C. K., Nickell, C. D., and Gray, L. E. (1988). Effects of soil compaction on the incidence of *phytophthora megasperma* F.Sp. *Glycinea* in soybeans. *Plant Disease* 72 (10), 896–900.

Morse, R. D. (1999). No-till vegetable production—Its time is now. *HortTechnology* 9 (3), 373–9.

Morse, R. D. (1989). Strip tillage is a good conservation method. *American Vegetable Grower* 37 (2), 32–3.

Mundy, C., Creamer, N. G., Crozier, C. R., Wilson, L. G., and Morse, R. D. (1999). Soil physical properties and potato yield in no-till, subsurface-till, and conventional-till systems. *HortTechnology* 9 (2), 240–7.

Munkholm, L. J. (2011). Soil friability: A review of the concept, assessment, and effects of soil properties and management. *Geoderma* 167-68, 236–46.

Muresan, A. O., Indrea, D., Maniutiu, D., and Sima, R. (2011). The effect of minimum strip-tillage cover crop system on a few vegetable crops. *Journal of Horticulture, Forestry and Biotechnology* 15 (1), 35–9.

Parish, R. L., and McCoy, J. E. (1998). Inconsistency of metering with a precision vegetable seeder. *Journal of Vegetable Crop Production* 4 (2), 3–7.

Paul, B. K., Vanlauwe, B., Ayuke, F. *et al.* (2013). Medium-term impact of tillage and residue management on soil aggregate stability, soil carbon, and crop productivity. *Agriculture Ecosystems & Environment* 164, 14–22.

Pedersen, H. H., Sørensen, C. G., Oudshoorn, F. W. *et al.* (2013). A wide span tractor concept developed for efficient and environmental friendly farming. In *Land Technik AgEng 2013*, 47–53. Hannover: VDI-Verlag, Dusseldorf.

Pedersen, H. H., McPhee, J. E., and Chamen, W. C. T. (2016). Wide span—Re-mechandising vegetable production. *Acta Horticultural*. XXIX IHC—Proc. Int. Symposia on the Physiology of Perennial Fruit Crops and Production Systems, and Mechanisation, Precision Horticulture and Robotics, 551–558.

Pedersen, Hans Henrik. (2011). Harvest capacity model for a wide span onion bunker harvester. *Automation and System Technology in Plant Production, CIGR Section V & NJF Section*. (Ed. Ibrahim Abd El-Hameed) 44–46. Herning, Denmark: NJF, Nordic Association of Agricultural Scientists.

Perdok, U. D., and Lamers, J. G. (1985). Studies of controlled agricultural traffic in the Netherlands. *Journal of Terramechanics* 22 (3), 187–8.

Petersen, H. (2002). Effects of Non-inverting deep tillage vs. conventional ploughing on collembolan populations in an organic wheat field. *European Journal of Soil Biology* 38 (2), 177–80.

Phatak, S. C., Dozier, J. R., Bateman, A. G., Brunson, K. E., and Martini, N. L. (2002) *Cover Crops and Conservation Tillage in Sustainable Vegetable Production*. Making Conservation Tillage Conventional: Building a Future on 25 Years of Research. (Ed. by E. Van Santen) 24–26. Proc. of 25th Annual Southern Conservation Tillage Conference for Sustainable Agriculture, Auburn, AL.

Pikul, J. L., Chilom, G., Rice, J. *et al.* (2009). Organic matter and water stability of field aggregates affected by tillage in South Dakota. *Soil Science Society of America Journal* 73 (1), 197–206.

Radford, B. J., Yule, D. F., McGarry, D., and Playford, C. (2007). Amelioration of soil compaction can take five years on a vertisol under no-till in the semi-arid subtropics. [In English]. *Soil & Tillage Research* 97 (2), 249–55.

Raghavan, G. S. V., Taylor, F., Vigier, B., Gauthier, L., and McKyes, E. (1982). Effect of compaction and root rot disease on development and yield of peas. [In English]. *Canadian Agricultural Engineering* 24 (1), 31–34.

Reicosky, D. C., Reeves, D. W., Prior, S. A. *et al.* (1999). Effects of residue management and controlled traffic on carbon dioxide and water loss. [In English]. *Soil & Tillage Research* 52 (3–4), 153–65.

Reyes, M. R., Raczkowski, C. W., Reddy, G. B., and Gayle, G. A. (2005). Effect of wheel traffic compaction on runoff and soil erosion in no-till. *Applied Engineering in Agriculture* 21 (3), 427–33.

Roberts, W., and Cartwright, B. (1991). *Vegetable Production with Conservation Tillage, Cover Crops, and Raised Beds*. Proc. of the 1991 Southern Conservation Tillage Conference, 148, 72–6.

Rogers, G. S., Little, S. A., Silcock, S. J., and Williams, L. F. (2004). No-Till Vegetable Production Using Organic Mulches. Paper presented at the Sustainability of Horticultural Systems in the 21st Century. Proc. of the XXVI International Horticultural Congress, Toronto, Canada.

Schrader, S., and Lingnau, M. (1997). Influence of soil tillage and soil compaction on microarthropods in agricultural land. *Pedobiologia* 41 (1–3), 202–9.

Servadio, P. (2013). Compaction effects of green vegetable harvester fitted with different running gear systems and soil-machinery relationship. *Journal of Agricultural Science and Applications* 2 (2), 72–9.

Spoor, G., and Godwin, R. J. (1978). Experimental investigation into deep loosening of soil by rigid tines. *Journal of Agricultural Engineering Research* 23 (3), 243–58.

Stalham, M. A., Allen, E. J., Rosenfeld, A. B., and Herry, F. X. (2007). Effects of soil compaction in potato (*solanum tuberosum*) crops. *Journal of Agricultural Science* 145 (4), 295–312.

Stirling, G. R. (2008). The impact of farming systems on soil biology and soilborne diseases: Examples from the Australian sugar and vegetable industries—The case for better integration of sugarcane and vegetable production and implications for future research. *Australasian Plant Pathology* 37 (1), 1–18.

Stirzaker, R. J., Bunn, D. G., Cook, F. J., Sutton, B. G., and White, I. (1995). *No-Tillage Vegetable Production Using Cover Crops and Alley Cropping*. Soil Management in Sustainable Agriculture. (Ed. by H. F. Cook and H. C. Lee). Proc. of the Third International Conference on Sustainable Agriculture, Wye College, University of London, United Kingdom.

Stirzaker, R. J., Sutton, B. G., and Collisgeorge, N. (1993). Soil-management for irrigated vegetable production. 1. The growth of processing tomatoes following soil preparation by cultivation, zero-tillage and an in situ grown mulch. *Australian Journal of Agricultural Research* 44 (4) 817–29.

Sutherland, R. A., Crisp, P., and Angell, S. M. (1989). The effect of spatial arrangement on the yield and quality of 2 Cultivars of Autumn Cauliflower and Their Mixture. *Journal of Horticultural Science* 64 (1), 35–40.

Taylor, J. H. (1994). Development and benefits of vehicle gantries and controlled traffic systems. In *Soil Compaction in Crop Production*. 521–37.

Tillett, N. D., and Holt, J. B. (1987). The use of wide span gantries in agriculture. [In English]. *Outlook on Agriculture* 16 (2), 63–7.

Titmarsh, G., Waters, D., Wilson, A. *et al.* (2003). Experiences in runoff and erosion of controlled traffic farming systems, Southern Queensland. *ISTRO 16: Soil Management for Sustainability, Brisbane*. 235–40.

Tu, J. C., and Tan, S. C. (1988). Soil compaction effect on photosynthesis, root rot severity, and growth of white beans. [In English, French]. *Canadian Journal of Soil Science* 68 (2), 455–9.

Tullberg, J. N. (1994). Controlled Traffic in Rainfed Grain Production. Paper presented at the XII World Congress on Agricultural Engineering: Volume II. Milan, Italy, 1295–1301.

Tullberg, J. N., Yule, D. F., and McGarry, D. (2007). Controlled traffic farming: From research to adoption in Australia. *Soil & Tillage Research* 97 (2), 272–81.

Tullberg, J. N., Ziebarth, P. J., and Li, Y. X. (2001). Tillage and traffic effects on runoff. *Australian Journal of Soil Research* 39 (2), 249–57.

Unger, Paul W. (1996). Soil bulk density, penetration resistance, and hydraulic conductivity under controlled traffic conditions. *Soil and Tillage Research* 37 (1), 67–75.

Upadhyaya, S. K., Lancas, K. P., Santos-Filho, A. G., and Raghuwanshi, N. S. (2001). One-pass tillage equipment outstrips conventional tillage method. *California Agriculture* 55 (5), 44–7.

Vermeulen, G. D., and Mosquera, J. (2009). Soil, crop, and emission responses to seasonal-controlled traffic in organic vegetable farming on loam soil. *Soil & Tillage Research* 102 (1), 126–34.

Voorhees, W. B., Senst, C. G., and Nelson, W. W. (1978). Compaction and soil structure modification by wheel traffic in the northern corn belt. *Soil Science Society of America Journal* 42 (2), 344–9.

Wang, X., Gao, H., Tullberg, J. N. *et al.* (2008). Traffic and tillage effects on runoff and soil loss on the loess plateau of northern china. *Australian Journal of Soil Research* 46 (8), 667–75.

Westcott, M. P., and Callan, N. W. (1990). Modeling plant-population and rectangularity effects on broccoli head weights and yield. *Journal of the American Society for Horticultural Science* 115 (6), 893–97.

4 Development in Energy Generation Technologies and Alternative Fuels for Agriculture

Janusz Piechocki, Piotr Sołowiej, Maciej Neugebauer, and Guangnan Chen

CONTENTS

4.1 INTRODUCTION

Agricultural productivity largely depends on energy, water, and land resources. The increasing use of energy resources is currently one of the major challenges to agriculture (Chen *et al.*, 2015; Bundschuh and Chen, 2014; Bundschuh *et al.*, 2017). Continuous high fuel and electricity prices and the need for significant reductions in greenhouse gas emissions make the improvement of farming energy efficiency essential. Exploration of new alternative and renewable energy sources is also vital.

Agricultural production relies heavily solar energy, which is indispensable in the photosynthetic process to transform inorganic compounds into organic substances that give rise to living organisms. Light, including visible, infrared, and partially ultraviolet light, is required for plant and animal production. The sun is the cheapest and the most effective source of light in the wavelength range required for the growth of all living organisms. Sunlight best meets the developmental needs of plants and animals in all stages of growth (per year) and times of day, and it induces growth phases in plants (Folta and Maruhnich, 2007; Franklin, 2009). Contemporary agriculture consumes significant amounts of energy, and it relies on both direct and indirect energy sources. Direct sources of energy include fuel for agricultural vehicles and machines, whereas indirect sources involve energy accumulated in fertilizers (Arizpe, 2011). At present, fossil fuels, in their various forms, supply most of

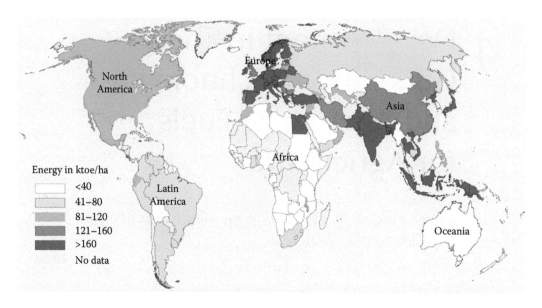

FIGURE 4.1 Demand for energy in agriculture. (From http://na.unep.net/geas/articleImages/Apr-12-figure-4.png)

the energy required by agriculture that feeds the world. The demand for energy per 1 ha of farmland is presented in Figure 4.1. It is determined by the level of technical advancement and population density in a given country.

Energy is one of the most important factors for business, and it significantly influences the development of modern agriculture. The progressing automation and mechanization of agricultural production increases the demand for energy in the form of heat, electricity, and fuel for powering agricultural machines.

The growing demand for energy spurs the development of new solutions for energy generation in agriculture, including the use of heat from biomass processing, milk cooling, or the composting of biological wastes. The growing popularity of renewable sources of energy in agriculture also stems from technological progress, including small biogas plants, which are fueled with biodegradable wastes from various types of farms. Advanced technological solutions rely on solar energy to dry agricultural produce; heat buildings, water, and greenhouses; pump water; or generate electricity with the involvement of photovoltaic cells. Some agricultural processes generate heat that is irreversibly lost. Those processes are optimized to reduce energy losses and increase their economic efficiency, which contributes to more rational energy generation and more sustainable agricultural performance in the face of limited resources. By relying on renewable sources of energy, energy consumption in agriculture can be reduced without compromising performance.

This chapter will discuss some of the selected energy generation technologies and alternative fuels in agriculture, including the recovery of heat from biomass composting; production technology; and uses of biogas and biodiesel, solar energy technologies, and optimization of production processes to minimize energy losses in agriculture.

4.2 RECOVERY OF HEAT FROM BIOMASS COMPOSTING AND ITS USE IN AGRICULTURAL PRODUCTION

Biomass is obtained from plant and animal sources and can be converted from stored chemical energy (originally from solar energy captured during photosynthesis) into bioheat, biopower, biofuels, and biomaterials. Biomass feedstocks are wide ranging, but can be broadly classified into forest residues, crop residues, animal wastes, and dedicated energy crops. The challenge is to develop environmentally sustainable and economically viable practices to produce, collect, process, store, transport, and deliver the biomass to bioenergy conversion plants.

Bioenergy uses biomass to generate either heat, electricity, or transport fuels. Bioenergy can be regarded as a form of solar energy, since photosynthesis combines atmospheric carbon dioxide with water in the presence of sunlight to form biomass, while also producing oxygen.

Increasingly, agriculture is being looked to as a source of energy. Bioenergy crops, or agricultural products, can be converted to solid or liquid fuel, and offer the potential of a lower carbon-emitting source of energy. Owing to the concern over global warming, unstable diesel fuel prices in the world market, and a limited supply in the future, many farmers have been looking for alternative fuels or growing their own. To achieve the best outcomes, there are many factors to be taken into account for each bioenergy resource, such as moisture content, resource location and distribution, and the type of conversion process.

Composting is a biological degradation process which mineralizes organic matter and releases energy as heat. Organic matter is produced from inorganic substances under exposure to sunlight in the process of photosynthesis (Equation 4.1).

$$6CO_2 + 6H_2O + \text{solar energy} \rightarrow C_6H_{12}O_6 + 6O_2 \qquad \text{(Equation 4.1)}$$

During photosynthesis, glucose and oxygen are produced from carbon dioxide and water in plant cells exposed to sunlight.

Organic material is biodegraded by microorganisms. When oxygen is available, biomass is broken down in the process of aerobic decomposition, which is the reverse of photosynthesis. Composting is the sum of microbiological processes, during which biological material is transformed by microorganisms into humus (Alexander, 1977; Ros *et al.*, 2006; Moreno *et al.*, 2013; López-González *et al.*, 2014). Microorganisms also produce significant amounts of carbon dioxide, which are released into the atmosphere, and heat (Liang *et al.*, 2003; Miyatake and Iwabuchi, 2006). Microbial biomass increases significantly during composting. The processes that take place during aerobic decomposition of organic matter can be described with the use of the Equation 4.2 (Alexander, 1977):

$$\text{Organic matter} + O2 + \text{aerobic microorganisms} \rightarrow$$
$$CO2 + NH4 + PO4 + \text{microbial biomass} + \text{heat} + \text{humus} \qquad \text{(Equation 4.2)}$$

Temperature is the most important parameter during composting (Liang *et al.*, 2003; Miyatake and Iwabuchi, 2006). The composting process can be divided into three phases: mesophilic, thermophilic, and stabilization. The mesophilic phase is characterized by the high activity of mesophilic microorganisms that feed on readily digestible organic matter, mainly sugars and amino acids. Temperature in this phase ranges from 25 to 45°C. The thermophilic phase begins when temperature exceeds 45°C. The activity of mesophilic microorganisms is slowed down, the thermophilic microorganisms are activated, and their metabolic processes raise the substrate temperature from 70 to 80°C. Biodegradation processes are most intense in the thermophilic phase. The third phase involves substrate cooling and maturation. The population size of thermophilic bacteria decreases, temperature drops to 35–40°C, mesophilic microorganisms are activated, and they decompose the remaining biomass. The thermophilic phase can be prolonged by aerating the substrate and keeping its moisture content at a stable level. This maximizes the effectiveness of biomass degradation, shortens composting time, and reduces methane emissions to the atmosphere. Air supply should be optimized to provide thermophilic microbes with the required levels of oxygen without excessively cooling or drying the substrate, which could slow down or even completely inhibit the composting process.

In addition to oxygen and water (the recommended moisture content of composted material is 60 to 70%), microorganisms use nitrogen to increase their mass, and they rely on carbon as a source of energy. The optimal carbon to nitrogen ratio (C:N) in a compost heap is 30:1 (with a tolerance limit of 25 to 35 parts carbon to 1 part nitrogen). When the C:N ratio exceeds 35:1, the process is significantly slowed down and the composted material is partially decomposed, and when the C:N ratio falls below 20:1, nitrogen can be released into the atmosphere. Various substances are added to composted matter to maintain the optimal C:N ratio (urea or liquid manure) and substrate porosity (cereal straw) (Adhikari *et al.*, 2008; Chang and Chen, 2010; Estevez *et al.*, 2012).

The rates of carbon dioxide release, oxygen uptake, and biomass decomposition have been analyzed in several studies (Finstein, 1975; Strom, 1978; Rothbaum, 1961; Wiley, 1957) which demonstrated the highest levels of microbial activity at a temperature of around 60°C, which creates optimal conditions for most thermophilic microorganisms. The production of heat with a temperature of 60–65°C inside a compost heap contributes to pasteurization and pathogen elimination, and promotes aeration and decomposition of organic matter in deeper layers (Macgregor *et al.*, 1981). Due to the specific structure and physical characteristics of composted biological material, heat is accumulated in the heap whose temperature can exceed 80°C. When optimal composting conditions are maintained and fresh biomass is continuously fed into the system, the high temperature achieved in the thermophilic phase of the process can be used to heat farm buildings or water.

A pioneering method for recovering heat and biogas from compost was developed 70 years ago in the previous century by Jean Pain, a French farmer who relied on the composting process to heat his home, prepare hot water, and recover biogas (Figure 4.2). A compost heap can be a source of heat for up to 18 months (Poulain, 1981). The compost heap generated approximately 500 m^3 of gas which was used to supply two heating stoves and a combustion engine in a power generator which charged batteries for household lamps. Jean Pain developed a sustainable method for recovering low-temperature heat and safely managing biological wastes.

In 1992, Japanese scientists Hirakazu Seki and Tomoaki Komori proposed a novel method for recovering heat from exhaust air leaving the compost heap (Seki and Komori, 1992). Exhaust air was

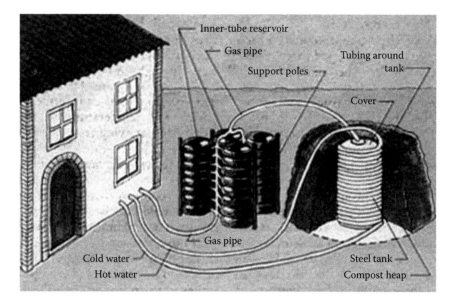

FIGURE 4.2 Diagram of the heat and biogas recovery system developed by Jean Pain. (From http://journeytoforever.org/biofuel_library/methane_pain.html)

passed through a specially designed column where it was used to heat water. Water was heated to a temperature of up to 30°C, and up to 72% of composting heat was effectively recovered on average (Figure 4.3).

In the article entitled "Extracting Thermal Energy from Composting," Truckner described one of the first systems for recovering heat from composted cattle manure and organic farm wastes (Truckner, 2006). The recovered heat served two purposes: to heat water which was then used to prepare calf feed, and to supply the floor heating system in the calf barn. Heat meters were installed in the system to optimize heat production, the composition of the compost heap, and the structure of the compost container.

In an experiment conducted in 2005, Sołowiej investigated the effectiveness of compost heat for heating a vegetable greenhouse in northern Poland in early spring (Sołowiej, 2007). The test stand comprised two plastic tunnels with an area of 120 m^2 each (an experimental tunnel with preheated soil and a control tunnel without heating), compost heaps, a system of pipes (PVC, Ø 16 mm, wall thickness 1.5 mm) connecting compost heaps and soil in the experimental tunnel, a circulation pump distributing water in the pipe system, an expansion vessel and thermometers for measuring compost, supply water, and return water temperature. The diagram of the test stand is presented in Figure 4.4.

The compost heap had five layers. Each layer was composed of dry straw (15–20 cm), fresh organic matter (cabbage leaves, carrot, beetroot discards, etc.), dry straw, and soil (10–15 cm). During the construction of the compost heap, every layer was watered to obtain the required moisture content. A system of pipes collecting heat was placed inside the heap. The highest demand

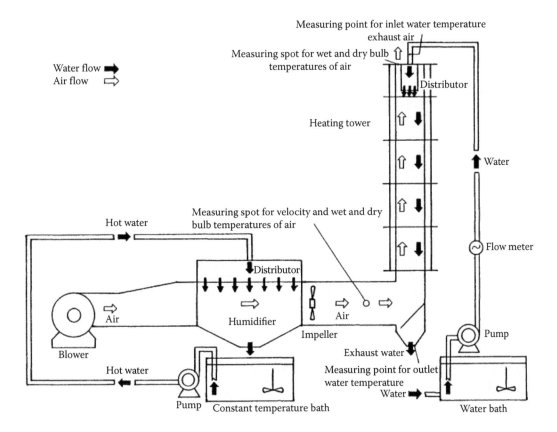

FIGURE 4.3 Schematic diagram of the experimental apparatus. (From Seki, H. and Komori, T., 1992. *Journal of Agricultural Meteorology* 48 (3), 246–273.)

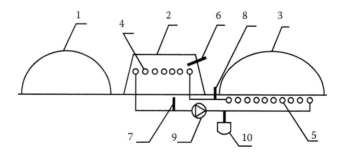

FIGURE 4.4 Testing station diagram: (1) control tunnel, (2) compost heap, (3) tunnel with preheated soil, (4) system of pipes collecting heat, (5) system of pipes preheating soil, (6) thermometer measuring compost temperature, (7) thermometer measuring supply water temperature, (8) thermometer measuring return water temperature, (9) circulating pump, (10) expansion vessel.

for energy was noted at the beginning of the experiment, because soil in the experimental tunnel was frozen after winter. This demand was met by utilizing the highest composting temperature, which is noted at the beginning of composting. Soil temperature was stabilized after 10 days and remained constant at 9–11°C until the end of the experiment. The energy generated by the heap was used only to maintain constant soil temperature. Lettuce grown in the heated tunnel was harvested on experimental day 34, i.e., 22 days after planting. The experiment was concluded on day 42, when lettuce from the control tunnel was harvested. Lettuce grown in the heated tunnel was harvested six days earlier than the crops grown in the control tunnel. Compost produced during the experiment was used as fertilizer.

Scientists from the University of New Hampshire developed a heat recovery system for pre-heating water in a farm (Smith and Aber, 2014). The UNH heat exchange system operates by blowing compost vapor (110–170°F) against an array of two-phase, super-thermal conductor heat pipes which were developed by a Canadian company named Acrolab. The six heat pipes (isobars) are 30 feet long, with 22 feet contained within a 24-inch diameter vapor duct, and another 8 feet contained within a 295-gallon water tank. The isobars provide thermal uniformity across the entire length of the pipe, meaning if one end is heated, the energy is immediately distributed evenly across the entire length of the pipe (Acrolab, 2013). More specifically, when compost heated vapor is applied to the evaporator side of the pipe (portion contained within the 24-inch diameter pipe), the refrigerant inside the isobar heats up and vaporizes. The vapor stream within the isobar travels up the pipe, condensing on the cooler side, releasing its energy in the bulk storage water tank through the latent heat of condensation. After condensing, the refrigerant is returned to the warm end of the pipe through gravity, repeating the process without any moving parts (Figure 4.5).

Compost heaps are also used inside greenhouses. The energy produced by the composting process heats the greenhouse, and the released carbon dioxide is used up by plants during photosynthesis (manure compost as passive greenhouse heating; 2009). Some passive heating systems rely on vermicompost, whose temperature reaches 40°C (Figure 4.6).

For the composted waste to generate the optimal quantities of heat and carbon dioxide, the biomass has to be suitably aerated and kept moist. The use of a compost heap as a source of low-temperature heat:

- Improves the cost-effectiveness of greenhouse vegetable production by significantly speeding up plant growth and harvest
- Reduces the demand for conventional sources of energy and minimizes air pollution associated with traditional heating methods
- Promotes the reuse of organic wastes from the field and contributes to sustainable agriculture
- Produces valuable organic fertilizer which can be used on the farm or sold

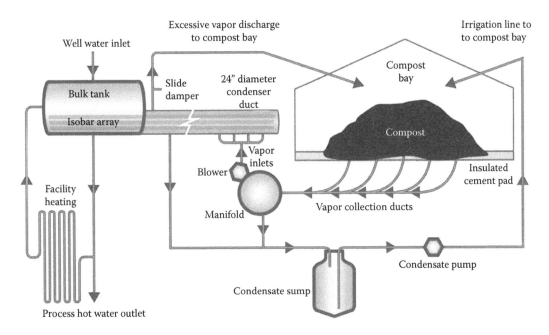

FIGURE 4.5 Flow diagram of a heat recovery system. (From https://www.biocycle.net/2014/02/21/heat-recovery-from-compost)

FIGURE 4.6 Vermicompost bin. (From http://www.permaculture.co.uk/articles/heating-greenhouse-compost-and-manure)

4.3   PRODUCTION TECHNOLOGY AND USE OF BIOGAS

Agricultural biogas is a gaseous fuel which is produced from farm wastes, mostly agricultural biomass and by-products of agricultural production, including liquid or solid manure, plant residues from food processing, as well as forest biomass (Deng *et al.*, 2012; Kafle and Kim, 2013; Deng *et al.*, 2014). Farm wastes can be fermented to produce methane. Agricultural biomass is increasingly being obtained from farms that specialize in the production of energy crops (Figure 4.7).

Biogas is also produced naturally from organic matter that decomposes under anaerobic conditions. This process takes place already at temperatures higher than 10°C when organic compounds are fermented by symbiotic microbial communities in the natural environment.

The process by which gas is produced during anaerobic decomposition of biological matter has been long known, but it was first described in the seventeenth century by Von Helmont (1630). In the following years and centuries, this phenomenon was investigated by Shirley (1667), Volta (1776), Priestly (1790), and Dalton (1804). In 1806, the first laboratory experiment which resulted in the production of methane from organic waste was conducted by Davy. In 1868, Bechamp demonstrated that sediments from the production of starch and sucrose were decomposed by microorganisms, which led to the release of methane and carbon dioxide. Anaerobic decomposition of cellulose was investigated in a series of experiments conducted by Mitscherlich, Hoppe-Seyeler, Popoff, and Tappeiner in the nineteenth century (Marchaim, 1992).

Further scientific inquiries into the methane-generating processes were made in the 1930s in Europe, and research aiming to harness methane for energy generation was also undertaken in China and India. At present, those countries operate the highest number of biogas plants which play a significant role in their energy production systems. In Europe, the interest in biogas plants peaked in two distinctive periods. The first period covered the late 1940s and the early 1950s, which resulted from energy shortages and economic hardships after World War II. The interest in biogas plants was revived in the 1970s in the face of the global energy crisis, and it paved the way to research work into biogas production in the United States and other countries.

The recent interest in methane production can be attributed not only to its energy generating potential, but mainly to the fact that this process can be used to effectively manage organic matter which accumulates waste and sewage sludge. This method can help solve environmental problems, particularly in rural areas where the sludge that remains after digestion can be used as organic fertilizer which is much more available for plants than non-digested sludge. In municipal wastewater

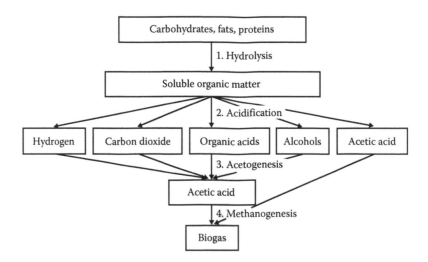

FIGURE 4.7   Stages of biogas production during methane fermentation. (From https://en.wikipedia.org/w /index.php?title=Biogas&oldid=709899992)

treatment plants, energy can be recovered from sewage sludge through methane production during anaerobic digestion (Figure 4.8).

Methane production processes can be conducted on an industrial scale to control biogas production in biogas plants. Biogas plants supplied with farm waste are referred to as agricultural biogas plants, and the generated biogas is known as agricultural biogas (Figure 4.9).

Methane is produced by mesophilic bacteria which readily proliferate under the conditions found in biogas plants. Mesophilic bacteria thrive at a temperature of 37–40°C. Biogas production relies on living microorganisms which have to be provided with optimal physical and chemical conditions, as well as strictly controlled quantities of biomass as substrate for the fermentation process. When those conditions are not met, the biogas is not produced, and the generation process is not energy efficient (Deublein and Steinhauser, 2008).

The quantity of biogas produced from different substrates during methane fermentation is different (Table 4.1). The C:N ratio of composted substrate is also a very important parameter which determines the rate at which organic matter is decomposed. The optimal C:N ratio is 20:30:1. A narrow C:N ratio contributes to the excessive release of ammonia (NH_3), whereas a very wide C:N ratio prolongs the time between fermentation and biogas production. Both narrow and wide values of this parameter lower the pH, which inhibits the growth of mesophilic bacteria (Pang *et al.*, 2008).

The composition of biomass fed to biogas plants can be varied, but it should be noted that fermentation can only take place within a given range of physicochemical parameters, and that bacteria have a limited capacity for processing biomass components. Not all biomass is decomposed in the digester tank. In addition to biogas, the fermentation process also generates substantial amounts of anaerobically digested biomass which constitutes valuable fertilizer alone, or in combination with other compounds. Nitrogen found in fermented biomass is converted to the ammonium form that is easily available for plant uptake and less readily leached from soil (Table 4.2).

Agricultural biomass can be used to produce biogas and vehicle fuel, which is a valuable resource in agricultural production. Most importantly, fuel can be used locally in agricultural production, including processes that produce biomass for the generation of biofuels. Agricultural biomass is also used in other fermentation processes to obtain substances for the production of biofuels. Such substances include methanol and ethanol which, together with liquid hydrocarbons, are used as

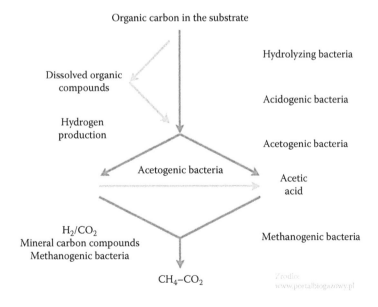

FIGURE 4.8 Bacterial communities involved in biogas production. (From https://en.wikipedia.org/w/index .php?title=Biogas&oldid=709899992)

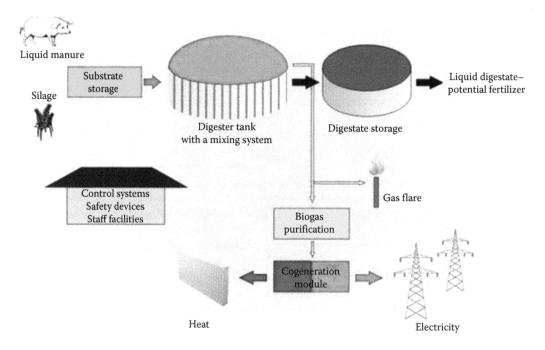

FIGURE 4.9 Diagram of a biogas plant in a farm. (From https://en.wikipedia.org/w/index.php?title =Biogas&oldid=709899992)

TABLE 4.1

Quantity of Biogas Produced from Different Substrates during Methane Fermentation

Substrate	Biogas Quantity (m³/Mg)
Liquid cattle manure	25
Liquid pig manure	36
Whey	55
Sliced beetroot	75
Brewer's spent grain	75
Dried distiller's grains with soluble	80
Green waste	110
Biological waste	120
Maize silage	200
Fat	800

Source: https://en.wikipedia.org/w/index.php?title=Biogas&oldid=709899992

---

biofuels for powering tractors and agricultural machines. This is an important problem related to the use of renewable energy sources in farming systems, and requires separate discussion.

The content of methanol or ethanol in biofuel varies considerably across countries. It usually ranges from several to less than 20%, but in some countries, such as Brazil, it can be as high as 30%. For combustion engines to burn biofuel with high methane content, their compression ratio has to be modified. Biogas can also be used directly in the fuel combustion engines of power generators.

TABLE 4.2
Typical Composition of Agricultural Biogas

Typical Composition of Biogas		
Compound	Formula	%
Methane	CH_4	50–75
Carbon dioxide	CO_2	25–50
Nitrogen	N_2	0–10
Hydrogen	H_2	0–1
Hydrogen sulfide	H_2S	0–3
Oxygen	O_2	0–0.5

Source: https://en.wikipedia.org/w/index.php?title=Biogas&oldid=709899992

Generators produce electricity, and the heat recovered from exhaust gas can also be used in agricultural production (Murphy and Baxter, 2013).

The principles of energy generation and use of renewable energy sources have been laid down in the EU Directive, and the relevant regulations are applicable in all Member States (EU Directive, 2009).

4.4   PRODUCTION TECHNOLOGY AND USE OF BIODIESEL

The tractor is a mobile power unit and one of the most widely used pieces of farm equipment. Modern tractors are often powered by internal combustion (IC) engines due to their high reliability and efficiency and ability in implementing heavy workloads. The engine's reliance on fossil fuels and the emissions in the atmosphere from the combustion process has contributed to global climate change. Biodiesels present a genuine opportunity as the future of renewable fuel for agricultural and other machinery, both to eliminate the further depletion fossil fuels and to provide a significant reduction in greenhouse gas emissions.

Biodiesel is a liquid fuel made from processing of either tallow (animal fat) or vegetable oil in a process called "esterification." Overall, biodiesel production is a relatively simple process, basically consisting in putting together the animal tallow or vegetable oil with an alcohol in a catalyst to have the process of transesterification, in which the oil is separated from the glycerin. This can be possibly done in a small scale on farms to provide fuel for diesel-powered farm machinery. By comparison, ethanol production involves a fermenting process and it is more expensive to set up a processing plant.

Biodiesel is renewable, and can be used as a fuel in diesel engines either as biodiesel or even as straight oil that has been filtered. The energy content of biodiesel is approximately 36.2 MJ/l. Compared to petro-diesel, biodiesel gives off considerably lower emissions of particulate matter (PM), carbon monoxide (CO), and hydrocarbon (HC). For example, it is found that switching to a B20 fuel will reduce greenhouse gas emissions by some 17%, together with reductions in other aspects of air pollution (Chen *et al.*, 2015).

Although biodiesel has better combustion efficiency and lower emissions, there are a number of issues that could potentially influence the future production and utilization of biodiesel, including relatively poor performance at low temperatures, fuel quality standards, and a decrease in power and torque generated by biodiesels (Sadeghinezhad *et al.*, 2013). Others may include carbon deposit formation, fuel filter clogging, and engine wear. Higher costs of maintenance may thus be likely for biodiesel.

Biodiesel blends up to 5% by volume in North America and 7% in Europe are now routinely used in agriculture machines without impacting machine performance or durability. Other research showed that biodiesel blends of 20% (B20) or less would also not change the engine performance in a noticeable way. Higher biodiesel blends, however, require additional care to ensure performance and durability. For higher blending, it may also be necessary to modify the machinery, particularly the fuel delivery system. A second fuel tank and other modifications to machinery may be required.

Food versus fuel balance is an important issue when using food crops for the production of biodiesel (Girard and Fallot, 2006). Excessive uses of food for fuel could either decrease the food supply around the world, or increase deforestation to provide more farmland, both of which are undesirable. It was said that filling one tank of biodiesel could use one year's worth of food for poor people. It was also estimated that the biofuel production needed to replace just 10% of fossil fuels in transport in the United States, Canada, and the European Union could require between 30 to 70% of existing crop areas. The coproducts from the production of biodiesel, canola, and soybean meals may be used in pig and poultry diets.

Alternative sources of new second or third generation feed stocks for biodiesel (for example, from "industrialized" sources such as agricultural and food wastes) or algae oil are being actively investigated (Girard and Fallot, 2006; Bhuiya *et al.*, 2016). It is predicted that the technology for the second generation biofuels may still take several years to become viable to compete with "cheap" mineral oils.

4.5 SOLAR ENERGY TECHNOLOGIES IN AGRICULTURE

Solar energy is a renewable resource that can be harnessed for agricultural production. The potential solar energy around the world is presented in Figure 4.10. According to Devabhaktuni *et al.* (2013), there are three main methods of capturing solar energy for agricultural use:

- Photovoltaic systems integrated with buildings: Photovoltaic systems offer a promising solution for sites that are located remotely from power grids, including farms. They can be used off-grid, which means that stables and other buildings can be situated far from the main farm building and do not have to be connected to the grid. In buildings connected to the grid, the electricity generated by PV panels can be distributed to specific circuits or systems. Alternatively, two-way meters can be installed to sell excess power back to the grid during the day. This solution does not differ from other PV installations, and it is not limited to agricultural use only.

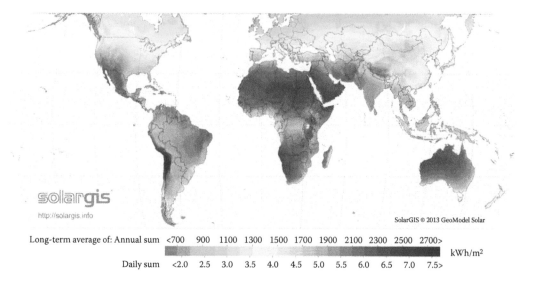

FIGURE 4.10 Annual and daily sum of solar energy in the world in kilowatts per square meter. (From https://upload.wikimedia.org/wikipedia/commons/9/9d/SolarGIS-Solar-map-World-map-en.png)

FIGURE 4.11 Photovoltaic system in Broadwater Farm. (From http://photonenergy.co.uk/agriculture/case
-studies/58-case-study-broadwater-farm)

Photovoltaic systems have been described by many researchers (Tudisca *et al.*, 2013;
Moosavian *et al.*, 2013), and various commercial solutions are available on the market
(Viessmann; PhotoEnergy http://photonenergy.co.uk/agriculture). A model PV system is shown
in Figure 4.11. A system with the annual energy output of 67,000 kWh can reduce CO_2
emissions by 38 tons. The system has rated power of a 84 kWp, and all PV panels are connected to the grid via six inverters and two-way electricity meters. The presented system is
used in agriculture, but identical solutions are used in urban areas. Agricultural facilities can be
supplied with electricity not only directly from the grid, but also from PV panels. Additionally,
due to annual and daily fluctuations in photovoltaic generation, such solutions are not used to
power farm machines. Such installations are referred to as "on-grid" or "grid-tied" systems,
which means that all of the generated energy is sold to the grid, and the power required for
machine operation is purchased from the grid. Farms that generate electricity can also balance
their energy output. The produced electricity is used by the farm, and only excess power is sold
to the grid. When the farm needs more energy or when the local generation system is down,
electricity is purchased from the grid. Such solutions are cost-effective, and they have been
used for many years outside the agricultural sector. Tudisca *et al.* (2013) described efficient PV
systems in Sicilian farms. According to the authors, the payback period in such projects,
including PV installations that were financed solely by farmers, ranged from 5 to 7.5 years.
However, the results of economic analyses evaluating the cost-effectiveness of PV systems
may vary depending on both sun exposures in a given region and electricity prices. Legal
regulations are another important consideration, including whether private entities and natural
persons are allowed to sell energy or not. This can significantly affect the profitability of PV
installations even in regions receiving similar amounts of sunlight.
- Off-grid systems are yet another popular solution. They differ from on-grid systems in that
  excess energy is stored and used when the PV installation is down (e.g., at night). Such
  solutions are relatively rarely used outside the farming sector because they require expensive
  and ineffective electrical batteries. Akikur *et al.* (2013) demonstrated that off-grid systems
  (in this case, hybrid systems combining PV panels and wind farms) are cost-effective only
  when the distance between the farm and the nearest power grid exceeds 3 km. According to
  Ghafoor and Munir (2015), off-grid systems are profitable in regions with high solar potential
  (the Middle East, Africa, and tropical countries) due to the steady decrease in the prices of PV
  installations.
- Strictly agricultural uses of PV systems: For example, solar energy can be used to dry agricultural produce and in other farming operations.

The above points two and three will be discussed in greater detail below.

4.5.1 The Use of PV Systems

The main rationale for the use of PV systems in farming is to increase the percentage of energy from renewable sources in agricultural production (Bardi *et al.*, 2013). At present, the farming sector relies primarily on energy from fossil fuels. Tractors and agricultural machinery are powered by petroleum products derived from crude oil. In systems that directly generate electricity, this problem can be addressed in two ways:

- Electricity can be converted to fuel for powering combustion engines (or other engines). The only solution of the type that has been implemented in practice involves water electrolysis and vehicles powered by hydrogen. However, it is not used in agriculture due to problems with hydrogen storage and distribution (Bardi *et al.*, 2013).
- Electricity can be used directly in agricultural production: This solution is not related to on-grid or off-grid systems for powering ordinary electrical devices, which were described in previous sections, but it involves supply of electricity to equipment which is powered by other conventional energy carriers.

An electric vehicle (low power tractor) developed as part of the RAMSES project is an example of the second solution (Faircloth *et al.*, 2013; Mousazadeh *et al.*, 2009b). The vehicle differs from conventional electrical tractors. In addition to performing standard operations, it acts as a power source for other agricultural machines and mechanical devices, including watering, sowing, planting, and harvesting equipment. The vehicle weighs 1700 kg, and it has a maximum speed of 45 km/h and a carrying capacity of up to 1000 kg. The battery is charged via a PV system, and it can power the tractor for a range of around 80 km on roads or 4 ha of work in the field. The batteries can be connected in series (96 V) or in parallel (48 V). The tractor is presented in Figure 4.12.

A greenhouse where a PV installation is used for passive cooling is yet another example of a system which directly uses electricity from PV cells (Figure 4.13). Photovoltaic louvres installed on the roof can be rotated at different angles to control the amount of energy generated by each panel and to create shade (and lower temperatures) inside the greenhouse. The amount of solar energy reaching the greenhouse has to meet the crops' energy requirements. The energy balance is calculated based on the amount of solar radiation reaching the greenhouse and all energy losses (Figures 4.14 and 4.15).

FIGURE 4.12 Battery-powered RAMSES agricultural vehicle. (From Mousazadeh, H. *et al.*, 2009a, *Journal of Cleaner Production* 17 (9), 781–90.)

FIGURE 4.13 Greenhouse with a PV system. (From Marucci, A. and Cappuccini, A., 2016, *Energy* 102, 302–12.)

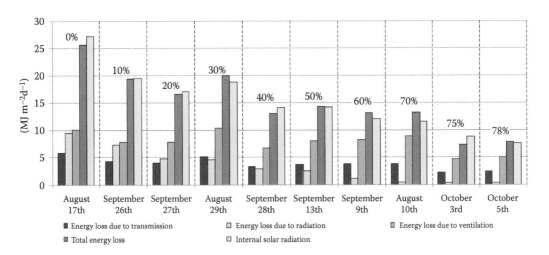

FIGURE 4.14 Energy loss in a greenhouse on selected days. (From Marucci, A. and Cappuccini, A., 2016, *Energy* 102, 302–12.)

Based on the calculated values of solar irradiation and energy losses inside the greenhouse (Figures 4.14 and 4.15), photovoltaic louvres are manipulated to achieve the optimal degree of shading which meets the energy requirements of crops at each growth stage (Marucci and Cappuccini, 2016).

4.5.2 HEAT USE

Heat from solar radiation can also be used in a variety of practical applications:

- In solar thermal collectors for heating water; this solution is widely used outside agriculture (Viessmann, 2016).

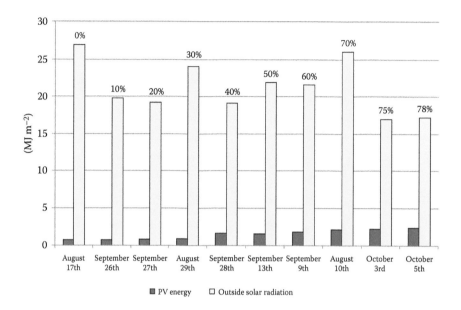

FIGURE 4.15 Outside solar radiation and energy generated by PV panels. (From Marucci, A. and Cappuccini, A., 2016, *Energy* 102, 302–12.)

- In conventional greenhouses to create optimal conditions for thermophilic plants grown in colder climates.
- For drying agricultural produce (Vijaya Venkata Raman *et al.*, 2012); according to El-Sebaii and Shalaby (2012), this is one of the key uses of solar energy in agriculture. Losses in food drying processes can be as high as 30 to 40%. Freshly harvested agricultural produce can be quickly and effectively dried with the use of freely available solar energy to substantially minimize those losses.

The latter solution is still rarely used in agriculture. According to El-Sebaii and Shalaby (2012), there are four main drying processes that rely on solar energy:

- Natural drying: The material intended for drying is placed in a well-ventilated location with ample sunlight exposure. A photograph of naturally dried raisins is presented in Figure 4.16.
- Direct drying: The material is placed in containers (with or without side walls) covered with transparent material. The heat from solar radiation evaporates moisture from the dried product, and hot air is evacuated by convection. The drying process is presented in Figure 4.17, and a photograph of various fruit dried with the use of this technology is shown in Figure 4.18.
- Indirect drying: In this technology, air is heated by solar energy, and it is directed to drying chambers (Figure 4.19). The continuous flow of hot air dries the food inside the chamber. A photograph of an indirect drying system is shown in Figure 4.20.
- Mixed drying: The mixed drying system combines direct and indirect drying processes described in points two and three. A diagram of the mixed drying process is presented in Figure 4.21.

FIGURE 4.16   Naturally dried raisins. (From https://cdn.comsol.com/wordpress/2016/01/Sun-drying-process .jpg)

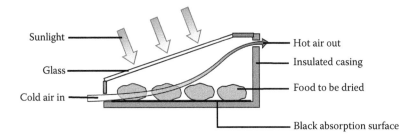

FIGURE 4.17   Diagram of a direct drying process. (From https://upload.wikimedia.org/wikipedia/commons /thumb/2/27/Direct_Solar_dryder.svg/2000px-Direct_Solar_dryder.svg.png)

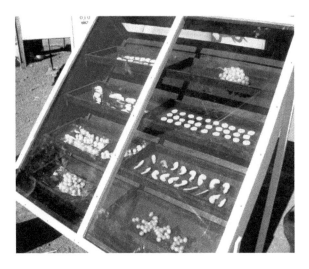

FIGURE 4.18   Direct drying of fruit. (From http://www.siffordsojournal.com/uploaded_images/food_dryer _016-778693.jpg)

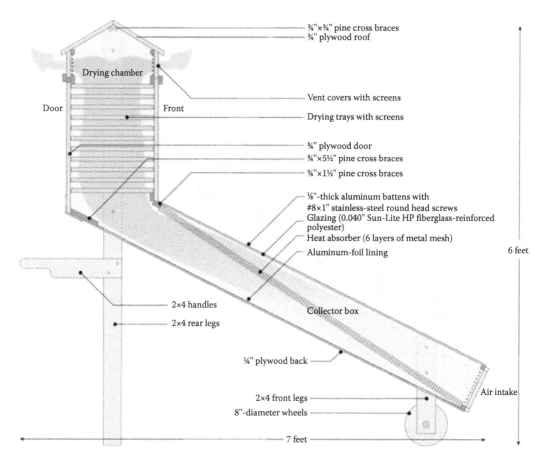

FIGURE 4.19 Diagram of an indirect drying process. (From http://www.motherearthnews.com/~/media /Images/MEN/Editorial/Special%20Projects/Issues/2014/06-01/Best%20Ever%20Solar%20Food%20Dehydrator %20Plans/Lead-Chart%20jpg.jpg?la=en)

Sun drying is one of the oldest food preservation methods known to man, but according to the literature (Vijaya Venkata Raman *et al.*, 2012; Kalogirou, 2014), it is not widely used in industrial production. Forced air-drying methods that rely on heat from fossil fuels are much more popular in the food processing industry, including in countries with a high solar potential. Sun-drying methods are generally regarded as outdated and are used only in small farms.

4.5.3 OTHER EXAMPLES OF SOLAR ENERGY USE IN AGRICULTURE— DESALINATION OF SEAWATER

Water is one of the major resources in agricultural production of both crops and livestock. Global water resources are being depleted at an alarming rate, which necessitates the search for a new source of water. The desalination of seawater offers a viable solution (Kalogirou, 2014). The existing desalination methods are presented in Figure 4.22.

FIGURE 4.20 Indirect food dryer. (From http://www.activistpost.com/wp-content/uploads/2016/02/solar
_dehydrator.png)

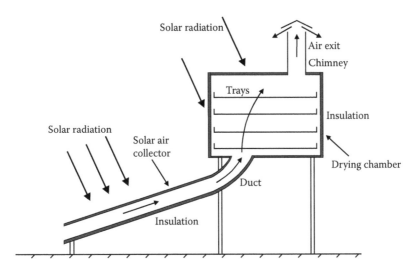

FIGURE 4.21 Diagram of a mixed drying process. (Reprinted from *Solar Energy Engineering: Processes and
Systems*, Second Edition, Kalogirou, S. A., p. 423, Copyright 2014, with permission from Elsevier.)

The percentage of renewable energy sources involved in the process of seawater desalination is
shown in Figure 4.23.

An example of a stationary system for water desalination is presented in Figure 4.24.

According to Shatat *et al.* (2013), desalination methods that rely on renewable energy sources
have reached technological maturity and can successfully compete with conventional solutions.

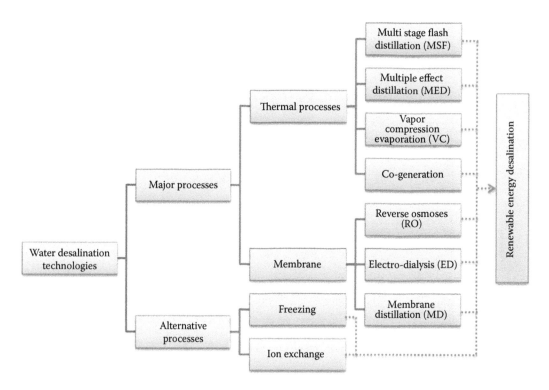

FIGURE 4.22   Water desalination methods. (From Shatat, M. and Riffat, S., 2012, *International Journal of Low-Carbon Technologies*. Oxford University Press, 1–19.)

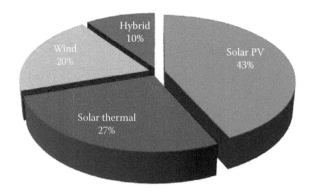

FIGURE 4.23   Percentage of renewable energy sources involved in seawater desalination. (From Shatat, M. *et al.*, 2013, *Review, Sustainable Cities and Society* 9, 67–80.)

They are highly recommended for regions which have extensive access to renewable energy sources, but have poorly developed infrastructure for transmitting conventional energy carriers, in particular the Middle East and Africa where sunlight is ample most of the year (Figure 4.10). The development of water desalination plants in those regions, including in agriculture, will increase the supply of potable water, reduce the consumption of energy from conventional sources, increase agricultural output, and minimize CO_2 emissions.

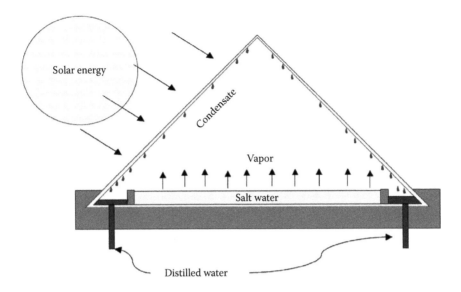

FIGURE 4.24 Stationary system for water desalination. (From Ali, M. T. *et al.*, 2011, *Renewable and Sustainable Energy Reviews* 15 (8), 4187–99.)

4.6 IMPROVING ENERGY EFFICIENCY IN AGRICULTURE BY OPTIMIZING PRODUCTION PROCESSES AND MINIMIZING ENERGY LOSSES

According to Bardi *et al.* (2013), the global renewable energy resources that can be used to implement modern technological solutions are estimated at 8.5 EJ. The present demand for energy in agriculture is approximately 30 EJ. The main renewable sources of energy with the greatest potential for agricultural production are wind and solar power. If energy were to be supplied by PV panels only, photovoltaic cells would have to cover the area of 30,000 km^2 (with only 20% conversion efficiency and average solar energy levels of 8×10^9 J/m^2; Figure 4.10) to fully meet the energy demand of the farming sector. This accounts for only 0.2% of farmland which is presently used to grow cereals worldwide. An environmental impact assessment, including changes in the intensity of sunlight reaching the Earth (reflected radiation) and effects on biodiversity, should also be carried out (Statistical Review of World Energy, 2012; Hernandez *et al.*, 2014).

The costs associated with the desalination of seawater have been decreasing steadily in recent years and are presently estimated at (Shatat *et al.*, 2013):

- With energy from fossil fuels: Estimated ultimate recovery 0.35–2.70 /m^3
- With wind energy: Estimated ultimate recovery 1.0–5.0 /m^3
- With energy from PV panels: Estimated ultimate recovery 3.14–9.0 /m^3
- With energy from solar thermal collectors: Estimated ultimate recovery 3.5–8.0 /m^3

Methods for generating energy from biomass, particularly biological wastes from agricultural production (such as livestock manure or a mixture of chicken manure and straw), deserve special attention. Biogas production and composting utilize waste by turning it into methane and heat—energy carriers that can be used locally by the farm to significantly improve its production efficiency.

The Energy Returned on Energy Invested (EROEI) ratio is a highly useful indicator for evaluating the efficiency of renewable energy sources. It measures the relationship between the amount of usable energy delivered by a particular source or device and the total amount of energy that was invested to obtain that source or manufacture that device. When the EROEI is less than one, the energy source is inefficient, and the analyzed process does not generate usable energy. According to Bardi *et al.* (2013), energy generation methods are not profitable when the EROEI is lower than four or five. The EROEI of fossil fuels has been decreasing steadily due to their depletion and increasing mining costs. In contrast, renewable sources of energy are characterized by increasing EROEI values. At present, wind energy and solar energy can effectively compete with fossil fuels as reliable sources of high-quality power. The progress made in renewable energy technology has driven down the prices and increased the efficiency of green solutions. Unfortunately, the above does not apply to energy generated from biofuels. According to Bardi *et al.* (2013), this can be explained by the low efficiency of photosynthesis as well as the relatively high cost and complexity of processing substrates for biofuel production. The EROEI of biofuels can be improved by using the generated energy as close as possible to its generation site. This goal should not be very difficult to accomplish in agriculture.

4.7   CONCLUSION

A significant amount of energy is consumed in agriculture. Most of this energy comes from non-renewable sources. At present, nearly all the tractors and agricultural machinery available run on petroleum products such as diesel and petrol. Significant research has been done in the past to reduce our dependency on petroleum products. A number of alternatives like biodiesel and biogas have also been investigated. However, there are still a number of cost and technical issues related to their usage to be overcome.

Energy use is now seen as one of the key indicators of sustainable development. Renewable energy sources should contribute to the energy security. Agriculture can also play a dual role as an energy user and as an energy supplier in the form of bioenergy. Advanced technologies and alternative fuels can be used to improve the efficiency of agricultural production, while minimizing energy consumption in the farming sector.

Despite the current limitations, it is believed that the long-term future for renewable energy is positive, since the price of fossil fuels will continue to rise as the resources are depleted, while the price of renewable energy will continue to decrease. Currently, many farms are set remotely from power grids, and they produce organic materials that can be converted into energy. Those materials include wastes which have to be sustainably managed without causing harm to the natural environment. Biological wastes constitute biomass which is an excellent substrate for composting, a process that generates heat for agricultural production. Biomass can also be fermented to produce biogas which is used in hybrid systems that cogenerate electricity and heat. Biomass can also be used to produce other biofuels for powering combustion engines in tractors and agricultural machines. Advanced technologies that convert solar energy into both heat and electricity are also increasingly used in the farming sector.

Another important problem related to the use of renewable energy sources in farming systems is energy storage. A variety of technologies are being developed. However, due to its extensiveness and complexity, this problem should be thoroughly discussed in a separate paper. Overall, considerable technology demonstration will be required to prove the technical performance, understand implementation requirements, and build local knowledge and capability. In addition, the farming practices of precision agriculture, controlled traffic farming (CTF), direct drilling, and minimum tillage could also be used to reduce energy use.

REFERENCES

Adhikari, B. K., Barrington, S., Martinez, J., and King, S. (2008). Characterization of food 406 waste and bulking agents for composting. *Waste Manage.* 28, 795–804.

Acrolab. (2013). Isobar heat pipe. www.acrolab.com/products/isobars-heat-pipes.php.

Akikur, R. K., Saidur, R., Ping, H. W., and Ullah, K. R. (2013). Comparative study of stand-alone and hybrid solar energy systems suitable for off-grid rural electrification: A review. *Renewable and Sustainable Energy Reviews* 27, 738–52.

Alexander, M. (1977). *Introduction to Soil Microbiology.* J. Wiley & Sons, New York, U.S.A.

Ali, M. T., Fath, H. E. S., and Armstrong, P. R. (2011). A comprehensive techno-economical review of indirect solar desalination. *Renewable and Sustainable Energy Reviews* 15 (8), 4187–99.

Arizpe, N., Giampietro, M., and Ramos-Martin, J. (2011). Food security and fossil energy dependence: an international comparison of the use of fossil energy in agriculture (1991–2003). *Critical Reviews in Plant Sciences* 39, 45–63.

Bardi, U., Asmar, T. E., and Lavacchi, A. (2013). Turning electricity into food: The role of renewable energy in the future of agriculture. *Journal of Cleaner Production* 53, 224–31.

Bhuiya, M. M. K., Rasul, M. G., Khan, M. M. K., Ashwath, N., Azad, A. K., and Hazrat, M. A. (2016). Prospects of second generation biodiesel as a sustainable fuel. Part II. Properties, performance and emission characteristics. *Renewable and Sustainable Energy Reviews* 55, 1129–46.

Bundschuh, J. and Chen, G. (Eds.) (2014). *Sustainable Energy Solutions in Agriculture,* CRC Press, Taylor & Francis Books.

Bundschuh, J., Chen, G., Chandrasekharam, D., and Piechocki, J. (Eds.) (2017). *Geothermal, Wind and Solar Energy Applications in Agriculture and Aquaculture,* CRC Press, Taylor & Francis Books.

Chang, J. I. and Chen, Y. (2010). Effects of bulking agents on food waste composting. *Bioresource Technol.* 101, 5917–24.

Chen, G., Maraseni, T., Bundschuh, J., and Zare, D. (2015). Agriculture: Alternative Energy Sources. (Ed. S. Anwar). *Encyclopedia of Energy Engineering and Technology.* Taylor & Francis Books, London, UK.

Deng, L., Chen, Z., Yang, H., Zhu, J., Liu, Y., Dlugi, Y., and Zheng, D. (2012). Biogas fermentation of swine slurry based on the separation of concentrated liquid and low content liquid. *Biomass and Bioenergy* 45, 187–94.

Deng, L., Li, Y., Chen, Z., Liu, G., and Yang, H. (2014). Separation of swine slurry into different concentration fractions and its influence on biogas fermentation. *Applied Energy* 114, 504–11.

Deublein, D. and Steinhauser, A. (2008). *Biogas from Waste and Renewable Resources.* WILEY-VCH Verlag GmmH & Co. KGaA DOI: 10.1002/9783527632794.ch8.

Devabhaktuni, V., Mansoor, A., Depuru, S. S. S. R., Green, R. C., Nims, D., and Near, C. (2013). Solar energy: Trends and enabling technologies. *Renewable and Sustainable Energy Reviews* 19, 555–64.

El-Sebaii, A. A. and Shalaby, S. M. (2012). Solar drying of agricultural products: A review. *Renewable and Sustainable Energy Reviews* 16, 37– 43.

Estevez, M. M., Linjordet, R., and Morken, J. (2012). Effects of steam explosion and codigestion in the methane production from Salix by mesophilic batch assays. *Bioresour. Technol.* 104, 749–56.

EU Directive. (2009). Renewable Energy Sources Directive.

Faircloth, W. H., Rowland, D. L., and Lamb, M. C. (2013). *Evaluation of Peanut Cultivars for Suitability in Biodiesel Production Systems.* University of Georgia, College of Agricultural and Environmental Sciences. http://caes2.caes.uga.edu/commodities/fieldcrops/peanuts/pins/documents/EvaluationofPeanutCultivarsforSuitability.pdf.

Finstein, M. S. and Morris, M. L. (1975). Microbiology of municipal solid waste composting. *Advan. Appl. Microbiol.* 19, 113–51.

Folta, K. M. and Maruchnich, S. A. (2007). Green Light: A signal to slown down or stop. *J. Exp. Bot.* 58, 3099–111.

Franklin, K. A. (2009). Light and temperature signal crosstalk in plant development. *Curr. Op. Plant. Bio.* 12, 63–8.

Hernandez, R. R., Easter, S. B., Murphy-Mariscal, M. L., Maestre, F. T., Tavassoli, M., Allen, E. B., Barrows, C. W., Belnap, J., Ochoa-Hueso, R., Ravi, S., and Allen, M. F. (2014). Environmental impacts of utility-scale solar energy. *Renewable and Sustainable Energy Reviews* 29, 766–79.

Ghafoor, A. and Munir, A. (2015). Design and economics analysis of an off-grid PV system for household electrification. *Renewable and Sustainable Energy Reviews* 42, 496–502.

Girard, P. and Fallot, A. (2006). Review of existing and emerging technologies for the production of biofuels in developing countries. *Energy for Sustainable Development* 10 (2), 92–108.

Kafle, G. K. and Kim, S. H. (2013). Anaerobic treatment of apple waste with swine manure for biogas production: batch and continuous operation. *Applied Energy* 103, 61–72.

Kalogirou, S. A. (2014). *Solar Energy Engineering: Processes and Systems.* Second Edition, Elsevier. ISBN-13: 978-0-12-397270-5.

Liang, C., Das, K. C., and McClendon, R. W. (2003). The influence of temperature and moisture content regimes on the aerobic microbial activity of a biosolids composting blend. *Bioresource Technology* 86 (2), 131–7.

López-González, J. A., Vargas-García, M. C., López, M. J., Suárez-Estrella, F., Jurado, M., and Moreno, J. (2014). Enzymatic characterization of microbial isolates from lignocellulose waste composting: Chronological evolution. *J. Environ. Manage.* 145, 137–46.

Macgregor, S. T., Miller, F. C., Psarianos, K. M., and Finstein, M. S. (1981). Composting process control based on interaction between microbial heat output and temperature. *Appl. Environ. Microbiol.* 41, 1321–30.

Manure compost as a passive greenhouse heating. (2009). http://www.growbetterveggies.com/growbetterveggies /2009/03/manure-compost-as-passive-greenhouse-heating.html.

Marchaim, U. (1992). Biogas processes for sustainable development. FAO. ISBN 95-5-103126-6.

Marucci, A. and Cappuccini, A. (2016). Dynamic photovoltaic greenhouse: Energy balance in completely clear sky condition during the hot period. *Energy* 102, 302–12.

Miyatake, F. and Iwabuchi, K. (2006). Effect of compost temperature on oxygen uptake rate, specific growth rate and enzymatic activity of microorganisms in dairy cattle manure. *Bioresource Technology* 97, 961–5.

Moreno, J., López, M. J., Vargas-García, M. C., and Suárez-Estrella, F. (2013). Recent advances in microbial aspects of compost production and use. *Acta. Horticult. (ISHS)* 1013, 443–57.

Mousazadeh, H., Keyhani, A., Mobli, H., Bardi, U., Lombardi, G., and El Asmar, T. (2009a). Environmental assessment of RAMseS multipurpose electric vehicle compared to a conventional combustion engine vehicle. *Journal of Cleaner Production* 17 (9), 781–90.

Mousazadeh, H., Keyhani, A., Mobli, H., Bardi, U., and El Asmar, T. (2009b). Sustainability in agricultural mechanization: Assessment of a combined photovoltaic and electric multipurpose system for farmers. *Sustainability* 1 (4), 1042–68.

Pang, Y. Z., Liu, Y. P., Wang, K. S., and Yuan, H. R. (2008). Improving biodegradability and biogas production of corn stover through sodium hydroxide solid state pretreatment. *Energy & Fuels* 22 (4), 2761–6.

Poulain, N. (1981). *Jean Pain: France's King of Green Gold.* Reader's Digest, 76–81.

Ros, M., Klammer, S., Knapp, B., Aichberger, K., and Insam, H. (2006). Long term effects of compost amendment of soil in functional and structural diversity and microbial activity. *Soil Use Manage.* 22, 209–18.

Rothbaum, H. P. (1961). Heat output of thermophiles occurring on wool. *J. Bacteriology* 81, 165–71.

Sadeghinezhad, E., Kazi, S. N., Badarudin, A., Oon, C. S., Zubir, M. N. M., and Mehrali, M. (2013). A comprehensive review of bio-diesel as alternative fuel for compression ignition engines. *Renew. Sustain. Energy Rev.* 28, 410–24.

Seki, H. and Komori, T. (1992). Packed-column-type heating tower for recovery of heat generated in compost. *Journal of Agricultural Meteorology* 48 (3), 246–273.

Shatat, M. and Riffat, S. (2012). Water desalination technologies utilizing conventional and renewable energy sources. *International Journal of Low-Carbon Technologies.* Oxford University Press, 1–19.

Shatat, M., Worall, M., and Riffat, S. (2013). Opportunities for solar water desalination worldwide. *Review, Sustainable Cities and Society* 9, 67–80.

Smith, M. and Aber, J. (2014). Heat recovery from compost. *BioCycle* 55 (2), 27.

Sołowiej, P. (2007). The example of using a compost heap as a low-temperature source of heat. [In Polish]. *Inżynieria Rolnicza* 8 (96), 247–53.

Statistical Review of World Energy. (2012). http://www.bp.com/assets/bp_internet/globalbp/globalbp_uk_english /reports_and_publications/statistical_energy_review_2011/STAGING/local_assets/pdf/statistical_review _of_world_energy_full_report_2012.pdf.

Strom, P. F. (1978). The thermophilic bacterial populations of refuse composting as affected by temperature. Ph.D. Thesis. Rutgers University, New Brunswick, NJ.

Truckner, M. F. (2006). Extracting thermal energy from composting. *BioCycle* 47 (8), 38.

Tudisca, S., Di Trapani, A. M., Sgroi, F., Testa, R., and Squatrito, R. (2013). Economic analysis of PV systems on buildings in Sicilian farms. *Renewable and Sustainable Energy Reviews* 28, 691–701.

Viessmann. (2016). Heating with Solar Energy. http://www.viessmann-us.com/content/dam/vi-brands/CA /pdfs/solar/heating_with_solar_energy.pdf/_jcr_content/renditions/original.media_file.inline.file/file.pdf.

Vijaya Venkata Raman, S., Iniyan, S., and Goic, R. (2012). A review of solar drying technologies. *Renewable and Sustainable Energy Reviews* 16, 2652–70.

Wiley, J. S. III. (1957). Progress report of high rate composting studies. *Proc. Ind. Waste Conf.* 12, 596–603.

5 Applied Machine Vision in Agriculture

Matthew Tscharke

CONTENTS

5.1 INTRODUCTION

Food production has been steadily increasing globally to maintain supply relative to the growth in the human population. Along with this increased demand for food, variability in climate and reduced availability of fertile land presents additional pressures on farming systems in the future. Improved food quality, food security, and sustainable agricultural practice are objectives intertwined in a global marketplace, where goods are distributed internationally between countries with different rules, regulations, and resources. Producing higher quality and reliable yields in a smaller area with less money requires well-informed decisions and improved risk management. Measurements need to be recorded throughout the food production and supply chain to provide a better understanding of the existing inefficiencies and opportunities for improvement. Sensing has always been an integral part of agricultural production, with farmers utilizing sight, smell, touch, and taste to analyze their produce and inform management decisions. Of these senses, vision is arguably the most common input into on-farm decision-making processes. Artificial sensing technologies, such as machine vision systems, enhance a farmer's ability to gather visual information from their farm and extend their sensing capability outside of their visible range. When integrated into agricultural systems appropriately, machine vision systems make important on-farm measurements readily available to farm managers in order to assist in decision making and enhance the functionality and precision of tasks undertaken by machines. Machine vision systems overcome limitations in human observations, as they can continuously perform objective measurements without fatigue, and therefore increase the efficiency, accuracy, and understanding of farm operations and food production.

5.2   MACHINE VISION SYSTEMS

Machine vision systems are comprised of camera and computer hardware, software algorithms, and memory that interact in a conceptually similar manner to our human visual system to extract information from images and provide decision support or automation within a system. The following section presents an overview of the common processes, hardware and software technologies, and machine learning and image analysis techniques that are the foundation of machine vision systems.

Machine vision systems aim to solve engineering problems through the application of image analysis techniques that isolate and extract target information from images. In agricultural settings, target information may be an object such as a livestock animal, plant, or machine (or a subcomponent thereof, like a leg, leaf, or tire). Target information characterizes the knowledge that the machine vision system is required to extract from an image to analyze and present meaningful results. For example, the target may be a combination of objects that provide the description of a scene, the movement of an object in a series of images that describes an object's behavior, or the target object may simply be a particular color in the image. Ultimately, the target is the image information required by the machine vision system to contribute an informed and reliable decision or measurement within a process. To extract target information, the unique descriptive features of the target object need to be isolated and identified through segmentation and feature extraction processing tasks. The task of separating target information from background information is known as segmentation. Segmentation effectively removes redundant information from the image so that target information is easier to isolate and extract. For example, the color green is a common feature of most plants, therefore green objects in an image may initially be segmented from the image to help identify a target plant amongst several different plants. A feature extraction task follows segmentation, which identifies and extracts metrics describing the features of the remaining objects (i.e., plants and plant components) in the segmented information. Then a further recognition task is required to identify the unique descriptive features of the target object (i.e., target plant) from the descriptive features of the non-target objects (i.e., remaining plants). The next section considers the implications of imaging hardware in machine vision systems, as the descriptive features of a target object are limited by the dynamic range of the imaging hardware that is available to sense them.

5.3   IMAGING SENSORS AND FEATURES

Imaging devices are evolving continuously, with currently available sensors able to sample photon energy from our surrounding environment across electromagnetic spectrum bands, ranging in wavelengths of 10^{-1} m (12×10^3 Hz; sonar images derived from propagated soundwaves) to wavelengths less than 10^{-12} m (10^{19} Hz; gamma camera images). To highlight the superiority of machine vision systems, our human visual sensing capacity which is sensitive to wavelengths between 4×10^{-7} and 7×10^{-7} m, is 0.0003% of the range of wavelengths (10^{-2} m to 10^{-12} m) across the electromagnetic spectrum. We do sense other photons with wavelengths in this range such as sound (20 Hz to 20×10^3 Hz), however, we are unable to construct an image to visualize them in two-dimensional space. Figure 5.1 illustrates the various imaging technologies at particular wavelengths across the electromagnetic spectrum.

Imaging devices consist of an array of small sensing elements (called pixels) that are designed to sample photon energy across specific wavelength ranges. The majority of cameras available in the market have pixels that sense the photon energy in the visible range of the electromagnetic spectrum, with wavelengths between 4×10^{-7} and 7×10^{-7} m and into the start of the near infrared (NIR) range. To fuel plant growth, the photosynthesis process in plants requires photon energy in the blue and red bands of the visible range to be absorbed by the leaves' green pigment, consisting of chlorophyll and several other accessory pigments. Chlorophyll reflects green and gives plants their color. A healthy plant will reflect more near infrared (NIR) due to increased chlorophyll. Hence, imaging devices can be used to capture the interaction (absorption and reflection) between plant material and the visible

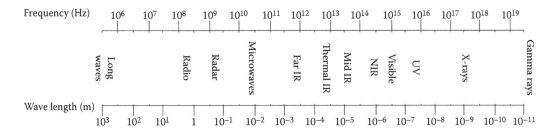

FIGURE 5.1 Various sensing technologies at particular wavelengths across the electromagnetic spectrum.

and near infrared photons to determine the health status of the vegetation. At either side of the visible range, cameras tend to become increasingly more expensive, but provide additional feature information about our surroundings that can be used in specialized applications. For example, cameras with pixels sensitive to long-wavelength infrared light (between 3.5×10^{-6} and 1.5×10^{-5} m) are used to generate images called thermographs that contain a digital record of the scene's temperature.

The availability and affordability of imaging technologies that sense photon energies across the ranges (shown in Figure 5.1) provide limits on the features of an object or scene that a machine vision system can utilize. As a pixel array captures a two dimensional grid of sample measurements from a physical space, textural patterns between the intensities of pixels in images can be identified and used as descriptive features of target objects. The geometry of an object in an image scene can also be recovered from a two-dimensional image, but is limited due to the camera capturing a single plane perspective view of the object in each image frame. A camera can be strategically positioned alongside a conveyor system in a production line to identify planar object geometry in this instance, however, when the application demands the identification of more sophisticated and variable object geometries, a camera system and/or additional image processing techniques may be required to determine the scene depth (Wu *et al.*, 2004; Moreda *et al.*, 2012). Camera systems capable of sensing depth include time-of-flight (Klose *et al.*, 2009), stereo (Wu *et al.*, 2004; Si *et al.*, 2015; Reina *et al.*, 2012), stereo and structure from motion (Turner *et al.*, 2012), light field (Polder *et al.*, 2014), and structured light (Rosell-Polo *et al.*, 2015). Depth images provide a volumetric representation of the captured scene which, when overlaid with feature information, can provide feature information relative to positions in three-dimensional space (Barone *et al.*, 2006; Prakash *et al.*, 2006).

To help identify unique feature information, the perspective view of the camera and its distance relative to the target object should also be considered. Camera lenses can be selected to provide different levels of magnification and detail of an image scene. For example, machine vision systems with different perspective views can be used to identify humans by facial or fingerprint recognition. The boundaries of physical imaging areas range from electron microscopes that can capture images of object features as small as 10^{-10} m to telescopes that can extract features of distant galaxies. Other imaging hardware, such as omnidirectional cameras, enable complete surround views to be captured in a single frame (Ericson *et al.*, 2010). When time is required to describe a target, such as during behavior recognition, sequential movement of an object's geometry from video recordings can be used to determine motion features such as distance, velocity, and acceleration (Eerens *et al.*, 2014). Some imaging devices also infer feature information, such as pixel intensities in X-ray and ultrasound images, which correlate to the density of an object (Wang *et al.*, 2015). Examples of feature information that can be extracted from image objects using image analysis techniques include color, depth, texture, size, shape, motion, chemical properties, and temperature, among others. These target features are either used directly as inputs into decision support or control systems and/or stored as feature data sets. This information provides a raw knowledge base for a learning method to identify the target information in subsequent images.

The main consideration to make when selecting and configuring an imaging device is to ensure that the sensitivity (bit depth or sample resolution) of the device is sufficient to identify and extract unique object features against background information, so that the feature measurements sampled by the device are repeatable. An image is made up of pixels, with each pixel capable of being assigned a value, depending on the bit depth of the imaging device. The pixels of a basic monochrome camera sensor array have a bit depth of eight, and can record one of 2^8 (256) values from 0 (Black) to 255 (White) during image capture. Basic color cameras have pixel elements with a bit depth of eight that are covered with a color filter array (CFA) that allows photons with wavelengths corresponding to either red, green, or blue to pass through. This CFA enables visible light to be recorded in red, green, and blue image channels that are combined to construct a color image with 24-bit color pixels that can each be one of 16,777,216 different color combinations. More specialized cameras have pixels that have bit depths of 12, 14, or 16 bits which have data ranges 0 to 4095, 0 to 16383, and 0 to 65535, respectively. Imaging devices that record with greater sensitivity to changes in the photon energies of the image scene being observed have a greater potential to record unique feature information of a target object. Hyperspectral imagers can capture hundreds of channels of intensity information at nanometer increments, creating the increased potential to record feature information. These channels provide a higher spectral resolution and can be used to isolate features relating to the color and chemical composition of the objects within the scene (Dale *et al.*, 2013).

Understanding and control of camera operation is essential to obtain repeatable measurements and minimize variation in feature data. Often cameras have built-in automatic settings which can cause variation in recorded intensity-based features. To eliminate unknowns, either the automatic adjustments should become features themselves, paired with the resulting intensity feature data, or the automatic settings should be disabled, so that intensity features always maintain values relative to a known base setting. For some applications, automatic settings can have a beneficial effect, such as using automatic exposure for the identification of shape features which rely on detecting changes between intensities (gradients and edges) within the image. Following these considerations ensures that the camera setup is known, and the variability in extracted feature information is minimized. Only after the camera hardware configuration is understood and set appropriately should the extracted feature data be input into a learning method to compile the feature knowledge into a computer model that receives feature inputs and outputs a decision.

5.4   MACHINE LEARNING USING FEATURE DATA

Learning methods are required in machine vision systems, since initially they have no knowledge of the target object. Hence, these methods are required to formulate a model that is learned from a target object's extracted features. This model is subsequently used in recognition and decision-making tasks performed by the machine vision system. There are a range of different learning approaches applied during machine vision system development. On one side of the spectrum, much of the heavy lifting can be achieved from the programmer's experience and learning by experimenting with different image processing techniques, and refining the techniques that sufficiently isolate the target object features from a test set of images. This approach is most useful in controlled environments, where lighting and other variability can be minimized. Consequently, the variability in the target object's features are also minimized, well-defined, and generally more easily learned, making it more difficult for the target to be mistaken for something else. However, a machine vision system operating in an outdoor environment generally requires a complex knowledge base that consists of the variation in the target object's features in respect to the external conditions it is subjected to. Temperature, natural lighting, and motion create sources of variability during image capture, which can decrease a machine vision system's target-identification performance and require increased algorithm complexity to counteract the poor repeatability in the image appearance. In the case of plants and livestock, the color and morphology may vary considerably due to condition, nutrition, wind, or animal movement. These sources of variability in the feature appearance of target and non-target objects can overlap, making it difficult to isolate unique features of a target object in outdoor conditions. The greater the variability in the scene, the more likely that a non-target object with similar features is falsely identified as a target object.

The machine vision programmer should aim to "capture the variance" of the feature measurements of the target object into the machine vision system's knowledge base, while (from a hardware application perspective) trying to minimize it to ensure the best chance of repeatability in measurements. The most common learning approach used to acquire this knowledge involves supervised training, which involves labeling target information. The variability in the labeled target and non-target feature information can be used to determine how much each non-target and target feature overlap, and the strength and uniqueness of features that can be utilized during identification. Commonly used techniques to interface between computer readable knowledge (feature sets) and output decision actions are machine learning techniques like neural networks, support vector machines, Bayesian networks, decision tress, and K-means clustering methods. These machine learning techniques generally oversee two main processes called clustering and classification. Clustering algorithms are responsible for grouping similar feature information into clusters (i.e., knowledge) and classification algorithms are responsible for determining which class (set of clusters) that the information of an object belongs to (i.e., decision). Figure 5.2 illustrates the flow of an adaptive clustering and classification procedure, where the feature set of a target or non-target object is compared to an existing model containing clusters of prior knowledge. If the feature set matches a class within the knowledge model, then the classifier can output the result which is analogous to a decision. This output can be information reported to a farmer to assist with making a decision, or provided as an input into the control system of a machine. In order to update the system's knowledge, a clustering step is performed, where the feature set is compared with the clusters that currently form the knowledge model. If the feature set formed is part of a class (group of feature clusters attributed to a target object) and is significantly similar to that class, then the respective class can be updated. If the feature set does not significantly match an existing class and cluster, it can form a new cluster and be introduced into a class over time.

Classes could be types of diseases on apples for example, with clusters consisting of color and texture feature information of different diseases, or the classes could simply be defective and salable apples, irrespective of whether they are diseased or damaged (Zhang *et al.*, 2015). Both clustering and classification algorithms have learning rules which make the decision of whether feature information belongs to a cluster, or what combination of clusters best represents a class. The classifier provides a weighted response across the classes and a decision whether the feature information provided to the classifier during classification is target or non-target information. A test set of labeled

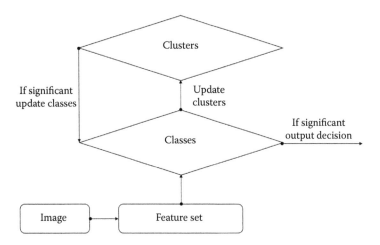

FIGURE 5.2 Example of an adaptive clustering and classification process.

feature data is often used after training a classifier to determine how well the trained model performed the identification of the target object. Object recognition can occur after sufficient knowledge of the target has been programmed into the system that yields test results that are satisfactory for the application. After the target information has been identified, additional metrics can be determined, such as counting the number of target objects identified at the scene.

Machine vision systems have superior decision-making potential compared to humans, due to their ability to sense outside the human sensory range and store and analyze huge quantities of feature information. This fact makes machine vision systems an attractive methodology for use in agricultural operations, where data gathered manually via visual inspection is commonly used in decision making, and more frequent and widespread visual observations are desirable.

5.5 APPLICATIONS OF MACHINE VISION IN AGRICULTURE

Machine vision systems enhance farm surveillance through the aggregation, analysis, and reporting of relevant target information prior, during, or after an on-farm operation takes place. Machine vision systems are utilized throughout agricultural production systems to assist automating tasks and making decisions. An agricultural production system can be generalized into seven top-level management themes that group particular on-farm operations.

These seven management themes in the agricultural production cycle can be defined as: *Preparation* of the field or facility and infrastructure for planting or rearing livestock; *protection*, where the inputs (e.g., nutrition) and external threats (e.g., pests, weeds, and diseases) to the cycle are monitored to help maintain quality in line with market demands; *planting or rearing* livestock, where the often vulnerable starting growth period for the plant or animal is managed; *productivity*, where the performance output and health of the plant or animal is monitored and optimized; *harvesting*, where the productivity and protection have met market demands and selection is viable for a market; *processing*, where the plants or animals are processed to suit storage, transport, and sale to a market; and *surveillance*, which provides the link between the tools monitoring these operations and the data used by management to assist in optimizing production. The relationships among these themes is shown in Figure 5.3, where surveillance tools monitor and extract data during preparation, planting or rearing, productivity, harvest, and processing operations to inform protection operations across the themes.

To provide the most viable, reliable, and repeatable machine vision solutions, a detailed understanding of the target object and the variation in its features over time is required. The farm operations need to be understood by consulting with farmers, industry organizations, and industry service providers (such as agronomists and veterinarians) to determine the most suitable point in the production cycle to integrate the system. Generally, when the task is to minimize a production loss or to enhance efficiency, it is beneficial to detect and track the target object from an early stage, as an operation can be planned to maximize its effectiveness and minimize costs. For example, from a protection perspective, weeds should be removed before they create competition with the surrounding crops, and additional mechanical energy or chemical application is required to control them (Steward *et al.*, 1999); or a disease should be detected early, so treatment is more likely to be effective and the spread of the disease can be minimized. Suitable timing and methodology for the operation can be established during surveillance using information that quantifies the scale of the problem and the rate in which it is developing.

Machine vision systems can operate continuously to perform surveillance tasks that would otherwise not have been possible due to labor requirements and inaccurate analysis caused by fatigue or inexperienced human inspectors. Future machine vision applications are likely to extend the surveillance tasks of an agricultural business, providing opportunities for expansion, and lead to job creation during their development and commercialization from supporting service industries. Many of the surveillance tasks machine vision systems now capable of performing were previously never considered economically viable to perform manually. These systems aim to have short payback periods through improved on-farm decision making, efficiency, and productivity.

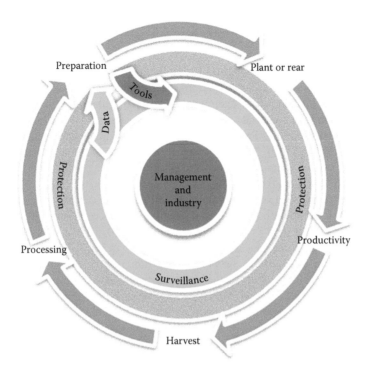

FIGURE 5.3 Relationship between management themes in the agricultural production cycle.

Data collected by machine vison systems will increasingly be integrated into farm management tools, such as the AgX® platform, which is a platform as a service (PaaS) that provides a centralized point for the distribution of agricultural data collected from sensors to inform on-farm decisions and operations (AgX® SST Software, Stillwater, Oklahoma, and GEOSYS, Plymouth, Minnesota). While machine vision systems have the capability to automatically make decisions from the analysis of image information, the perspective in which they observe a target object may be insufficient to make a correct classification. In many applications, vehicles are required to position the camera sensor in the range of the target, where the detail is sufficient to make an identification with high confidence. Aerial- and ground-based robots commonly referred to as unmanned aerial vehicles (UAVs) and unmanned ground vehicles (UGVs) are emerging as suitable vehicles to act as a body for maneuvering a machine vision system into unique ground and aerial vantage points to extract a wide variety of information from images. Emerging and future machine vision systems have the potential to increase product yield and quality, profit margins, farmer and consumer awareness, and confidence. Machine vision systems can enhance the scale of observations and increase the frequency of measurements over these seven themes, as presented in Section 5.5.1.

5.5.1 Machine Vision in Preparation

Preparation involves aligning available and predicted farm resources, knowledge, and experience with a market demand, and then performing the necessary operations in preparation for planting or rearing livestock, with the aim to satisfy that market demand. Many applications of machine vision systems can be found in the literature that perform on-farm preparation tasks, including variety selection based on phenotyping information, soil analysis, and weed management operations prior to planting.

The production cycle generally starts with reproduction from a seed or semen, which creates the initial potential for the agricultural production process. At this important stage, machine vision systems are used to assist with selection of the seed or semen to reduce the variability in the end product and the likelihood of inefficiencies and failure. For many plant types the geometry, mass, and size of seeds are important indicators of the their quality, which can lead to improved emergence rates and growth performance (Sankaran *et al.*, 2016; Ureña *et al.*, 2001). Machine vision systems are well suited for performing repetitive tasks, such as grading and sorting. For this reason, a variety of seeds or grains are inspected using machine vision techniques before sale or planting (Szczypiński *et al.*, 2015; LeMasurier *et al.*, 2014; Donis-González *et al.*, 2014; Wang *et al.*, 2011). In animal production systems semen is graded using machine vision systems to determine suitable samples for artificial insemination (Elsayed *et al.*, 2015). Machine vision systems have also been developed to automatically determine the behavior of animals *in estrus* to ensure that artificial or natural insemination can be timed effectively (Tsai *et al.*, 2014).

The environment that the seeds or young animals will be introduced into needs to be prepared. Management of the soil condition for pasture and planting requires assessment of the nutrients and physical properties of the soil such as compaction level (Głąb, 2014). The hydrology of the site in relation to the soil also needs to be considered. Image analysis can help with mapping the terrain for leveling and determining relationships between production parameters, such as yield and site topography, so improvements to water and nutrient distribution can be made prior to planting (Rovira-Más *et al.*, 2008; Abdel Rahman *et al.*; McKinion *et al.*, 2010; Godwin *et al.*, 2003). Several commercially available software tools can recover depth information and site topography from images, one example is Pix4D software (Pix4D, Pix4D SA, Lausanne, Switzerland).

Machine vision systems are key components in phenotyping products and services. Sources of variation experienced outdoors (such as temperature, natural lighting, and motion) from wind can be controlled in indoor phenotyping laboratories to help compare and analyze plant responses. A number of companies offer phenotyping services that utilize a range of sensors, including RGB, hyperspectral, NIR, Thermal, 3D imaging devices, and chlorophyll fluorescence kinetic imaging. A challenge of phenotyping plants indoors is translating the results experienced under controlled, indoor conditions to the conditions the plants will be subject to when planted outdoors. To understand and quantify how different external conditions influence plant development, the sources of variation experienced outdoors can be recorded and used to control the conditions in the indoor environment. Once this control is achieved, the plant's response to each variable can be analyzed independently while the remaining variables are fixed. Companies that provide phenotyping services or products include WPS (De Lier, the Netherlands), LemnaTec GmbH (Aachen, Germany), Phenotype Screening Corporation (Knoxville, Tennessee, USA), Photon Systems Instruments (*PSI*, Brno, Czech Republic), QUBIT Phenomics (Kingston, Ontario, Canada), and WIWAM (Partnership between SMO, Eeklo, Belgium and VIB, Gent, Belgium) among others.

A number of machine vision technologies have been used in commercially available weed spot-spray systems to prepare the field before planting and crop protection is required. The WeedSeeker® (Trimble Navigation Limited, Sunnyvale, California, USA) and WEEDit (Rometron, Steenderen, the Netherlands) vision-based weed control systems are primarily used during preparation to perform weed control in cases where there is no crop present, such as for fallow or post-harvest spraying operations. These weed-control systems do not utilize a camera to identify weeds, and instead work by propagating light onto the ground and actuating spray solenoids if a change in the reflected light is detected due to the presence of green plant material (chlorophyll). These machine vision systems can be retrofitted to UGV robotic platforms to automate the operation. For example, the Australian company SwarmFarm Robotics Pty Ltd have a robotics platform that automatically traverses the field and can spray weeds using either the WEEDit or WeedSeeker® technology.

Automatic spur pruning implements have been developed to manage the growth of grape vines. This implement utilizes a machine vision system to position its robotic pruning arms and instruct where and when it is suitable to cut the vines. The implement is towed behind a tractor that moves the implement forward autonomously (Vision Robotics Corporation, San Diego, California, USA).

The WALL-YE company has developed a compact UGV robot that identifies and prunes grape vines (WALL-YE, Mâcon, France).

5.5.2 Machine Vision in Planting or Rearing Livestock

Young plants and animals are vulnerable during their early stages of life and often require more frequent monitoring to ensure they maintain a healthy trajectory for their journey ahead. Machine vison systems are becoming increasingly important phenotyping tools, as they can record and report the physiological and biochemical measurements of plants and animals to inform future breeding programs and selection from past production phenotyping feature data (Minervini *et al.*).

To ensure good emergence rates and uniform distribution of plants, some crops are sown with additional seed, followed by a thinning operation that determines satisfactory plant density relative to soil condition, which reduces inter-plant competition for light and nutrients (Moghaddam *et al.*, 2016; Nakarmi *et al.*, 2012; Klassen *et al.*, 2003). Sowing operations can be guided using machine vision techniques that detect the planting lines of previously sown rows (Leemans *et al.*, 2007). Seedlings are cultivated in an environment where higher germination and growth rates can be experienced. To improve the viability of seedlings and plant density, machine vision systems have been used to establish whether the condition of seedlings are satisfactory prior to transplanting in the field (Tong *et al.*, 2013; Ali Ashraf *et al.*, 2011; Koenderink *et al.*, 2010; Ureña *et al.*, 2001). Machine vision systems have also been used to assist with pregnancy detection, monitoring animal behavior, and potential problems during birth (Scully *et al.*, 2014; Scully *et al.*, 2015; Cangar *et al.*, 2008; Rutten *et al.*, 2013).

Machine vison systems have been utilized in lettuce-thinning operations to identify the most viable lettuce plants germinating from seeds. Once a viable plant is determined on the row and at the correct inter-plant spacing, the machine vision system sends a command to spray the surrounding lettuces to increase the nutrients available for the selected plant to grow (Blue River Technology, Sunnyvale, California; Vision Robotics Corporation, San Diego, California).

5.5.3 Machine Vision in Protection

Identifying both good and bad situations around the farm and managing trends in the market and climate can lower risk and lead to improved productivity and business growth. The farming environment increasingly has insufficient and unreliable sources of key inputs, such as nutrients, feed, and water to sustain growth throughout the entire production cycle. To ensure that productivity is maintained, farm resources need to be stored and managed to smooth out variability in resources. Operations such as irrigation and fertilization are often required to maintain acceptable levels for productivity in pasture and cropping systems. Feed allocation is required to maintain livestock body conditions in line with market demands to avoid penalties at sale (Jørgensen *et al.*, 2007). Additional stressors such as weeds, pests, and diseases cause problems and losses throughout the production cycle. Machine vision systems are emerging surveillance tools that can identify weeds, pests, and diseases; help manage resource allocation; and assist with undertaking the operations themselves.

To protect crops from water stress, water usage and distribution can be monitored and managed through irrigation control driven by image analysis, and nutrient and water deficiencies can be detected and rectified (Hendrawan *et al.*, 2011; García-Mateos *et al.*, 2015; Hsieh *et al.*, 2007; Vrindts *et al.*, 2003; Story *et al.*, 2010). The machinery undertaking operations can also be guided by machine vision control systems (Meng *et al.*, 2015; Xue *et al.*, 2012; Billingsley *et al.*, 1997).

Both vertebrate and invertebrate pests can be found on farms. Identifying and eliminating these pests while maintaining a healthy population of beneficial vertebrates and invertebrates remains an active research and development area. With the assistance of trapping devices, machine vision systems can identify pests through direct observation, and can also infer a particular pest species from the type of damage and interference with the plant material that the pest has left behind (Ding *et al.*, 2016).

Diseases also often create symptoms on plant material that can be captured by machine vision systems for disease identification purposes, and used to apply pesticide at a variable rate according to the level of infection (Chung *et al.*, 2016; Barbedo, 2016; Sarangi *et al.*, 2016; Sannakki *et al.*, 2011; Qin *et al.*, 2009; Du *et al.*, 2008; Siripatrawan *et al.*, 2015; Jones, 2014; Bauriegel *et al.*, 2011). At the microscopic level, image analysis has been used to count viable nematode worms for use in biological control for soil-borne pest insects (Kurtulmuş *et al.*, 2014; Slusarewicz *et al.*). Protection of the product during storage is of equal importance, as a range of invertebrate pests can infest and contaminate grain and other seeds, which can be detected using machine vision techniques (Fornal *et al.*, 2007; Neethirajan *et al.*, 2007). Vertebrate pests also cause significant amounts of loss to farming communities. Wild animals (such as pigs) trample and eat crops, causing a significant amount of damage which can be quantified using image analysis (Bengsen *et al.*, 2014). Machine vision systems have been designed to identify vertebrate pests and control drafting gates to separate native and livestock animals and pest animals into two fenced areas (Finch *et al.*, 2006).

In cropping systems, isolating green plant material from the soil background in images is a necessary segmentation step before extracting the features of the plant used for identification (Romeo *et al.*, 2013; Hernández-Hernández *et al.*, 2016). Once isolated, the green plant material can be used for the detection of weeds in crop and fallow situations and machine and implement guidance (Berge *et al.*, 2012; Burgos-Artizzu *et al.*, 2011; Tang *et al.*, 2016, Tellaeche *et al.*, 2008; Tellaeche *et al.*, 2011; Wu *et al.*, 2011; Wiles, 2011; Billingsley *et al.*, 1997; Meng *et al.*, 2015). Figure 5.4 shows the result of an image analysis algorithm that detects the crop row and determines weeds that are not on the row, so that they can be spot-sprayed. The Garford RoboCrop inRow system utilizes machine vision algorithms to identify plants and plant rows to control weeds, using hoeing mechanisms and implement guidance (Garford Farm Machinery Ltd, Peterborough, England). The RoboCrop inRow technology operates well under conditions where plant and row spacings are consistent, the plant foliage is clearly separated from the next plant, and the crop is the dominate feature in the image. The Robovator by F. Poulsen Engineering uses a machine vision system to detect weeds based on the crop plant size. The system detects when a plant is smaller than the crop size and triggers either a short burst of heat or a hoeing mechanism to thermally or mechanically kill the plant (F. Poulsen Engineering ApS, Hvalsø, Denmark). The Robovator also has reduced performance when the plant's foliage overlaps and the row is closed. Smaller weeding UGVs are commercially available. Naïo Technologies (Ramonville-Saint-Agne, France) have developed a weeding system that navigates down crop rows and uses a range of mechanical weed-removal tools between the row and in-row to kill weeds on small-scale vegetable plots. The ecoRobotix (Yverdon-les-Bains, Switzerland) system is a lightweight and efficient robot being field tested to weed large sugar beet crops. The ecoRobotix system utilizes a machine vison system to determine the weeds from the crop and spray or mechanically remove them.

The Rowbot™ UGV is under development to assist with applying nutrients (such as nitrogen) to crops in an operation known as side-dressing. The Rowbot™ system uses laser, 3D stereo, and video information acquired from the tri-modal sensor to identify corn stalks and navigate itself between rows of corn.

5.5.4 Machine Vision in Productivity

Productivity monitors the plant or animal's response to inputs (i.e., water, food) to inform decisions that assist management achieving the optimal yield and quality product as output. As productivity monitoring requires the observable characteristics of plants and animals to be recorded and measured within their environment, the information gathered may also be used to complement the plant or animal's phenotype.

Machine vision systems have been developed to estimate the weight and monitor the growth of several plant and livestock species (Shimizu *et al.*, 1995; Kataoka *et al.*, 2003) including pigs, sheep, cattle, and chickens (Koenig *et al.*, 2015; Tasdemir *et al.*, 2011; Kongsro, 2014; Schofield *et al.*,

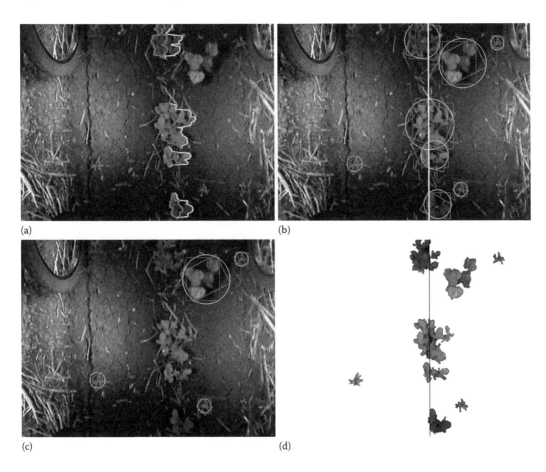

(a)　　　　　　　　　　　　　　(b)

(c)　　　　　　　　　　　　　　(d)

FIGURE 5.4 Example of weed detection algorithm that detects the crop row and determines weeds that are not on the row so they can be sprayed. (a) Finding the middle points of green objects in an image to identify a row of a crop. (b) Determining the green objects in an image. (c) Removing green objects that are assumed to be the crop and identifying potential weeds. (d) Image after segmentation and row detection with only the green objects remaining in the image.

1991; Menesatti *et al.*, 2014; Wang *et al.*, 2006; Mortensen *et al.*, 2016; Frost *et al.*, 1997). Particular body measurements of livestock animals provide an indication of a condition. For example, researchers have used machine vision to track a pig's movements and guide an ultrasonic probe to positions on the pig's body to perform a series of back-fat thickness measurements (French *et al.*, 2003; Frost *et al.*, 2004; Tillett *et al.*, 2002). Figure 5.5 shows the result of an image analysis algorithm that automatically determines the body measurements of a pig, which are then used to estimate its weight and make a decision if it is ready for market. Monitoring the behavior of livestock animals and determining appropriate environmental conditions and enrichment is also an important consideration to ensure that the animals remain healthy, comfortable, and happy. Figure 5.6 shows the result of an image analysis algorithm that determines the volume of activity of chickens around a feeder and the feeder utilization by the birds (Neves *et al.*, 2015). Fighting and injury can occur when insufficient space is allocated to animals and opportunities are limited to escape from confrontation (McFarlane *et al.*, 1995). Livestock animals' environments can also be a cause of injury and discomfort, so researchers have performed assessments using vision based tools (Stefanowska *et al.*, 2001; Gronskyte *et al.*, 2015). Gait scores can be used to indicate the suitability of an animal's walking surface (Aydin *et al.*, 2010; Magee *et al.*, 2002;

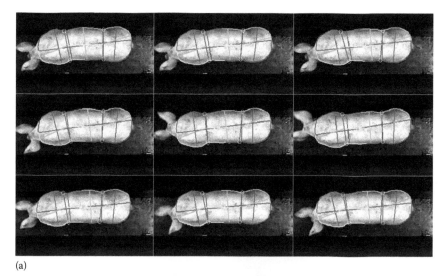

(a)

(b)

FIGURE 5.5   Example of a condition monitoring algorithm that determines the body measurements of a pig which are used to estimate its weight and determine whether it is ready for market. (a) Colored overlay of a pig's body measurements that were extracted for the purpose of formulating a weight estimation model from image data. (b) Colored overlay of a pig's body measurements that were used in a weight estimation model to estimate a pig's body weight in a commercial environment.

Rutten *et al.*, 2013). The lying patterns and distribution of the animals can be assessed and are used to control the climate in housed animal environments (Nasirahmadi *et al.*, 2015; Shao *et al.*, 2008). Animal stress and inflammation of joints can be detected using thermal machine vision applications (Stewart *et al.*, 2007; Pampariene *et al.*, 2016; Westin *et al.*, 2010). Plant stress indicators, such as wilting and thermal properties, have also been detected using machine vison systems (Klassen *et al.*, 2003; Cai *et al.*, 2013; Murase *et al.*, 1997; Leinonen *et al.*, 2004). Crop load estimation in apple orchards has been performed using machine vision technologies mounted on UGV scouting tools, enabling the number and size of the apples to be estimated. These statistics can be used to determine yield and inform harvesting requirements and timing (Stajnko *et al.*, 2004; Stajnko *et al.*, 2009; Hung *et al.*, 2015).

Historical records of plant and animal responses to their production environment provide benchmarks that can be analyzed and reviewed periodically to inform adjustments to the production cycle's inputs. Companies are developing services and tools to record these benchmarks using

FIGURE 5.6 Example of a behavior monitoring algorithm that determines the number and activity of chickens around a feeder and the feeder utilization by the birds.

machine vision systems. The company GAMAYA captures hyperspectral images of fields using UAVs and performs image analysis on them to detect and classify weeds, nematodes, and plant nutrient content; predict yield and monitor growth; and detect soil erosion and planting errors (GAMAYA, Lausanne, Switzerland).

5.5.5 Machine Vision in Harvest

Harvest involves making well-timed decisions of when to stop the production cycle and harvest, based on the quantity and quality of the product in respect to market conditions and available farm resources. Machine vision in harvest flows from monitoring productivity, which provides information to assist with forecasting yield and deciding when to harvest. Most agricultural plants' flowers and fruit provide early visual cues that can indicate harvest timing and approximations of yield (Aquino *et al.*, 2015; Zhu *et al.*, 2016; Wang *et al.*, 2014; Ye *et al.*, 2007; Rezaei Kalaj *et al.*, 2016; Jadhav *et al.*, 2014). During the harvest operation, machine vision systems assist with sorting, selecting, and guiding the machinery or robot to pick the product (Benson *et al.*, 2003; Bulanon *et al.*, 2004; Rajendra *et al.*, 2009; Hannan *et al.*, 2007). Quantifying the mass or numbers of fruits or vegetables distributed around a field or orchard enables yield maps to be created that can be used to identify weak performing areas for improvement, or strong performing areas to learn from. Energid Technologies (Cambridge, Massachusetts) have a citrus-harvesting robot under development that utilizes machine vision systems to locate the coordinates of the fruit, and guide the robot manipulator to retrieve it. The AGROBOT SW6010 adopts the AGvision® system, which is a machine vison system that enables the farmer to configure the quality standards of the robotic picker based on the strawberries' ripeness (AGROBOT, Huelva, Spain).

5.5.6 Machine Vision in Transport, Processing, and Storage

A wide range of machine vision applications can be found in the transport, processing, and storage of agricultural products. The animal or plant often needs to be stored or transported to a processing facility post-harvest. Welfare problems can arise during animal transport (Frost *et al.*, 1997; Petherick, 2005; Foster *et al.*, 2014; Phillips *et al.*, 2013) so vision systems have been used to assist in recording and assessing incidents to guide layout and design improvements (Gronskyte *et al.*, 2015). Many species of fruit and vegetables, cuts of meat, and grain kernels are graded and sorted using machine vision systems, and cut and packaged for sale using machine vision guidance systems (Cubero *et al.*, 2011; Polkinghorne *et al.*, 2010; Davies, 2013; Pothula *et al.*, 2014; Lee *et al.*, 2008; Benalia *et al.*, 2016; Baigvand *et al.*, 2015; Unay *et al.*, 2011; Li *et al.*, 2002; Zhang *et al.*, 2015; Zhang *et al.*, 2015; Sampson *et al.*, 2014; Garrido-Novell *et al.*, 2012; Keresztes *et al.*, 2016; Zareiforoush *et al.*, 2015; Rezaei Kalaj *et al.*, 2016; Abdullah *et al.*, 2006; Yu *et al.*, 2014; Nielsen *et al.*, 2014; Misimi *et al.*, 2016). The SmartSortD system from Smart Vision Works (Orem, Utah) is a date-sorting system that uses machine vision technology to identify and sort into categories such as maturity, size, and skin separation. The Shibuya Group (Shibuya Corporation, Mameda-honmachi, Kanazawa, Japan) have sorting machines that utilize machine vision hardware, including black and white cameras, color cameras, NIR cameras, and electron beam tomography equipment. Analysis of the images collected from these devices can be used to inspect the quality related to sweetness; acidity; internal defects (skin separation or hollow parts); size, shape, and diameter; and the color and blemishes of various fruits and vegetables on packing lines. Machine vision is also used increasingly in automated meat processing systems where machine vision is used to determine body measurements to grade carcasses, and identify reference points to guide robotic arms to make cuts (SCOTT®, Dunedin, New Zealand).

5.6 CONCLUSION

Since their inception into the agricultural system, machine vision systems have increased the automation and accuracy of farm operations, and have provided additional sources of information for decision support and control. Many machine vision systems in agriculture are still in their infancy, and while they have been proved conceptually there are practical hurdles that need to be overcome to provide robust, commercially-ready systems. This is mainly due to the variability of temperature, weather conditions, natural lighting, and motion in the outdoor environment. There are also a number of examples of vision systems that have proven themselves in the marketplace, especially when used to grade and sort produce. The commercially-ready systems that have been presented in this chapter provide a snapshot of the intelligent systems powered by machine vison technology that will help drive efficiency and productivity improvements in agriculture in the years to come. Future opportunities come from the ground-, air-, and water-based mobile and robotic platforms that have been designed to maneuver imaging devices into positions where they can capture data of animals and crops from new perspectives to provide additional details to inform decision-making in agricultural systems and processes.

REFERENCES

Abdel, Rahman, Mohamed, A. E., Natarajan, A., and Hegde, Rejendra. Assessment of land suitability and capability by integrating remote sensing and GIS for agriculture in Chamarajanagar district, Karnataka, India. *Egyptian Journal of Remote Sensing and Space Science.*

Abdullah, M. Z., Mohamad-Saleh, J., Fathinul-Syahir, A. S. *et al.* (2006). Discrimination and classification of fresh-cut starfruits (*Averrhoa Carambola L.*) using automated machine vision systems. *Journal of Food Engineering* 76 (4), 506–23.

Ali Ashraf, Muhammad, Kondo, Naoshi, and Shiigi, Tomoo. (2011). Use of machine vision to sort tomato seedlings for grafting robot. *Engineering in Agriculture, Environment and Food* 4 (4), 119–25.

Aquino, Arturo, Borja, Millan, Gutiérrez, Salvador *et al.* (2015). Grapevine flower estimation by applying artificial vision techniques on images with uncontrolled scenes and multi-model analysis. *Computers and Electronics in Agriculture* 119, 92–104.

Aydin, A., Cangar, O., Eren Ozcan, S. *et al.* (2010). Application of a fully automatic analysis tool to assess the activity of broiler chickens with different gait scores. *Computers and Electronics in Agriculture* 73 (2), 194–9.

Baigvand, Mehrdad, Ahmad Banakar, Saeed Minaei *et al.* (2015). Machine vision system for grading of dried figs. *Computers and Electronics in Agriculture* 119, 158–165.

Barbedo, Jayme Garcia Arnal. (2016). A review on the main challenges in automatic plant disease identification based on visible range images. *Biosystems Engineering* 144, 52–60.

Barone, S., Paoli, A., and Razionale, A. V. (2006). A biomedical application combining visible and thermal 3D imaging. XVIII Congreso internactional de Ingenieria Grafica Barcelona.

Bauriegel, E., Giebel, A., Geyer, M. *et al.* (2011). Early detection of fusarium infection in wheat using hyperspectral imaging. *Computers and Electronics in Agriculture* 75 (2), 304–12.

Benalia, Souraya, Cubero, Sergio, Manuel Prats-Montalbán, José *et al.* (2016). Computer vision for automatic quality inspection of dried figs (*Ficus Carica L.*) in real-time. *Computers and Electronics in Agriculture* 120, 17–25.

Bengsen, Andrew J., Gentle, Mattew N., Mitchell, James L. *et al.* (2014). Impacts and management of wild pigs (*Sus scrofa*) in Australia. *Mammal Review* 44 (2), 135–47.

Benson, E. R., Reid, J. F., and Zhang, Q. (2003). Machine Vision–based Guidance System for Agricultural Grain Harvesters Using Cutting-edge Detection. *Biosystems Engineering* 86 (4), 389–98.

Berge, T. W., Goldberg, S., Kaspersen, K. *et al.* (2012). Towards machine vision based site-specific weed management in cereals. *Computers and Electronics in Agriculture* 81, 79–86.

Billingsley, J., and Schoenfisch, M. (1997). The successful development of a vision guidance system for agriculture. *Computers and Electronics in Agriculture* 16 (2), 147–63.

Bulanon, D. M., Kataoka, T., Okamoto, H. *et al.* (2004). Development of a real-time machine vision system for the apple harvesting robot. SICE 2004 Annual Conference.

Burgos-Artizzu, Xavier P., Ribeiro, Angela, Guijarro, Maria *et al.* (2011). Real-time image processing for crop/weed discrimination in maize fields. *Computers and Electronics in Agriculture* 75 (2), 337–46.

Cai, X., Sun, Y., Zhao, Y. *et al.* (2013). Smart detection of leaf wilting by 3D image processing and 2D Fourier transform. *Computers and Electronics in Agriculture* 90, 68–75.

Cangar, Ö. T., Leroy, M., Guarino *et al.* (2008). Automatic real-time monitoring of locomotion and posture behavior of pregnant cows prior to calving using online image analysis. *Computers and Electronics in Agriculture* 64 (1), 53–60.

Chung, Chia-Lin, Huang, Kai-Jyun, Chen, Szu-Yu *et al.* (2016). Detecting Bakanae disease in rice seedlings by machine vision. *Computers and Electronics in Agriculture* 121, 404–11.

Cubero, Sergio, Aleixos, Nuria, Moltó, Enrique *et al.* (2011). Advances in machine vision applications for automatic inspection and quality evaluation of fruits and vegetables. *Food and Bioprocess Technology* 4 (4), 487–504.

Dale, Laura M., Thewis, André, Boudry, Christelle *et al.* (2013). Hyperspectral imaging applications in agriculture and agro-food product quality and safety control: A review. *Applied Spectroscopy Reviews* 48 (2), 142–59.

Davies, E. R. (2013). Machine vision in the food industry. (Ed. Darwin G.) *Robotics and Automation in the Food Industry*, 75–110. Woodhead Publishing.

Ding, Weiguang, and Taylor, Graham. (2016). Automatic moth detection from trap images for pest management. *Computers and Electronics in Agriculture* 123, 17–28.

Donis-González, Irwin R., Guyer, Daniel E., Fulbright, Dennis W. *et al.* (2014). Post-harvest noninvasive assessment of fresh chestnut (*Castanea* spp.) internal decay using computer tomography images. *Postharvest Biology and Technology* 94, 14–25.

Du, Qian, Chang, Ni-Bin, Yang, Chenghai *et al.* (2008). Combination of multispectral remote sensing, variable rate technology and environmental modeling for citrus pest management. *Journal of Environmental Management* 86 (1), 14–26.

Eerens, Herman, Haesen, Dominique, Rembold, Felix *et al.* (2014). Image time series processing for agriculture monitoring. *Environmental Modelling & Software* 53, 154–62.

Elsayed, Mohamed, El-Sherry, Taymour M., and Abdelgawad, Mohamed. (2015). Development of computer-assisted sperm analysis plugin for analyzing sperm motion in microfluidic environments using Image-J. *Theriogenology* 84 (8), 1367–77.

Ericson, Stefan and Åstrand, Björn. (2010). Row-detection on an agricultural field using an omnidirectional camera. Intelligent Robots and Systems (IROS), IEEE/RSJ International Conference.

Finch, N. A., Murray, P. J., Dunn, M. T. *et al.* (2006). Using machine vision classification to control access of animals to water. *Animal Production Science* 46 (7), 837–9.

Fornal, Józef, Jeliński, Tomasz, Sadowska, Jadwiga *et al.* (2007). Detection of granary weevil (*Sitophilus granarius L.*) eggs and internal stages in wheat grain using soft X-ray and image analysis. *Journal of Stored Products Research* 43 (2), 142–8.

Foster, Susan F. and Overall, Karen L. (2014). The welfare of Australian livestock transported by sea. *The Veterinary Journal* 200 (2), 205–9.

French, A. P., Frost, A., Pridmore, T. P. *et al.* (2003). An image analysis system to guide a sensor placement robot onto a feeding pig. Irish machine vision and image processing conference, Portrush, Co. Antrim, Northern Ireland.

Frost, A. R., Schofield, C. P., Beaulah, S. A. *et al.* (1997). A review of livestock monitoring and the need for integrated systems. *Computers and Electronics in Agriculture* 17 (2), 139–59.

Frost, A. R., French, A. P., Tillett, R. D. *et al.* (2004). A vision guided robot for tracking a live, loosely constrained pig. *Computers and Electronics in Agriculture* 44 (2), 93–106.

García-Mateos, G., Hernández-Hernández, J. L., Escarabajal-Henarejos, D. *et al.* (2015). Study and comparison of color models for automatic image analysis in irrigation management applications. *Agricultural Water Management* 151, 158–66.

Garrido-Novell, Cristóbal, Pérez-Marin, Dolores, Amigo, Jose M. *et al.* (2012). Grading and color evolution of apples using RGB and hyperspectral imaging vision cameras. *Journal of Food Engineering* 113 (2), 281–8.

Głąb, Tomasz. (2014). Effect of soil compaction and N. fertilization on soil pore characteristics and physical quality of sandy loam soil under red clover/grass sward. *Soil and Tillage Research* 144, 8–19.

Godwin, R. J. and Miller, P. C. H. (2003). A review of the technologies for mapping within-field variability. *Biosystems Engineering* 84 (4), 393–407.

Gronskyte, Ruta, Clemmensen, Line Harder, Hviid, Marchen Sonja *et al.* (2015). Pig herd monitoring and undesirable tripping and stepping prevention. *Computers and Electronics in Agriculture* 119, 51–60.

Hannan, M. W., Burks, T. F., and Bulanon, D. M. (2007). A real-time machine vision algorithm for robotic citrus harvesting. 2007 ASAE Annual Meeting.

Hendrawan, Yusuf and Murase, Haruhiko. (2011). Neural-intelligent water drops algorithm to select relevant textural features for developing precision irrigation system using machine vision. *Computers and Electronics in Agriculture* 77 (2), 214–28.

Hernández-Hernández, J. L., García-Mateos, G., González-Esquiva, J. M. *et al.* (2016). Optimal color space selection method for plant/soil segmentation in agriculture. *Computers and Electronics in Agriculture* 122, 124–32.

Hsieh, Y. D., Gau, C. H., Kung Wu, S. F. *et al.* (2007). Dynamic recording of irrigating fluid distribution in root canals using thermal image analysis. *International Endodontic Journal* 40 (1), 11–7.

Hung, Calvin, Underwood, James, Nieto, Juan *et al.* (2015). A feature learning based approach for automated fruit yield estimation. *Field and Service Robotics: Results of the 9th International Conference*. (Ed. Luis Mejias, Peter Corke and Jonathan Roberts) 485–98. Cham: Springer International Publishing.

Jadhav, U., Khot, L. R., Ehsani, R. *et al.* (2014). Volumetric mass flow sensor for citrus mechanical harvesting machines. *Computers and Electronics in Agriculture* 101, 93–101.

Jones, R. A. C. (2014). Trends in plant virus epidemiology: Opportunities from new or improved technologies. *Virus Research* 186, 3–19.

Jørgensen, Rasmus Nyholm, Sørensen, Claus Grøn, Jensen, Helle Frank *et al.* (2007). Feeder Ant: An autonomous mobile unit feeding outdoor pigs. 2007 ASAE Annual Meeting.

Kataoka, Takashi, Kaneko, Toshihiro, and Okamoto, Hiroshi. (2003). Crop growth estimation system using machine vision. Advanced Intelligent Mechatronics, 2003. AIM 2003. Proc. 2003 IEEE/ASME International Conference.

Keresztes, Janos C., Goodarzi, Mohammad, and Saeys, Wouter. (2016). Real-time pixel based early apple bruise detection using short-wave infrared hyperspectral imaging in combination with calibration and glare correction techniques. *Food Control* 66, 215–26.

Klassen, S. P., Ritchie, G., Frantz, J. M. *et al.* (2003). Real-time imaging of ground cover: Relationships with radiation capture, canopy photosynthesis, and daily growth rate. *Digital imaging and spectral techniques: Applications to precision agriculture and crop physiology*, 3–14.

Klose, Ralph, Penlington, Jaime, and Ruckelshausen, Arno. (2009). Usability study of 3D time-of-flight cameras for automatic plant phenotyping. *Bornimer Agrartechnische Berichte* 69 (12), 93–105.

Koenderink, Nicole J. J. P., Broekstra, Jeen, and Top, Jan L. (2010). Bounded transparency for automated inspection in agriculture. *Computers and Electronics in Agriculture* 72 (1), 27–36.

Koenig, Kristina, Höfle, Bernhard, Hämmerle, Martin *et al.* (2015). Comparative classification analysis of post-harvest growth detection from terrestrial LiDAR point clouds in precision agriculture. *ISPRS Journal of Photogrammetry and Remote Sensing* 104, 112–25.

Kongsro, Jørgen. (2014). Estimation of pig weight using a Microsoft Kinect prototype imaging system. *Computers and Electronics in Agriculture* 109, 32–5.

Kurtulmuş, Ferhat and Ulu, Tufan C. (2014). Detection of dead entomopathogenic nematodes in microscope images using computer vision. *Biosystems Engineering* 118, 29–38.

Lee, Dah-Jye, Schoenberger, Robert, Archibald, James *et al.* (2008). Development of a machine vision system for automatic date grading using digital reflective near-infrared imaging. *Journal of Food Engineering* 86 (3), 388–98.

Leemans, V. and Destain, M. F. (2007). A computer vision–based precision seed drill guidance assistance. *Computers and Electronics in Agriculture* 59 (1–2), 1–12.

Leinonen, Ilkka and Jones, Hamlyn G. (2004). Combining thermal and visible imagery for estimating canopy temperature and identifying plant stress. *Journal of Experimental Botany* 55 (401), 1423–31.

LeMasurier, L. S., Panozzo, J. F., and Walker, C. K. (2014). A digital image analysis method for assessment of lentil size traits. *Journal of Food Engineering* 128, 72–8.

Li, Qingzhong, Wang, Maohua, and Gu, Weikang. (2002). Computer vision–based system for apple surface defect detection. *Computers and Electronics in Agriculture* 36 (2–3), 215–23.

Magee, Derek R. and Boyle, Roger D. (2002). Detecting lameness using "re-sampling condensation" and "multi-stream cyclic hidden Markov models." *Image and Vision Computing* 20 (8), 581–94.

McFarlane, Nigel J. B. and Schofield, C. Paddy. (1995). Segmentation and tracking of piglets in images. *Machine vision and applications* 8 (3), 187–93.

McKinion, J. M., Willers, J. L., and Jenkins, J. N. (2010). Comparing high density LIDAR and medium resolution GPS-generated elevation data for predicting yield stability. *Computers and Electronics in Agriculture* 74 (2), 244–9.

Menesatti, Paolo, Costa, Corrado, Antonucci, Francesca *et al.* (2014). A low-cost stereovision system to estimate size and weight of live sheep. *Computers and Electronics in Agriculture* 103, 33–8.

Meng, Qingkuan, Qiu, Ruicheng, He, Jie *et al.* (2015). Development of an agricultural implement system based on machine vision and fuzzy control. *Computers and Electronics in Agriculture* 112, 128–38.

Minervini, Massimo, Fischbach, Andreas, Scharr, Hanno *et al.* Finely-grained annotated datasets for image-based plant phenotyping. *Pattern Recognition Letters.*

Misimi, Ekrem, Øye, Elling Ruud, Eilertsen, Aleksander *et al.* (2016). GRIBBOT–Robotic 3D vision-guided harvesting of chicken fillets. *Computers and Electronics in Agriculture* 121, 84–100.

Moghaddam, Parviz Ahmadi, Arasteh, Amir Sheykhi, Komarizadeh, Mohammad Hasan *et al.* (2016). Developing a selective thinning algorithm in sugar beet fields using machine vision systems. *Computers and Electronics in Agriculture* 122, 133–8.

Moreda, G. P., Muñoz, M. A., Ruiz-Altisent, M. *et al.* (2012). Shape determination of horticultural produce using two-dimensional computer vision: A review. *Journal of Food Engineering* 108 (2), 245–61.

Mortensen, Anders Krogh, Lisouski, Pavel, and Ahrendt, Peter. (2016). Weight prediction of broiler chickens using 3D computer vision. *Computers and Electronics in Agriculture* 123, 319–26.

Murase, H., Nishiura, Y., and Mitani, K. (1997). Environmental control strategies based on plant responses using intelligent machine vision techniques. *Computers and Electronics in Agriculture* 18 (2–3), 137–48.

Nakarmi, A. D. and Tang, L. (2012). Automatic inter-plant spacing sensing at early growth stages using a 3D vision sensor. *Computers and Electronics in Agriculture* 82, 23–31.

Nasirahmadi, Abozar, Richter, Uwe, Hensel, Oliver *et al.* (2015). Using machine vision for investigation of changes in pig group lying patterns. *Computers and Electronics in Agriculture* 119, 184–90.

Neethirajan, S., Karunakaran, C., Jayas, D. S. *et al.* (2007). Detection techniques for stored-product insects in grain. *Food Control* 18 (2), 157–62.

Neves, Diego Pereira, Mehdizadeh, Saman Abdanan, Tscharke, Matthew *et al.* (2015). Detection of flock movement and behaviour of broiler chickens at different feeders using image analysis. *Information Processing in Agriculture* 2 (3–4), 177–82.

Nielsen, J. U., Madsen, N. T., and Clarke, R. (2014). Automation in the meat industry: Slaughter line operation. *Encyclopedia of Meat Sciences.* (Second Edition), 43–52. Oxford: Academic Press.

Pampariene, I., Veikutis, V., Oberauskas, V., Zymantiene, J., Zelvyte, R., Stankevicius, A., Marciulionyte, D., and Palevicius, P. (2016). Thermography based inflammation monitoring of udder state in dairy cows: Sensitivity and diagnostic priorities comparing with routine California mastitis test. *Journal of Vibroengineering* 18 (1), 511–521.

Petherick, J. Carol. (2005). Animal welfare issues associated with extensive livestock production: The northern Australian beef cattle industry. *Applied Animal Behaviour Science* 92 (3), 211–34.

Phillips, Clive J. C., and Santurtun, Eduardo. (2013). The welfare of livestock transported by ship. *The Veterinary Journal* 196 (3), 309–14.

Polder, Gerrit and Hofstee, Jan Willem. (2014). *Phenotyping large tomato plants in the greenhouse using a 3D light-field camera.* Montreal, Quebec, Canada.

Polkinghorne, R. J. and Thompson, J. M. (2010). Meat standards and grading: A world view. *Meat Science* 86 (1), 227–35.

Pothula, Anand Kumar, Igathinathane, C., Kronberg, S. *et al.* (2014). Digital image processing based identification of nodes and internodes of chopped biomass stems. *Computers and Electronics in Agriculture* 105, 54–65.

Prakash, Surya, Lee, Pei Yean, Caelli, Terry *et al.* (2006). Robust thermal camera calibration and 3D mapping of object surface temperatures. Defense and Security Symposium.

Qin, Jianwei, Burks, Thomas F., Ritenour, Mark A. *et al.* (2009). Detection of citrus canker using hyperspectral reflectance imaging with spectral information divergence. *Journal of Food Engineering* 93 (2), 183–91.

Rajendra, Peter, Kondo, Naoshi, Ninomiya, Kazunori *et al.* (2009). Machine vision algorithm for robots to harvest strawberries in tabletop culture greenhouses. *Engineering in Agriculture, Environment and Food* 2 (1), 24–30.

Reina, Giulio and Milella, Annalisa. (2012). Towards autonomous agriculture: Automatic ground detection using trinocular stereovision. *Sensors* 12 (9), 12405–23.

Rezaei Kalaj, Yousef, Kaveh, Mollazade, Werner Herppich *et al.* (2016). Changes of backscattering imaging parameters during plum fruit development on the tree and during storage. *Scientia Horticulturae* 202, 63–9.

Romeo, J., Pajares, G., Montalvo, M. *et al.* (2013). A new expert system for greenness identification in agricultural images. *Expert Systems with Applications* 40 (6), 2275–86.

Rosell-Polo, Joan R., Cheein, Fernando, Auat, Gregorio, Eduard *et al.* (2015). Advances in structured light sensors applications in precision agriculture and livestock farming. *Advances in Agronomy*. (Ed. L. Sparks Donald) 71–112. Academic Press.

Rovira-Más, Francisco, Zhang, Qin, and Reid, John F. (2008). Stereo vision three-dimensional terrain maps for precision agriculture. *Computers and Electronics in Agriculture* 60 (2), 133–43.

Rutten, C. J., Velthuis, A. G. J., Steeneveld, W. *et al.* (2013). Invited review: Sensors to support health management on dairy farms. *Journal of Dairy Science* 96 (4), 1928–52.

Sampson, David Joseph, Chang, Young Ki, Vasantha Rupasinghe, H. P. *et al.* (2014). Corrigendum to "A dual-view computer vision system for volume and image texture analysis in multiple apple slices drying" [J. Food Eng. 127 (2014), 49–57]. *Journal of Food Engineering* 130, 62.

Sankaran, Sindhuja, Wang, Meng, and Vandemark, George J. (2016). Image-based rapid phenotyping of chickpeas' seed size. *Engineering in Agriculture, Environment and Food* 9 (1), 50–5.

Sannakki, Sanjeev S., Rajpurohit, Vijay S., Nargund, V. B. *et al.* (2011). Leaf disease grading by machine vision and fuzzy logic. *Int J* 2 (5), 1709–16.

Sarangi, Sanat, Umadikar, Jayalakshmi, and Kar, Subrat. (2016). Automation of agriculture support systems using Wisekar: Case study of a crop-disease advisory service. *Computers and Electronics in Agriculture* 122, 200–10.

Schofield, C. Patrick, and Marchant, John A. (1991). Image analysis for estimating the weight of live animals. *Fibers' 91*, Boston, MA.

Scully, S., Butler, S. T., Kelly, A. K. *et al.* (2014). Early pregnancy diagnosis on days 18 to 21 post-insemination using high-resolution imaging in lactating dairy cows. *Journal of Dairy Science* 97 (6), 3542–57.

Scully, S., Evans, A. C. O., Carter, F. *et al.* (2015). Ultrasound monitoring of blood flow and echotexture of the *corpus luteum* and uterus during early pregnancy of beef heifers. *Theriogenology* 83 (3), 449–58.

Shao, Bin, and Xin, Hongwei. (2008). A real-time computer vision assessment and control of thermal comfort for group-housed pigs. *Computers and Electronics in Agriculture* 62 (1), 15–21.

Shimizu, H. and Heins, R. D. (1995). Computer vision–based system for plant growth analysis. *Transactions of the ASAE* 38 (3), 959–64.

Si, Yongsheng, Liu, Gang, and Feng, Juan. (2015). Location of apples in trees using stereoscopic vision. *Computers and Electronics in Agriculture* 112, 68–74.

Siripatrawan, U., and Makino, Y. (2015). Monitoring fungal growth on brown rice grains using rapid and nondestructive hyperspectral imaging. *International Journal of Food Microbiology* 199, 93–100.

Slusarewicz, Paul, Pagano, Stefanie, Mills, Christopher *et al.* Automated parasite fecal egg counting using fluorescence labeling, smartphone image capture, and computational image analysis. *International Journal for Parasitology*.

Stajnko, Denis, Lakota, Miran, and Hočevar, Marko. (2004). Estimation of number and diameter of apple fruits in an orchard during the growing season by thermal imaging. *Computers and Electronics in Agriculture* 42 (1), 31–42.

Stajnko, Denis, Rakun, Jurij, and Blanke, Michael. (2009). Modelling apple fruit yield using image analysis for fruit color, shape, and texture. *European Journal of Horticultural Science*, 260–7.

Stefanowska, J., Swierstra, D., Braam, C. R. *et al.* (2001). Cow behavior on a newly grooved floor in comparison with a slatted floor, taking claw health and floor properties into account. *Applied Animal Behaviour Science* 71 (2), 87–103.

Steward, B. L. and Tian, L. F. (1999). Machine vision weed density estimation for real-time, outdoor lighting conditions. *Transactions of the ASAE* 42 (6), 1897.

Stewart, M., Webster, J. R., Verkerk, G. A. *et al.* (2007). Noninvasive measurement of stress in dairy cows using infrared thermography. *Physiology & Behavior* 92 (3), 520–5.

Story, David, Kacira, Murat, Kubota, Chieri *et al.* (2010). Lettuce calcium deficiency detection with machine vision computed plant features in controlled environments. *Computers and Electronics in Agriculture* 74 (2), 238–43.

Szczypiński, Piotr M., Klepaczko, Artur, and Zapotoczny, Piotr. (2015). Identifying barley varieties by computer vision. *Computers and Electronics in Agriculture* 110, 1–8.

Tang, Jing-Lei, Chen, Xiao-Qian, Miao, Rong-Hui *et al.* (2016). Weed detection using image processing under different illumination for site-specific areas. *Computers and Electronics in Agriculture* 122, 103–11.

Tasdemir, Sakir, Urkmez, Abdullah, and Inal, Seref. (2011). Determination of body measurements on Holstein cows using digital image analysis and estimation of live weight with regression analysis. *Computers and Electronics in Agriculture* 76 (2), 189–97.

Tellaeche, Alberto, Burgos-Artizzu, Xavier P., Pajares, Gonzalo *et al.* (2008). A vision-based method for weed identification through the Bayesian decision theory. *Pattern Recognition* 41 (2), 521–30.

Tellaeche, Alberto, Pajares, Gonzalo, Burgos-Artizzu, Xavier P. *et al.* (2011). A computer vision approach for weed identification through support vector machines. *Applied Soft Computing* 11 (1), 908–15.

Tillett, R. D., Frost, A. R., and Welch, S. K. (2002). AP—Animal production technology: Predicting sensor placement targets on pigs using image analysis. *Biosystems engineering* 81 (4), 453–63.

Tong, Jun H., Jiang, B. Li, and Huan, Y. Jiang. (2013). Machine vision techniques for the evaluation of seedling quality based on leaf area. *Biosystems Engineering* 115 (3), 369–79.

Tsai, Du-Ming, and Huang, Ching-Ying. (2014). A motion and image analysis method for automatic detection of estrus and mating behavior in cattle. *Computers and Electronics in Agriculture* 104, 25–31.

Turner, Darren, Lucieer, Arko, and Watson, Christopher. (2012). An automated technique for generating georectified mosaics from ultra–high resolution unmanned aerial vehicle (UAV) imagery, based on structure from motion (SfM) point clouds. *Remote Sensing* 4 (5), 1392–410.

Unay, Devrim, Gosselin, Bernard, Kleynen, Olivier *et al.* (2011). Automatic grading of bi-colored apples by multispectral machine vision. *Computers and Electronics in Agriculture* 75 (1), 204–12.

Ureña, R., Rodríguez, F., and Berenguel, M. (2001). A machine vision system for seeds quality evaluation using fuzzy logic. *Computers and Electronics in Agriculture* 32 (1), 1–20.

Vrindts, E., Reyniers, M., Darius, P. *et al.* (2003). Analysis of soil and crop properties for precision agriculture for winter wheat. *Biosystems Engineering* 85 (2), 141–52.

Wang, Chenglong, Li, Xiaoyu, Wang, Wei *et al.* (2011). Recognition of worm-eaten chestnuts based on machine vision. *Mathematical and Computer Modelling* 54 (3–4), 888–94.

Wang, Laigang, Tian, Yongchao, Yao, Xia *et al.* (2014). Predicting grain yield and protein content in wheat by fusing multisensor and multitemporal remote-sensing images. *Field Crops Research* 164, 178–88.

Wang, Weilin and Li, Changying. (2015). A multimodal machine vision system for quality inspection of onions. *Journal of Food Engineering* 166, 291–301.

Wang, Y., Yang, W., Winter, P. *et al.* (2006). Noncontact sensing of hog weights by machine vision. *Applied Engineering in Agriculture* 22 (4), 577–82.

Westin, R., and Rydberg, A. (2010). Thermal imaging for early detection of shoulder lesion development in sows. International Conference on Agricultural Engineering–AgEng: Towards Environmental Technologies, Clermont-Ferrand, France.

Wiles, L. J. (2011). Software to quantify and map vegetative cover in fallow fields for weed management decisions. *Computers and Electronics in Agriculture* 78 (1), 106–15.

Wu, Jiahua, Tillett, Robin, McFarlane, Nigel *et al.* (2004). Extracting the three-dimensional shape of live pigs using stereo photogrammetry. *Computers and Electronics in Agriculture* 44 (3), 203–22.

Wu, Xuewen, Xu, Wenqiang, Song, Yunyun *et al.* (2011). A detection method of weed in wheat field on machine vision. *Procedia Engineering* 15, 1998–2003.

Xue, Jinlin, Zhang, Lei, and Grift, Tony E. (2012). Variable field-of-view machine vision based row guidance of an agricultural robot. *Computers and Electronics in Agriculture* 84, 85–91.

Ye, Xujun, Sakai, Kenshi, Manago, Masafumi *et al.* (2007). Prediction of citrus yield from airborne hyperspectral imagery. *Precision Agriculture* 8 (3), 111–25.

Yu, Keqiang, Zhao, Yanru, Li, Xiaoli *et al.* (2014). Identification of crack features in fresh jujubes using Vis/NIR hyperspectral imaging combined with image processing. *Computers and Electronics in Agriculture* 103, 1–10.

Zareiforoush, Hemad, Minaei, Saeid, Alizadeh, Mohammad Reza *et al.* (2015). A hybrid intelligent approach based on computer vision and fuzzy logic for quality measurement of milled rice. *Measurement* 66, 26–34.

Zhang, Baohua, Huang, Wenqian, Gong, Liang *et al.* (2015). Computer vision detection of defective apples using automatic lightness correction and weighted RVM classifier. *Journal of Food Engineering* 146, 143–51.

Zhang, Baohua, Huang, Wenqian, Wang, Chaopeng *et al.* (2015). Computer vision recognition of stem and calyx in apples using near-infrared linear-array structured light and 3D reconstruction. *Biosystems Engineering* 139, 25–34.

Zhu, Yanjun, Cao, Zhiguo, Lu, Hao *et al.* (2016). In-field automatic observation of wheat heading stage using computer vision. *Biosystems Engineering* 143, 28–41.

6 Advances in Unmanned Aerial Systems and Payload Technologies for Precision Agriculture

Felipe Gonzalez, Aaron Mcfadyen, and Eduard Puig

CONTENTS

6.1 INTRODUCTION

Today's farmers have to deal with an increasingly complex industry and international competition to market their products. Issues such as water use, climate change, regulations, soil quality, commodity prices, and input prices are common concerns, to name a few. As a result, growers are turning to precision agriculture (PA) to address some of these challenges. PA can be defined as the use of spatial and temporal information of crops in order to perform site-specific management. Its main aim is to increase crop yield and farm profitability through a more efficient use of resources. It is highly beneficial for the economy to have a competitive agricultural industry with high quality standards. Nonetheless, the benefits of PA to society also include the creation of high-tech jobs, for instance, for remote sensing systems and machinery guidance, as well as the mitigation of negative environmental impacts arising from an excessive application of chemical inputs (Mulla, 2013).

Conventional farm management techniques are often based on applying uniform quantities of crop inputs, such as seeds, water, fertilizers, pesticides, and herbicides. On the contrary, PA consists of delivering customized inputs based on georeferenced crop information and the partition of fields into zones with particular treatment requirements. With the advent of miniaturized sensor technologies and the ever-increasing number of PA applications, a variety of agricultural equipment of increasing complexity is progressively being developed. Since the 1990s, variable-rate technology (VRT), which describes any technology that enables the variable application of inputs, is one of the most popular PA techniques (Zhang and Kovacs 2012). Some of the most common inputs are fertilizer, pesticides (herbicide, insecticide, and fungicides), manure, seeding, tillage, and irrigation. By optimizing their usage, farm managers can increase productivity while minimizing the environmental and health risks of chemical inputs.

6.1.1 REMOTE SENSING IN PRECISION AGRICULTURE

Precision agriculture relies on actionable information obtained from sensors and data analysis software in order to achieve efficiency in farm practices. Actionable information is the result of integrating different sources of approximate real-time sensor information into a decision support system. Some of the most common platforms from which the sensors acquire the data include satellites, manned aircraft, unmanned aircraft, tractors, and handheld devices. Remote sensing (RS) is the science of obtaining information and measuring properties of objects on the Earth's surface from a distance, typically from aircraft or satellites; as opposed to proximal sensing, which refers to sensor data acquisition from ground vehicles and handheld devices (Mulla, 2013).

Objects on the Earth's surface reflect, absorb, transmit, and emit electromagnetic energy from the sun. Digital sensors have been developed to measure all types of electromagnetic energy as they interact with objects. Figure 6.1 shows the visible and infrared portions of the electromagnetic spectrum, which are used predominantly for remote sensing in PA.

Traditional RS technologies based on satellite and aircraft platforms are continuously improving in terms of spatial and temporal resolution, thus enhancing their suitability for PA applications. As the distance between the sensor and the crop increases from an aircraft to a satellite, the surveyed area will increase from a farm scale to a regional scale. Each combination of sensor and platform has pros and cons for a given application, which involve technological, operational, and economic considerations. Satellite surveys typically survey larger areas and the user can access the imagery in the following days or weeks after, however, satellite data providers typically require a large minimum surface in square kilometers, which may not be cost-effective for the farmer to purchase on a regular basis for PA applications. Additionally, satellite imagery may suffer from cloud cover, atmospheric distortion, and lack of flexibility to capture imagery following the phenology phases. On the other hand, surveys carried out by general aviation aircraft can be scheduled more flexibly than satellite imagery, but they often require complex and costly campaign organization efforts (Berni et al. 2009).

The operational success of VRT relies heavily on timely sensor data collection and the accurate computation of prescription maps describing crop vigor, deficiencies, disease, weeds, and pests, as well as soil variables, such as moisture and nutrient content. The data can be obtained in a variety of ways, such as the traditional remote sensing platforms described previously; more recently other platforms, such as unmanned aerial vehicles (UAVs), or sensors are mounted on vehicles can provide higher sensor proximity to the crops as well as near real-time information. Only in the last 10 years have UAVs and miniaturized sensor technologies become widely available at cost-effective prices and are demonstrating a significant suitability for PA applications. In fact, UAVs are not only facilitating remote sensing data acquisition, but also present an interesting business case for deployment as VRT platforms, where only a small quantity of input is required.

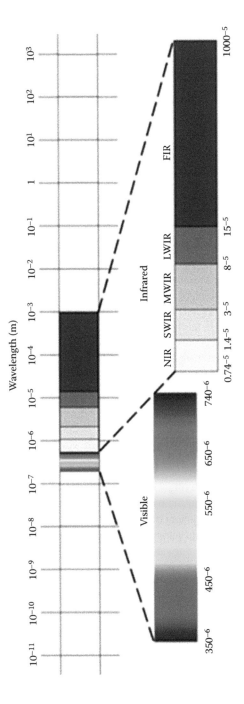

FIGURE 6.1 Visible and infrared portions of the electromagnetic spectrum. The infrared is populated by near-infrared (NIR), short-wave infrared (SWIR), mid-wave infrared (MWIR), long-wave infrared (LWIR), and far-infrared (FIR).

6.1.2 UAVs in Precision Agriculture

There are a variety of terms and acronyms used in the context of unmanned aircraft in addition to the UAV, which generally refers only to the vehicle as a platform including all systems necessary to fly. The term "unmanned aerial system" (UAS) is also very common in the aviation industry to refer not only to the UAV hardware but the sensor payload, data processing units, and ground station equipment. From a regulatory perspective, the International Civil Aviation Organization (ICAO) uses the terms "remotely piloted aircraft" (RPA) and "remotely piloted aircraft systems" (RPAS) roughly with the same meanings as UAV and UAS, respectively. National aviation regulators are expected to converge and adopt the terms established by ICAO in official documents. Nonetheless, the most widely-used term for unmanned aircraft by the general public is "drone," particularly for consumer-level UAVs.

UASs often contain a more cost-effective and flexible sensor platform than satellite or general aviation aircraft, particularly for farm-scale areas up to hundreds of hectares. A relevant factor on the use of UASs by farm managers, agricultural consultants, and researchers is their relatively lower cost, either when purchased directly from the market or alternatively as an on-demand service provided by a specialized company. For that reason, UASs can be deployed as frequently as required based on the phenology phase and weather conditions, often in a more cost-effective way than manned aircraft or satellites. The use of UASs allows the end-user to plan in advance the type of sensor that will be used to survey the farm, as well as the spatial and temporal resolutions requirements for a given application. In fact, commercial UASs involved in agricultural operations are typically required to fly below the ceiling of 400 ft above ground level. For any given imaging sensor, the closer the UAV flies from the crop, the higher the spatial resolution is. Similarly, the more surveys conducted along the crop season, the more temporal resolution the dataset will contain. Multi-temporal studies require datasets along one or multiple seasons of a crop field. In order to obtain comparable data from multiple dates, it is helpful to carry out the survey in similar light conditions. For that purpose, the optimal time is usually around peak sunlight and clear skies to avoid cloud shades. In terms of the type of sensor, each PA application may benefit the most from one or a combination of sensor technologies, such as such as high-resolution RGB (visible spectrum), thermal, multispectral, or hyperspectral cameras. In general, UAVs and most of the sensor technologies require a modest capital investment when compared to most farm equipment. However, cost-recovery relies not only on the data capture, but also on the ease of use and the data processing costs to generate accurate actionable information. Operating UASs is relatively simple and becoming easier with the increased autonomy in every new generation of unmanned aircraft.

One of the main advantages of deploying UASs in farm environments is their capability to perform autonomous flights without being actively controlled by a ground-based operator. As a result, UASs can be deployed to survey an area of interest by accurately following predefined flight parameters. Some of the most critical flight parameters are the flight altitude, which determines the spatial resolution of an imaging sensor; the forward velocity; and frontal and lateral overlaps between images, which may be required to facilitate data processing. For a farm containing a few hundred hectares of land, UAV operations can be performed within visual line-of-sight (VLOS). In optimal light conditions with clear skies and maximum visibility, VLOS may be up to 1 km from the human operator.

The lifecycle representation of a PA application that relies on UAS for data acquisition to generate prescription maps is presented in Figure 6.2. Initially, a UAS service starts with the definition of the area of interest, the sensor technology required for that particular application, as well as relevant data acquisition parameters such as the spatial and spectral resolutions. If the application requires monitoring the crop along a period of time, the frequency in which the surveys are carried out is described as temporal resolution. Once the flight parameters are inserted into the ground station software, the flight can be performed autonomously, following a set of predefined waypoints to maximize the accuracy of the data collection campaign. An increasing number of countries have

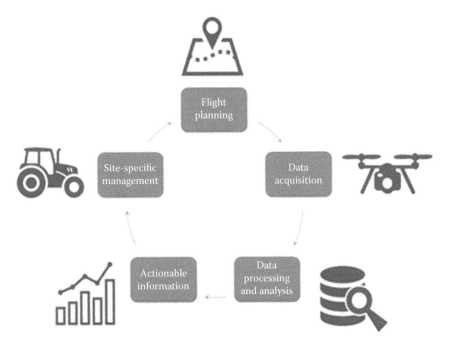

FIGURE 6.2 Life cycle of a sample PA application with UAS as a support tool.

developed or are in the process of developing regulations to operate UAV platforms for commercial operations. Therefore, all flights performed on the farm will be subject to the applicable rules, and often will require a preflight safety assessment. Once the data acquisition is completed, the post-processing and analysis can be performed with specialized software onboard the UAV's computer or by using a separate ground-based computer or cloud service. The resulting actionable information will feed into the VRT system as a prescription map that delivers customized inputs to the crop. The VRT system is typically a ground vehicle that can carry significant volumes of input. However, in certain applications, such as pest hotspots or weeds, only modest quantities of pesticide may be required. In those cases, UAS may be more cost-effective and accurate in delivering custom amounts of inputs.

In regards to the data post-processing and analysis, some UAV and sensor manufacturers include a fully or partially automated data processing service as part of their product package. In the instances which the farm manager lacks the expertise to collect and manage the data, it is likely that the same UAV company that performs the aerial data collection also has the capability to deliver the actionable information in a short timeframe.

6.1.3 ADOPTION AND ECONOMIC IMPACT OF UAV TECHNOLOGY

The first UAV designs can be traced back decades ago with applications mostly in the defense domain. However, only in the last 10 years did UAVs take off as a widely available technology and with an ever-expanding international market. The rapid growth can be attributed primarily to the miniaturization and progressive cost reductions of electronic devices, such as GPS receivers and inertial sensors, which lower the market's entry barriers for companies with new ideas. Moreover, open source communities, such as ardupilot.org or diydrones.com, are also having a major impact in facilitating the understanding and manageability of complex flight systems to the general public.

The UAV industry encompasses not only the hardware that allows the platform to perform flight operations and the sensor technologies that collect aerial data, but also the software, data

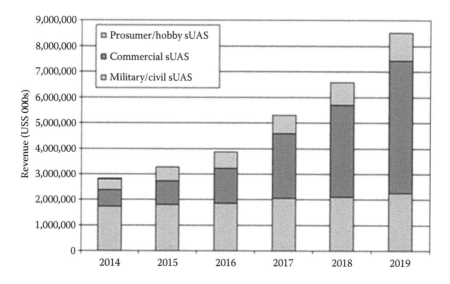

FIGURE 6.3   World sUAS revenue by segment between 2014 and 2019.

processing services, licensing, and legal services that comprise the commercial use of UAVs. Figure 6.3 presents a forecast released in 2015 of the UAV industry, with revenue in the global market for small platforms that weigh less than 25 kg, also known as small UASs (sUASs). The forecast reflects an exponential increase of revenue from less than 1$US billion in 2014 to 5$US billion in 2019.

Unlike other UAV applications (such as real estate photography or mail delivery) that can take place in urban areas, agricultural applications mostly take place in low population density areas, with few structural obstacles and few privacy concerns. Consequently, the use of UASs in agriculture is a primary candidate for minimal regulation and easier adoption by industry end-users. As a result, UASs operated in crop fields and livestock rangelands represent a minimal safety threat to people on the ground or other aircraft, provided that the UASs are operated following the applicable aviation rules and guidelines.

Considering the technical, economic, and safety aspects discusses in previous sections, it is reasonable to argue that the use of UASs in agricultural environments will keep increasing in the upcoming years. A recent research report published by the Association for Unmanned Vehicle Systems International (AUVSI), concluded that by 2025 agriculture will make up about 80% of the commercial UAS industry (AUVSI, 2015). In more specific terms, the American Farm Bureau has forecast that farmers using drone services to monitor their crops could see a return on investment of $12 per acre for corn, $2.60 per acre for soybeans, and $2.30 per acre for wheat. Also in the near future, farmer managers and companies will start adopting UASs for targeted application of herbicides and pesticides (AFBF, 2015).

6.2   UAV PLATFORMS AND REGULATIONS

6.2.1   Overview

UASs represent a technological opportunity for PA applications by improving safety, reducing liability, increasing accuracy, and saving time and money when used in a systematic and effective way. An important aspect when compared to manned aircraft is that the risk to an aircraft pilot and crew during an incident or accident is eliminated. Many types of UAVs are available today for precision agriculture and related fields, such as environmental monitoring. Types of UAVs can be classified into three categories: fixed-wing, multi-rotors, and helicopters. Similarly, UAVs are often classified in terms of weight for regulation reasons. Some countries, like Australia and Canada, have been international pioneers in defining a specific regulation for the use of UASs in their airspace. Both countries have followed a similar approach in dividing weight types into: very small (<2 kg), small (2.1 kg to 25 kg) as well as medium and large with different weight limits. In both cases an exemption is available for commercial operators to not apply for special flight operation certificates. Considering the work, investment, and training required to obtain certification, the sub–two kilograms exemption is likely to increase the adoption of UAS technology by farm managers and contractors, but also for agricultural and environmental research in universities.

6.2.2 UAV Platforms

The UAV designs mostly used for commercial applications are fixed-wing and rotary-wing aircraft. Among rotary-wing designs, helicopter and multirotor configurations are the most common. Multirotors are the most popular and easiest UAV type to manually control using radio transmitters. There are a number of multirotor configurations generally named after the quantity of rotors that form the propulsion system, such as quadcopters (four-rotor), hexacopters (six-rotor), and octacopters (eight-rotor). Taking into consideration recent statistics on the UAV platforms used in the United States for commercial operations (AUVSI, 2016), Table 6.1 shows the approximate percentage of each type of platform. Across all applications, multirotors represent nearly 90% of the entire market, demonstrating UAV operators have a clear preference for multirotor over fixed-wing. For agriculture in particular, nearly one-third of UAV platforms are fixed and two-thirds are multirotor.

Among the UASs typically used in agriculture, helicopters and multirotors tend to have less payload bay limitations, allowing for an increased variety of on-board sensors to be integrated. While fixed-wing aircraft can also be large enough to accommodate heavy payloads, large fixed-wing aircraft also present the need for a relatively flat runway to land and take off. Therefore, lightweight (less than 15 kg) fixed-wing UASs that can be hand launched are the most common type of fixed-wing aircraft used in agricultural applications.

In terms of power requirements, multirotors rely mostly on LiPo batteries, which allow for a typical flight endurance of 10 to 25 minutes. Hand launched fixed-wing aircraft are usually made of foam

TABLE 6.1

Percentage of Each Type of UAV Platform in the US Commercial Industry

Design Type	% of Platforms
Across commercial applications	
Rotary-wing	90%
4-Rotor (Quadcopter)	67.23%
8-Rotor (Octocopter)	17.01%
6-Rotor (Hexacopter)	4.96%
Helicopter	0.54%
12-Rotor and other	0.27%
Fixed-wing	10%
Hand launch	8.30%
Launcher	1.70%
Agricultural applications	
Rotary-wing	72.70%
Fixed-wing	28.30%

or other light materials and are also powered by LiPo batteries, usually enduring for 30 to 60 minutes. On the other hand, larger fixed-wing aircraft that require runways for takeoff and landing, usually require more robust airframes and fuel engines. The endurance in the latter case is typically above one hour and up to several days if the aircraft is designed for military applications. However, these UAVs are rarely used for agriculture due to high costs and operational complexities. Depending on the sensors used, multiple data sets may be collected with a high spatial and temporal resolution. However, with more complexity and capability comes higher operational requirements, and additional specialist skills may be required. Usually larger platforms are costly and a significant financial investment is required. Additionally, and perhaps more importantly, are the safety implications of using such platforms in commercial applications. They have the potential to cause considerable damage (to humans and property) and, as such, fall under stricter operating guidelines than smaller UAVs (see Section 6.2.3).

In the latter discussion, no discrimination between aircraft type was made. It is important to recognize the differences in capability between aircraft type in the context of plant biosecurity. Fixed-wing aircraft typically can cover more areas over a given time interval and provide flexibility in sensor mounting points. As they are unable to hover and have minimum operating height requirements, high spatial diversity can be achieved at the cost of decreased spatial resolution.

Unmanned aircraft systems can be used in standalone operations involving a single platform or more advanced systems utilizing multiple aircraft. In each case a ground station is usually required for remote piloting and mission command. Multiple UAVs can be flown in a swarm or coordinated to fly separate with complementary trajectories for a given application. This requires advanced centralized or decentralized control and guidance algorithms, but has the potential to increase quality and quantity of data collected at reduced operator workload. Currently, the use of multiple UAVs has been demonstrated for a range of related applications, but also in agriculture (Techy, Schmale III, and Woolsey 2010).

The market is rapidly evolving and new innovations are presented every year. According to AUVSI (AUVSI, 2016), more than two-thirds of all UAV platforms used by companies operating in the USA are manufactured by DJI Innovations (Shenzhen, China). The DJI Phantom, the most common UAV, is followed only by other models of the same company. In the agricultural domain, 76 operated fixed-wing platforms while 192 operated rotary-wing platforms (AUVSI, 2016). The USA market is a good representative of world trends, a prudent interpretation of the study is that is that UAS users tend to favor low-cost (smaller platforms) and multirotor aircraft (easier use). However, when endurance is a limitation, fixed-wing platforms such as the popular Sensefly eBee series (Cheseaux-sur-Lausanne, Switzerland) or the PrecisionHawk models (Raleigh, North Carolina) are also regularly being used.

6.2.3 REGULATION

In many countries, the operation of UAVs for both commercial and research requires certification and needs to be carried out under regulated conditions. That is a consequence of UAVs operating in the same airspace as manned aircraft. Aviation regulatory bodies define the rules and restrictions that govern who can access the national airspace and under what conditions. They can be thought of as road rules for aircraft aimed at protecting the general public and other airspace users by ensuring that safety standards are met. Some rules may only apply to a particular type of aircraft, while some may apply to all aircraft operating under certain weather conditions or flight types. For example, consider commercial road vehicles. They may have relaxed parking restrictions, but, depending on their size, they may be restricted from operating in particular areas at certain times of the day. On the contrary, small personal vehicles are relatively free to access public road networks, regardless of the time of operation, and require a different driver's license class.

From an international perspective, ICAO published the first edition of the remotely piloted aircraft systems manual (2015). The manual shows how the existing regulatory framework that was developed for manned aviation applies to unmanned aircraft. Moreover, it serves as an educational tool for states, industries, service providers, and other stakeholders on most of the topics that comprise the regulatory framework. Rulemaking on a new technology is a difficult task, especially when the implications of UAS rules impact critical aspects of a country such as national airspace, public safety, and privacy. In Sections 6.2.3.1 and 6.2.3.2, we'll be presenting an outline of the current regulation in the United States and Australia. The United States is a good representative of the countries that are developing the regulations at a slow pace, while Australia has been one of the most proactive countries since the early 2000s.

6.2.3.1 United States

In February 2012, President Obama signed the FAA Modernization and Reform Act of 2012 (FMRA; P.L.112-95). The legislation mandated that the Federal Aviation Administration (FAA) develop a comprehensive plan to integrate unmanned aircraft systems (UASs) into the national airspace and begin implementing the plan starting in October 2015.

Under a special rule established by the FMRA, model aircraft and hobby UAVs operated strictly for noncommercial, recreational purposes are permitted to fly below 400 feet, so long as they remain within sight of the operator, outside of restricted airspace, and away from airports, unless appropriate prior notification has been given to airport operators and air traffic control towers. Under this rule, operations of hobby UAVs have proliferated, creating significant enforcement challenges for the FAA (Elias, 2016).

Meanwhile, the FAA has proceeded slowly and cautiously in complying with the FMRA mandate related to government and commercial operations. It has allowed government agencies and operators of small commercial drones to obtain permits on a case-by-case basis under section 333 of the FMRA. As of September 2015, the FAA approved 1407 applications out of 2650 petitions, and the agency approves about 50 new operations a week, a process expedited by the FAA rolling out a summary grant process, whereby similar petitions are batched and analyzed together, rather than individually (AUVSI, 2016). In February 2015, the FAA released its "Notice of Proposed Rulemaking for Small UAS[s]," a set of rules that, once finalized, are expected to govern the commercial drone industry for platforms up to 55 pounds. Until this set of rules is reviewed and completed, the Section 333 exemption process remains the most effective way for commercial entities to gain access to the airspace for UAS operations (AUVSI, 2016).

Civil operators are authorized via Section 333 that grants of exemption are automatically issued a "blanket COA" to conduct civil UAS operations nationwide. The blanket COA authorizes flights at or below 400 feet to any UAS operator with a Section 333 exemption for aircraft that weigh less than 55 pounds, operate during daytime Visual Flight Rules (VFR), operate within visual line of sight (VLOS) of the pilots, and stay away from airports or heliports at the following distances:

- 5 nautical miles (NM) from an airport having an operational control tower
- 3 NM from an airport with a published instrument flight procedure, but not an operational tower
- 2 NM from an airport without a published instrument flight procedure or an operational tower
- 2 NM from a heliport with a published instrument flight procedure

6.2.3.2 Australia

In 2002, Australia was the first country in the world to regulate UAS with Civil Aviation Safety Regulation (CASR) Part 101. In April 2016, CASA is reviewing CASR Part 101 in two phases, and will eventually modernize it into CASR Part 102. Up until 2016, with CASR Part 101, commercial and research operations with UAVs required the companies to have an operator certificate and their pilots a remote pilot license. Despite these requirements, there are 550 companies currently registered in Australia as of May 2016. Taking effect on September 2016, the amendment to CASR Part 101

TABLE 6.2

Classes of RPA as Defined by CASA to Take Effect in Late 2016

Subclass	Very Small	Small	Medium	Large
Max Weight (kg)	<2 kg	2–25 kg	25–150 kg	>150 kg

will open new opportunities for farmers and companies to use UAS without the need of certification. Firstly, five new weight classes of remotely piloted aircraft (RPA) have been created (Table 6.2).

Secondly, commercial operators flying very small RPAs, i.e., weighing less than two kilograms, will not require an operator certificate or a remote pilot license. Operators will only have to provide one notification to CASA at least five days before their first commercial flight and operate by the standard operating conditions. Moreover, private land owners will be allowed to carry commercial-like operations on their own land with a small RPA without needing an operator certificate or a remote pilot license, provided that they follow the standard operating conditions and none of the parties involved receive remuneration:

- You must only fly during the day and keep your RPA within visual line-of sight.
- You must not fly your RPA higher than 120 meters (400 ft) AGL.
- You must keep your RPA at least 30 meters away from other people.
- You must keep your RPA at least 5.5 km away from controlled aerodromes.
- You must not fly your RPA over any populous areas. These include beaches, parks, and sporting ovals.
- You must not fly your RPA over or near an area affecting public safety or where emergency operations are underway (without prior approval).
- This could include situations such as a car crash, police operations, fire and associated firefighting efforts, and search and rescue.
- You can only fly one RPA at a time.

Autonomous flight is prohibited under the current amendments. CASA is still developing suitable regulations for autonomous flight. However, there is scope for CASA to approve autonomous flight on a case-by-case basis.

6.3 PAYLOAD TECHNOLOGIES

6.3.1 OVERVIEW

Some inexpensive consumer UASs can be used out of the box to take a video or still photo from above a field, which may be useful to some extent for those applications that only require high-resolution RGB imagery. However, to obtain valuable data for PA applications it is often necessary to use specialized payloads, as well as an autopilot system to survey a field following a predefined flight path. There are a wide variety of payload technologies currently used for PA applications with different levels of consolidation and adoption in commercial environments. In this chapter, we briefly describe the essentials of both imaging and non-imaging payload technologies.

6.3.2 IMAGING SENSORS

The ability of sensors to measure the reflected energy coming from land and water allows us to use remote sensing to quantify features and changes on the Earth's surface. The amount of energy reflected from these surfaces is usually expressed as a percentage of the amount of energy striking the objects. Reflectance is 100% if all of the light striking an object bounces off and is detected by the sensor. If none of the light returns from the surface, reflectance is said to be 0%. In most cases, the reflectance value of each object for each area of the electromagnetic spectrum is somewhere between these two extremes. Across any range of wavelengths, the percent of reflectance values for landscape features such as water, sand, roads, forests, etc. can be plotted and compared. Such plots are called "spectral response curves" or "spectral signatures." Differences among spectral signatures are used to help classify remotely sensed data into different classes. Figure 6.4 shows a typical spectral signature for healthy vegetation across visible and infrared bands. The photosynthetically active region of the wavelength is between 400 nm and 700 nm, with the rest of the wavelengths being related to cell structure and water content. The valleys in the signature are the primary absorption areas which are related to the chlorophyll and water content in plants. For each plant species, significant reflectance variations from the healthy signature will reflect the presence of a deficiency. The type of deficiency can be inferred from the individual wavelengths where the difference is most significant. The rapid change region between the wavelengths at 680 nm and 730 nm is known as "red edge" and represents a critical portion of the spectrum to measure plant qualities such as chlorophyll and nitrogen content.

Imaging sensors measure the radiation emitted or reflected from the object of interest, which in PA are mostly crops, weeds, and soil. The objective of imaging surveys is to estimate crop attributes at the leaf or canopy level with the reflectance values at various wavelengths. A spectral variation index (VI) is typically defined as an arithmetic operation of various reflectance values applied to each pixel of an area of interest. From a biological perspective, VIs are used to explain the biological phenomena that the crop is experimenting. From a PA perspective, VIs are also used to segment and classify the surveyed area into various partitions that share similar characteristics. The spatial information generated can be used by the farmer to assess the crop status or the progress of a crop treatment. When maps are converted to a suitable format for the VR equipment, the farmer can deliver site-specific treatments to the crop with GPS assistance.

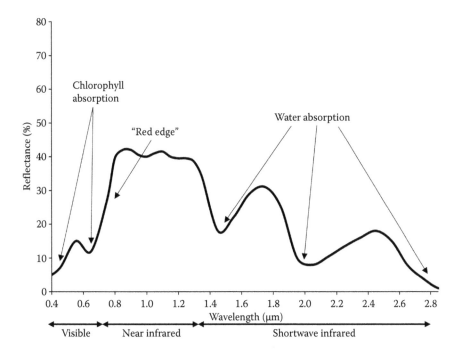

FIGURE 6.4 Typical spectral signature of healthy vegetation across the visible, near-infrared, and short-wave infrared portions of the spectrum. Highlighted are the areas of the signature affected by chlorophyll and water absorption, as well as the steep climb known as "red edge."

VIs are often classified as broadband indices when the bands used have a bandwidth higher than 10 nm, while narrowband VIs are often defined as requiring bands of bandwidths below 10 nm. Some of the most widely used broadband VIs in aerial and satellite remote sensing are presented in Table 6.3.

Relatively small RGB (visible spectrum), multispectral and hyperspectral cameras are currently available for UAVs. Depending on the type of camera, multiple wavelengths may be measured simultaneously to varying degrees of spectral resolution. A lower ground resolution, also known as ground sampling distance (GSD), in centimeters per pixel will deliver more detail of the target crop that is being surveyed. However, achieving higher ground detail (lower GSDs) implies flying lower, which, in turn, leads to surveying less area for a given forward overlap between consecutive images. Therefore, the flight planning for optimal data collection will require a balance between the area to cover and the GSD obtained. Figure 6.5 shows an instance of a graph relating the flight height (in meters) and ground resolution (in centimeters per pixel) as well as the ground speed (in meters per second) necessary to achieve an 80% overlap between forward images.

An important part of the data analysis is producing an orthoimage (also known as an orthophoto) in order to visualize the entire crop field in a single image. The orthoimage can be described as the result of geometrically correcting each individual image and then stitching them together into a larger mosaic. This is a fairly complex algorithm that requires significant computational resources and specialized software. In the context of drone surveys, orthoimages are often created by using specialized, structure-from-motion (SfM) software, such as Pix4D or Agisoft Photoscan. To effectively create the orthoimage, it is recommendable to plan for a flight path that maintains around 80% forward and a 60% lateral overlap between images. Thus, it is generally imperative to fly the UAV in autonomous mode using the autopilot ground station software, rather than flying manually with a radio transmitter, which wouldn't be as accurate.

It is generally accepted that a finer spectral resolution can provide more information. By quantifying light reflectance in narrower bands of the spectrum, it is possible to measure subtler changes and confirm the presence, absence, or severity of a particular deficiency, pest, or disease. Often the most relevant bands to evaluate crop health are the NIR and red edge, which are not perceived by the human eye. Sections 6.3.2.1 through 6.3.2.3 describe RGB (visible spectrum), multispectral, and hyperspectral cameras in further details.

TABLE 6.3

List of VIs, Their Formula, and Reference

Index	Formula	Reference
NDVI	(NIR − Red)/(NIR + RED)	Rouse *et al.* (1973)
NDRE	(REDedge − RED)/(REDedge + RED)	Barnes *et al.* (2000)
GNDVI	(NIR − GREEN)/(NIR + GREEN)	Gitelson *et al.* (1998)
DVI	NIR − RED	Tucker *et al.* (1979)
EVI2	2.5*(NIR-RED)/(NIR + 2.5*RED + 1)	Jiang *et al.* (2008)
GRVI	NIR/GREEN	Sripada *et al.* (2006)
IPVI	NIR(NIR + RED)	Crippen *et al.* (1990)
MSR	[(NIR/RED) − 1]/[(NIR/RED)^(1/2) + 1]	Chen *et al.* (1994)
SAVI	1.5*(NIR − RED)/(NIR + RED + 0.5)	Huete *et al.* (1988)
ExR	1.4*REDedge − GREEN	Meyer *et al.* (1998)
GVI	(GREEN − REDedge)/(GREEN + REDedge)	Gitelson *et al.* (2002)

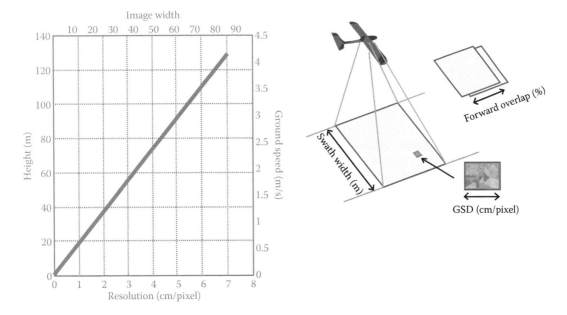

FIGURE 6.5 Sensor performance for the Micasense MCA multispectral camera with Ground sampling distance, ground speed and ground swath in relation to flight height.

6.3.2.1 Visible Spectrum Cameras

Visual imaging devices, such as consumer-level digital cameras, can be a much cheaper alternative to multispectral and hyperspectral sensors. Although they offer high resolutions between 10 and 50 megapixels, the spectral resolution is limited to three bands of the visible spectrum, namely red, green, and blue (RGB). In order to assess crop health, the most relevant bands are NIR and red edge, which are outside the visible spectrum. However, consumer-level digital cameras may be suitable to detect crop diseases that appear in green or yellowish tones. Thanks to the high spatial resolution, the imagery provided by these cameras may also be useful to visually identify weeds or other unwanted elements in the field, or to assess crop damage.

In Laliberte *et al.* (2010) two visual imaging (RGB) cameras were used on-board UAVs to monitor and classify the vegetative composition of rangelands and arid landscapes. Spatial resolutions less than 2 m/pixel were observed, highlighting the potential for crop or district level monitoring depending on available flight time and supported data storage and transmission capability. A further increase in spatial resolution was reported in a recent publication on pest damage assessment (Puig, 2015). The spatial resolution achieved in this study was below 2 cm/pixel with a Sony NEX-5. The results presented demonstrate how various levels of crop damage and the areas they occupy can be automatically mapped from RGB imagery.

6.3.2.2 Multispectral Imaging

Typically, multispectral images provide intensity information from up to 10 wavelengths of approximately 10 nm bandwidths, usually discontinuous and not overlapping. This is achieved by a set of board level cameras each with a particular band filter or a modified RGB camera that presents sufficient sensitivity on the NIR bands by changing the manufacturer's original filter. In the latter approach, various filters are used to isolate spectral regions of interest (such as NIR reflectance for vegetation monitoring) (MaxMax). This provides a lower cost solution for obtaining specific spectral data without using a specialized multispectral camera (Micasense or Tetracam). Such camera arrangements may also have faster shutter times, simplifying image rectification. These cameras are often called modified NIR cameras and have become quite common due to lower costs. However, depending on the specific setup, novel image correction methods (due to ambient conditions) may be required to utilize various vegetation indices. In terms of actual multispectral cameras, in some cases, it is possible to replace filters in the field or in the lab, allowing these systems to be repurposed to detect different wavelengths at different times for different applications. Also, some cameras can be used off the shelf and others have been designed explicitly for a particular application (Yang, 2012).

Airborne, satellite, or ground-based imagery has been used for disease detection in cereal crops (Franke and Menz, 2007). A project in Colorado, for example, conducted field experiments to measure the effects of aphid infestations on cereal crops using an airborne multispectral camera (Pearson, Golus, and Hammon, 2012). In Oklahoma, multispectral imaging was able to accurately discriminate between plant stress caused by Russian aphids and other factors using airborne imagery (from manned aircraft) (Backoulou et al., 2013). This is an important result, highlighting the benefit of aerial imagery for disease discrimination. Using a UAV at lower altitude is expected to increase the spatial resolution, and therefore improve detail and thus accuracy of such results.

Using fixed-wing UAVs, fruit crop health was monitored in Suarez et al. (2010) whilst various vegetation monitoring tasks were studied in Berni et al. (2009). Similarly, water stress detection in cereal crops has also been detected from UAVs using the thermal band. Leaf area index (LAI) was measured for winter wheat under varied fertilization schemes using a single, modified EO camera onboard a small UAV (Hunt et al., 2012). NIR-green-blue imagery was obtained and was able to provide comparable results to NIR-red-green color infra-red cameras with considerably less post processing required. Flights above and below 400 ft were conducted demonstrating that this type of system would be useful for plant and crop level monitoring onboard small UAVs.

In Lucieer et al. (2012); Turner, Lucieer, and Watson; and Nebiker, Annen, and Scherrer (2008) multispectral imaging was used onboard rotary wing UAVs to detect moss and grape vine health with relatively high spatial resolution. In Merz and Chapman (2011), only the near infrared band was used onboard an autonomous, custom-built helicopter to measure LAI on wheat plots for phonemics and plant level monitoring, with a resolution of 1–2 cm that was recorded over multiple flights covering up to 1 ha. In Inoue, Morinaga, and Tomita (2000) four CCD cameras and associated filters were used on a blimp to monitor rice and soy leaf area index from low altitude between 30 and 400 m. Later, a blimp was used for wheat field monitoring with respect to nutrient levels using two CCD cameras for visual and NIR imagery of a 2 ha plot (Jensen et al., 2007). Parasails and multispectral cameras have also been used for ecological monitoring (Clark, Woods, and Oechsle, 2010) and agriculture (Antic, Culibrk, Crnojevic, and Minic, 2010).

6.3.2.3 Hyperspectral Imaging

Hyperspectral images typically provide low spatial resolution compared to consumer level cameras, but very high spectral resolution is typically more than 100 bands. For instance, the Nano-Hyperspec (Headwall) delivers 272 spectral bands in the visible and NIR spectrum from 400 nm to 100 nm, while the spatial resolution is limited to 640 spatial bands. They are often continuous, overlapping bands or spread over a wide bandwidth spanning the visible to near-infrared (NIR) and long-wavelength infra-red (LWIR) regions. This means subtler changes may be detected in the context of plant stress and disease lifecycles. A single camera is required but can be prohibitively large. Recently, smaller hyperspectral cameras amendable to UAV implementation have become available at a relatively low cost. Many such hyper-spectral cameras have GPS/IMU included (Eaglet, Headwall) providing synchronized data for effective image rectification and improving data post-processing.

As with other imaging, hyperspectral imaging (Figure 6.6) can be used in a passive or active manner. In the former, the plant reflectance due to sunlight is measured. In the latter, a light source temporarily illuminates the observed structure and the reflectance is measured. This is also referred to as fluorescence. Additionally, it may be difficult to adapt fluorescence-based imaging approaches to

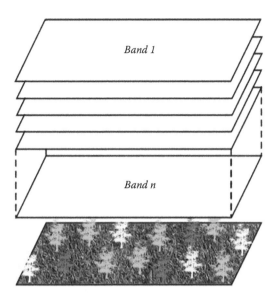

FIGURE 6.6 Basic principle for hyperspectral imaging, fine spectral partitions.

UAV. Payload restrictions may limit the activating sources that can be used and stability of the camera platform may be inadequate to achieve comparable results.

Although some recent results are based on ground or lab-based experiments, they suggest sampling in a subregion of the EM spectrum specific to the particular disease is only required, relaxing the need for a hyperspectral camera. This implies that cheaper multispectral imaging may suffice; pending reliable indices have been developed. The focus then is on collecting good quality images and ensuring the indices have been comprehensively tested against a range of in-field conditions. UAV can provide a means to gather large amounts of the required test data for such an analysis.

6.3.3 Non-Imaging Sensors

Volatile organic compounds (VOCs) are emitted throughout the lifecycle or plants. Different compounds are emitted from leaves and stems depending on plant species, growth stages, and relative health condition. By measuring VOCs emitted from healthy plants and those under stress, it may be possible to distinguish between various diseases and provide early stress detection. It can be considered a maturing technology that provides a relatively reliable and non-invasive approach to detection and monitoring. Although suggested as a research opportunity in recent literature reviews (Sankaran *et al.*, n.d.), it has not been applied to UAVs. The difficulty arises due to the way in which samples must be taken, and the weight of the compounds to be sampled. Close proximity to the plant or leaf is required and the volatile compounds are typically very light. This implies a rotorcraft type UAV would be required, however, the downwash from the rotors may disperse the compounds away from the sensor. As such, this type of sensor may be better suited to ground robots that could perhaps work in coordination with UAVs. A UAV could sample the broader area using optical sensors and relay information regarding site-specific points of interest. The data from both platforms can then be fused to provide greater depth of information. Alternatively, the sensor could be suspended and a slow, hover-like flight would be required to move through the field and obtain reliable measurements.

In physical sampling methods, a tissue or spore sample is collected, and by microscope observation and/or DNA identification, a pest may be detected. They are useful for detecting invertebrate pests such as fruit flies, or fruit and spores from disease-causing fungal pathogens such as *Fusarium*. More importantly, collecting a spore as opposed to infected tissue or an insect directly has the advantage of early detection and, as such, improves the chances of effective preventive measures.

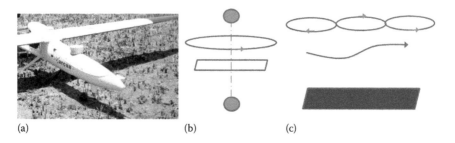

(a)   (b)   (c)

FIGURE 6.7 Example aerial spore trap and sampling trajectories (gray and red). In (a) an aerial spore trap is shown. In (b) example sampling trajectories about the focus of an infestation is shown. In (c) sampling along an arbitrary trajectory (red) and along a predicted plume line (gray) is shown.

Many physical sampling methods for invertebrates, such as vacuum aspirators, water sumps, nets, and small aperture funnels are not amendable to UAV implementation due to size, weight, and aerodynamic restrictions.

Smaller mobile spore traps using adhesive surfaces or substances (agar type jellies) can be used (Savage *et al.*, 2012). Originally used on remotely piloted vehicles (Gottwald and Tedders, 1985), they have since been demonstrated onboard UAVs in a number of applications locally (Gonzalez *et al.*, 2011) and overseas (Schmale III, Dingus, and Reinholtz, 2008; Aylor *et al.*, 2011; Schmale III *et al.*, 2012). Given the dispersal of the spores is relatively unknown and difficult to determine, it is unclear what flight path should be flown for data collection, so a number of solutions have been proposed (Schmale III, Dingus, and Reinholtz, 2008; Lin *et al.*, 2013; Techy, Woolsey, and Schmale III, 2008; Wang *et al.*, 2007). A zero sample is not enough to confirm the absence of a pathogen and large presence is not enough to assume broad scale exposure or infection. Recent simulation work provides insight as to more efficient flight path with respect to spore sampling (Savage *et al.*, 2012). Depending on the UAV type, specific flight paths may be favored, but will result in different data, and thus make comparative analysis difficult. Multiple UAVs may help to alleviate these problems. Two UAVs have been used simultaneously with advanced flight control to improve the sample size on a given sortie and, along with previous results, has allowed researchers to consider modeling dispersal and arrival paths of particular fungal pathogens (Figure 6.7) (Tallapragada, Ross, and Schmale III, 2011).

6.4 UAS APPLICATIONS IN PA

Most of the current agricultural applications of UAS involve either the acquisition of aerial data from a variety of sensors or the mechanical release of different inputs for site-specific crop treatment. In terms of the operation. Digital cameras are the most commonly used data collection sensors aboard UASs. Additionally, UASs may be equipped with infrared sensors that provide information on crop and soil temperature, more specialized imaging sensors such light detection and ranging (LIDAR) systems that use laser scans to capture high-resolution contour maps and images, or hyperspectral and multispectral imaging systems that capture a broad spectral range of light reflectance both within and beyond the limits of human vision. With the development of new sensor technologies, there is an ever-increasing number of UAS applications for PA. Some of the most common applications are:

- Crop health assessments, such as nitrogen content and plant vigor
- Crop disease detection and severity assessments
- Soil moisture content and irrigation efficiency assessments
- Field drainage assessments with digital elevation models

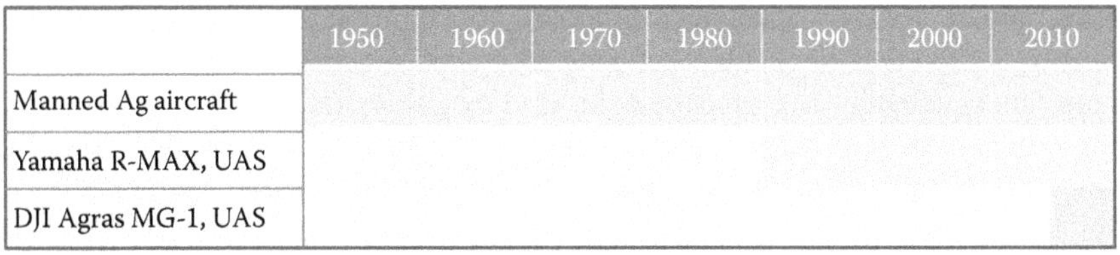

	1950	1960	1970	1980	1990	2000	2010
Manned Ag aircraft							
Yamaha R-MAX, UAS							
DJI Agras MG-1, UAS							

FIGURE 6.8 Timeline of crop spraying technologies from manned agricultural aircraft to remotely controlled aircraft by a pilot on the ground, such as the Yamaha R-MAX, and, more recently, waypoint flight capable UASs, such as the DJI Agras MG-1.

- Yield estimate assessments
- Aerial surveys for weed detection
- Aerial application of crop inputs such as fertilizers and pesticides
- Aerial surveys to detect the presence and damage from insect pests
- Direct release of benign insects for weed control or insect pest mitigation
- Collection of pathogen samples with spore traps
- Farm surveillance to monitor livestock

6.4.1 CASE STUDY: UAS AS A VR TOOL

A successful transition from a manned aircraft agricultural practice to a modern unmanned approach is progressively taking place in the practice of crop dusting (Figure 6.8). The first manned aircraft designed specifically for aerial application were introduced in the early 1950s. Beginning in Japan and South Korea in the late 1990s, UAVs, such as the Yamaha R-MAX, became common in mountainous terrain and were used by relatively small, family-owned farms, where lower-cost and higher precision spraying was required. More recently the R-MAX has been used in other countries such as the United States and Australia as the regulation for UAS has been developed. Despite featuring more than 20 kg of payload capacity and an endurance of up to one hour, the R-MAX appears to be a good alternative to agricultural aircraft. However, there are some limitations such as the required pilot on the ground flying the aircraft with a radio transmitter as well as the limitation of 150 m distance between the pilot and the aircraft. The ability for low-cost UAS to fly autonomously has only become available in the last decade.

A good representative of the new generation of crop spraying UAS aimed at commercial use is the DJI Agras MG-1 (Figure 6.8). Although 10 kg payload capacity is nearly half the capacity of the R-MAX, the price of the R-MAX costs $100,000, while the price of the Agras MG-1 is $10,000. Most importantly, unlike the Yamaha R-MAX, the Agras MG-1 is capable to fly autonomously a waypoint flight and deliver custom volumes of liquid in predefined GPS locations. This technology is better aligned with the principles of PA, and it represents a further step towards facilitating the adoption of cost-effective UAS technologies to farm managers, agricultural consultants, and researchers.

6.5 DISCUSSION AND CONCLUSIONS

As PA practices are increasingly becoming a more cost-effective approach to grow crops, farm managers are called to adopt new technologies and become more competitive. PA practices involve using only the necessary volumes of inputs such as water, fertilizers, herbicides, and pesticides, which helps reduce costs but also minimizes the ecological impact of chemicals on the soil, aquifers, and rivers. The product cycle of a PA application generally consists of data collection, the development of prescription maps, and delivering custom input applications.

UAVs have become a widely accessible technology for the past 10 years, while aviation regulators are converging towards facilitating commercial operations. Farm managers and agricultural consultants are slowly adopting UAV technologies lead by a sharp increase in research publications by universities worldwide. UAVs provide an aerial platform for specialized sensors to collect aerial data, and often represent a cost-effective alternative to manned aircraft and satellite imagery. Additionally, UAVs have can also be deployed with spraying mechanisms to deliver custom volumes of input in predefined locations over the crop.

The use of UAVs for remote sensing, has reached a relatively high level of maturity. By acquiring aerial imagery with multispectral and hyperspectral sensors, the farmer can obtain quantitative information of the crop. As these sensors can collect spectral information in multiple bands, it is possible to generate maps of crop vigor, nitrogen content, chlorophyll content, and many other crop properties. Also, the use of thermal sensors provides valuable information on the moisture content in the crops and soil, which can be used to optimize the use of irrigation.

The UAS market is growing internationally and many companies are focusing on developing integrated UASs for agriculture. Currently, it is fairly common for UAV service providers and knowledgeable farm managers to acquire a low-cost UAV from a company, purchase a sensor from another company, and have the data processed yet by another company. This trend is likely to continue and remain a strong option for clients that require minimum investment. However, in the near future we can expect integrated products that will meet the need to simplify the adoption of this technology. Also, as single-board computers become small, lighter, and more powerful, it will become possible to process the data onboard the UAV, practically delivering real-time, actionable information for immediate farm management actions.

From a technological perspective, the introduction of PA represented an important step towards managing a farm, based not only on experience but also on actual quantitative information. Subsequently the introduction of GPS-guided VRT tractors and more recently the expansion on the use of UASs, reflect the fact that farming practices are gearing towards a fully-automated industry. The farm of the future is likely to present ground and air vehicles that will monitor the crop and communicate with each other to efficiently perform the necessary tasks. In order to remain competitive, humans may no longer be required to execute manual labor but progressively will become supervisors and controllers of technology systems.

REFERENCES

AFBF. (2015). Fact Sheet: Quantifying the Benefits of Drones. American Farm Bureau Federation and Measure/ Informa Economics.

AgWeb. (2011). Attack on Plant Health. Retrieved April 22, 2013, from http://www.agweb.com/article/attack _on_plant_health/.

Antic, B., Culibrk, D., Crnojevic, V., and Minic, V. (2010). An efficient UAV-based remote sensing solution for precision farming. Retrieved May 30, 2013, from http://www.ktios.net/stari/images/stories/clanovi _katedre/borislav_antic/biosense_borislav01.pdf.

Ascending Technologies. (n.d.). Retrieved from http://www.asctec.de/uav-applications/research/products.

Australian Research Centre for Aerospace Automation. (n.d.). Enhanced flight assist system EFAs for automated aerial survey of powerline networks. Retrieved April 22, 2013, from http://www.arcaa.net /research.

Australian Bureau of Statistics. (2012). Characteristics of Australian exporters, 2010–11.

Australian Government. (2013). Department of Agriculture, Fisheries and Forestry. Retrieved June 14, 2013, from www.daff.gov.au/.

Australian Government. (n.d.). Radiocommunications (Aircraft and Aeronautical Mobile Stations) Class Licence 2006. Retrieved from http://www.comlaw.gov.au/Details/F2012C0058.

Australian Government. (n.d.). Radiocommunications (Citizen Band Radio stations) Class Licence 2002. Retrieved from http://www.comlaw.gov.au/Details/F2011C00307.

Australian Government. (n.d.). Radiocommunications (Low Interference Potential Devices) Class License Variation Notice (No. 1). Retrieved from http://www.comlaw.gov.au/Details/C2013G00637.

Australian Research Centre for Aerospace Automation (ARCAA). (n.d.). Enhanced Flight Assist System (eFAS) for automated aerial survey of powerline networks. Retrieved from http://www.arcaa.net/research /enhanced-flight-assist-system-efas-for-automated-aerial-survey-of-powerline-networks/.

AUVSI. (2013). Precision agriculture will lead civil UAS. Retrieved May 20, 2013, from http://www .aviationweek.com.

AUVSI. (2016). The first 1000 commercial UAS exemptions. Association for Unmanned Vehicle Systems International.

AUVSI. (2016). UAS buyer's guide. Association for Unmanned Vehicle Systems International.

Aylor, D., Schmale III, D., Shields, E., Newcomb, M., and Nappo, C. (2011). Tracking potato blight pathogens in the atmosphere using unmanned aerial vehicles and lagrangian modelling. *Agricultral and Forest Meteorology.*

Backoulou, G., Elliot, N., Giles, K., and Rao, M. (2013). Differentiating stress of wheat fields induced by *diuraphis noxia* from other stress causing factors. *Computers and Electronics in Agriculture*, 90, 47–53.

Basiri, M., Schill, F., Lima, P., and Floreano, D. (2012). Robust acoustic source localization of emergency signals from micro air vehicle. IEEE/RSJ Int. Conf. Intelligent Robots and Systems, 4737–42. Villamoura.

Baurigel, E., Giebel, A., Geyer, M., Schmidt, U., and Herppich, W. (2011). Early detection of Fusarium in wheat using hyperspectral imaging. *Computers and Electronics in Agriculture.*

Berni, J., Zarco-Tejada, P., Saurez, L., and Fereres, E. (2009). Thermal and narrowband multispectral remote sensing for vegetation monitoring from an unmanned aerial vehicle. *IEEE Transactions Geoscience and Remote Sensing*, 47 (3), 722–38.

Berni, J., Zarco-Tejada, P., Sepulcre-Canto, G., Fereres, E., and Villalobos, F. (2009). Mapping conductance and CWSI in olive orchards using high resolution thermal remote sensing imagery. *Remote Sensing of Environment*, 113, 2380–8.

Burling, K., Hunsche, M., and Noga, G. (2011). Use of blue-green and chlorophyll fluorescence measurements for differentiation between nitrogen and pathogen infection in winter wheat. *Journal of Plant Pathology.*

Canis, B. (2015). Unmanned Aircraft Systems (UAS): Commercial outlook for a new industry. Congressional Research Service.

Choi, H., Geeves, M., Alsalam, B., and Gonzalez, L. F. (2016). Open source computer vision–based guidance system for UAVs on-board decision making. 2016 IEEE Aerospace Conference. Yellowstone Conference Center, Big Sky, Montana.

Civil Aviation Safety Authority. (n.d.). Retrieved May 23, 2013, from http://www.casa.gov.au/scripts/nc.dll ?WCMS:STANDARD::pc=PC_100375.

Civil Aviation Safety Authority. (1988). Civil Aviation Regulations 1988, Volume 3, Part 12.

Civil Aviation Safety Authority. (2002). Advisory Circular 101-1(0).

Clark, A., Woods, J., & Oechsle, O. (2010). A low-cost airborne platform for ecological monitoring. *Int. Archives of Photogrammetry, Remote Sensing and Spatial Information Science*, 38 (5), 167–72.

Crookston, K. (2006). A top 10 list of developments and issues impacting crop management and ecology during the past 50 years. *Crop Science*, 46, 2253–62.

Department of Agriculture, Fisheries and Forestry (DAFF). (2012). Agricultural commodity statistics.

Division of Agriculture, University of Arkansas. (n.d.). Retrieved April 22, 2013, from http://www .insectphotos.uark.edu/smallgrains.index.htm.

Ecosure. (2009). Aerial mapping of riparian vine weeds. West Burleigh: Tweed Shire Council.

Elias, B. (2016). Unmanned aircraft operations in domesting airspace: U.S. policy perspectives and regulatory landscape. Congressional Research Service.

Europena Commision Seventh Framework Program. (n.d.). Plant and Food Biosecurity. Retrieved May 2, 2013, from http://www.plantfoodsec.eu.

Everaerts, J. (2008). The use of unmanned aerial vehicles (UAVs) for remote sensing and mapping. *Int. Archives of Photogrammetry, Remote Sensing and Spatial Information Science*, 37.

Ferry, N., Stavroulakis, S., Guan, W., Davison, G., Bell, H., Weaver, R. *et al.* (2011). Molecular interactions between wheat and cereal aphid (*Sitobion avenae*): Analysis of changes to the wheat proteome. *Proteomics.*

Fletcher, A., and Erskine, P. (2012). Mapping of a rare plant species (*Boronia deanei*) using hyper-resolution remote sensing and concurrent ground observation. *Ecological Management and Restoration*, 13 (2).

Fletcher, A., and Erskine, P. (2012). Mapping of a rare plant species (*Boronia deanei*) using hyper-resolution remote sensing and concurrent ground observation. *Ecological Management and Restoration*, 13 (2).

Franke, J., and Menz, G. (2007). Multi-temporal disease detection by multispectral remote sensing. *Precision Agriculture*.

Gatewing. (n.d.). Retrieved April 22, 2013, from http://www.gatewing.com.

Gonzalez, F., Castro, M., Narayan, P., Walker, R., and Zeller, L. (2011). Development of an autonomous unmanned aircraft to collect time-stamped samples from the atmosphere and localise potential pathogen sources. *Journal of Field Robotics*.

Gonzalez, F., Heckmann, A., Notter, S., Zürn, M., Trachte, J., and McFadyen, A. (2015). Nonlinear model predictive control for UAVs with slung/swung load. *International Conference on Robotics and Automation*. Washington State Convention and Trade Center (WSCC), Seattle, Washington.

Gottwald, T., and Tedders, W. (1985). A spore trap and pollen trap for use on aerial remotely piloted vehicles. *The American Phytopathological Society*, 75 (7), 801–7.

Government of Western Australia. (2013). Department of Agriculture and Food. Retrieved April 22, 2013, from http://www.agric.wa.gov.au.

Grains Research and Development Corporation. (2010). *GRDC Annual Report 2011–2012*.

Hardin, P., and Jensen, R. (2011). Small-scale unmanned aerial vehicles in environmental remote sensing: challenges and opportunities. *GIScience and Remote Sensing*, 48 (1), 99–111.

Hazel, B. A. (2015). In commercial drones the race is on. Oliver Wyman, Aviation Aerospace & Defense.

Huang, W., Lamb, D., Niu, Z., Zhang, Y., Liu, L., and Wang, J. (2017). Identification of yellow rust in wheat using *in situ* spectral reflectance measurements and airborne hyperspectral imaging. *Precision Agriculture*, 8, 187–97.

Hunt, E., Hively, W., Fujikawa, S., Linden, D., Daughtry, C., and McCarty, G. (2012). Aquisition of NIR-green-blue digital photographs from unmanned aircraft for crop monitoring. *Remote Sensing*.

HyVista. (n.d.). Retrieved April 22, 2013, from http://hyvista.com/?page_id=440.

ICAO. (2015). Manual on remotely piloted aircraft systems (RPAS). Montreal, Canada. International Civil Aviation Organization.

Inoue, Y., Morinaga, S., and Tomita, A. (2000). A blimp-based remote sensing system for low-altitude monitoring of plant variables: A preliminary experiment for agricultural and ecological applications. *Int Journal of Remote Sensing*, 21 (2), 379–85.

Jackson, S., and Bayliss, K. (2011). Spore traps need improvement to fulfil plant biosecurity requirements. *Plant Pathology*, 6, 801–10.

Jensen, T., Apan, A., Young, F., and Zeller, L. (2007). Detecting the attributes of a wheat crop using digitial imagery acquired from a low-altitude platfom. *Computers and Electronics in Agriculture*, 59, 66–77.

Johnson, L., Herwitz, S., Dunagan, S., Lobitz, B., Sullivan D, and Slye, R. (2003). Collection of ultra-high spatial and spectral resolution image data over Californian vineyards with a small UAV. *Proc. Int. Symposium on Remote Sensing of Environment*.

Kazmi, W., Bisgaard, M., Garcia-Ruiz, F., Hansen, K., and Cour-Harbo, A. (2011). Adaptive surveying and early treatment of crops with a team of autonomous robots. *Proc. Fifth European Conference on Mobile Robots ECMR 2011*, 253–8.

Kelcey, J., and Lucieer, A. (2012). Sensor correction of six-band multispectral imaging sensor for UAV remote sensing. *Remote Sensing* 4, 1462–93.

Kuckeberg, J., Tarachnyk, I., and Noga, G. (2009). Temporal and spatial changes of chlorophyll fluorescence as a basis for early and precise detection of leaf rust and powdery mildew infection in wheat leaves. *Precision Agriculture*.

Lai, J., Mejias, L., and Ford, J. (2011). Airborne vision-based collision detection system. *Journal Field Robotics*, 28 (2), 137–57.

Laliberte, A., Goforth, M., Steele, C., and Rango, A. (2011). Multispectral remote sensing from unmanned aircraft: Image processing workflow and applications for rangeland enviornments. *Remote Sensing*.

Laliberte, A., Herrick, J., Rango, A., and Winters, C. (2010). Aquisition, or theorectification, and object-based classifiaction of unmanned aerial vehicle imagery for ranglenad monitoring. *GIScience and Remote Sensing*, 76 (6), 661–72.

Lin, B., Bozorgmagham, A., Ross, S., and Schmale III, D. (2013). Small fluctuations in the recovery of *fusaria* across consecutive sampling intervals with unmanned aircraft 100 meters above ground level. *Aerobiologica*.

Lua, J., Dacheng, W., Yingying, D., Wenjiang, H., and Jindi, W. (2011). Developing an aphid damage hyperspectral index for detecting aphid (*Hemiptera: Aphididae*) damage levels in winter wheat. *Geoscience and Remote Sensing Symposium.*

Lucieer, A., Robinson, S., Turner, D., Harwin, S., and Kelcey, J. (2012). Using a micro UAV for ultra-high resolution multisensor observations of antarctic moss beds. *Int. Archives of Photogrammetry, Remote Sensing and Spatial Information*, 39, 429–33.

Mahlein, A., Oerke, E., Steiner, U., and Dehne, H. (n.d.). Recent advances in sensing plant diseases for precision crop protection. *European Journal of Plant Pathology.*

MaxMax. (n.d.). Retrieved June 24, 2013, from http://maxmax.com/vegetation_stress_mkii.htm.

Mcfadyen, A., Corke, P., and Mejias, L. (2012). Rotorcraft collision avoidance using spherical image-based visual servoing and single point features. *IEEE/RSJ Conference Intelligent Robots and Systems.* Villamoura.

Mcgwire, K., Weltz, M., Finzel, J., Morris, C., Fenstermaker, L., and Mcgraw, D. (2013). Multiscale assessment of green leaf cover in a semi-arid rangeland with small unmanned aerial vehicle. *International Journal of Remote Sensing.*

Merz, T., and Chapman, S. (2011). Autonomous unmanned helicopter system for remote sensing missions in unknown environments. *Int. Archives of Photogrammetry, Remote Sensing and Spatial Information Sciences* 38, 143–8.

Mewes, T., Franke, J., and Menz, G. (Precision Agriculture). (2011). Spectral requirement on airborne hyperspectral remote sensing for wheat disease detection.

Mikrocopter. (n.d.). Retrieved from http://www.mikrokopter.de/ucwiki.

Mirik, M., Ansley, R., Michels, G., and Elliott, N. (2012). Spectral vegetation indices selected for quantifying russian wheat aphid (*Diuraphis noxia*) feeding damage in wheat (*Tricum aestivum L.*). *Precision Agriculture.*

Mississippi State University. (n.d.). Mississippi Crop Situation. Retrieved April 22, 2013, from http://www.mississippi-crops.com.

Moshou, D., Bravo, C., Oberti, R., West, J., Bodria, L., McCartney, A. *et al.* (2005). Plant disease detection based on data fusion of hyperspectral and multispectral fluorescence imaging from Kohonen maps. *Real Time Imaging*, 11, 75–83.

Moshou, D., Garvalos, I., Kateris, D., Bravo, C., Oberti, R., West, J. *et al.* (2012). Multisensor fusion of remote sensing data from crop disease detection. In *Geospatial Techniques for Managing Environmental Resources.*

Mulla, D. (2013). 25 years of remote sensing in precision agriculture: Key advances and remaining knowledge gaps. *Biosystems Engineering.*

Murray, M., and Brennan, P. (2010). Estimating disease losses to the Australasian barley industry. *Australasian Plant Pathology*, 38 (6), 558–70.

Murray, M., and Brennan, P. (2009). Estimation disease losses to Australasian wheat industry. *Australasian Plant Pathology*, 39 (1), 85–96.

Murray, M., Clarke, B., and Ronning, A. (2013). Estimation invertebrate pest losses in six major Australian crops. *Australian Journal of Entomology.*

Nansen, C., Macedo, T., Swanson, R., and Weaver, D. (2009). Use of spatial structure analysis of hyperspectral data cubes for detection of insect-induced stress in wheat plants. *International Journal of Remote Sensing.*

Navia, D., Santos de Mendonca, R., Skoracka, A., Szydlo, W., Knihinicki, D., Hein, G. *et al.* (2013). Wheat curl mite (*Aceria tosichella*) and transmitted viruses: An expanding pest complex affecting cereal crops. *Exp. Application Acarol* 59, 95–143.

Nebiker, S., Annen, A., and Scherrer, M. (2008). A lightweight multispectral sensor for micro UAV opportunities for very high resolution airborne remote sensing. *Int Archives of the Photogrammetry, Remote Sensing and Spatial Information Sciences* 37, 113–1199.

Pearson, C. H., Golus, H. M., and Hammon, R. W. (2012). Fifty years of agronomic research in western colorado. Retrieved April 16, 2013, from http://www.colostate.edu/programs/wcrc/pubs/information/fiftyyears.htm.

Perry, E., Brand, J., Kant, S., and Fitzgerald, G. (2012). Field-based rapid phenotyping with unmanned aerial vehicles. *Australian Society of Agronomy.*

Plaza, A., Plaza, J., Paz, A., and Sanchez, S. (2011). Parallel hyperspectral image and signal processing. *IEEE Signal Processing Magazine.*

Puig, E., Gonzalez, F., Hamilton, G., and Grundy, P. (2015). Assessment of crop insect damage using unmanned aerial systems: A machine learning approach. 21st International Congress on Modelling and Simulation, Gold Coast, Australia.

Queensland Times. (2013, May 27). Drones fly on cutting edge of ten-meter farm grants. Retrieved May 29, 2013, from http://www.qt.com.au/news.

Radiocommunications (Low Interference Potential Devices) Class Licence 2000. (n.d.). Retrieved from http://www.comlaw.gov.au/Details/F2011C00543.

Rango, A., and Laliberte, A. (2010). Impact of flight regulations on effective use of unmanned aircraft systems for natural resource applications. *Journal of Applied Remote Sensing* 4.

Rango, A., Laliberte, A., Herrick, J., Steele, C., Destelmeyer, B., and Chopping, M. (2006). Use of UAVs for remote measurement of vegetation canopy variables. *AGU Fall Meeting Abstracts.*

RapidEye. (n.d.). Retrieved April 22, 2013, from http://www.rapideye.com/solution/agriculure.htm.

Saari, H., Pellikka, I., Pesonen, L., Tuominen, S., Heikkila, J., Holmlund, C. *et al.* (2011). Unmanned aerial vehicle (UAV) operated spectral camera system for forest and agriculture applications. *Remote Sensing for Agriculture, Ecosystems and Hydrology* XIII, SPIE, 8174.

Sankaran, S., Mishra, A., Ehsani, R., and Davis, C. (n.d.). A review of advanced techniques for detecting plant diseasees. *Computers and Electronics in Agriculture.*

Savage, D., Barbetti, M., MacLeod, W., Salam, M., & Renton, M. (2012). Mobile traps are better than stationary traps for surveillance of airborne fungal spores. *Crop Protection*, 36, 23–30.

Schmale III, D., Ross, S., Fetters, T., Tallapragada, P., Wood-Jones, A., and Dingus, B. (2012). Isolates of *Fusarium graminearum* collected 40 to 320 meters above ground level cause *Fusarium* head blight in wheat and produce *trichothecene mycotoxins. Aerobiologica.*

Schmale III, D., Dingus, B., and Reinholtz, C. (2008). Development and application of an autonomous aerial vehicle for precise aerobiological sampling above agricultural fields. *Journal of Field Robotics.*

SenseFly. (n.d.). Retrieved from http://www.sensefly.com/operations/overview.html.

Singh, C., Jays, D., Paliwal, J., and White, N. (2012). Fungal damage detection in wheat using short-wave near-infrared hyperspectral and digital color imaging. *Int. Journal of Food Properties.*

Suarez, L., Zarco-Tejada, P., Gonzalez-Dugo, V., Berni, J., Sagardoy, R., Morales, F. *et al.* (2010). Detecting water stress effects on fruit quality in orchards with time-series PRI airborne imagery. *Remote Sensing in Environment* 114, 286–98.

Sullivan, D., Fulton, J., Shaw, J., and Bland, G. (2007). Evaluating the sensitivity of unmanned thermal infrared aerial system to detect water stress in cotton canopy. *American Society of Agriculture and Biological Engineers* 50 (6), 1955–62.

Tallapragada, P., Ross, S., and Schmale III, D. (2011). Lagragian coherent structures are associated with fluctuations in airborne microbial populations. *Chaos: An Interdisciplinary Journal of Nonlinear Science* 21 (3).

Techy, L., Schmale III, D., and Woolsey, C. (2010). Coordinated aerobiological sampling of a plant pathogen in the lower atmosphere using two autonomous unmanned aircraft vehicles. *Journal of Field Robotics.* 27 (3), 335–343.

Techy, L., Woolsey, C., and Schmale III, D. (2008). Path planning for efficient coordination in aerobiological sampling missions. *IEEE Int. Decision and Control.*

Tetracam. (n.d.). Retrieved from http://www.tetracam.com.

Texas A&M. (2013). Wheat insect monitoring beginning to spot aphids and mites. Retrieved April 22, 2013, from http;//today.agrilife.org/2013/04/04/wheat-insect-monitoring-beginning-to-spot-aphids-and-mites/.

Trachte, J., Gonzalez, F., and Mcfadyen, A. (2014). Nonlinear model predictive control for a multirotor with heavy slung load. Proc. of the 2014 International Conference on Unmanned Aircraft Systems, 1105–10. Orlando, Florida.

Turner, D., Lucieer, A., and Watson, C. (n.d.). Development of an unmanned aerial vehicle (UAV) for hyper resolution vineyard mapping on visible, multispectral and thermal imagery.

United States Department of Agriculture. (n.d.). Retrieved April 22, 2013, from http://www.ars.usda.gov/is/ar/archive/apr04/aphid0404.htm.

University of Tasmania. (n.d.). Terraluma. Retrieved from http://www.terraluma.net.

Victorian Government. (n.d.). Department of environment and primary industries. Retrieved April 22, 2013, from http://www.dpi.vic.gov.au/agriculture/about-agriculture/biosecurity/biosecurity-strategy-implementation-plan/plant-biosecurity.

Wang, J., Patel, V., Woolsey, C., Hovakimyan, N., and Schmale III, D. (2007). Adaptive L1 control of a UAV for aerobiological sampling. Proc. of American Control Conference.

West, J., Bravo, C., Oberti, R., Moshou, D., Ramon, H., and McCartney, A. (2012). Detection of fungal diseases optically and pathogen inoculum by air sampling. *Precision Crop Protection—The Challenge of Heterogeneity.* Springer.

Woodhouse, I. (2011). Remote Sensing System. Patent Application Publication.

Yamaha. (n.d.). Retrieved from http://rmax.yamaha-motor.com.au/

Yang, C. (2012). A high-resultion airborne four-camera imaging system for agricultural remote sensing. *Computers and Electronics in Agriculture*, 88, 13–24.

Zarco-Tejada, P., Gonzalez-Dugo, V., and Berni, J. (2012). Fluorescence, temperature and narrow-band indices acquired from UAV platform for water stress detection using a micro-hyperspectral imager and thermal camera. *Remote Sensing of Environment*, 117.

Zarco-Tejada, P., Guillen-Climent, M., Hernandez-Clemente, R., Catalina, A., Gonzalez M, and Martin, P. (2013). Estimating leaf carotenoid content in vineyards using hyper resolution hyperspectral imagery from unmanned aerial vehicles (UAVs). *Agriculture and Forest Meteorology*.

Zhang, C. and Kovacs, J. (2012). The application of small unmanned aerial systems for precision agriculture: A review. *Precision Agriculture* 13, 693–712.

Zürn, M., McFadyen, A., Notter, S., Heckmann, A., Morton, K., and Gonzalez, L. F. (2016). MPC–controlled multirotor with suspended slung load: System architecture and visual load detection. 2016 IEEE Aerospace Conference, Yellowstone Conference Center, Big Sky, Montana.

7 Agricultural Robotics

Santosh K. Pitla

CONTENTS

7.1 INTRODUCTION TO AGRICULTURAL ROBOTICS

Advances in electronics and computing is influencing agriculture in numerous ways. Present day agricultural machines utilize sensors and high-speed computers for efficient application of crop inputs (e.g., seeds, fertilizers, and chemicals), and facilitate bio material transfer (e.g., grain and biomass) from fields to processing facilities. Modern agricultural tractors require human supervision to warrant safe operation, even though they are equipped with numerous sensors and controls. In contrast, the next generation of agricultural machines, in addition to the control hardware and software, will comprise of intelligence to learn, react, and make decisions for optimized crop management in the absence of an operator. Agricultural machinery is evolving from simple mechanical agricultural devices to state-of-the-art autonomous machines, capable of performing tasks with potentially unlimited autonomy. This path of technological advancement in agricultural machinery automation is poised to continue, as we will be tasked with producing food, fiber, and fuel for approximately nine billion people in the year 2050 with the same amount of land and limited resources. Efficient automated machine system solutions, including agricultural robotics, will play a key role in performing field operations in the most efficient and productive way possible for increasing farm yields while minimizing environmental impacts.

Robotic equipment is used in a multitude of applications, ranging from shop floor assembly tasks in manufacturing plants to space exploration. Some applications require the robots to perform monotonous repetitive activities, whereas others demand the exploration of unknown hazardous environments to accomplish complex tasks. In any case, the demand for robotics is increasing across various sectors and agricultural production is no exception; researchers, equipment manufacturers, and producers around the world are recognizing that conventional automated agricultural equipment and robotics can be profitable in lieu of labor shortages, and the need for efficient resource management. Worldwide, automated conventional machinery and agricultural robots (ag-robots) are seen as key solutions for performing precise, efficient field operations (e.g., planting, fertilizer spreading, spraying, weed management, and harvesting) to increase productivity while reducing environmental impacts. In Asian countries like China, Japan, and India, the increasing average age of farmers, and the migration of the younger generation to developing large cities is driving the demand for agricultural machine automation. The increasing need for productivity gains, emerging demand for organic crops, and the impending obligation to transform agriculture into a sustainable and environment-friendly practice are some of the major driving forces for the research, development, and adoption of highly automated systems and ag-robots. Economic feasibility, socioeconomics, and the liability aspects of the ag-robots have to be addressed before they can be fully realizable in agricultural production.

7.1.1 PARADIGM SHIFT IN MACHINE SIZE (SMALL VERSUS BIG)

One of the major trends that can be observed in present day agriculture is to utilize bigger and faster machines to increase field capacity (hah^{-1}) and cover more field area in a given time window. Contemporary field machinery may have reached a saturation point in terms of their size (width and mass), and further increases in the size without negatively affecting soil structure and field efficiency will mean significant modifications to the vehicle construction, transport, and power management systems. Wider implements have higher field capacities (hah^{-1}), but take more preparation and transport time, which leads to lower effective field efficiencies (Taylor *et al.*, 2001). For example, producers with wide implements and multiple moderately-sized fields might expend more time transporting the equipment in between the fields and preparing them for field operations, compared to the time spent in the field doing actual field operations. Wider equipment poses problems for targeted accurate metering and placement of seeds, fertilizer, and chemicals. Planter and sprayer errors attributed to overlap and underlap of coverage areas are typical where significant loss of crop inputs are reported (Luck *et al.*, 2009; Shearer and Pitla, 2014). However, application errors caused by overlap and underlap are reduced using additional instrumentation, GPS-based automatic steering, and map-based control on conventional machinery. Coming to the gross vehicle weight of the equipment, we might have reached a saturation point. A 55-ton grain cart with a total ballasted weight of approximately 70 tons is common in some of the large producer fields for harvested grain transfer (Shearer and Pitla, 2013). The high axle loads of the machinery cause soil compaction, which adversely impacts crop yields, and yield reductions up to 50% are possible for severely compacted soils (Wolkowski and Lowery, 2008). Autonomous operation of equipment of this size may only serve to magnify the existing problems. Even so, there is a desire for one person to supervise multiple large agricultural machines to reduce labor costs while increasing productivity. The rapidly evolving agricultural equipment industry is moving towards the deployment of large autonomous vehicles. One important consideration is that with the increase in engine power and size of the autonomous machines, the issues concerning safety of the surroundings and humans are exacerbated.

Liability and soil compaction associated with larger ag-robots are some of the most important driving factors for thinking small. Blackmore *et al.* (2004) discussed the concept of replacing conventional large agricultural machines with multiple, small autonomous machines. Small ag-robots are believed to work 24 hours a day to compensate for the reduced work rate while potentially minimizing compaction. From a field deployment perspective, the ag-robots would need refilling of fuel and recharging time depending on whether the power source is an internal combustion engine, or battery operated, respectively. The eventual goal of these small ag-robots is to reduce the scale of management to a level where plants can be treated individually (Shibusawa, 1996). The hypothetical concept of utilizing multiple small ag-robots to replace conventional vehicles is an effective strategy, however the practical deployment of a swarm of ag-robots is challenging. Standardized control architectures addressing specific agriculture applications, intelligence to react and mitigate unknown environments, inter-machine communication methodologies, and data transfer infrastructure have to be made feasible. Also, the capital costs for small ag-roots is too high due to the added electronics, sensors, and computing hardware costs. One encouraging observable trend is that the increased use of sensors and intelligent systems in the automotive sector is driving down the costs of sensors and electronic hardware which is aiding the feasibility of ag-robots in practical applications. Technology (hardware, software, and sensors) costs might not be an issue in the future, instead, a major consideration for the deployment of multiple ag-robots in agriculture could be the issue of mitigating liability. Liability is a result of unpredictable behavior of ag-robots which can be addressed by developing intelligent control architectures for ensuring safe, predictable operation and accomplishment of assigned tasks. Efforts are underway in universities, research institutions, and technology startup companies to build intelligence in ag-robots both large and small for performing specific agricultural applications in both broad acre crops and orchard environments.

Ag-robots can be classified into two broad categories: manipulators and unmanned agricultural ground vehicles (UAGVs). Section 7.2 of this chapter discusses ag-robots that are under development in universities and research institutions, whereas Section 7.3 presents trends in the industry of UAGVs. Majority of the chapter's discussion is focused on UAGV development and their applications, and factors affecting the advancement and widespread acceptance of UAGVs. Insights on the need for further developmental work on the techno-economics of ag-robots, machine data acquisition and telematics, and the inevitability of a standardized control framework for swarms of UAGVs is discussed towards the end of the chapter.

7.2 AG-ROBOT PROTOTYPES AND APPLICATIONS

7.2.1 AG-ROBOTS: MANIPULATORS

Manipulator type ag-robots are predominantly used in the food processing, dairy, horticulture, and orchard industries. In dairy industry, robotic milking stations are now commercially available worldwide (Kuipers, 1996; Hogewerf et al., 1992) and as of 2012, around 10,000 farms in 25 different countries are using robotic milking stations (Broucek and Tongel, 2015). In a typical stationary robot milking system with manipulators, cows visit a milking cell where the robot attaches teat cups to the cow with the aid of a vision system. While robotic manipulators for milking and some horticulture applications are commercially available, robotic fruit harvesting in orchards (e.g., apples, pears, citrus, and grape orchards) is currently in its research and development phase. Researchers are developing robotic arms with integrated cameras (e.g., stereovision cameras) for identification of the fruit's three-dimensional location to facilitate automated harvest (Davinia et al., 2014). Varying lighting conditions in outdoor environments and fruit clustering can be noted as some of the main challenges for deploying robotic harvesting successfully. An example of a robotic arm mounted on a multi-utility vehicle for harvesting apples in an orchard can be seen in Figure 7.1 (Silwal et al., 2014; Davidson et al., 2015). Extensive literature is available on robotic citrus harvesting and other specialty crops (Hannan et al., 2004) similar to apple-harvesting robots. Robotic harvesters for cherry tomatoes (Kondo et al., 1996), cucumbers (Van Henten et al., 2002), mushrooms (Reed et al., 2001), cherries (Tanigaki et al., 2008), and other fruit (Kondo et al., 1995; 1996) have also been developed. A multidisciplinary approach is required to address the technical, horticultural, economical, and producer acceptance issues for making significant progress toward the adoption of fruit harvesting robots in various fruit orchard markets (Burks et al., 2005).

(a)   (b)

FIGURE 7.1   (a) Robotic manipulator installed on a utility vehicle in orchards; (b) Robotic manipulator harvesting the apples. (Courtesy of Manoj Karkee, Washington State University.)

7.2.2   AG-ROBOTS: UNMANNED AGRICULTURAL GROUND VEHICLES (UAGVS)

UAGVs navigate fields autonomously to complete an assigned agricultural production task, and at a basic level, are comprised of an instrumented mobile platform with onboard sensors, and computing hardware with algorithms and software. UAGVs typically can support attachments for performing field operations like planting, spraying, and harvesting. In some other instances, the UAGVs could be ground-based sensing platforms that are equipped with an array of sensors and cameras for scouting soil conditions, disease pressure, and crop health evaluation. Field operable UAGVs that can handle harsh field conditions have been developed successfully at various research institutes (Katupitiya and Eaton, 2008; Oksanen, 2012; Guivant *et al.*, 2012). Researchers at the Technical University of Denmark developed an autonomous UAGV prototype for mapping weeds (Madsen and Jakobsen, 2001). The overall goal was to map glyphosate-resistant weed species, such as waterhemp, and manage them with targeted spraying. The autonomous Demeter system is a retrofitted New Holland 2550 self-propelled windrower developed at Carnegie Mellon University (Pilarski *et al.*, 2002). The robotic machine harvested more than 40 hectares of crop without human intervention.

While the Demeter System was developed specifically for harvesting, Agribot, a multi-purpose UAGV, was developed for site-specific crop management. The robot shown in Figure 7.2 was

FIGURE 7.2   AgriBot. (Courtesy of Eduardo Godoy, UNESP, Brazil.)

developed as part of a public-private partnership between research organizations and industries in Brazil. This four-wheel steered UAGV uses a a diesel engine as the prime mover for its hydrostatic transmission (Godoy *et al.*, 2012). The robotic platform consisted of a controller area network (CANbus) system, an industry standard for in-vehicle communication of off-road equipment (Stone *et al.*, 1999). Unlike Agribot, which is a high clearance robotic platform with a considerable physical footprint, UAGVs with a smaller footprint that can go in between the rows are also under development. AgBo is a small UAGV with a CAN bus system that is capable of navigating in between crop rows using laser scanners (Grift, 2007). Management of glyphosate-resistant weeds was one of the primary applications of AgBo (see Figure 7.3). This UAGV had independent four-wheel steering, which allowed for crabbing, spin turns, front/rear, and all-wheel steering. A track type inter-row UAGV that navigates with two LIDAR sensors can be seen in Figure 7.4. The inter-row UAGV uses a fuzzy logic navigation controller, which takes inputs only from one-dimensional (1D) ranging sensors without any need for any additional instrumentation. The UAGV's control system was

FIGURE 7.3   AgBo. (Courtesy of Tony Grift, University of Illinois Urbana–Champaign.)

FIGURE 7.4   Inter-row robot for under canopy measurements. (a) Embedded controllers and Lidar sensors; (b) laboratory evaluation of autonomous navigation; (c) field evaluation of autonomous navigation. (Courtesy of Santosh K. Pitla, University of Nebraska–Lincoln.)

implemented on basic hardware and operated on a commonly available microcontroller development platform (e.g., Arduino) and open-source software libraries. The primary application of this inter-row robot was to count plant populations, and perform under-canopy soil and microclimate measurements. Insect-inspired UAGVs have been explored as a concept to perform swarm applications in agriculture. AgAnts (see Figure 7.5) are small robotic platforms with mechanical linkages which work in groups (Grift, 2007). The physical structure of these UAGVs resemble insects, and have replaced conventional wheels with multiple legs. AgAnts work in groups where wireless communication and cooperative behavior strategies play an important role in the operation of swarm systems. Pitla (2012) developed a multi-robot system (MRS) consisting of three autonomous vehicle platforms. The group of UAGVs seen in Figure 7.6 were developed to simulate various levels of cooperation required among UAGVs while planting, baling, and harvesting operations in non-field conditions. The UAGVs are battery-operated electrical machines and consist of a CAN bus system for data acquisition from multiple sensors and distributed control. Important applications of these

FIGURE 7.5   AgAnt. (Courtesy of Tony Grift, UIUC.)

FIGURE 7.6   MRS for Cooperative behavior evaluation. (Courtesy of Santosh K. Pitla, University of Nebraska–Lincoln.)

UAGVs are to evaluate the deliberative and reactive robotic behaviors, and the coordinated and task-sharing abilities among multiple UAGVs (Pitla, 2012).

UAGVs with integrated manipulators or robotic arms are a combination of both categories of ag-robots, and are found in limited, specialized agricultural applications. Their demand, however, is anticipated to increase in plant phenotyping and fruit-harvesting applications, where ag-robots have to navigate in between the crop/tree rows to collect leaf samples or harvest fruits, respectively. UAGVs with robotic arms for conventional crop production might also become available in the future with the maturity of various technologies. Ag-robot research is progressing swiftly in universities and research institutes, and a similar trend can be found in the industry. The motivation behind this trend is the increasing demand for automating processes in greenhouses, test plots, orchards, high-throughput phenotyping applications, and production agriculture. In the last few years there has been a proliferation of technology start-up companies that are developing ag-robots, specifically UAGVs, for a wide variety of agricultural applications. Traditional agricultural equipment manufacturers, in addition to providing conventional automated solutions, are also focusing their efforts on the development of UAGVs for fully autonomous field operations.

7.3 TRENDS IN UNMANNED AGRICULTURAL GROUND VEHICLE (UAGV) DEVELOPMENT

The most sophisticated tractors available today feature automation of numerous machine functions, but require an operator to closely monitor the performed tasks. Advancements in sensing, communication, and control technologies, coupled with Global Navigation Satellite Systems (GNSS) and Geographical Information Systems (GIS), are aiding the progression of agricultural equipment from simple mechanical machines to intelligent machines of the future. Specifically, ISOBUS and GNSS systems are contributing to the accelerated transition of tractors to highly automated machines (Lenz *et al.*, 2007; Shearer and Pitla, 2013). Even with the use of substantial technology on the equipment today, an operator is still needed inside the cab to monitor operations and attend to emergency situations. For example, planting operations still require the tractor operator to raise the planter row units at the end of the rows, and lower the row units for planting in the next row. Additionally, layers of intelligence have to be added to the tractor control system to enable situational and environmental awareness for managing moving and static obstacles, and automatically cycle through different optimal tractor states. In the case of planting, these tractor states could correspond to stopping, planting, and not planting, based on whether the tractor is going through obstacle avoidance, traveling in straight passes, or traversing at end-of-the-row turns, respectively. Efforts to develop this type of intelligence for field-usable UAGVs is underway, and some equipment manufacturers are progressing toward making fully autonomous tractors commercially available. Shearer and Pitla (2013) categorized autonomous agricultural machinery into Gen-I and next generation UAGVs. Additional categorization is envisioned, based on the recent evolving industry trends and clear differences in the autonomy levels and the functional attributes of the UAGVs. The UAGVs are categorized in to three types: First generation (Gen-I), second generation (Gen-II), and third generation (Gen-III). Detailed discussion on each type of the UAGV with an example, their differences, and applications are discussed.

7.3.1 FIRST GENERATION (GEN-I)

Gen-I UAGVs are contemporary agricultural equipment with additional hardware and software, which enables autonomous operation with minimal changes to the vehicle's overall mechanical and control structure. These UAGVs can be operated with or without the operator inside the tractor cab. A good example of Gen-I UAGV is the autonomous grain cart system developed by Kinze (Kinze Manufacturing Inc., Williamsburg, Iowa) in partnership with Jaybridge Robotics (Jaybridge

Robotics, Cambridge, Massachusetts). The autonomous grain cart system (see Figure 7.7) can be summoned by the operator of the combine during the grain unloading operation. Additionally, the combine operator can initiate four autonomous grain cart functions (follow, unload, park, and idle) remotely (Held, 2014). The Kinze autonomous grain cart system is commercially available to selected users and can work with any row crop-tractor model and combine.

Similar efforts to transform existing conventional tractors to fully autonomous tractors can be observed with equipment manufacturers. Deere and Company (Moline, Illinois) developed a fleet of autonomous peat moss harvesters (Johnson *et al.*, 2009). These peat moss harvesters were modified, full-scale tractors with three-point hitch-mounted peat harvesting equipment. A fleet of semiautonomous tractors for mowing and spraying in an orchard were also developed (Zeitzew, 2006; Moorehead *et al.*, 2009). Autonomous Tractor Corporation (Fargo, North Dakota), a startup company, is developing an autonomous tractor that uses two proprietary autonomous technologies called eDrive^{TM} and AutoDrive^{TM}. This ag-robot can be considered as a Gen-I UAGV with significant changes made to the drive transmission and the power source of the vehicle (see Figure 7.8). The UAGV resembles a conventional, articulated, four-wheel drive tractor from the exterior, however the entire transmission system is replaced by four electric motors which are powered by a diesel-electric system. The AutoDrive^{TM} technology is the proprietary autonomous navigation component of this UAGV that doesn't use a GPS. Instead, a Laser Radio Navigation System (LRNS) with sub-inch positioning accuracy is used for autonomous navigation. The majority of the tractor auto guidance systems use GPS navigation, and not having to rely on the GPS is one of the benefits of using the AutoDrive^{TM} over conventional auto guidance systems. Thus, Gen-I UAGVs are modern day, high-horsepower tractors with autonomous hardware and software. In the event of a malfunction, managing this high-horsepower equipment can be tremendously difficult. Even though redundant safety features are programmed into the Gen-I UAGVs to mitigate liability, the ability to effectively stop a faulted, large autonomous machine in unpredictable situations is a challenge that has to be overcome with repeated testing and evaluation. Testing of this equipment is ongoing at a swift pace in the research and development facilities, and Gen-I UAGVs could be available for commercial use in the next few years.

FIGURE 7.7 Kinze Autonomous Grain Cart System. (Courtesy of Dustan Hahn, Kinze.)

FIGURE 7.8 Autonomous Tractor with eDrive™ and AutoDrive™ system. (Courtesy of Kraig Schulz, Autonomous Tractor Corporation.)

7.3.2 SECOND GENERATION (GEN-II)

Gen-II UAGVs are "autonomous only" machines, and cannot be driven by an operator, as they typically do not have a tractor cab. These machines, however, can be controlled using a remote in addition to autonomous mode, and are substantially different from Gen-I UAGVs. A good example of a Gen-II UAGV is presented in Figure 7.9. This UAGV named GreenBot™ is developed by Precision Makers (Giessen, the Netherlands). It is evident from Figure 7.9 that there is no operator's cab, and that this UAGV is a scaled-down version of a full-scale tractor. The liability is considerably reduced because of the smaller size, and in the event of a malfunction this UAGV can still be contained with high-tensile strength fences. This type of mitigation might not be possible with Gen-I UAGVs because of their high gross vehicle weights. GreenBot™ is a 1.8-m wide four-wheel steering vehicle with a lifting capacity of 1500 kg. Potential markets for the use of GreenBot™ are fruit farming, horticulture, and non-agricultural applications, such as roadside construction work. For safe operation in the field, the GreenBot™ is programmed to alert the owner of the machine in the event of an unrecognizable obstacle in its path. Another good example of a Gen-II UAGV is Spirit™, an autonomous vehicle that is under development by Autonomous Tractor Corporation (Fargo, North Dakota). Spirit™ operates with the same eDrive™ and AutoDrive™ systems (see Figure 7.10) used on the Gen-I UAGV developed by ATC that was discussed earlier. Spirit™ is a multipurpose autonomous tractor that provides the ability to attach implements to the front, rear, and

(a)   (b)

FIGURE 7.9 GreenBot™. GreenBot performing (a) autonomous soil preparation; (b) autonomous mowing operation. (Courtesy of Allard Martinet, Precision Makers.)

FIGURE 7.10 Spirit™. (Courtesy of Kraig Schulz, Autonomous Tractor Corporation.)

in the middle of the tractor frame. It can be observed from Figure 7.10 that the UAGV has a track-style design to minimize compaction. The liability of the equipment is somewhat reduced, due to the decreased size when compared to conventional large tractors. This UAGV will not be able to traverse through the field once the crop reaches a certain growth stage, and post-emergence field operations, such as spraying and anhydrous side dress application, might not be possible because of the low ground clearance of this UAGV.

7.3.3 Third Generation (Gen-III)

Gen-III UAGVs include greater adjustability in their mechanical structure and control architecture, enabling them to perform multiple field operations throughout the working season. The instrumentation hardware and software will allow the user to easily change the equipment's mechanical configuration with sophisticated automation features. A good example of Gen-III UAGV that is under development can be seen in Figure 7.11. BoniRob™ is developed by Deepfield Robotics (Germany), a Bosch startup company. Built-in exchangeable application modules of BoniRob™ can be customized for various field operations. The adjustable mechanical structure, multifunctionality, and the ability of this UAGV to work in swarm configuration with other UAGVs places this autonomous machine in Gen-III category. Various communication interfaces, both wired (Gigabit-Ethernet) and wireless (WiFi and Bluetooth) are provided for establishing communication with external devices (e.g., other ag-robots or monitoring stations). BoniRob™ can also work with unmanned air vehicles in coordinated ground/air mission settings. Thus, multifunctionality and the ability to work in a group of UAGVs are some of the primary differentiating aspects of Gen-III UAGVs relative to Gen-II and Gen-I UAGVs. Some important design and control architectural requirements of Gen-III UAGV's are summarized in the following subsections.

7.3.3.1 Multifunctionality

Gen-III UAGVs have to be multi-utility vehicles which can perform different field operations. The multifunctionality enables these UAGVs to operate throughout the working season, thereby allowing for maximum utilization of the machine's working life. The approach of using an UAGV specifically for one task might not be economically feasible, unless the UAGV is deployed in a high-value crop where high capital costs can be justified. Embedding multifunctionality is a challenging task.

FIGURE 7.11   BoniRob. (Courtesy of Albert Amos, Deepfield Robotics.)

For example, in addition to providing adequate power for various operations demanding different levels of power, the UAGV's control architecture has to be designed with appropriate flexibility for providing diverse control functions. Planting and spraying are low draft applications, whereas tillage operation is a high draft application. To serve the diverse power demands of these operations, the UAGV should employ a power source that can provide a range of power outputs for operating different implements. A Gen-III UAGV can be envisioned to contain a base robotic platform, onto which hardware and software modules can easily be added and removed to address different implement requirements for multifunctionality. The ability to perform multiple functions by UAGVs will greatly facilitate coordinated tasks in a swarm configuration. Depending on the cooperative task, the multifunctional Gen-III UAGV can be assigned different types of roles (e.g., leader or follower) to complete a collective task.

7.3.3.2   Scale Neutral Systems

Availability of the same level of technology to producers both large and small is significant for maximizing productivity and profitability, irrespective of farm size across the producer spectrum. Currently, only producers with large fields can afford technological solutions, as most of the advanced technology features come with high capital cost machinery. Scale-neutral autonomous systems that can be used in both large and small producer farms have to be developed. These scale-neutral systems can be a swarm of GEN-III UAGVs, where large fields will use a bigger swarm compared to small fields, with the technology aspects remaining the same for both large and small farms. Assuming a linear trend, a producer with 100-hectare field size might require 5 UAGVs, versus a producer with a 1000-hectare field who will need 50 UAGVs. Both producers in this case can access GEN-III UAGVs with the only difference being the number of UAGVs in the swarm.

7.3.3.3   Cooperative Behavior in UAGV Swarms

Swarms of UAGVs are required to match the field capacities of larger conventional equipment, and, therefore, the Gen-III UAGVs must incorporate intelligence for working in groups. Control frameworks needed to deploy UAGV swarms specifically in agricultural field settings have to be researched extensively for facilitating the progression of the UAGV swarms. The nature of the field operation will dictate the form of cooperation needed among UAGVs. For example, a group of UAGVs might be doing the same field operation or, they can each be doing a different operation that requires some form of cooperation with the other UAGVs in the group to finish the task. Gen-III UAGV technologists and researchers have to take in to account this variability in coordination requirements, and build intelligence in the control architectures of the UAGV swarms to address the various communication requirements. Control architectures for both individual and multi-robot systems specifically for agricultural production are required for the deployment and progression of Gen-III UAGVs.

7.4   CONSIDERATIONS FOR THE ADVANCEMENT OF AG-ROBOTICS

Manipulator-type ag-robotics and UAGVs both have their unique challenges which need to be addressed for their advancement, commercial deployment, and wide-scale adoption. In the case of manipulator-type ag-robotics, the success of robotic milking stations already demonstrated their social acceptance and economic feasibility for profitable farm operations. The field of UAGVs, however, require few more years of technological maturity before they can be commercially viable. While stationery robotic milking stations work in a fairly controlled environment, robotic arms deployed in field crops and orchards operate in more uncontrolled settings leading to engineering challenges. These challenges include: inability to work in varying lighting conditions and dusty environments, and long processing times associated with the accurate localization of fruit/crop for robotic picking. Established robotic technologies used in industrial automation, automotive manufacturing, and manipulator robotic research efforts in citrus and orchard crops will aid the progression of manipulator type ag-robotics.

Additional challenges exist for UAGVs when compared to manipulator type ag-robotics primarily because of their operational ground speeds, varying terrain, and the requirement for performing a field operation on the go (e.g., planting or tilling of the soil). UAGVs require additional sensors, software, and control algorithms for safe, autonomous operation. Some of the major issues that could impede the emergence and wider adoption of UAGVs are cost of technology, safety, and limited availability of standardized control architecture frameworks. Relative to their industrial and non-agricultural robotic counterparts, UAGVs have to be cost-effective, given their seasonal use and narrow profit margins. However, with the advancements in automotive industry in the last few years; sensors, computing hardware, and control components are now produced on a large scale; resulting in reduced costs, which could make UAGVs economically viable in the future. Remaining concerns to be resolved include making the UAGVs safe, fault-tolerant and intelligent enough to perform precise field operations with minimal human supervision. From an engineering design perspective, the safety, efficacy, and functionality of an UAGV greatly depends on the control framework/ architecture. To effectively interpret, arbitrate, and prioritize information obtained from sensors and produce desired actions, effective UAGV control architectures are required. The task becomes even more challenging when multiple UAGVs are working with each other in cooperative settings, where multi-robot system control architectures applicable specifically for agricultural production are needed. In addition to the economic feasibility of UAGVs, standardized control architectures of individual and swarms of UAGVs are critical for their progression. Techno-economics, control architectural work on UAGVs, and the importance of data infrastructure and telematics are discussed in this section.

7.4.1   TECHNO-ECONOMICS

Upstream and downstream field operations could be affected with the introduction of UAGVs in a conventional production system, which subsequently impacts the economics of production. A systems approach is important for understanding the economic impacts of deploying UAGVs in production agriculture. As an example, when a conventional planter is replaced with an UAGV or multiple UAGVs, the approach for refilling the seed might be different from conventional methods of refilling. The UAGVs have to autonomously navigate toward a movable seed tender parked at the edge of the field to refill the seed and resume planting after the refill. This process requires optimized route planning and task scheduling, which directly impacts the operational costs. Assuming a more efficient path taken by a UAGV planter relative to a conventional planter system, to quantify the reduction in overall cost of operation, techno-economic analysis of the production system has to be performed. Techno-economic modeling is a well-established process of business case modeling that takes into account the technical dependencies and constraints during the process of cost and revenue calculations (Knoll, 2012). This type of modeling is prominently found in the biofuel production analysis to evaluate the cost-effectiveness of a product (Nagarajan *et al.*, 2013). Some researchers are applying this modeling technique for precision agricultural machinery with some success (Toledo *et al.*, 2014). However, further research specifically focused on the techno-economic modeling of UAGV swarms is needed.

The operational width, depth, and speed of operation of an implement are some of the major factors that determine the power demand (kW), fuel consumption (Lh^{-1}), and the field capacity (hah^{-1}) of agricultural machinery. These parameters (e.g., power demand, fuel consumption, and field capacity) which have a significant impact on the economics of crops, have to be considered in the techno-economic modeling phase of UAGVs. The machinery size impacts the soil structure which indirectly impacts yields and the profit margins of the operation. UAGVs with low gross vehicle weights provide yield benefits, and these benefits resulting from the use of UAGVs have to be included in the techno-economic modeling. The cost of operation per hectare (ha^{-1}) is a function of the field efficiency, effective field capacity (hah^{-1}), and the hourly machinery cost (ha^{-1}). In this case, the hourly machinery operation cost of the UAGV is determined primarily from both the fixed costs (e.g., capital costs, and machinery repair and maintenance costs) and variable costs (e.g., fuel costs; Beaton *et al.*, 2005). Machinery management standards available from the American Society of Agricultural and Biological Engineers (ASABE, 2011) provides data on the cost of operations of conventional agriculture. With the appropriate assumptions, this data from the ASABE standards can be used to perform preliminary techno-economic analysis of UAGVs.

Cost-benefit analysis can be done with various sensitivity variables to evaluate the advantages of UAVGs versus conventional machinery using the techno-economic modeling techniques. Technical obsolescence is another important aspect of UAGVs that need to be researched further and considered during the design phase of UAGVs, in addition to techno-economic considerations. "Obsolescence" refers to the non-use of available technology even though the technology is in good condition. Because of the availability of newer technologies that are more advantageous, the available technology is rendered obsolete. Toledo *et al.* (2014) discussed the importance of techno-economics and the obsolescence of small robotic weed machines. Their study concluded that, compared to conventional weed control techniques, using a small robotic weed control system could reduce the energy requirements by approximately 20% indicating decreased cost of operation ($/ha^{-1}$) of robotic weeding. With UAGVs, long hours of operation with reduced labor costs is possible, which could minimize the effect of technical obsolescence. Long hours of operation shorten the mechanical life of UAGVs because of continuous use, and will allow for matching the mechanical life with the technology life (e.g., electronics and computer life). In this case, the question is, "What is the ideal mechanical life of an UAGV before it becomes technologically obsolete?" Researchers are working on the question of technical obsolescence and furthering efforts that are required in this domain (Barreca, 2000; Shearer and Pitla, 2013). A multidisciplinary effort involving engineers, economists, agronomists, and manufacturers is needed to better understand the subject of technical obsolescence of UAGVs in production agriculture settings.

7.4.2   CONTROL ARCHITECTURES—INDIVIDUAL UAGVs AND UAGV SWARMS

Artificial intelligence research laboratories working in the areas of space exploration and military applications spearheaded the work on control architectures of mobile robots (Brooks, 1986; Arkin,

1998; Arbib, 1981; Yavuz et al., 1999; Yavuz and Bradshaw, 2002). Contrasting that with the field of agriculture, a reduced amount of research is available focused on UAGV control architectures. Given the complex nature of agricultural field environments, UAGV control architecture development is significant for achieving substantial progress in field automation. Uncertain and unpredictable situations are prevalent in agricultural fields where behavior based (BB) robotic architecture principles can be utilized. Use of BB robotic systems could be effective where the real world cannot be accurately modeled or characterized. Using BB robotic principles, Blackmore et al. (2002) developed system architecture for the behavioral control of an autonomous tractor. Blackmore et al. (2004) identified graceful degradation as a key element for a robust autonomous ground vehicle. Graceful degradation is the process in which the robot's actions are degraded in steps as opposed to an abrupt ending when an emergency occurs. Graceful degradation ensures predictable behavior of the robot and minimizes the chances of damage to the environment or the robot. An object-oriented approach was taken where the physical hardware of the autonomous robot was assumed to match the logical design. These research efforts identified important features of the UAGVs that are required for a robust and fault-tolerant operation.

A more practical approach of control architectures was proposed by Torrie et al. (2002), who suggested that to effectively support Unmanned Ground Vehicle (UGV) systems, a standard architecture is required. A Joint Architecture for Unmanned Ground Systems (JAUGS) that was independent of vehicle platform, mission, tasks, or computer hardware was proposed. According to Torrie et al. (2002), an autonomous agricultural system should consist of different Operational Control Units (OCUs); base Station OCU, remote OCU, laptop OCU, and the UGV. In a similar attempt, a system architecture which connected high-level and low-level controllers of a robotic vehicle was proposed by Mott et al. (2009). In addition to the aforementioned levels, a middle level was introduced to improve the safety of the autonomous vehicle. The middle level enforced timely communication and provided consistent vehicular control. When the high level was not transmitting in the desired way, the middle level recognized this condition and transitioned to a safe mode, where the vehicle stops. Ultimately, the middle level acted as a communication bridge, integrating the high- and low-level controllers providing robustness to the robotic vehicles. This concept was successfully deployed on a fully-autonomous stadium mower (Zeitzew, 2006), and a large-scale peat moss harvesting operation (Mooerehead et al., 2009). A five-layered Individual Robot Control Architecture (IRCA) was developed to address the intelligence aspect of an individual UAGV (Pitla, 2012; Posselius et al., 2016). The five functional layers methodically interpreted and translated sensor information into useful actions for the UAGV. Two essential facets of an UAGV (goal oriented autonomous navigation, and reactive intelligence to mitigate unpredictable situations) were incorporated into the design of the IRCA. Similar efforts to develop safety architecture to guide the design and deployment of autonomous agricultural vehicles can be found in literature (Kohanbash et al., 2012). Further development of standardized control architectures for individual UAGVs has to be continued to support the advancement and adoption of UAGVs.

Multiple UAGVs are required to match the high field capacity of large conventional equipment for finishing a field operation. However, managing and deploying multiple UAGVs is a challenging task, as each UAGV in the swarm need to be aware of their roles and tasks to execute a coordinated task. Algorithms for coordinating multi-machines have been developed for leader-follower types of situations where, typically, one manned tractor acts as a leader and the other tractor acts as a follower (Noguchi et al., 2004; Vougioukas, 2009; Held, 2014). The task becomes even more challenging when both the leader and the follower have to be fully autonomous with minimal human intervention in the control loop. Taking a step forward from leader-follower systems, fleets of fully autonomous tractors might have to be used for a highly efficient and productive operation. Fleets of semi-autonomous tractors were programmed to operate in orchard and harvesting applications (Moorehead et al., 2009; Johnson et al., 2009). These efforts clearly indicate that the field of agriculture is evolving and transforming to a point where swarms of UAGVs could be used. Parker (2002) suggested that a system consisting of multiple simpler robots will most likely be cost-effective and more robust than a system employing a single complex robot. In this type of a system, dedicated functions are delegated to multiple robots instead of one single robot with complex architecture requiring to perform all the functions. Reliability of the system as a whole is improved, because if one robot fails, the remaining robots continue to work to finish the given task. Complete dependence on one complex robot is thereby avoided, which improves the reliability of the system (Parker, 2002).

In a Multi-Robot System (MRS), the controlled architecture design becomes increasingly challenging with increase in number of robots in the swarm. Inter-robot communication and the levels/extent of communication among the UAGVs form a crucial part of the multi-robot architecture. Research on inter-robot communication for non-agricultural applications can be found readily (Balch et al., 1994; Rude et al., 1997; Wilke and Braunl, 2001). Various communication levels between robots were researched, and it was found that, in some cases, communication significantly increased task performance, but in other cases, the inter-agent communication was unnecessary. Since establishing communication resources among UAGV swarms come at substantial cost and complexity, the communication requirements among the UAGVs need to be researched and considered during the development of these control architectures. When a swarm of UAGVs is performing a task as a group, the first aspect that needs to be evaluated is whether the swarm requires some form of inter-robot communication for finishing the assigned task. Even though UAGVs are working as a group, there might be some cases where inter-robot communication is not required, as long as these UAGVs transmit their basic states (such as speed and location) to a remote station for monitoring purposes. This situation where UAGVs do not need communication is possible if a field can be divided in to multiple working zones, and each UAGV is constrained to work in its respective working zone. In this case, inter-robot communication is needed only when the UAGVs are required to perform different, dedicated functions to accomplish a cooperative task. The variety of coordination strategies required for specific agricultural operations will affect the communication requirements of the UAGVs.

The coordination strategy differs for homogeneous versus heterogeneous UAGV swarms. A homogeneous UAGV swarm is a group of functionally equivalent robots (e.g., swarm of planters, swarm of sprayers, etc.) that perform similar actions utilizing the same levels of sensing and control capabilities, whereas heterogeneous UAGV swarm is a group of functionally different robots (e.g., baler and bale picker, harvester and grain carts) that perform unique tasks to accomplish a collective assignment. Thus, the homogeneity and heterogeneity of the robots in a UAGV swarm affects the coordination strategies that, in turn, determine the type of inter-robot communication required. A classification of UAGV swarms based on homogeneity and heterogeneity and the levels of communication associated with each type of swarm is presented in Figure 7.12 (Posselius et al., 2016). A classification, such as the one shown in Figure 7.12, could provide basis for the development of the multi-robot system control architecture (MRSCA). MRSCA of UAGV swarms will provide a framework that will facilitate inter-machine communication in support of cooperative behaviors (no cooperation, modest, and absolute).

Although a UAGV swarm compels us to think that some form of communication between robots is necessary, in reality, no-communication between homogeneous UAGVs is as good as when there is some form of inter-robot communication. A major requirement for a homogeneous UAGV swarms to be robust with no inter-robot communication is that individual UAGVs must be intelligent and should react to the local environment. Homogeneous UAGVs will divide the task and work in their respective unique work areas without interfering with each other's operations. Thus, cooperation may not be required, and inter-robot communication becomes insignificant for homogeneous UAGV swarms. A swarm involving heterogeneous UAGVs would require inter-robot communication, as each UAGV performs a different function that necessitates coordinated actions. The extent of cooperation between two heterogeneous UAGVs will vary and be of two types; modest cooperation and absolute cooperation (Pitla et al., 2014; Posselius et al., 2016). For modest cooperation, the robots do not interfere with each other's work, however, short-term cooperation is established at the

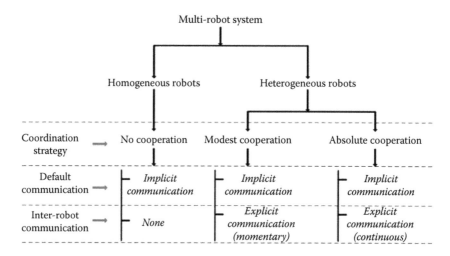

FIGURE 7.12 Classification of UAGV swarms based on coordination and communication requirements. (From Posselius, J. H. *et al.*, 2016, Multi-Robot System Control Architecture. Patent Number: 9,527,211.)

instant when UAGVs need to work together. For absolute cooperation, uninterrupted communication between UAGVs is required at all times. Implicit and explicit types of communication forms the communication framework for both homogeneous and heterogeneous UAGV swarms. Implicit communication is the unintentional communication of the UAGVs, where states of the UAGVs are transmitted over the wireless channel to a central monitoring station. Intentional communication directed at a specific UAGV constitutes explicit communication. While some tasks require implicit communication, a combination of implicit and explicit communication may be important for specific tasks (see Figure 7.12). In a homogeneous UAGV swarm, the UAGVs utilize implicit communication to transmit their states, whereas for heterogeneous UAGV swarms, explicit momentary and explicit continuous types of communication must be established for modest and absolute cooperation, respectively.

A good example of absolute cooperation behavior among UAGV swarms can be demonstrated by the harvesting operation. In this example, the harvester and the grain carts have to communicate continuously to transfer the harvested grain without any loss. Figure 7.13a shows three UAGVs, the UAGV in screen 1 on the front left assumes the role of a harvester, whereas the UAGV in the front right assumes the role of Grain Cart–I. Grain Cart–II follows behind Grain Cart–I to synchronize with the harvester after the Grain Cart–I is full with grain. Some form of synchronization flags have to be created among the UAGVs, so that each grain cart will know when to synchronize with the harvester to unload grain and drop off the grain at the field edge. The paths of the harvester, Grain Cart–I and Grain Cart–II in Figure 7.13b illustrates the synchronized coordinated operation of the UAGVs. All UAGVs communicate continuously throughout the operation in absolute cooperation behavior. The tasks, roles, and synchronization flags of the MRSCA framework allows for this complex, coordinated field operation of the UAGVs.

In a similar effort to develop multi-robot coordination, a system architecture for both individual and fleets of robots for improved reliability and decreased complexity was developed for a fleet of UAGV sprayers by Emmi *et al.* (2014). This architecture presented a centralization of the main controller and the principal sensory systems, which allowed for modularity and flexibility to add or remove sensor systems (RHEA, 2013). The development of control frameworks of UAGV swarms have to be expanded and researched extensively for the successful deployment of UAGV swarms for various operations under different field conditions. UAGV swarm complexity increases with the number of machines in the swarm, and the type of operations they perform. Field cultivation

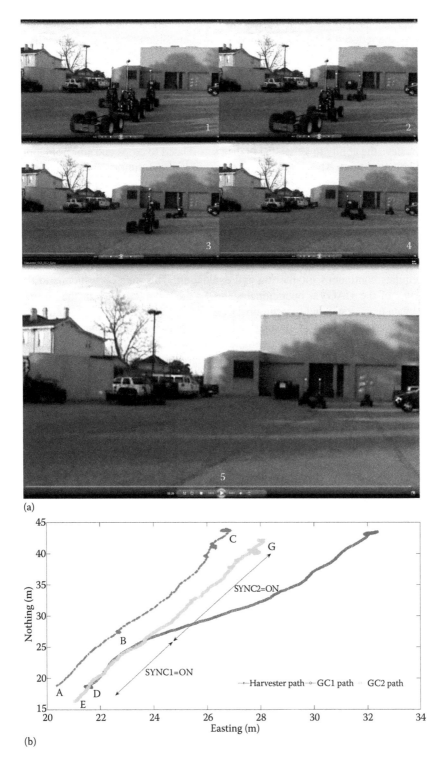

(a)

(b)

FIGURE 7.13 Absolute Cooperation Simulation (a) Three UAGVs simulating the internal states of harvesting operation. (b) Paths of the UAGVs and the states of the synchronization flags. (Courtesy of Santosh K. Pitla, University of Nebraska–Lincoln.)

and planting operations will be some of the first applications where UAGV swarms might be deployed, due to the reduced complexity of the inter-robot communication and established implement control technologies. Automated biomass (bale) retrieval and transfer of the biomass to the field edges or on-farm bioprocessing plants are some of the other immediate applications of the UAGV swarms.

7.4.3   Data and Connected Systems

UAGV swarms have the ability to generate big data sets related to crops and soil from long hours of unsupervised field operations. These big datasets can be of high resolution, both spatial and temporal, and could be in the order of per square meter scale, or, in some cases, even per plant basis. Some of these datasets have to be analyzed in real time for varying crop input application rates on the go (e.g., nitrogen application or chemical spraying), while others have to be stored and transferred to remote locations for post-processing and analysis. To manage and harness these big datasets effectively, infrastructure related to telematics and appropriate software tools for data analytics have to explored and investigated. The absence of the operator on the UAGVs will also lead to another major category of data that requires continuous monitoring and evaluation. This data corresponds to the internal operational states of the UAGVs, operational states of the other UAGVs in the swarm, and the environmental data (e.g., static and moving obstacles or terrain variation) in which the UAGVs are working. If the UAGVs are used as sensing and phenotyping platforms, plant data (e.g., spectral image data) have to be stored and sent to a remote location for post-processing. Thus, infrastructure to support the data requirements of the next-generation agricultural machinery have to be developed and installed for realizing the full potential of the UAGV swarms. UAGV swarms, in addition to establishing inter-robot communication, must be internet connected for high-bandwidth data transmissions, remote monitoring, remote software updates, and debugging. Internet-connected UAGV swarms will simplify the aspects of software downloads, repair and maintenance, resolving technical issues, and for providing firmware updates from a remote location.

7.5   CONCLUSION

Over the last few decades, significant progress has been made in the areas of plant sensing, environment perception, image processing, GNSS and machine vision based auto guidance, path planning, site-specific crop input application control, and machine performance improvements. These efforts have aided the progression of agricultural machinery into highly automated individual robotic systems and UAGV swarms. UAGV swarms are scale-neutral solutions that can be used in both large and small farms to significantly improve the crop production efficiency while minimizing environmental impacts. However, liability issues associated with UAGVs have to be addressed for their widespread acceptance and advancement. Liability here refers to property damage and losses incurred due to wastage of crop inputs (seeds, fertilizers, and chemicals) resulting from the improper operation of the UAGV.

Researchers and industry technologists are working on different robotic platforms (Gen-I, Gen-II, and Gen-III) and intelligent control architectures for mitigating the liability. Open source access to the hardware, algorithms, and software of the UAGVs has to be created for sharing these developmental efforts through a consortium of universities and industry partners for the progression and advancement of UAGVs. Further, a UAGV testing facility can be envisioned similar to conventional official tractor performance testing facilities. Currently, the Organization of Economic Co-operation and Development (OECD, 2016) approved test stations evaluate new model tractors for drawbar, power takeoff (PTO), and hydraulic power claims. Similar testing and benchmarking of UAGVs has to be performed for their ability to operate safely in field conditions. A controlled outdoor field facility with infrastructure to test and evaluate the UAGVs performing typical row production operations (e.g., cultivation, planting, and spraying) will facilitate efforts to advance and deploy

UAGV swarms for field automation. Research institutes, universities, technology companies, and equipment manufacturers have to take a public-private partnership approach in developing the standard test procedures for the testing and bench marking of UAGVs.

REFERENCES

Arbib, M. A. (1981). Perceptual structures and distributed motor control. *Handbook of Physiology—The Nervous System II: Motor control*. (Ed. V. B. Brooks). American Physiological Society, Bethesda, MD, 1449–80.

Arkin, R. C. (1998). *Behavior-Based Robotics*. The MIT Press. Massachusetts Institute of Technology, Cambridge, Massachusetts.

ASABE. (2011). ASAE D497.7 MAR2011 Agricultural Machinery Management Data. ASABE Standards 2011. St. Joseph, Michigan.

Balch, T., and Arkin, R. C. (1994). Communication in reactive multiagent robotic systems. *Autonomous Robots* 1, 27–52.

Barreca, S. L. (2000). Technology life-cycles and technological obsolescence. Available at: http://www.bcri .com/Downloads/Valuation%20Paper.PDF.

Beaton, A. J., Dhuyvetter, K. C., Kastens, T. L., and Williams, J. R. (2005). *Journal of Agricultural and Applied Economics* 37, 131–44.

Blackmore, B. S., Have, H., and Fountas, S. (2002). A proposed system architecture to enable behavioral control of an autonomous tractor (Keynote Address). Proc. of Automation Technology for Off-Road Equipment. (Ed. Q. Zhang.) ASAE: St. Joseph, Michigan, 13–23.

Blackmore, B. S., Fountas, S., Vougioukas, S., Tang, L., Sorensen, C. G., and Jorgensen, R. (2004). A method to define agricultural robot behaviors. Proc. of the Mechatronics and Robotics Conference (MECHROB), 1197–200.

Broucek, J. and Tongel, P. (2015). Adaptability of dairy cows to robotic milking: A review. *Slovak J. Anim. Sci.* 48 (2), 86–95.

Brooks R. A. (1986). A robust layered control-system for a mobile robot. *IEEE Journal of Robotics and Automation* 2, 14–23.

Burks, T., Villegas, F., Hannan, M., Flood, S., Sivaraman, B., Subramanian, V. and Sikes, J. (2005). Engineering and horticultural aspects of robotic fruit harvesting: Opportunities and constraints. *HorTechnology* 15 (1), 79–87.

Davinia F., Pallejà, T., Tresanchez, M., Runcan, D., Moreno, J., Martínez, D., Teixidó, M., and Palacín, J. (2014). A proposal for automatic fruit harvesting by combining a low-cost stereovision camera and a robotic arm. *Sensors* 14, 11557–79; doi:10.3390/s140711557.

Davidson, J. R., Silwal, A., Karkee, M., Li, J., Xia, K., Zhang, Q., and Lewis, K. (2015). Extended abstract: Human-machine collaboration for the robotic harvesting of fresh market apples. IEEE ICRA 2015 Workshop on Robotics in Agriculture. doi:10.1155/2012/368503 10.13140/RG.2.1.3457.2880.

Godoy, E. P., Tangerino, G. T., Tabile, R. A., Inamasu, R. Y., and Porto, A. J. V. (2012). Networked control system for the guidance of a four-wheel steering agricultural robotic platform. *Journal of Control Science and Engineering* 2012. doi:10.1155/2012/368503.

Grift, T. E. (2007). Robotics in crop production. *Encyclopedia of Agricultural, Food, and Biological Engineering*. doi:10.1081/E-EAFE-120043046.

Guivant, J., Cossell, S., Whitty, M., and Katupitiya, J. (2012). Internet-based operation of autonomous robots: The role of data replication, compression, bandwidth allocation and visualization. *Journal of Field Robotics* 29 (5), 793–818. http://dx.doi.org/10.1002/rob.21432.

Hannan, M. W. and Burks, T. F. (2004). *Current developments in automated citrus harvesting*. ASAE Paper No. 04–3087. ASAE, St. Joseph, Michigan.

Hogewerf, P. H., Huijsmans, P. J. M., Ipema, A. H., Janssen, T., and Rossing, W. (1992). Observations of automatic teat cup attachment in an automatic milking system. Proc. International Symposium on Prospects for Automatic Milking, 80–90. Wageningen, Netherlands, Pudoc Scientific Publishers.

Johnson D. A., Naffin, D. J., Puhalla, J. S., Sanchez, J., and Wellington, C. K. (2009). Development and implementation of a team of robotic tractors for autonomous peat moss harvesting. *Journal of Field Robotics* 26, 549–71.

Held, J. (2014). Kinze Autonomous Harvest System. Available at: http://www.kinze.com/article.aspx?id=341.

Katupitiya J. and Eaton, R. (2008). *Precision Autonomous Guidance of Agricultural Vehicles for Future Autonomous Farming*. ASABE Paper No. 084687. ASAE, St. Joseph, Michigan.

Knoll, T. M. (2012). Techno-Economic Modelling of Mobile Access Network Alternatives 41, Treffen der VDE/IT G-Fachgruppe 5.2.4, Berlin.

Kohanbash, D., Bergerman, M., Lewis, K. M., and Moorehead, S. J. (2012). A Safety Architecture for Autonomous Agricultural Vehicles. Proc. of ASABE Annual International Meeting. Paper No. 12–1337110.

Kondo, N., Monta, M., Fujiura, T. (1995). Fruit harvesting robots in Japan. *Physical, Chemical, Biochemical and Biological Techniques and Processes* 18, 181–4.

Kondo, N., Nishitsuji, Y., Ling, P. P., Ting, K. C. (1996). Visual Feedback Guided Robotic Cherry Tomato Harvesting. Transactions of the ASAE 39, 2331–8.

Kuipers, A. (1996). The milking robot. *Agri-Holland* 1–1996, 9–12. Department for Trade and Industry, Ministry of Agriculture, Nature Management and Fisheries, the Netherlands.

Lenz, J., Landman, R., and Mishra, A. (2007). Customized Software in Distributed Embedded Systems: ISOBUS and the Coming Revolution in Agriculture. Agricultural Engineering International: The CIGR Ejournal. Manuscript ATOE 07 007 (9).

Luck, J. D., Pitla, S. K., Mueller, T. G., Dillon, C. R., Higgins, S. F., Fulton, J. P., and Shearer, S. A. (2009). Potential for pesticide and nutrient savings via map-based automatic boom section control of spray nozzles. *Computers and Electronics in Agriculture* 70 (1), 19–26.

Madsen, T. E., and Jakobsen, H. L. (2001). *Mobile robot for weeding*. M.S. Thesis. Lynby, Denmark: Danish Technical University.

Moorehead, S., Ackerman, C., Smith, D., Hoffman, J., and Wellington, C. (2009). Supervisory Control of Multiple Tractors in an Orchard Environment. Proc. of the Fourth IFAC International Workshop on Bio-Robotics, Information Technology, and Intelligent Control for Bio-Production Systems. Champaign, Illinois.

Mott, C., Ashley, G., and Moorehead, S. (2009). Connecting High-Level and Low-Level Controllers on Robotic Vehicles Using a Supporting Architecture. Fourth IFAC International Workshop on Bio-Robotics, Information Technology, and Intelligent Control for Bio-Production Systems. Champaign, Illinois.

Nagarajan, S., Chou, S. K., Cao, S., C. Wu, Zhou Z. (2013). An updated, comprehensive techno-economic analysis of algae biodiesel. *Bioresource Technology* 145 (2013), 150–6.

Noguchi, N., Will, J., J. Reid, J., and Zhang, Q. (2004). Development of a master-slave robot system for farm operations. *Computers and Electronics in Agriculture* 44, 1–19.

OECD. 2016. Code 2 OECD Standard Code for the Official Testing of Agricultural and Forestry Tractor Performance. Paris, France.

Oksanen, T. (2012.) Embedded control system for large-scale unmanned tractors. Fifth Automation Technology for Off-Road Equipment Conference (ATOE), Valencia, Spain, 3–8.

Pilarski, T., Happold, M., Pangels, H., Ollis, M., Fitzpatrick, K., Stentz, A. (2002). The demeter system for automated harvesting. (Reprinted from Proc. of the American Nuclear Society: Eighth International Topical Meeting on Robotics Remote Systems, Pittsburgh, PA, 1999.) *Autonomous Robots* 13, 9–20.

Pitla, S. K. (2012). *Development of Control Architectures for Multi-Robot Agricultural Field Production Systems*. Dissertation, University of Kentucky Libraries: Lexington Kentucky.

Pitla, S. K., Luck, J. D., and Shearer, S. A. (2014). Multi-Robot System Control Architecture (MRSCA) For Agricultural Mobile Robots. Proc. of the Second International Conference on Robotics and Associated High-Technologies and Equipment for Agriculture and Forestry (RHEA—2014), Madrid, Spain.

Posselius, J. H., Foster, C. H., Pitla, S. K., Shearer, S. A., Luck, J. D., Sama, M. P., Zandonadi, R. S. (2016). Multi-Robot System Control Architecture. Patent Number: 9,527,211.

Reed, J. N., Miles, S. J., Butler, J., Baldwin, M., and Noble, R. (2001). Automatic mushroom harvester development. *Journal of Agricultural Engineering Research* 78, 15–23.

RHEA, 2013. A robot fleet for highly effective agriculture and forestry management. Available at: http://www.rhea-project.eu/.

Rude, M., Rupp, T., Matsumoto, K., Sutedjo, S., and Yuta, S. (1997). IRoN: An inter-robot network and three examples of multiple mobile robots' motion coordination. *IEE97*, 1437–44.

Shearer, S. A. and Pitla, S. K. (2013). Field production automation. *Agricultural Automation: Fundamentals and Practices*. (Ed. Q. Zhang and F. J. Pierce.) CRC Press, 97–124.

Shearer, S. A., and Pitla, S. K. (2014). Precision planting and crop thinning. *Automation: The Future of Weed Control in Cropping Systems*. (Ed. S. L. Young and F.J. Pierce.) Springer, 99–124. ISBN: (Print) 978-94-007-7512-1.

Shibusawa, S. (1996). *PhytoTechnology: An introduction to the concept and topic of a new project*. School of Bio-Applications and Systems Engineering, Tokyo University of Agriculture and Technology.

Silwal, A., Gongal, A., and Karkee, M. (2014). Identification of red apples in field environments with over-the-row machine vision system. *Agric Eng Int: CIGR Journal* 16 (4), 66–75.

Stone, M. L., McKee, K. D., Formwalt, C. W., and Benneweis, R. K. (1999). ISO 11783: Electronic communication protocols for agricultural vehicles. *ASAE Distinguished Lecture Series* (23), 1–17. ASABE, St. Joseph, Michigan.

Tanigaki, K., Fujiura, T., Akase, A., Imagawa, J. (2008). Cherry-harvesting robot. *Computers and Electronics in Agriculture* 63, 65–72. doi 10.1016/j.compag. 2008.01.018.

Taylor, R. K., Schrock, M. D., and Staggenborg, S. A. (2001). *Using GPS technology to assist machinery management decisions.* ASAE Paper No. MC01-24. ASABE, St. Joseph, Michigan.

Toledo, O. M., Steward, B. L., Gai, J., Tang, L. (2014). Techno-economic analysis of future precision field robots. 2014 ASABE and CSBE/SCGAB Annual International Meeting. Montreal, Quebec Canada. Paper Number: 141903313.

Torrie, M. W., Cripps, D. L., and Swensen, J. P. (2002). Joint Architecture for Unmanned Ground Systems (JAUGS) Applied to Autonomous Ground Vehicles. Automation Technology for Off-Road Equipment, Proc. of the Conference, Chicago, Illinois, USA. ASAE Publication Number 701P0502 (Ed. Qin Zhang), 1–12.

Van Henten, E. J., Hemming, J., van Tuijl, B. A. J., Kornet, J. G., Meuleman, J., Bontsema, J., van Os, E. A. (2002). An autonomous robot for harvesting cucumbers in greenhouses. *Autonomous robots* 13, 241–58.

Vougioukas, S. G. (2009). Coordinated master-slave motion control for agricultural robotic vehicles. Proc. 4th IFAC International Workshop on Bio-Robotics, Information Technology, and Intelligent Control for Bio-Production Systems, Champaign, Illinois.

Wilke, P., and Braunl, T. (2001). Flexible wireless communication networks for mobile robot agents. *Industrial Robot: An International Journal* 28 (3), 220–32.

Wolkowski, R., and Lowery, B. (2008). *Soil compaction: Causes, concerns, and cures (A3367).* University of Wisconsin-Extension. Available at: http://www.soils.wisc.edu/extension/pubs/A3367.pdf.

Yavuz, H., Chandler, A., Bradshaw, A., and Seward, D. (1999). Conceptual design and development of a navigation system for a mobile robot. Computer Aided Conceptual Design Workshop, Lancaster University, Lancaster, UK, 65–85.

Yavuz. H., and Bradshaw, A. (2002). A new conceptual approach to the design of hybrid control architecture for autonomous mobile robots. *Journal of Intelligent and Robotic Systems* 34, 1–26.

Zeitzew, M. (2006). *Autonomous Utility Mower.* Autonomous Technology for Off-Road Equipment, Bonn, Germany.

8 Development of Precision Livestock Farming Technologies

Thomas Banhazi and Marcus Harmes

CONTENTS

8.1 BACKGROUND TO THE PRECISION LIVESTOCK FARMING (PLF) CONCEPT

While the intention behind the Precision Livestock Farming (PLF) research and development efforts was to introduce Process Control principles in agricultural/animal production (Frost *et al.*, 1997), there have, in fact, been a number of incentives for management systems on farms. These incentives have stemmed from challenges for farmers and opportunities for both farmers and engineers. Precision Livestock Farming is possible because of technological advances, however, challenges for farmers (such as larger herds and animal identification, as well as demands for efficiency and, more recently, sustainability and welfare) have all created opportunities for technologies to be tested and applied. Therefore, the definition of PLF developments was that automation and increased information technology (IT) use would make animal production more efficient due to the increased control it affords (Banhazi and Black, 2009a). In turn, it was believed that such control would result in better animal welfare and environmental outcomes, including greater efficiency in using resources and limiting environmental pollution (Frost *et al.*, 2003; Stacey *et al.*, 2004). The introduction of process control procedures did result in significant improvements in other industries (Wathes *et al.*, 2001; Banhazi *et al.*, 2012b). A suggested data management and control diagram for livestock industries can be seen in Figure 8.1.

Predictions for a rise in the human population of 2.3 billion people by the middle of the twenty-first century is an additional factor scientists and demographers have considered when evaluating the global environmental impact of agriculture and land clearing (Tilman *et al.*, 2011). The intention to make animal production more efficient is a response to trends and demands in the overall development of the world livestock industry and global meat/protein demand. The World Health Organization (WHO, 2017) has forecasted global meat and milk consumption up to 2030, and they predict that the 218 million tons of meat consumed in 1999 will become 376 million in 2030 (WHO, 2017).

FIGURE 8.1   Data management and control diagram for a livestock enterprise.

Other predictions include a major rise in environmental pollution by 2050 from agriculture (Tilman and Clark, 2014). Particular regions, especially Southeast Asia, feature increasing consumption of and therefore demand for animal protein, a demand also driven by rising levels of per capita income (Huynh *et al.*, 2006). These predictions have prompted a range of responses, including changes to dietary trends and developing substitutes for sources of nutrition and protein (Boland *et al.*, 2013), but also trying to increase efficiency and control via PLF. The combined factors of an increasing population, a decreasing quantity of available arable land, and predicted changes to established seasonal patterns are factors that have all registered the necessity in the livestock farming industry for PLF (Banhazi, 2011).

The first step in implementing PLF is to automatically collect management information in a strategic fashion online. The available information then needs to be managed efficiently. To achieve this, the use of available information technology (IT) tools are required. The collected data need to be analyzed and interpreted, so the resulting information can be used for decision making (Banhazi and Black, 2009a). The main purpose of any PLF system should be to turn data into information and then into knowledge to improve operational outcomes (Banhazi *et al.*, 2012a; Banhazi and Black, 2009a). In parallel, implementation of the "System Approach"—or, the integration of currently independently controlled processes—is needed. Finally, more advanced production control methods need to be introduced on farms (Banhazi *et al.*, 2012b). The promoters of PLF technologies originally believed that the utilization of modern "production management" principles in livestock industries would create a framework which would ensure the identification of inefficiencies, and allow for continuous improvement of animal production, which, in turn, will improve profitability (Wathes *et al.*, 2008; Berckmans, 2011).

8.2 EARLY PLF HISTORICAL DEVELOPMENTS

Like many other developments in the associated fields of agriculture, computing, and engineering, the early development of PLF was driven by a range of variables in the care of livestock, the growth of herd size and the consequent inability of farmers to care for individual animals, the economic efficiency of farming and, increasingly, environmental considerations. As these variables have created a greater complexity in the work of farmers, it has become necessary that farmers be able to monitor variables in relation to the key livestock production processes (Frost *et al.*, 1997). A core principle in the development of PLF has been the development and deployment of technologies, which provide a farmer with information that is accurate and relevant (Frost *et al.*, 1997). Corresponding developments in agriculture have followed a similar trajectory of providing a farmer with up-to-date information on crops and fields (Zhang and Kovacs, 2012). Therefore, the development of PLF over time has not just included technology and its applications, but also evolving definitions and clarifying the idea of what PLF is. It can be understood as a set of technologies, but also processes of decision making (Banhazi and Black, 2009a).

The early PLF developments were mainly instigated in Europe and the United Kingdom. Early pioneers of the PLF concept included researchers at the Silsoe Research Institute, United Kingdom and Leuven University, Belgium, which still remain active in the PLF research area in Europe. Additional developments took place in other European countries, such as Germany, Denmark, the Netherlands, and Finland; and, outside of Europe, at the Volcani Research Centre, Israel (Brandl and Jorgensen, 1996; Lawson *et al.*, 2011; Meen *et al.*, 2015; Niemi *et al.*, 2010; Halachmi *et al.*, 1998; Morag *et al.*, 2001; Hartung *et al.*, 2017; Busse *et al.*, 2015). Australian PLF developments started in 2002 with involvement of scientists from the United Kingdom and Belgium (Banhazi *et al.*, 2003) but, earlier in 1999, the National Livestock Identification System (NLIS) began to be used on Australian farms (Souza-Monteiro and Caswell, 2004). Most pig industry-related PLF developments were led by scientists in South Australia (Banhazi *et al.*, 2007; Banhazi and Black, 2009b) while researchers attached to the University of New England pioneered spatially enabled livestock management systems. Commonwealth Scientific and Industrial Research Organization (CSIRO) researchers extensively investigated virtual fencing technologies (Bishop-Hurley *et al.*, 2007), while researchers at Sydney University developed a number of field robotics applications in addition to the same university's Future Dairy program. The Sheep Cooperative Research Centre (CRC) also developed different PLF-like applications (Dyall *et al.*, 2004). PLF developments are also accelerating in the United States of America (USA) and in Asia (Zhang *et al.*, 2002; Gates *et al.*, 2001).

Initially it was believed that precise control of animal production could be quickly achieved via PLF developments, but there was perhaps a general underestimation of the complexity of development tasks ahead (Banhazi *et al.*, 2015). Eastwood, Klerkx, and Nettle (2017) note a number of factors including cost, support structures, and the relationships between developers and users as being areas of complexity. It is also important to consider that the developments in Europe, the United Kingdom, and Australia are researcher-led developments and tools, rather than farmer-led developments, hence there will inevitably be disconnect between both sets of expectations. Further complicated sets of relationships are between public and private research, interactions to ensure that on-farm experience informs technology development, and researcher-to-farmer data sharing (Eastwood, Klerkx, and Nettle, 2017).

Historical development of PLF and its emerging and developing applications can be assessed using case studies. The emergence of PLF has made clear that the system itself comprises a number of components and farmers may use some but not all components (Batte and Arnholt, 2003), meaning that there are a variety of ways PLF system might be implemented. A study of Australian dairy farmers (Eastwood *et al.*, 2012) presented case studies of farmers' engagement with the technological potentials of decision support systems, which reinforced the significance of support as farmers moved from early to advanced stages in the use of technology and the importance of communication between developers and users. It showed that the development of knowledge is intrinsic to the use of the technology. Ohio case studies of farmers (Batte and Arnholt, 2003) revealed that farmers who had adopted component technologies of precision farming had strongly articulated reasons for doing so, and could identify benefits for profit, risk management, and environmental compliance. These case studies showed a variety of uses, including for monitoring soil for crops and, in one case, with pigs. As expected, these case studies revealed differences in emphasis, although, in terms of technology, the tools to measure yield and to sample soil could be used by farmers engaged in other types of farming besides livestock. In terms of information that could be obtained from technology, information which contributed to management and decision making were important. Busse's various studies on German agriculture uncovered what the terms both the hard and soft skill sets required for further development, from improvements to the control systems facilitated by technology, to the willingness of stakeholders to be innovators (Busse *et al.*, 2015).

8.3   HUMAN ASPECTS OF TECHNOLOGY DEVELOPMENT AND USE

Much has been written about the scientific and commercial aspects of PLF (Banhazi *et al.*, 2012a) but there is also an emerging literature on the human aspects of introducing and then consolidating PLF systems to the real-world environments of commercial farms. However, much literature related to farming efficiency is more general than specifically focused on PLF, but valuable points are made about the attitudes of the farmers who are the key stakeholders in the implementation of these systems.

While quantitative data has long been acknowledged by developers in the fields of science and engineering as a significant and effective source of information, researchers are increasingly aware that perceptions, attitudes, and what Jansen *et al.* (2009) refer to as the "normative frame of refer-ence," are important qualitative measures of the uptake and consolidation of innovations and developments in systems and technologies. These sources of information or "constructs" are com-monplaces in the "soft" fields (Jansen *et al.*, 2009) but they also have their significance to the field of PLF and farming management. As is discussed in Section 8.6, the sustainability of PLF and the possibility that there is a gap between reality and expectations are indications of the need to take note of the human participation in the development and consolidation of new systems. No matter how good or effective PLF may be, it is essential that these merits be communicated and accepted. The farmer can be centrally positioned as the stakeholder who is the most important decision-maker (Arbuckle *et al.*, 2013) in a range of current and compelling issues facing agriculture, including climate change and animal health. Using PLF is another dimension of farmer decision making, and, given that farmers may choose to use new technology, it is one of many decisions they will have to make about the management of their farms.

The literature provides a broad coverage of the uptake of farm and animal management. Jansen *et al.'s* (2009) study of farm management (specifically mastitis control in dairy farming) argued that effective farming practices could depend on the "human factor," which means the preferences of the farmers for management systems, as well as how they prioritize or feel about developments. The human factor includes what and how the farmers will decide about the management of their farms and their adherence to risk averse behaviors (Willock *et al.*, 1999). The "technology diffusion process" has long been part of the scholarly literature since the research of Beale and Bolen in 1955 (cited in Daberkow and McBride, 2003). Since then, multiple variables have been identified that cascade out from the attitudes of the farmers, including their awareness of new technologies and their desire for solid economic return (Daberkow and McBride, 2003). Thus, no matter how good developers or producers think their systems are, these benefits must be taught in meaningful ways.

One example is the approach taken by the Upper Midwest Aerospace Consortium (UMAC) in its endeavors to inculcate technology use for precision farming, whereby UMAC consciously thought of farmers as partners rather than as clients (Seelan *et al.*, 2003).

8.4   EXAMPLES OF PLF APPLICATIONS

Examples of PLF applications can be outlined according to specific and individual technologies' image analysis and sound recording. Section 8.4 will consider the wider uses of a mixture of technologies which facilitate estrus detection, herd management, behavior simulation, and production. Finally, this section considers the tools which facilitate data interrogation. Currently, no fully-integrated PLF systems exist, but there are examples of commercial companies, such as Farmex, offering semi-commercial PLF systems (Bird *et al.*, 2001) or PLF system components. Companies, such as Fancom, DeLaval, Petersime, Skov, and Big Dutchman have all developed PLF components (Guarino *et al.*, 2008; Kashiha *et al.*, 2013; Vranken and Berckmans, 2017). In 2010, PLF Agritech was established in Australia, and a number of PLF focused spin-off companies were established in Belgium. Australian researchers developed a number of key sensors systems that have been tested under commercial conditions, demonstrating the benefits of the sensors system (Banhazi *et al.*, 2009; Banhazi, 2009; Banhazi *et al.*, 2011). Various versions of these systems were tested on Australian and European farms (Figures 8.2 and 8.3), although it is also important to note that the Australian dairy industry has used EID since the 2000s, while early Animal Management Systems were in use in the Netherlands by 1992.

Specific technological developments have been the success stories of PLF developments, such as the commercially-available cleaning robots, milking robots, and automated heat-detection system (Braithwaite *et al.*, 2005; Zhang *et al.*, 2006; Bull *et al.*, 1996; Saumande, 2002).

Image analysis has several applications, including observation of animal behavior (such as if they are lying down, standing up, consuming food, or fighting), and evaluation of the cleanliness of body surfaces, for example, the teats of cows before robotic milking. The potential of detecting dirt on animal skin (teats) using information from images was assessed (Bull *et al.*, 1996). It was found that classification of the teats into "clean" and "dirty" categories was in reasonable agreement with human assessment.

Sound recordings are being used to either utilize the data (1) as an early warning system for impending respiratory problems in pigs and indirectly as an indicator of the level of airborne pollution; or (2) as a warning system to reduce crushing of piglets by the sow (Friend *et al.*, 1989). Belgian researchers studying the coughing sound of pigs suggested that incorporating the sound-analysis into building ventilation systems could improve the functionality of the ventilation systems by reducing the effects of airborne pollution (Guarino *et al.*, 2008).

Different electronic technologies have been developed to improve estrus detection in a number of species (Brehme *et al.*, 2008; De Mol *et al.*, 1997; Maatje *et al.*, 1997; Ostersen *et al.*, 2010; Saumande, 2002; Van Asseldonk *et al.*, 1998). Intravaginal probes and mount sensors are mainly used to monitor beef and dairy cattle. However, simple infrared movement detectors, first used by Danish research workers, were evaluated as potential sensors for automated estrus detection in sows. The infrared sensors were mounted above the sows and the difference in body movements was evaluated. Up to 80% of the sows were classified correctly as being in estrus based on the level of daily activity. However, in dairy cows the best estrus detection rate was achieved when different traits were analyzed in combination using multivariate fuzzy logic models (Krieter *et al.*, 2003). This research is a good example of how advanced and innovative data analysis, combined with the use of simple and cost-effective sensors can solve real production management problems. The potential benefits are not gained as a result of operating expensive hardware, but more from improved computer analysis of the data. This principle is probably true for the vast majority of PLF applications.

A few innovative management programs have also been produced in recent years. A system called PKS SCHWEIN has been reported by Krause (1990), which was designed to assist producers

FIGURE 8.2   Early version of the Weight-Detect™ system, designed to non-intrusively measure the weight of pigs on livestock farms.

managing pig breeding herds. The system comprised a series of modules to be used for sow organization, breeding, and reproduction management. LOGIPORK (Cordier and Gaudin, 1990) and CHESS (Dijkhuizen and Huirne, 1990) are similar and commercially-available systems designed to analyze productivity levels of livestock. These systems use comparative trend analysis as a tool to evaluate and contrast historical performance data of similar farms.

A behavior-based simulation model has been developed to enable agricultural engineers to optimize dairy farm design specifically for robotic milking (Halachmi, 2000). As robotic milking is a mature technology, this model was created to overcome the lack of available experience and to facilitate the design of "optimal milking sheds," while taking into consideration the circumstances of individual farmers. While this system was designed for dairy farmers, it is a good example how

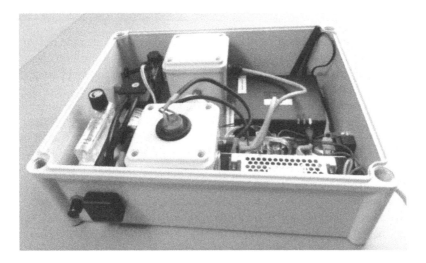

FIGURE 8.3 Early version of the Enviro-Detect™ system, designed to measure various environmental parameters in livestock buildings.

Decision Support Systems (DSS) can be used to collect, preserve, and disperse management experience/knowledge which might be in short supply.

An expert system for managing important aspects of production, such as feeding, shed environment, and disease, was developed for layer birds (Lokhorst and Lamaker, 1996). Knowledge of a number of experts was stored in the database. Based on the advice of these experts, different quantitative and qualitative data helped managers to detect irregularities in the production process and respond to it. However, it was highlighted that one of the critical aspects of developing an expert, knowledge-based system is the acquisition of appropriate and relevant knowledge.

Signal processing is another useful method for data interrogation (Marchant, 2003). It is the science of understanding and interpreting different "signals," for example, measurements of the heart rate of an animal in order to assess its well-being or control aspects of the production process (Arieli *et al.*, 2002; Kettlewell *et al.*, 1997; Santurtun *et al.*, 2014). Signal processing issues are especially relevant for so-called complex, sensor-rich environments, such as animal buildings and greenhouses (Pasgianos *et al.*, 2003; Tong *et al.*, 2013).

PLF has special requirements in networking as a large amount of information is exchanged between different components of the system via networks such as the Internet (Alvarez and Nuthall, 2006; Batte, 2005). Emerging networking technologies and their integration into PLF systems can facilitate improved communication procedures and the development of integrated digital environment (Gloy and Akridge, 2000; Gray *et al.*, 2017).

Different control systems, such as conventional (staged) ventilation systems are commonly used in animal buildings to control internal temperatures. Strategies for improved heating and ventilation control have been reviewed previously (Zhang and Barber, 1995). However, American researchers claimed that existing ventilation systems could be significantly enhanced by incorporating fuzzy control methods (Gates *et al.*, 2001). These "intelligent" controllers enable users to select the level of tradeoff between energy use and control precision. Some of these new controllers have the ability to take the thermal history of both the air space and the animal into consideration when adjusting environmental temperature in animal buildings.

Behavior-based environmental control for animal buildings has inherent advantages over the conventional, temperature-based control methods (Shao *et al.*, 1998). It has been reported that postural images of pigs might be used to control environmental temperature in sheds. Cameras were used to capture behavioral pictures of animals, which were then analyzed to gain an understanding of their thermal comfort level (Wouters *et al.*, 1990).

8.5   ANIMAL IDENTIFICATION

Traceability offers an immediate benefit to producers while encouraging them to develop advanced information communications technology and data management systems on farms that can also be used for other PLF functions (Artmann, 1999). Individual identification and monitoring of animals is an important step towards enhancing the tractability of livestock products and state-of-the-art developments in livestock and building control are continuing to emerge (Voulodimos *et al.*, 2010; Busse *et al.*, 2015; Fournel *et al.*, 2017). The latest generation of animal ID devices could also incorporate sensors, store additional data, and provide authentication protocols (Brown-Brandl *et al.*, 2013). Identification Devices with built-in sensors could be used for tasks such as health and reproduction status monitoring (Caja *et al.*, 1999; Champion *et al.*, 2005). Individual animal ID technology will enable livestock managers to once again treat animals as individuals rather than as a herd or flock (Eigenberg *et al.*, 2008; Eradus and Jansen, 1999; Goedseels *et al.*, 1992). This could facilitate the provision of individually tailored diets and environment control, which has enormous potential to improve productivity and welfare. The use of miniature injectable transponders could improve the security and practicality of electronic identification, as this application reduces the risk of transponder loss and fraud (Klindtworth *et al.*, 1999; Lambooij *et al.*, 1999). However, injected transponders have to be biologically compatible, technically feasible, and easily recoverable in slaughterhouses. It is expected that electronic identification systems will be further enhanced by technologies used for smart-cards, and can potentially be used to control production processes (Porto *et al.*, 2014; Reiners *et al.*, 2009; Schwartzkopf-Genswein *et al.*, 1999). However, potential problems with compatibility between different systems have to be resolved.

A cutting-edge technology (the use of retinal vascular pattern recognition combined with Global Positioning Satellite (GPS) systems) was described recently as a secure source verification of livestock. A case study of Irish cattle in 2008 found positive results from the technology as a means for registering and following animals (Allen *et al.*, 2008) Whittier et al. suggested that this technology offers a "fraud-free" verification of livestock in food safety or food retail systems (Whittier *et al.*, 2003), while GPS can also be used in order to track animals on farms.

Radio telemetry systems are available to remotely monitor and record physiological parameters, such as the heart rate and deep body temperature of animals (Kettlewell *et al.*, 1997). The transmitters are either ingested, fixed in the ear canal by expandable foam, or surgically implanted (Kyle *et al.*, 1998; Shipka, 2000). It is reported that these devices do not interfere with normal physiological function and behavior of the animals. These systems can be particularly useful in monitoring physiological responses of animals to environmental stressors in outdoor settings or during transport (Mitchell *et al.*, 2001).

8.6   EXPECTATIONS AND THE REALITY OF PLF IMPLEMENTATION

Complete PLF systems still need to be developed and delivered to the farming community, therefore, it is of ongoing value. At the moment, there is a limited existing support structure that can be used to deliver and maintain PLF applications on farms. Different hardware and perhaps software components are typically supplied to primary producers without any significant and ongoing technical assistance. However, empirical research indicates that the effective uptake of developments in farming systems can be influenced by the "human factor" of the farmer. One element of that factor is the farmer's confidence or feeling of control over a system or technology (Jansen *et al.*, 2009). Studies of differences in the health of livestock (a specific focus, and not explicitly focused on in PLF, but relevant as a comparative point) has accumulated empirical data pointing to a farmer's feeling of control as one factor in effective farm management.

Part of the "human factor" also includes support structures and effective communication. While new technologies may seem to be manifestly advantageous or valuable, Jansen *et al.* (2010) noted that "effective communication with farmers is essential in order to change their behavior and to improve their farm management." Part of effective communication includes the implementation of strategies that will motivate change in an industry that is reluctant to instigate change, and whose key stakeholders need strong proofs that a suggested change is viable (Banhazi *et al.*, 2003). Jansen *et al.* (2010) refer to the "cognitive dissonance," meaning that different stakeholders' perceptions and priorities will vary as a potent factor that can limit enthusiasm or motivation for change. Farmers may feel they already know enough or may make private judgments about what is or is not relevant and these attitudes, which can limit their support for innovation. It is also important to note that such perspectives are valid, since communication of expertise is two-way. No matter how well-informed an engineer may be, the farmers will best know their own needs, and communication with farmers needs to include recognition of the importance of their perspectives. Major studies of farmers in the United Kingdom (Willock *et al.*, 1999) and in Austria (Vogel, 1996) have already established that factors such as farmers' "practical knowledge and personal experience" will condition their behavior (Vogel, 1996). As information communication technologies are a main component of PLF, those who will use them must not only pay for them, but put in the time and effort to learn how to use them to their operational potential (Adrian *et al.*, 2005).

In 2005, Adrian *et al.* argued that data was needed on the "perceptions of usefulness," the ease of use, and "attitudes of confidence" among farmers toward PLF. Even if useful data is collected, such information is rarely used at its maximum potential, and producers are infrequently assisted to make good use of the collected information. Significantly, a limited amount of work was done on the systematic evaluation of the benefits of a fully integrated PLF system (cost/benefit, rate of return) has taken place on farms (Willis *et al.*, 2016; Black and Banhazi, 2013; Gregersen, 2011). However, any systematic evaluation would need to be undertaken with an underlying awareness of the essential

TABLE 8.1
Subjective SWOT Analysis of the PLF Research Area

Strengths	Weaknesses	Opportunities	Threats
PLF systems are unique products	PLF systems are new products and need to be accepted by the market People need to learn how to use them	Potential to enhance production	Lack of reliable cost vs. benefits analysis of PLF systems
Use of PLF systems on farms could increase precision/control of livestock production	Support structure needs to be developed Communication strategies needed	Further "precision tests" need to be conducted and used as marketing material	Lack of clearly demonstrated benefits of measuring key performance indicators and using the information to improve farm management
PLF systems could reduce cost	Service companies still need to be cultivated	Demonstrate the benefits of measuring key performance indicators frequently	
Identify animals in commercial settings with extreme Fuel Conversion Efficiency could facilitate genetic improvements		Develop a highly integrated system and thus create demand for system components	
Feed-intake and weight change data could identify sick, underperforming animals		Develop sensors for body composition measurement and welfare monitoring	

heterogeneity among farms; a point originally made about U.S. farms by Daberkow and McBride (2003), which has wider relevance. In addition, few service industry or support structures have been established, an absence that is further hindering the implementation of PLF systems on farms, even though it has been recognized for over a decade that the incorporation of PLF will be a "steep learning curve" involving what could be "difficult-to-use" technologies" (Adrian *et al.*, 2005). Research from the United States therefore indicates that a significant factor may be the age of the farmer, with better-educated, younger farmers being suggested as a demographic with a willingness to innovate via learning new skills (Daberkow and McBride, 2003). The involvement of larger commercial companies in PLF research and implementation is still limited, and, therefore, PLF research and development is still fragmented internationally. The involvement of larger commercial companies also bypasses concerns raised by Adrian et al. that smaller farms may struggle with the "large amounts of capital, time, and learning in order to use the technologies." The "Strength, Weakness, Opportunity, Threat" analysis of PLF technologies is presented in Table 8.1.

8.7 CONCLUSION: KEY IDEAS AND IDENTIFICATION OF FUTURE OPPORTUNITIES

A key challenge is to ensure that livestock industries are not left behind in the technological race (Wang *et al.*, 2006). A very careful selection of future research and development directions are needed, and the proposed selection criteria for establishing directions should include consideration of commercial importance, technical feasibility, and strategic advantage. Interlayered among these are further considerations of what the end-user themselves will need the technology to do, and how developers will respond to the way the needs of communities and users will shift and evolve. The common thread through all future research projects has to be the identification of production, animal welfare, and environmental key performance indicators, which can be measured on a real time basis to improve production efficiency (Banhazi and Black, 2009a).

Establishing good information management practices on farms can significantly improve the efficiency and profitability of livestock enterprises. To transform the information management systems currently used on farms into a more sophisticated system, data collection and, more importantly, data analysis needs to be automated (Banhazi and Black, 2009a). Automated data collection, management, and analysis, together with accessible data-warehouses would transform the currently used segregated systems into a powerful information-based PLF system. Centralized data management would enable uniform data analysis and the appropriate interpretation of available data. On-farm data processing could provide valuable support for farm managers in everyday management, but the real gains would come from using in-depth data analysis provided by remote data-warehouses via the Internet (Wathes *et al.*, 2008). A well-designed data collection system would reduce the need for manual recording of farm production data, and would present the producer with management data in the most efficient form. Achieving a well-designed data collection, presents numerous challenges, including the willingness of organizations to share commercial data. Initiatives such as Farm Data Standards in NZ (www.farmdatastandards.org.nz), aims to make key data available, such as breed and feed lists. Using automated data collection systems on farms, it would be possible to transfer details of animal performance, labor input, and environmental performance of farms throughout the supply chain (Frost *et al.*, 1997). A farm equipped with appropriate data-collection systems would provide data from the production sites for review and analysis. Applying in-depth analysis, artificial intelligence, and investigation of the available data could demonstrate direct financial benefits for producers. At all points, however, effective two-way channels of communication with the human stakeholders will be important. Research into what will or will not prompt a farmer to adopt a particular practice is long-standing but also ongoing (Baumgart-Getz *et al.*, 2012).

Farms that enabled research data to be collected would provide supplies of performance data at a minimal operating cost and would allow precise evaluation of the impact of changing feed formulations, veterinary drugs, housing, or management conditions on individual farms under specific environmental conditions. These envisaged research farms would also facilitate the economical evaluation of PLF systems, which is one of the most important tasks to be completed by the PLF research community.

REFERENCES

Adrian, A. M., Norwood, S. H. and Mask, P. L. (2005). Producers' perceptions and attitudes toward precision agriculture technologies. *Computers and Electronics in Agriculture* 48 (3), 256–71.

Allen, A., Golden, B., Taylor, M., Patterson, D., Henriksen, D. and Skuce, R. (2008). Evaluation of retinal imaging technology for the biometric identification of bovine animals in Northern Ireland. *Livestock Science* 116 (1–3), 42–52.

Alvarez, J. and Nuthall, P. (2006). Adoption of computer-based information systems: The case of dairy farmers in Canterbury, NZ, and Florida, Uruguay. *Computers and Electronics in Agriculture* 50 (1), 48–60.

Arbuckle, J. G., Morton, L. W. and Hobbs, J. (2013). Farmer beliefs and concerns about climate change and attitudes toward adaptation and mitigation: Evidence from Iowa. *Climatic Change* 118 (3–4), 551–63.

Arieli, A., Kalouti, A., Aharoni, Y. and Brosh, A. (2002). Assessment of energy expenditure by daily heart rate measurement—Validation with energy accretion in sheep. *Livestock Production Science* 78 (2), 99–105.

Artmann, R. (1999). Electronic identification systems: State of the art and their further development. *Computers and Electronics in Agriculture* 24 (1–2), 5–26.

Banhazi, T. (2009). User-friendly air quality monitoring system. *Applied Engineering in Agriculture* 25 (2), 281.

Banhazi, T. and Black, J. L. (2009a). Precision livestock farming: A suite of electronic systems to ensure the application of best-practice management on livestock farms. *Australian Journal of Multi-disciplinary Engineering* 7 (1), 1–14.

Banhazi, T., Black, J. L. and Durack, M. (2003). Australian precision livestock farming workshops. *Joint Conference of ECPA - ECPLF* 1, 675–684 (Eds A. Werner and A. Jarfe). Berlin, Germany: Wageningen Academic Publisher.

Banhazi, T., Dunn, M., Cook, P., Black, J., Durack, M. and Johnnson, I. (2007). Development of precision livestock farming (PLF) Technologies for the Australian pig industry. *Third European Precision Livestock farming Conference* 1, 219–28 (Ed. S. Cox). Skiathos, Greece: University of Thessaly.

Banhazi, T., Vranken, E., Berckmans, D., Rooijakkers, L. and Berckmans, D. (2015). Word of caution for technology providers: Practical problems associated with large-scale deployment of PLF technologies on commercial farms. In Seventh European Conference on Precision Livestock Farming 1, 105–11 (Ed I. Halachmi). Leuven, Belgium: Wageningen Academic Publishers.

Banhazi, T. M. (2011). The current state of PLF affairs. *Multidisciplinary Approach to Acceptable and Practical Precision Livestock Farming for SMEs in Europe and Worldwide* 1, 11–17 (Eds. I. G. Smith and H. Lehr). Halifax, UK: European Commission.

Banhazi, T. M., Babinszky, L., Halas, V. and Tscharke, M. (2012a). Precision Livestock Farming: Precision feeding technologies and sustainable livestock production. *International Journal of Agricultural and Biological Engineering* 5 (4), 54–61.

Banhazi, T. M. and Black, J. L. (2009b). Precision livestock farming: A suite of electronic systems to ensure the application of best practice management on livestock farms. *Australian Journal of Multi-disciplinary Engineering* 7 (1), 1–14.

Banhazi, T. M., Lehr, H., Black, J. L., Crabtree, H., Schofield, P., Tscharke, M. and Berckmans, D. (2012b). Precision Livestock Farming: An international review of scientific and commercial aspects. *International Journal of Agricultural and Biological Engineering* 5 (3), 1–9.

Banhazi, T. M., Rutley, D. L., Parkin, B. J. and Lewis, B. (2009). Field evaluation of a prototype sensor for measuring feed disappearance in livestock buildings. *Australian Journal of Multi-disciplinary Engineering* 7 (1), 27–38.

Banhazi, T. M., Tscharke, M., Ferdous, W. M., Saunders, C. and Lee, S.-H. (2011). Improved image analysis-based system to reliably predict the live weight of pigs on farms: Preliminary results. *Australian Journal of Multi-disciplinary Engineering* 8 (2), 107–19.

Batte, M. T. (2005). Changing computer use in agriculture: Evidence from Ohio. *Computers and Electronics in Agriculture* 47 (1), 1–13.

Batte, M. T. and Arnholt, M. W. (2003). Precision farming adoption and use in Ohio: Case studies of six leading edge adopters. *Computers and Electronics in Agriculture* 38 (2), 125–39.

Baumgart-Getz, A., Prokopy, L. S. and Floress, K. (2012). Why farmers adopt best management practices in the United States: A meta-analysis of the adoption literature. *Journal of Environmental Management* 96 (1), 17–25.

Berckmans, D. (2011).What can we expect from precision livestock farming and why? *Acceptable and Practical Precision Livestock Farming*, 1, 7–10 (Eds I. G. Smith and H. Lehr). Halifax, UK: European Commission.

Bird, N., Crabtree, H. G. and Schofield, C. P. (2001). Engineering technologies enable real-time information monitoring in pig production. *Integrated Management Systems for Livestock: Occasional Publication* (28), 105–12 (Eds. C. M. Wathes, A. R. Frost, F. Gordon and J. D. Wood). Edinburgh: British Society of Animal Science and Institution of Agricultural Engineers.

Bishop-Hurley, G. J., Swain, D. L., Anderson, D. M., Sikka, P., Crossman, C. and Corke, P. (2007). Virtual fencing applications: Implementing and testing an automated cattle control system. *Computers and Electronics in Agriculture* 56 (1), 14–22.

Black, J. L. and Banhazi, T. M. (2013). Economic and social advantages of precision livestock farming in the pig industry. *Sixth European Conference on Precision Livestock Farming* 1, 199–208 (Eds. D. Berckmans and J. Vandermeulen). Leuven, Belgium: Catholic University of Leuven.

Boland, M. J., Rae, A. N., Vereijken, J. M., Meuwissen, M. P., Fischer, A. R., van Boekel, M. A., Rutherfurd, S. M., Gruppen, H., Moughan, P. J. and Hendriks, W. H. (2013). The future supply of animal-derived protein for human consumption. *Trends in Food Science & Technology* 29 (1), 62–73.

Braithwaite, I., Blanke, M., Zhang, G. Q. and Carstensen, J. M. (2005). Design of a vision-based sensor for autonomous pig house cleaning. *EURASIP Journal on Applied Signal Processing* 2005 (13), 473–9.

Brandl, N. and Jorgensen, E. (1996). Determination of live weight of pigs from dimensions measured using image analysis. *Computers and Electronics in Agriculture* 15 (1), 57–72.

Brehme, U., Stollberg, U., Holz, R. and Schleusener, T. (2008). ALT pedometer—New sensor-aided measurement system for improvement in oestrus detection. *Computers and Electronics in Agriculture* 62 (1), 73–80.

Brown-Brandl, T. M., Rohrer, G. A. and Eigenberg, R. A. (2013). Analysis of feeding behavior of group housed growing–finishing pigs. *Computers and Electronics in Agriculture* 96 (0), 246–52.

Bull, C. R., McFarlane, N. J. B., Zwiggelaar, R., Allen, C. J. and Mottram, T. T. (1996). Inspection of teats by color image analysis for automatic milking systems. *Computers and Electronics in Agriculture* 15 (1), 15–26.

Busse, M., Schwerdtner, W., Siebert, R., Doernberg, A., Kuntosch, A., König, B. and Bokelmann, W. (2015). Analysis of animal monitoring technologies in Germany from an innovation system perspective. *Agricultural Systems* 138 (0), 55–65.

Caja, G., Conill, C., Nehring, R. and Ribo, O. (1999). Development of a ceramic bolus for the permanent electronic identification of sheep, goats, and cattle. *Computers and Electronics in Agriculture* 24 (1–2), 45–63.

Champion, R. A., Cook, J. E., Rook, A. J. and Rutter, S. M. (2005). A note on using electronic identification technology to measure the motivation of sheep to obtain resources at pasture. *Applied Animal Behaviour Science* 95 (1–2), 79–87.

Chao, K., Chen, Y.-R., Hruschka, W. R. and Gwozdz, F. B. (2002). Online inspection of poultry carcasses by a dual-camera system. *Journal of Food Engineering* 51 (3), 185–92.

Cordier, L. & Gaudin N. (1990). LOGIPORC—A Successful Example of Integration in Pig Production. *Third International Congress for Computer Technology—Integrated Decision Support Systems in Agriculture*, 228–32 (Ed. F. Kuhlmann). Frankfurt a. M. - Bad Soden: DLG.

Daberkow, S. G. and McBride, W. D. (2003). Farm and operator characteristics affecting the awareness and adoption of precision agriculture technologies in the United States. *Precision Agriculture* 4 (2), 163–77.

de Mol, R. M., Kroeze, G. H., Achten, J. M. F. H., Maatje, K. and Rossing, W. (1997). Results of a multivariate approach to automated oestrus and mastitis detection. *Livestock Production Science* 48 (3), 219–227.

Dijkhuizen, A. A. and Huirne, R. B. M. (1990). CHESS—An Integrated Decision Support and Expert System to Analyze Individual Sow-Herd Performance. *Third International Congress for Computer Technology—Integrated Decision Support Systems in Agriculture*, 221–7 (Ed. F. Kuhlmann). Frankfurt a. M. - Bad Soden: DLG.

Dyall, T. R., Eccles, D. A., Miron, D. J. and Nethery, R. D. (2004). Data standards for the Australian sheep industry. *Animal Production in Australia* 1, 49–52 (Eds. R. Stockdale, J. Heard and M. Jenkin). University of Melbourne, Victoria, Australia: CSIRO Publishing.

Eastwood, C., Chapman, D. and Paine, M. (2012). Networks of practice for co-construction of agricultural decision support systems: Case studies of precision dairy farms in Australia. *Agricultural Systems* 108, 10–18.

Eigenberg, R. A., Brown-Brandl, T. M. and Nienaber, J. A. (2008). Sensors for dynamic physiological measurements. *Computers and Electronics in Agriculture* 62 (1), 41–7.

Eradus, W. J. and Jansen, M. B. (1999). Animal identification and monitoring. *Computers and Electronics in Agriculture* 24 (1–2), 91–98.

Føre, M., Alver, M., Alfredsen, J. A., Marafioti, G., Senneset, G., Birkevold, J., Willumsen, F. V., Lange, G., Espmark, Å. and Terjesen, B. F. (2016). Modelling growth performance and feeding behaviour of Atlantic salmon (*Salmo salar L.*) in commercial-size aquaculture net pens: Model details and validation through full-scale experiments. *Aquaculture* 464, 268–78.

Fournel, S., Rousseau, A.N. and Laberge, B. (2017). Rethinking environment control strategy of confined animal housing systems through precision livestock farming. *Biosystems Engineering* 155, 96–123.

Friend, T., O'Connor, L., Knabe, D. and Dellmeier, G. (1989). Preliminary trials of a sound-activated device to reduce crushing of piglets by sows. *Applied Animal Behavior Science* 24, 23–9.

Frost, A. R., Parsons, D. J., Stacey, K. F., Robertson, A. P., Welch, S. K., Filmer, D. and Fothergill, A. (2003). Progress towards the development of an integrated management system for broiler chicken production. *Computers and Electronics in Agriculture* 39 (3), 227–40.

Frost, A. R., Schofield, C. P., Beaulah, S. A., Mottram, T. T., Lines, J. A. and Wathes, C. M. (1997). A review of livestock monitoring and the need for integrated systems. *Computers and Electronics in Agriculture* 17 (2), 139–59.

Gates, R. S., Chao, K. and Sigrimis, N. (2001). Identifying design parameters for fuzzy control of staged ventilation control systems. *Computers and Electronics in Agriculture* 31 (1), 61–74.

Gloy, B. A. and Akridge, J. T. (2000). Computer and internet adoption on large U.S. farms. *The International Food and Agribusiness Management Review* 3 (3), 323–38.

Goedseels, V., Geers, R., Truyen, B., Wouters, P., Goossens, K., Ville, H. and Janssens, S. (1992). A data-acquisition system for electronic identification, monitoring and control of group-housed pigs. *Journal of Agricultural Engineering Research* 52, 25–33.

Gray, J., Banhazi, T. M. and Kist, A. A. (2017). Wireless data management system for environmental monitoring in livestock buildings. *Information Processing in Agriculture* 4 (1), 1–17.

Gregersen, O. (2011). Economic aspects of PLF. *Acceptable and Practical Precision Livestock Farming* 1, 149–178 (Eds. I. G. Smith and H. Lehr). Halifax, UK: European Commission.

Guarino, M., Jans, P., Costa, A., Aerts, J. M. and Berckmans, D. (2008). Field test of algorithm for automatic cough detection in pig houses. *Computers and Electronics in Agriculture* 62 (1), 22–8.

Halachmi, I. (2000). Designing the optimal robotic milking barn, part II: Behavior-based simulation. *Journal of Agricultural Engineering Research* 77 (1), 67–79.

Halachmi, I., Edan, Y., Maltz, E., Peiper, U. M., Moallem, U. and Brukental, I. (1998). A real-time control system for individual dairy cow food intake. *Computers and Electronics in Agriculture* 20 (2), 131–44.

Hartung, J., Banhazi, T., Vranken, E. and Guarino, M. (2017). European farmers' experiences with precision livestock farming systems. *Animal Frontiers* 7 (1), 38–44.

Huynh, T. T. T., Aarnink, A. J. A., Drucker, A. and Verstegen, M. W. A. (2006). Pig production in Cambodia, Laos, the Philippines, and Vietnam: A review. *Asian Journal of Agriculture and Development* 4 (1), 69–90.

Jansen, J., Steuten, C. D. M., Renes, R. J., Aarts, N. and Lam, T. J. G. M. (2010). Debunking the myth of the hard-to-reach farmer: Effective communication on udder health. *Journal of Dairy Science* 93 (3), 1296–306.

Jansen, J., van den Borne, B. H. P., Renes, R. J., van Schaik, G., Lam, T. J. G. M. and Leeuwis, C. (2009). Explaining Mastitis incidence in Dutch dairy farming: The influence of farmers' attitudes and behaviour. *Preventive Veterinary Medicine* 92 (3), 210–23.

Kashiha, M., Pluk, A., Bahr, C., Vranken, E. and Berckmans, D. (2013). Development of an early warning system for a broiler house using computer vision. *Biosystems Engineering* 116 (1), 36–45.

Kettlewell, P. J., Mitchell, M. A. and Meeks, I. R. (1997). An implantable radio-telemetry system for remote monitoring of heart rate and deep body temperature in poultry. *Computers and Electronics in Agriculture* 17 (2), 161–75.

Klindtworth, M., Wendl, G., Klindtworth, K. and Pirkelmann, H. (1999). Electronic identification of cattle with injectable transponders. *Computers and Electronics in Agriculture* 24 (1–2), 65–79.

Krause, J. (1990). PKS SCHWEIN. *Third International Congress for Computer Technology—Integrated Decision Support Systems in Agriculture*, 174–181 (Ed. F. Kuhlmann). Frankfurt a. M. - Bad Soden: DLG.

Krieter, J., Firk, R., Stamer, E. and Junge, W. (2003). Improving estrus detection in dairy cows by a combination of different traits using fuzzy logic. *First EPLFC* 1, 99–104 (Ed. S. Cox). Germany: Wageningen Academic Publisher.

Kyle, B. L., Kennedy, A. D. and Small, J. A. (1998). Measurement of vaginal temperature by radiotelemetry for the prediction of estrus in beef cows. *Theriogenology* 49 (8), 1437–49.

Lambooij, E., van't Klooster, C. E., Rossing, W., Smits, A. C. and Pieterse, C. (1999). Electronic identification with passive transponders in veal calves. *Computers and Electronics in Agriculture* 24 (1–2), 81–90.

Lawson, L. G., Pedersen, S. M., Sørensen, C. G., Pesonen, L., Fountas, S., Werner, A., Oudshoorn, F. W., Herold, L., Chatzinikos, T., Kirketerp, I. M. and Blackmore, S. (2011). A four nation survey of farm information management and advanced farming systems: A descriptive analysis of survey responses. *Computers and Electronics in Agriculture* 77 (1), 7–20.

Lokhorst, C. and Lamaker, E. J. J. (1996). An expert system for monitoring the daily production process in aviary systems for laying hens. *Computers and Electronics in Agriculture* 15 (3), 215–31.

Maatje, K., de Mol, R. M. and Rossing, W. (1997). Cow status monitoring (health and estrus) using detection sensors. *Computers and Electronics in Agriculture* 16 (3), 245–54.

Marchant, B. P. (2003). Time-frequency analysis for biosystems engineering. *Biosystems Engineering* 85 (3), 261–81.

Meen, G. H., Schellekens, M. A., Slegers, M. H. M., Leenders, N. L. G., Van Erp-van der Kooij, E. and Noldus, L. P. J. J. (2015). Sound analysis in dairy cattle vocalization as a potential welfare monitor. *Computers and Electronics in Agriculture* 118, 111–5.

Mitchell, M. A., Kettlewell, P. J., Lowe, J. C., Hunter, R. R., King, T., Ritchie, M. and Bracken, J. (2001). Remote physiological monitoring of livestock—An implantable radio-telemetry system. *Livestock Environment VI*. Proc. of the Sixth International Symposium, 535–41. Louisville, Kentucky: The Society for Engineering in Agricultural, Food, and Biological Systems.

Morag, I., Edan, Y. and Maltz, E. (2001). An individual feed allocation decision support system for the dairy farm. *Journal of Agricultural Engineering Research* 79 (2), 167–76.

Niemi, J. K., Sevón-Aimonen, M.-L., Pietola, K. and Stalder, K. J. (2010). The value of precision feeding technologies for grow-finish swine *Livestock Science* 129, 13–23.

Ostersen, T., Cornou, C. and Kristensen, A. R. (2010). Detecting estrus by monitoring sows' visits to a boar. *Computers and Electronics in Agriculture* 74 (1), 51–8.

Park, B., Chen, Y. R. and Nguyen, M. (1998). Multispectral image analysis using neural network algorithm for inspection of poultry carcasses. *Journal of Agricultural Engineering Research* 69 (4), 351–63.

Park, B., Yoon, S. C., Lawrence, K. C. and Windham, W. R. (2007). Fisher linear discriminant analysis for improving fecal detection accuracy with hyperspectral images *Transactions of the ASABE* 50 (6), 2275–83.

Pasgianos, G. D., Arvanitis, K. G., Polycarpou, P. and Sigrimis, N. (2003). A nonlinear feedback technique for greenhouse environmental control. *Computers and Electronics in Agriculture* 40 (1–3), 153–77.

Porto, S. M. C., Arcidiacono, C., Giummarra, A., Anguzza, U. and Cascone, G. (2014). Localization and identification performances of a real-time location system based on ultra-wide band technology for monitoring and tracking dairy cow behavior in a semi-open free-stall barn. *Computers and Electronics in Agriculture* 108, 221–29.

Reiners, K., Hegger, A., Hessel, E. F., Böck, S., Wendl, G. and Van den Weghe, H. F. A. (2009). Application of RFID technology using passive HF transponders for the individual identification of weaned piglets at the feed trough. *Computers and Electronics in Agriculture* 68 (2), 178–84.

Santurtun, E., Moreau, V. and Phillips, C. J. C. (2014). A novel method to measure the impact of sea transport motion on sheep welfare. *Biosystems Engineering* 118, 128–37.

Saumande, J. (2002). Electronic detection of estrus in postpartum dairy cows: Efficiency and accuracy of the DEC(R) (showheat) system. *Livestock Production Science* 77 (2–3), 265–71.

Schwartzkopf-Genswein, K. S., Huisma, C. and McAllister, T. A. (1999). Validation of a radio frequency identification system for monitoring the feeding patterns of feedlot cattle. *Livestock Production Science* 60 (1), 27–31.

Seelan, S. K., Laguette, S., Casady, G. M. and Seielstad, G. A. (2003). Remote sensing applications for precision agriculture: A learning community approach. *Remote Sensing of Environment* 88 (1), 157–69.

Shao, J., Xin, H. and Harmon, J. D. (1998). Comparison of image feature extraction for classification of swine thermal comfort behavior. *Computers and Electronics in Agriculture* 19 (3), 223–32.

Shipka, M. P. (2000). A note on silent ovulation identified by using radiotelemetry for estrus detection. *Applied Animal Behaviour Science* 66 (1–2), 153–9.

Souza Monteiro, D.M. and Caswell, J.A. (2004). The Economics of Implementing Traceability in Beef Supply Chains: Trends in Major Producing and Trading Countries. University of Massachusetts, Amherst Working Paper, No. 2004-6.

Stacey, K. F., Parsons, D. J., Frost, A. R., Fisher, C., Filmer, D. and Fothergill, A. (2004). An automatic growth and nutrition control system for broiler production. *Biosystems Engineering* 89 (3), 363–71.

Tilman, D., Balzer, C., Hill, J. and Befort, B. L. (2011). Global food demand and the sustainable intensification of agriculture. *Proceedings of the National Academy of Sciences* 108 (50), 20260–4.

Tilman, D. and Clark, M. (2014). Global diets link environmental sustainability and human health. *Nature* 515 (7528), 518–22.

Tong, J. H., Li, J. B. and Jiang, H. Y. (2013). Machine vision techniques for the evaluation of seedling quality based on leaf area. *Biosystems Engineering* 115 (3), 369–79.

van Asseldonk, M. A. P. M., Huirne, R. B. M. and Dijkhuizen, A. A. (1998). Quantifying characteristics of information technology applications based on expert knowledge for detection of estrus and mastitis in dairy cows. *Preventive Veterinary Medicine* 36 (4), 273–86.

Vogel, S. (1996). Farmers' environmental attitudes and behavior. *Environment and Behavior* 28 (5), 591–613.

Voulodimos, A. S., Patrikakis, C. Z., Sideridis, A. B., Ntafis, V. A. and Xylouri, E. M. (2010). A complete farm management system based on animal identification using RFID technology. *Computers and Electronics in Agriculture* 70 (2), 380–8.

Vranken, E. and Berckmans, D. (2017). Precision livestock farming for pigs. *Animal Frontiers* 7 (1), 32–7.

Wang, N., Zhang, N. and Wang, M. (2006). Wireless sensors in agriculture and food industries—Recent development and future perspectives. *Computers and Electronics in Agriculture* 50 (1), 1–14.

Wathes, C. M., Abeyesinghe, S. M. and Frost, A. R. (2001). Environmental design and management for livestock in the twenty-first century: Resolving conflicts by integrated solutions. *Livestock Environment VI*. Proc. of the Sixth International Symposium, Louisville, Kentucky: The Society for Engineering in Agricultural, Food and Biological Systems.

Wathes, C. M., Kristensen, H. H., Aerts, J. M. and Berckmans, D. (2008). Is precision livestock farming an engineer's daydream or nightmare, an animal's friend or foe, and a farmer's panacea or pitfall? *Computers and Electronics in Agriculture* 64 (1), 2–10.

Whittier, J. C., Shadduck, J. A. and Golden, B. L. (2003). Secure identification, source verification of livestock—The value of retinal images and GPS. *1st EPLFC* 1, 167–172 (Ed S. Cox). Germany: Wageningen Acedemic Publishers.

WHO (2017). Global and regional food consumption patterns and trends. 2017, http://www.who.int/nutrition /topics/3_foodconsumption/en/index4.html. World Health Organization.

Willis, S., Black, J. and Banhazi, T. (2016). Estimation of production losses associated with short term growth rate reduction and sub-optimal thermal conditions on pig farms using Auspig simulation software. *Asian Conference on Precision Livestock Farming (PLF-Asia 2016)* 1, 45–52 (Eds G. Zhang, L. Zhao, C. Wang, W. Zheng, Q. Tong, D. Berckmans and K. Wang). Beijing, China: China Agricultural University.

Willock, J., Daery, I. J., McGregor, M. M., Sutherland, A., Edwards-Jones, G., Morgan, O., Dent, B., Grieve, R., Gibson, G. and Austin, E. (1999). Farmers' attitudes objectives, behaviors, and personality traits: The Edinburgh Study of decision-making on farms. *Journal of Vocational Behavior* 54 (1), 5–36.

Wouters, P., Geers, R., Parduyuns, G., Goossens, K., Truyen, B., Goedseels, V. and Van der Stuyft, E. (1990). Image-analysis parameters as inputs for automatic environmental temperature control in piglet houses. *Computers and Electronics in Agriculture* 4, 233–46.

Yoon, S. C., Park, B., Lawrence, K. C., Windham, W. R. and Heitschmidt, G. W. (2011). Line-scan hyperspectral imaging system for real-time inspection of poultry carcasses with fecal material and ingesta. *Computers and Electronics in Agriculture* 79 (2), 159–68.

Zhang, C. and Kovacs, J. M. (2012). The application of small, unmanned aerial systems for precision agriculture: A review. *Precision Agriculture* 13 (6), 693–712.

Zhang, G., Strom, J. S., Blanke, M. and Braithwaite, I. (2006). Spectral signatures of surface materials in pig buildings. *Biosystems Engineering* 94 (4), 495–504.

Zhang, N., Wang, M. and Wang, N. (2002). Precision agriculture—A worldwide overview. *Computers and Electronics in Agriculture* 36 (2–3), 113–32.

Zhang, Y. and Barber, E. M. (1995). An evaluation of heating and ventilation control strategies for livestock buildings. *Journal of Agricultural Engineering Research* 60 (4), 217–25.

Section II

Water and Irrigation Engineering

9 Efficient Use of Energy in Irrigation

J. I. Córcoles, A. Martínez-Romero, R. Ballesteros,
J. M. Tarjuelo, and M. A. Moreno

CONTENTS

9.1   INTRODUCTION

In countries that suffer water scarcity caused by the increase on water demands due to different uses, irrigated agriculture plays an important role in providing food for an increasing population under different climate change scenarios. With this aim, pressurized irrigation systems, such as sprinkler and drip irrigation systems, are essential to optimize water management and to maximize water productivity. In addition, it is important to improve water and energy uses simultaneously, focusing on sustainable farming practices and considering social, environmental, and economic constraints.

Efficient water and energy use are becoming more important in agriculture due to the wide-spread trends of reduced water availability, increasing energy costs, and accounting for carbon dioxide emissions. In these cases, energy prices determine the viability of irrigated agriculture in many areas of the world, especially where groundwater is the main water source. Moreover, water scarcity typical of arid and semiarid regions, together with an increasing trend in production costs (seeds, fertilizers, etc.), creates uncertainty about irrigated land viability, which is related to rural development. Indeed, one of the most important characteristics of irrigated land is that it should be a sustainable activity, in order to guarantee its economic viability.

The main factors influencing water and energy use include climate, irrigation system efficiency, pump efficiency, and total dynamic head (water lift and system pressure). The risk and sensitivity analysis of water, energy, and emissions in irrigated agriculture is imperative, so that areas for potential improvements and the most efficient use of inputs can be identified to ensure resilience at the farm level, maintain agricultural production, and minimize environmental impacts. Jackson *et al.* (2011) quantifies the risk and uncertainties that may influence water and energy use at the farm level by assigning and modeling a range of parametric values related to climatic and technical factors. Integrated modeling of the parameters driving risk and uncertainty in water and energy use will allow critical parameters to be identified to ensure more efficient use of both water and energy at farm level. Sensitivity analysis can help to identify which of these parameters has the biggest impact on energy and water use.

The main objective of this chapter is to contribute to a better understanding of the balance between energy and water use in irrigation. It is necessary, therefore, to provide an overview of the key technologies included in the literature that can contribute directly to improve the use of both water and energy in irrigation. These technologies, which are applied mainly in areas with water scarcity, high water prices due to energy costs, and a low gross margin for farmers, can be classified as follows:

1. Tools and models aiming at water saving and selecting the proper crop pattern at the farm level, to optimize economic water productivity, and to minimize the environmental impact. This action line can be performed with the use of precision agriculture, information and communication technologies (ICT), or remote sensing at different resolutions for crop status determination together with decision support system (DSS) models and tools.
2. Tools and models for improving irrigation infrastructure design and management as a whole, based on water and energy savings, such as: optimal design, size, and management of pressurized irrigation systems on the plot scale with low pressure sprinklers and emitters; design, sizing, and optimal operation of collective irrigation networks; and optimization of design, sizing and regulation of pumping systems.
3. Actions to reduce energy consumption and/or cost, such as the use of benchmarking techniques, energy audits, models for optimizing the electricity tariffs, telemetry, and remote-control systems, and renewable energy.

In addition to these three specific action lines, transversal activities can be included, such as: to promote the usage and usefulness of irrigation advisory services (IAS) to transfer and share real-time information with farmers; to create a network of leaders among farmers and technicians who can act as examples for farmers; and to create web-based geographic information system (GIS) platforms, or to use existing ones for information and technology transfer to end-users in a feedback process.

In this chapter, a general review considering the most relevant references about these topics is carried out. In addition, several developed models and tools to improve energy efficiency in irrigated areas are also described.

The results obtained in the different studies presented here should be analyzed in-depth for specific countries' cases, considering the relationships between energy and investment costs, which can modify the obtained results.

9.2 STRUCTURE OF THE CHAPTER

The most complex case in designing and sizing irrigation infrastructure workflow is the modernization of irrigable areas through collective irrigation networks. Figure 9.1 shows the key elements and management requirements in this process.

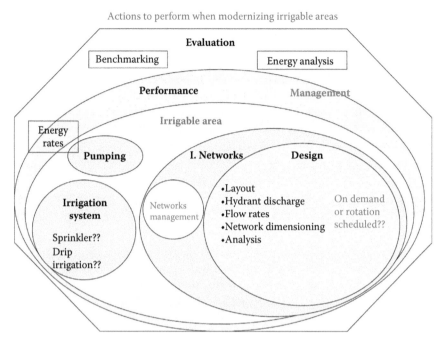

FIGURE 9.1 Modernization of irrigable areas workflow.

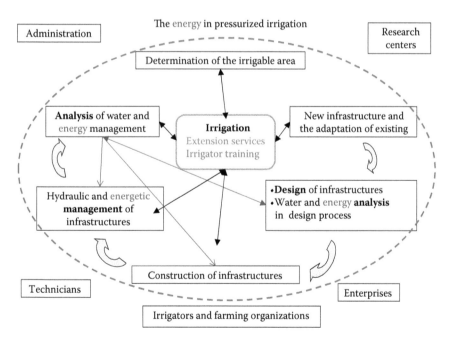

FIGURE 9.2 Key aspects to be considered to control the energy in irrigation.

Nowadays, in addition to warranty and proper performance, the energy scenario demands actions to reduce energy consumption and, therefore, the carbon footprint in the design, sizing, and management process of these infrastructures.

In this paper, all of these key elements will be analyzed and some solutions for reducing energy consumption will be presented and discussed in detail, using mainly the cases more directly related with our experience, given the great diversity of situations that can occur at the international level. Our objective is give an overview of the main aspects to be considered in the improvement of energy use in irrigation and not to make a revision paper over this subject. Thus, this chapter will describe the methodologies and tools to perform a proper design aiming at energy saving, perform energy analysis of already designed and existing infrastructures, and implement an optimal management regime aiming at minimum energy consumption. Each of these steps will be analyzed from the water source to the emitter of the irrigation system, from water abstraction systems, water storage infrastructure (reservoirs), pumping stations, piping systems for collective water distribution networks, and irrigation systems in plot. Figure 9.2 shows the key aspects to be considered to control the energy in irrigation.

9.3 ACCOUNTING FOR ENERGY SAVING IN THE IRRIGATION INFRASTRUCTURE DESIGN PROCESS

9.3.1 DESIGN OF WATER ABSTRACTION SYSTEMS

Pumping for groundwater abstraction and for water distribution are the main energy consumers in pressurized water networks in areas where the main water source is groundwater. In fact, several authors have developed different algorithms to minimize the energy and investment costs in pumping (Moradi-Jalal, Mariño, and Afshar, 2003; Moradi-Jalal, Rodin, and Mariño, 2004; Pulido-Calvo *et al.*, 2003; Planells *et al.*, 2005; Moreno *et al.*, 2007; Carrión *et al.*, 2016). For large irrigation areas that use groundwater, the most common option is to use of reservoirs to store water and pumping stations to provide pressure to the irrigation network (Moreno *et al.*, 2007; Izquiel *et al.*, 2015). Figure 9.3 illustrates different pumps for water abstraction (groundwater and surface water).

One of the main constraints found in this kind of infrastructure is the improper dimensioning of the pumping for wells that supply water to reservoirs. Moreno *et al.* (2010a) developed the DOS (Diseño Optimo de Sondeos) tool to optimize pumping for groundwater abstractions, which supply water to reservoirs by obtaining the minimum total cost (investment plus operation costs) by optimizing the pump characteristic and efficiency curves, together with the pumping pipe diameter. This methodology is based on the theoretical relationships between the characteristic and efficiency curves (Figure 9.4). It considers different variables, such as hydrologic and topographic variables (water table level, seasonal water table level variations, well flow rate, distance from well to discharge point, and elevation difference), hydraulic variables (head losses in pipes and demanded flow), and economic variables (energy rate and pump and pipe costs). In order to obtain general results, different combinations of these variables were studied, leading to a number of general conclusions.

The results showed that the steepness of the characteristic curve was mainly associated with the water table level variation throughout the year, and the pumping pipe diameter was mainly associated with the water demand (volume). Thus, when the water table level variation is high throughout the year, the steepness of the characteristic curve should also be high, to better fit the variable conditions.

(a)   (b)

FIGURE 9.3   Example of pumps for (a) groundwater abstraction and (b) surface water pumping.

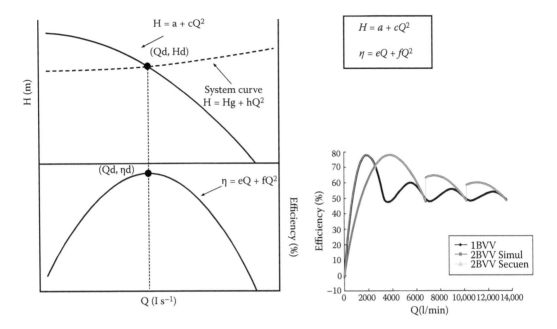

FIGURE 9.4 Example of characteristic and efficiency curves.

In addition, under this highly variable water table level condition, the operation point with the maximum efficiency should correspond to the month of highest demand. The flow rate and the optimal pumping pipe diameter remain practically constant for each considered demanded volume, and they are not affected by the variations of the water table level.

9.3.2 Design of Piping Systems

The process for designing the piping systems of collective irrigation networks is mainly concerned by the flow rates in each pipe and the differences in elevation. The differences in elevation are easily obtained with digital elevation models or global navigation satellite systems (GNSS) equipment. It is more difficult to determine the flow rate in each pipe because it depends on the individual farmer's irrigation behavior.

To determine the flow rate in each pipe, the following should be considered:

1. The discharge of each hydrant of the network, which depends on the crop pattern, area, and number of sectors to irrigate in the plot.
2. The layout of the water distribution network.
3. The crop growth stage in each plot and, therefore, the time of irrigation and the irrigation interval used.

The layout of the irrigation network is primarily concerned by the path's location and plot boundaries to avoid the purchase of right of way. However, many algorithms that optimally determine the layout in this type of network have been developed (Labye, 1988; Lamaddalena and Sagardoy, 2000; Planells, Ortega, and Tarjuelo, 2007).

Once the layout is determined, it is necessary to determine the distribution of the water volumes required through space and time by on-demand irrigation networks. With this aim, several statistical (Clément, 1960; Mavropoulos, 1997) and probabilistic (Lamaddalena and Sagardoy, 2000; Moreno *et al.*, 2010b) models have been developed.

In the process of sizing the pipes of the irrigation network, many different methods can be used (Lamaddalena and Sagardoy, 2000; Estrada *et al.*, 2009). The most utilized methods are Labye's iterative discontinuous method and the linear programming methods. From an energy point of view, it is necessary to determine the required pressure in the sizing process at the pumping station, if needed, to supply the required pressure to the hydrants. An optimum balance between the size of the pipes (investment cost) and the required pressure should be achieved for minimum total cost, considering the future trends of energy costs. GESTAR is an advanced computational hydraulic software tool specially adapted for the design, planning, and management of pressurized irrigation networks (Estrada *et al.*, 2009). Also, a holistic approach that considers the performance of the pumping station should be considered, as described in Section 9.3.3.

9.3.3 DESIGN OF WATER STORAGE SYSTEMS

Proper sizing and management of reservoirs has been the objective of numerous studies, which emphasized the need for this kind of research (Hirose, 1997; Pulido-Calvo *et al.*, 2006) (Figure 9.5). Izquiel *et al.* (2015) developed a tool to optimize the design and management of water abstraction and its application with center pivot systems seeking to minimize water application cost per unit area (C_T), including investment (C_a), energy (C_e), and maintenance costs (C_m); considering the abstraction and application process as a whole and differentiating the abstraction and application with the irrigation system costs. With this aim, two options were considered: to feed the center pivot directly from an aquifer or by using a regulation reservoir. DEPIRE, a software tool designed with a center pivot and regulating reservoir, was developed to determine the optimal flow rate, pipe diameters, pump power, and the volume of the regulation reservoir for any crop water requirement, different electricity rates, and water availability in the borehole. With this tool, the effect of the irrigated area (S), dynamic water level (DL) in the aquifer, and the pumping flow rate on the C_T was evaluated for a maize crop in the "Eastern Mancha" hydrogeological unit (HU 08.29) in the Júcar River Basin of the Castilla-La Mancha Region in Spain. The study area representing the minor C_T was 70 ha for direct pumping from the borehole and 100 ha when using an intermediate reservoir. Incorporating a regulation reservoir generates lower C_T than direct feed from the borehole for a S greater than 100 ha for any DL. C_T increased linearly with the DL due to a significant increase in

FIGURE 9.5 Examples of different types of reservoirs, water abstraction from a borehole, and booster stations.

C_e which primarily affects the cost of water abstraction from the aquifer, with a smaller effect on the application cost by the irrigation system. In this study, energy (C_{we}) is the most important component of water abstraction cost (C_w), reaching up to 70% of the total cost of water abstraction. For a DL greater than 30 m, C_{we} becomes more than five times the cost of investment (C_{wa}) for a DL equaling 200 m. However, the investment component (C_{Aa}) in the water application cost (C_A) is more important than energy (C_{Ae}), and this decreases with increasing size of the irrigated plot (S), ranging from C_{Aa} 600 to 350 € ha year when S increases from 30 to 150 ha, while C_{Ae} increases from 200 to 400 € ha year.

The optimal center pivot capacity was approximately $1.50 \text{ Ls}^{-1}\text{ha}^{-1}$, being mainly influenced by energy cost and the need for avoiding the use of the high energy rate period due to cost (the available time in off-peak period was six hours).

Izquiel et al. (2015) developed the additional model named DRODIN (Design of Reservoirs of Regulation in On-Demand Irrigation Networks) with similar approaches, considering the irrigation network demand throughout the irrigation season. The tool was applied to one on-demand irrigation network with 171 ha of drip irrigation for vineyard and olive crops. The results indicate that the C_e is the main component of C_T, with both the abstraction and the water supply to the irrigation network representing between 57% for DL equaling 0 m, to 78% for DL equaling 250 m. The operation of the pumping station, determined by the power, flow rate, and pumping time, determines the size of the reservoir and the annual costs of water supply to the network for a given water supply guarantee. The C_T increases linearly with DL, mainly due to increase in C_e, although there is a clear relationship between C_a, C_e and the reservoir size, which is only possible to analyze with tools as DRODIN.

Some advantages can be highlighted when one storage and regulation reservoir is utilized, such as the use of several wells to supply water to the reservoir (it is possible to pump water from the reservoir to several irrigation systems under different flow rates and pressure conditions) and, in case of a water well pump breakdown, irrigation water for one or several days can be stored. The use of a reservoir needs to be considered when the investments costs needed to build the reservoir are not compensated by the energy cost savings.

9.3.4 DESIGN OF PUMPING STATIONS

Different algorithms for minimizing the total cost of pumping stations (investment and operation costs) have been developed (Moradi-Jalal, Mariño, and Afshar, 2003; Moradi-Jalal, Rodin, and Mariño, 2004; Pulido-Calvo et al., 2003; Planells et al., 2005). They attempted to obtain the combination of variable and fixed-speed pumps that best minimized costs. However, none of these considered the calibrated pump characteristic curves. These studies considered the theoretical characteristic curves of the pumps, involving only the pump and motor efficiencies, not the efficiency of other components of the pumping station, such as wires or variable speed drivers (Figure 9.6). With these methodologies, energy saving was mainly obtained by means of the improvement of the sequence of activation of the pumps. In fact, Moradi-Jalal, Mariño, and Afshar (2003) stated that the major portion of the cost-reduction results from energy savings were through employing a better operation rule. This author obtained an energy saving of 32%, but it was necessary to use different types of pumps, involving maintenance and regulation problems. Pulido-Calvo, Roldán, López-Luque, and Gutierrez-Estrada (2003) obtained an energy saving of 41% by selecting the proper pumps and their regulation.

The performance of the irrigation networks can substantially vary from the initial conditions in which their design was based, therefore, the pumping station can cause a lack of energy efficiency when the pumping stations are working. Problems can be detected and solved by applying the developed model to each pumping station, and possible changes in the sequence of pump activation to improve energy efficiency can be proposed. Also, the use of frequency speed drives can mitigate these problems.

FIGURE 9.6   Example of frequency speed drives.

Monitoring the pumping stations and the irrigation networks is a key management strategy to optimize energy consumption. The instruments utilized to monitor pumping stations (electrical network analyzers, ultrasonic flowmeters, and pressure transducers) are easy to use. In addition, with the obtained energy savings, the investment cost for this kind of monitoring system is quickly recovered. The monitoring of pumping stations permits the detection of breakdowns in pumps, which improves the maintenance and quick detection of anomalies in the pumping station.

Moreno *et al.* (2007) developed a model for analyzing energy efficiency at pumping stations, which permitted the determination of the sequence of pump activation that minimized the energy cost for real demand scenarios. The model was calibrated for the pumping station of Tarazona de La Mancha (Spain) by measuring hydraulic and electrical parameters for each pump.

One of the main aspects to consider when carrying out the energy cost analysis at pumping stations is to estimate the discharge distribution throughout the irrigation season. The optimum sequence of activation of pumps was the option that best fitted the discharge distribution in terms of efficiency.

Moreno *et al.* (2007) illustrates that simple electrical and hydraulic measurements at pumping stations can help to improve their management, and, as a result, obtaining a high reduction in energy costs by carrying out a proper sequence of pump activation. In the case of this study, cost savings of 16% were obtained by changing the regulation of the pumping station from the current sequence of pump activation, in which two variable speed pumps worked simultaneously, to a sequence of pump activation that considered two variable speed pumps working sequentially. In only one season, the additional costs of applying this methodology (electrical network analyzers, flowmeters, and pressure transducers) would have been recovered with the energy efficiency improvement. Results could vary in other cases, depending on the discharge distribution and the working points of the pumps.

9.3.5   Design of Irrigation Systems in Plots

The main aspects that should be considered in the process of design and management of an irrigation system are maximizing uniformity, minimizing the drift and evaporation losses (Playán *et al.*, 2005), obtaining the minimum total cost of water application with the irrigation system, and determining the control and telemetry system requirements. Thus, aspects related to soil water redistribution and crop water use are outside the scope of this study. In order to optimize the design and management of in-plot irrigation systems (Figure 9.7), it is necessary to develop tools and models that act as decision support systems. Some of these tools are already available and will be briefly described in this section.

The optimum hydraulic design of a pressurized irrigation system is reached by determining the sizes of pump and distribution pipes that ensure a proper flow and intake pressure head in the sprinkler or the emitter, with minimum annual water application cost.

FIGURE 9.7 Examples of pressurizes irrigation systems.

Optimization of design and management is essential for improving the efficiency of water and energy use in pressurized irrigation systems, reducing CO_2 emissions contributing to the sustainable intensification of agriculture, and improving food production. Throughout the history of sprinkler and drip irrigation, it has always been of interest find the system characteristics that produce the most cost-effective results (Kumar *et al.*, 1992; Lamaddalena and Pereira, 2007b; Ortíz, De Juan, and Tarjuelo, 2010; Montero *et al.*, 2012; Carrión *et al.*, 2013; Carrión *et al.*, 2014). Kang, Yuan, and Nishiyama (1999) developed a method for hydraulic design of drip irrigation systems with minimum costs, using finite elements by analyzing the characteristics of water application uniformity as affected by lateral pipe lengths and diameters in both uniformly and non-uniformly sloping fields. Hassanli and Dandy (2000) applied genetic algorithms to develop hydraulic calculations and designs for rectangular drip irrigation units.

Daccache, Lamaddalena, and Fratino (2009) developed algorithms to analyze the relationship between irrigation uniformity and pressure at the hydrant level. To do so, this author developed models based on the COPAM (Combined Optimization and Performance Analysis Model) hydraulic model and applied these models to the Capitanata Water User Association (WUA; Italy). It describes how the strategies to reduce energy consumption could affect irrigation uniformity at the plot level, and therefore crop yield.

Moreno *et al.* (2012) developed a software for the optimal design of center pivot and moving lateral systems fed directly from wells, optimizing the characteristic and efficiency curves for the pump (Moreno *et al.*, 2009) as well as the types and diameters of pipes for pumping and distribution. This is preformed considering the energy rates and dynamic water table level conditions for each month of the irrigation season, for minimum total water application cost (investment plus operation costs). In addition, the tool calculates the conditions of system performance to avoid runoff. The method accounts for hydrological variables (water table level and its temporal variation), soil variables (infiltration parameters, surface storage capacity, and surface impermeability), hydraulic variables (head losses in pipes and flow demand), and economic variables (energy costs and pump and pipe costs) in the optimization process. Results show that the best options are timing irrigation to avoid periods of high energy costs, as well as increasing pumping power and pipe size with a greater system capacity (1.5 L s^{-1} ha^{-1}), and shorter operation time (18 h day). The minimum water application cost is obtained in all case studies in this paper for center pivot systems irrigating 75 ha, with lateral pipes of 254 mm (10 in.). These results should be analyzed in-depth in cases specific to each country, considering the energy and investment relationship, which could modify the results.

Carrión *et al.* (2013; 2014) developed a DSS tool, named PRESUD (Pressurized Subunit Design) (Figure 9.8), for optimal hydraulic design and sizing of solid set sprinkler and micro-irrigation systems with minimum total cost (operation plus investment) per unit of irrigated area, considering the proper type and size of pump together with the irrigation system. Use of these tools makes it possible to accurately determine the uniformity of the system, which is directly related to the crop yield and, therefore, the efficient use of water.

The investment cost of pressurized irrigation systems depends on the equipment and design, materials, and level of automation. This cost is also influenced by other factors such as shape, layout, and size of the plot or the distance from the water source to the plot (Van Der Gulik, 2003). The operation costs (mainly energy) can be reduced by selecting emitters and sprinklers that can operate with low pressure, and selecting the type of pump that best fits with the variations in flow rates and dynamic water lift (Moreno *et al.*, 2010b).

A diverse set of parameters influence the cost of abstraction and application of water by a pressurized irrigation system, and many partial studies of the issue have been developed, but the irrigation system should be analyzed as a whole, from the water source to the emitter, to avoid potentially significant errors. Carrión *et al.* (2016b) developed a decision support tool named DOPIR (Design of Pressurized Irrigation) that takes a holistic approach to optimize the process of water abstraction from one aquifer, analyzing pressurized irrigation systems as a whole, from the water source to the emitter, integrating the main factors involved in the process. The objective was to minimize the total cost of water abstraction and application with the irrigation system (C_T) investment (C_a) plus energy (C_e) plus maintenance cost (C_m) per unit of irrigated area according to the type of aquifer, crop water requirement, and electricity rate period during the irrigation season. The authors considered the type of aquifer through the main parameters that define it (type of borehole, transmissivity, etc.) and obtained the optimum drilling and pumping pipe diameters. The authors also determined the type of centrifugal pump to use based on the shape of their characteristic and efficiency curves, seeking maximum efficiency for the working interval of pressure and flow rate, considering the energy losses in cables normally not considered in these types of studies, and possible fluctuations in the dynamic water lift (DL) during the irrigation season. The optimum lateral

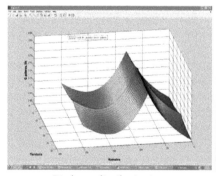

Discharge distribution in a
subunit with PRESUD

Optimal design of center
pivot (DOP)

FIGURE 9.8   Examples of tools for irrigation system design: PRESUD and DOP.

and manifold pipes were previously calculated using the PRESUD tool (Carrión *et al.*, 2013; Carrión *et al.*, 2014), which is linked to DOPIR, where any plot shapes, topography and borehole locations can be used as input for DOPIR. Results showed that:

1. Energy cost is the most important component of total water application cost (50% to 72%) in the case studies, for static water table (SWT) from 20 to 100 m, with a permanent sprinkler irrigation system in plot from 5 to 60 ha. Thus, it is very important to adapt the design and management of the irrigation and pumping system throughout the irrigation season to the energy rate periods. Moreover, the coefficients that define the suitability of water application in the subunit, such as emission uniformity (EU), total coefficient of variation of flow rate (CV_q) and emitter flow and pressure head variation in the subunit (Δq, Δh), are calculated.
2. The number of subunits by plot to minimize the C_T is obtained using most of the available hours of low and medium energy periods to irrigate, with a margin of security in the number of available hours for irrigation to recover any hours lost due to system failure, but not operating during high cost energy periods.
3. The criterion of limiting Δq to 10%, considering EU is equal to 90%, to calculate the irrigation application efficiency is widely used when designing a drip or sprinkler irrigation subunits, is not always the most convenient value from an economic point of view, and the PRESUD tool developed obtains better solutions.
4. The DOPIR tool developed for optimizing irrigation engineering in the specific conditions of investment and energy cost in each country or region of the world is useful for selecting the optimal type and size of the pump, pipe diameters, and irrigation subunit design under the specific conditions of an irrigated plot. It is also a valuable decision support system tool to transfer to the Irrigation Advisory Services (IAS) for helping farmers and technicians to design and size their pressurized irrigation systems.

9.4 ENERGY ANALYSIS OF IRRIGATION INFRASTRUCTURES

Several works have been published about how to implement performance analysis of irrigation systems. These studies present methodologies, models, and case studies that help to improve the water and energy efficiency of irrigation systems (Khadra and Lamaddalena, 2006; Lamaddalena and Pereira, 2007a; Lamadalena and Pereira, 2007b; Calejo *et al.*, 2008; Abadia *et al.*, 2008).

Many studies in recent years have focused on developing methodologies to quantify water and energy efficiency on land with improved irrigation systems. Lecina *et al.* (2010) analyzed modernization in the Alto Aragón irrigable area (Spain), replacing open-channel, gravity-based systems by pressurized distribution networks and switching from surface to pressurized irrigation systems. They highlighted some advantages and disadvantages of migrating from surface to sprinkler irrigation systems, quantified water savings, and some economic and social impacts of this transformation. Moreno *et al.* (2010c) analyzed the energy efficiency in more than 20 water user associations in Castilla-La Mancha and proposed measures to increase energy efficiency doing energy audits (Figure 9.9). Abadia *et al.* (2010) developed a comparative analysis of the energy efficiency in more than 30 collective irrigation networks in Murcia and Castilla-La Mancha, proposing some measures to improve energy efficiency and the main problems found in these infrastructures. Rodríguez Díaz *et al.* (2012b) analyzed some modernized areas in Andalucía (Spain) that migrated from surface irrigation systems to sprinkler and drip irrigation systems, highlighting some strengths and weaknesses of this type of modernization.

Countries such as Australia, Spain, Turkey, Malaysia, Sri Lanka, and others in Southern Asia are implementing a set of actions regarding the improvement of water and energy efficiency in irrigable areas using benchmarking techniques (Ghazalli, 2004; Molden, Sakthivadivel, and Habib, 2001;

FIGURE 9.9   Equipment to energy audits.

Cakmak *et al.*, 2004; Jayatillake, 2004; Rodríguez Díaz *et al.*, 2009; Córcoles *et al.*, 2012). These actions consist of analyzing the water and energy efficiency of Water User Associations (WUAs) by using energy indicators. They also propose measures to improve the use of energy to achieve an economic benefit. In order to calculate energy indicators, data obtained from field measurements, together with information supplied by the managers of the irrigated areas, were obtained and implemented in models. Thus, Moreno *et al.* (2007) developed the MAEEB model to analyze the performance of pumping stations, the AS model (Moreno, Córcoles, Moraleda, *et al.* 2010a) to analyze the performance of pumps and wells, the DOEB (Moreno *et al.*, 2009) model for optimal design of pumping stations, and the DOS (Moreno *et al.*, 2010a) model for optimal design of wells. The water-distribution networks can be analyzed with the EPANET (Rossman, 2000) or GESTAR (Estrada *et al.*, 2009) models.

These tools were validated by the energy analysis of 15 WUAs located in the Castilla-La Mancha Region during the 2007 irrigation season, and measures to improve the use of energy were proposed. These were monitored and evaluated in 7 of the 15 WUAs during the 2008 irrigation season.

The results of the energy analysis show that most of the analyzed WUAs operated with proper energy efficiency (higher than 50%) but they can be improved further. In the WUAs with high elevation differences, the energy efficiency is much lower, whereby the only way to solve this problem is the establishment of equal-elevation sectors for the management of the irrigation network.

The energy analysis and the use of these tools have made it possible to estimate the energy savings gained through application of the proposed measures. However, this is difficult to do accurately due to changes in water demand, in the management of the irrigation networks and pumping stations, and/or in the cropping patterns. In most of the study cases, an improvement of the energy efficiency after the implementation of the proposed measures was detected, with average energy savings of 10.2%. The proposed improvement measures that result in higher energy savings are those related to equipment improvements, mostly concerning the improvement of well pumps. From an economic point of view, an important proportion of the economic savings is gained through improvements in the electrical contract.

9.5   ENERGY MANAGEMENT IN COLLECTIVE IRRIGATION NETWORKS

Several methodologies have been developed to improve energy consumption in collective irrigation networks. In some cases, these networks are not designed properly. So, several tools to improve their design and management can be applied.

FIGURE 9.10 Examples of pumping stations for collective irrigation networks.

In irrigation networks, the energy is consumed by the pumping system for water distribution (Figure 9.10). Therefore, it is necessary to analyze the performance of pumping stations to determine the best regulation and management strategy and to enhance its energy efficiency.

The performance of pumping stations depends mostly on the real performance conditions, and not only on the conditions considered in the design process. Thus, it is necessary to develop pumping station analysis models that simulate the performance of the pumping stations and that help to optimize its regulation and management strategy. Moreno *et al.* (2007) developed a model for the energy analysis of pumping stations (MAEEB), with the aim of improving the energy efficiency of pumping stations. The MAEEB model is based on the fact that, in most cases, the design of pumping stations that feed irrigation water networks depends only on two parameters, the head pressure and the design flow (high discharges), while the rest of the discharges are not regarded.

The proper regulation of pumping systems is a key step in fitting energy consumption to the current energy demand. Lamaddalena and Khila (2011) demonstrated that from 27% to 35% of energy savings can be achieved using an appropriate average speed regulation in two Italian on-demand irrigation districts. However, most of these analyses do not consider the effect of the efficiency of the frequency speed drive on the final result and they assumed a high constant efficiency of the pumps. Therefore, it is possible to obtain a better approximation about pumping station efficiency, because the efficiency of the pumping systems can be poor for low discharges.

9.5.1 OPTIMAL MANAGEMENT OF PUMPING SYSTEMS

Management options for irrigation networks include rotational schedules or on-demand irrigation. The lack of flexibility, reliability, and predictability are some of the primary concerns highlighted in the scheduled rotation networks (Al-Abed, Shudifat, and Amayreh, n.d.). In most cases, the irrigation schedules are not followed, and the farmers tend to irrigate as much as possible. In fact, irrigation networks under a rotation schedule are more susceptible to being managed inefficiently than on-demand networks because of the lack of availability of tools for managers when selecting the configurations of open hydrants and choosing the proper pumping head for each of these configurations (Moreno *et al.*, 2010a, 2010b). An on-demand schedule provides users with flexibility in the frequency and duration of delivery (Burt and Plusquellec, 1990; Khadra and Lamaddalena, 2006) and delivers the exact quantity of water at the correct time, because the farmers decide when to irrigate based on crop water demands. In spite of the on-demand schedule, the energy efficiency may be low in some on-demand irrigation networks (Rodríguez Díaz *et al.*, 2009) because the upstream pressure head of the on-farm network can be subjected to high and continuous fluctuations, depending on the number of hydrants being simultaneously opened

(Daccache, Lamaddalena, and Fratino, 2009). Energy is a relevant factor to take into account, representing a high participation factor in the total management, operation, and maintenance costs. Moreover, the energy expenses are very high in comparison with the other types of irrigation management.

Several methodologies have been developed to improve energy consumption in irrigation networks, most of them based on the dimensioning and design of irrigation networks, such as sectoring.

Rodríguez Díaz et al. (2009) developed OPTIEN, an algorithm that simulates the potential energy savings obtained by sectoring the on-demand irrigation network, applied in Fuente Palmera (Spain). They showed that it would be possible to save more than 25% of the energy consumption in the maximum demand period, if the current water demand levels were operated in 12 ha rather than 24 ha. A further advance was achieved with WEBSO (Water and Energy Based Sectoring Operation; Carrillo Cobo et al., 2010; Navarro Navajas et al., 2012), an algorithm that includes a procedure for sectoring in pressurized irrigation networks. It was based on a topological characterization using a dimensionless coordinates system. This tool provides a monthly sectoring calendar for branched networks shown to achieve potential energy savings of up to 27% in Fuente Palmera. Fernández García et al. (2013) developed an optimization methodology aimed at minimizing energy consumption based on operational sectoring for one irrigation network with several source nodes. It was applied in the irrigation district of Palos de la Frontera (Spain) and tested in three pumping stations, showing that potential annual energy savings between 20% and 29% can be achieved.

None of the above-mentioned sectoring strategies considered the behavior of the pumping station when supplying a pressure head that was different from the one for which it was dimensioned, and a high constant efficiency was assumed. This fact might greatly affect the results of energy saving when defining new demand patterns (Moreno, Córcoles, Tarjuelo, et al. 2010a; Jiménez-Bello et al., 2015).

Other studies based on irrigation network sectoring have considered the performance of the pumping station. Jiménez-Bello et al. (2010) developed a methodology based on genetic algorithms, combined with hydraulic network modeling for irrigation network operating turns, achieving energy savings of nearly 36%. Fernández García, Moreno, and Rodríguez Díaz (2014) attempted to optimize the sectoring operation and pressure head, analyzing the performance of up to three of the frequency speed drives. These authors computed the energy savings obtained by means of sectoring, considering the effect of the pumping station efficiency (using the installation of frequency speed drives), depending on the generated flow rates. In that study, which was characterized by a steep topography with hydrant elevations ranging from 58 to 103 m, the authors obtained energy savings of up to 26%, while still guaranteeing a service pressure of 30 m at the hydrant level.

Irrigation network sectoring management can be analyzed using methodologies based on the location of hydrants with high energy requirements, which are defined as critical control points. The use of this tool to improve energy consumption has not been previously analyzed. Rodríguez Díaz, Montesinos, and Camacho Poyato (2012a) developed the WECP algorithm, useful in detecting critical points and designing improvement actions to minimize their impacts on energy demand. Application of the WECP to the El Villar irrigation district (Southern Spain) demonstrated energy savings of up to 30%. Other researchers have used critical points in the Palos de la Frontera irrigation district, showing that the control of critical points was more effective than sectoring, obtaining an additional annual energy saving of 10%. One study carried out focused on minimizing the energy costs of a pumping station in on-demand irrigation networks by means of an organization of the starting irrigation time for each hydrant (Córcoles et al., 2015). This study was applied in an on-demand irrigation network located in Tarazona de La Mancha (Spain). In this study, a tool developed in the MATLAB® environment with the EPANET® toolkit could minimize the energy cost of the pumping stations in on-demand irrigation networks, with the organization of the starting irrigation times for each hydrant. To achieve this objective, one of the novelties of this tool is that it is based on an analysis model that simulates the performance of the pumping station with consideration of the effect of the efficiency of the frequency speed drive. With this tool, using management of irrigation networks based on a variable pressure head at the pumping station, it was possible to reduce the pressure head at the pumping station and obtain values between 15% and 20% lower than with the current irrigation management (fixed pressure regulation).

The proposed methodology was used for a normal and a festivity day, where the total time is a low energy rate period, producing a higher concentration of water demand by the farmer. In comparison with fixed pressure regulation, and considering 18 hours as a daily operating time, energy savings close to 7% and 8% might be obtained for normal and feast days, respectively. In addition, the energy cost savings can be also relevant in comparison with the current management, with savings that ranged from 20% for a normal day and that were close to 11% for a feast day. In all the scenarios analyzed, the type of regulation that used two variable speed pumps activated sequentially, the rest of the pumps were fixed and had energy efficiencies that were slightly higher than the regulation. With regard to this outcome, the energy savings reached in this case ranged from 12% for a normal day and were close to 16% for a feast day. Additionally, a reduction in the energy costs might be achieved with this type of regulation, with energy cost savings that reached 17% (feast day) and 25% (normal day).

Other methodologies which can be useful for pumping stations' control regulation are focused on to estimate the pressure at all the nodes of an irrigation network using a minimum number of strategic nodes in the network (Córcoles, Tarjuelo, and Moreno, 2016b). Once the pressure head at all nodes in the network is estimated, the pressure head at the pumping station can be adjusted to supply the strictly necessary pressure to ensure the minimum pressure at the most restrictive of the open nodes of the network. This approach was applied in two on-demand, branched irrigation networks located in the provinces of Albacete and Cuenca, in the Castilla-La Mancha region. In both of the irrigation districts, the use of at least three control nodes showed high accuracy in estimating the pressure of all the hydrants. The use of three strategic nodes was identified as the most adequate solution because it simplifies the algorithm that regulates the pumping station and also reduces the cost of installing pressure transducers. According to the proposed methodology, the use of the minimum pumping head does not guarantee the lowest energy consumption rate, because it is necessary to consider the energy efficiency at the pumping station for each flow rate. Using this methodology, energy savings of approximate 3% to 5% were obtained relative to the average energy consumed with fixed pressure regulation.

9.5.2 OPTIMAL MANAGEMENT OF THE IRRIGATION NETWORKS: ON-DEMAND IRRIGATION OR FIXED ROTATION SCHEDULING

Another point to consider is the management of the irrigation networks, related to the use of on-demand irrigation networks or management under fixed rotation scheduling. Collective irrigation networks are often sized to work on-demand. However, the high investment cost of this kind of infrastructure often leads to the selection of scheduled rotational management, which reduces the investment cost.

Moreno *et al.* (2010c) analyzed the energy efficiency of both kinds of irrigation networks in combination with the use of performance indicators. This study was carried out in Castilla–La Mancha, analyzing four irrigation networks, paired ones with similar characteristics, with one of each pair working on-demand and the other operating under rotation scheduling. The four WUAs had similar infrastructures: water was pumped to a reservoir, where it was stored and then pumped to the irrigation network by means of a pumping station. The pumping stations of all the WUAs analyzed were composed of parallel, submergible pumps, regulated to a constant pumping head. Two of the irrigation networks irrigated vineyard crops using drip irrigation systems, one of them operated with on-demand, and the second one under rotation scheduling. The other two had common crop patterns (maize, barley, alfalfa, vegetables, etc.) and both were irrigated using sprinkler irrigation systems (solid set systems), one of them irrigated with on-demand irrigation and the other one with rotational schedule.

To analyze and compare the performance of these networks, several indicators were used to evaluate energy consumption, energy efficiency, and energy costs during two irrigation seasons. To calculate the indicators, it was necessary to obtain data from the managers about their networks and information from the WUA database, together with measurements of electric, hydraulic, and topographic data.

The results of this work showed that the irrigation networks working under rotational management were more susceptible to inefficient energy use than on-demand irrigation networks. However, if the pumping stations were in rotation scheduled irrigation networks that were properly managed, greater energy efficiency could be obtained than in on-demand irrigation networks. Rotation schedule irrigation networks are more susceptible to inefficient management because of the lack of available management tools to select the configurations of open hydrants and a proper pumping head for each of these configurations.

In the case studies examined, using the tools developed by the authors, improvements in energy efficiency achieved were between 3.5% and 24.9%, with higher potentials for improving energy efficiency occurring with irrigation networks that managed to operate under a rotation schedule.

9.5.3   OPTIMAL MANAGEMENT OF IRRIGATION SYSTEMS IN PLOTS

The evaluation of irrigation can be carried out using different performance evaluation tools, which can be useful to achieve efficient water and energy use to guarantee sustainable development in irrigable areas (Figure 9.11). One of these tools is called benchmarking technique (Malano, Burton, and Makin, 2004; Rodríguez-Díaz *et al.*, 2008; Córcoles *et al.*, 2012). This tool is based on the use of performance and energy indicators, which try to summarize the information for each of them (Kloezen, Kloezen, and Garces-Restrepo, 1998).

This methodology has been applied to evaluate irrigation systems performance. Thus, Córcoles *et al.* (2012) applied it during three irrigation seasons (2006–2008) in six irrigation districts belonging to the provinces of Albacete and Cuenca in Castilla-La Mancha Region. Four of them had sprinkler irrigation systems while the remaining had drip irrigation systems. These irrigation districts had pressurized irrigation networks that used groundwater resources. Using this technique, two of the most extended irrigation systems (sprinkler and drip irrigation systems) were compared by using performance indicators related to the management of the irrigated land. In this case, energy consumed by water abstraction from the borehole was not considered.

FIGURE 9.11   Examples of pressurized irrigation systems in plot.

The performance indicators proposed in this study were classified in four groups (system operation, financial, production efficiency, and environmental performance) (Malano and Burton, 2001). In addition, a new group was added based on energy indicators, and related to theoretical absorbed power, consumed active and reactive energy, and efficiency of the pumping system, among others.

Some indicators showed the differences between sprinkler and drip irrigation systems. Thus, the main system's water delivery efficiency, as the ratio between the total volume of irrigation delivery and the total volume of irrigation supply, reached higher values in sprinkler irrigation systems, with an average value close to 93%, which was slightly lower for a drip irrigation system (80%). In irrigation districts with drip irrigation systems, these lower main system water delivery efficiency values were due to common factors, such as water distribution losses, stored water evaporation, among others; but also to the cleaning process of the filtering equipment, which consumed water and energy. Although the volume of irrigation water supply is lower for drip irrigation systems than for sprinkler irrigation systems, it is important to control the main system water delivery efficiency by reducing irrigation water losses when irrigation pump filters are cleaned.

The irrigation water applied is slightly higher than crop requirements, mainly for drip irrigation systems which are used for tree crops, such as vineyards, olives, and almond trees. Some of the analyzed plots were managed under deficient irrigation regimes, but, in most cases, this management was not considered by farmers. Therefore, it is necessary to use new tools, such as the Irrigation Advisory Service of Castilla-La Mancha (http://crea.uclm.es/siar), to help farmers to improve crop management, mainly under regulated deficit irrigation management.

The values of the total management, operation, and maintenance costs per unit of irrigation delivery were for sprinkler irrigation systems 0.05 € m^3 lower than for drip irrigation systems (0.13 € m^3). It was related to the crop distribution for both irrigation systems, since there were crops with higher water requirements (usually higher than 6000 m^3 ha^{-1}) for sprinkler irrigation system districts than drip irrigation system districts (usually lower than 2000 m^3 ha^{-1}). Indeed, the total annual volume of irrigation water supply per unit of irrigated area (Vs) reached lower values for drip irrigation systems representing 20% of the volume required for sprinkler irrigation systems.

The average values of the energy cost per unit of irrigation delivery showed lower costs for sprinkler (0.022 € m^3) than drip irrigation systems (0.026 € m^3). These costs did not consider the costs of water abstraction from boreholes. If these costs were considered, the energy costs for sprinkler and drip irrigation systems will increase, reaching values close to 0.061 € m^3 and 0.071 € m^3, respectively. According to this, the high influence of water lift in irrigable areas which use groundwater resources can be noted.

The similarities between both irrigation systems might be explained, as the pumping head at the pumping station is very similar to fixed pressure regulation for both irrigation systems (58 m for sprinkler and 53 m for drip irrigation systems). In addition, the management and energy efficiency for each pump of the pumping stations were very similar for the analyzed areas.

The importance of energy cost is high for both irrigation systems, since it represents, for sprinkler irrigation system, 45% of the total management, operation, and maintenance costs, and 20% for drip irrigation system.

Indicators related to installed power of pumping stations were also really important. Considering the installed power per unit of irrigated area, there were not high differences between sprinkler and drip irrigation systems. Values ranged between 1.70 kW ha and 1.74 kW ha, respectively, without including the necessary power for water abstraction from boreholes.

The similarities between sprinkler and drip irrigation systems were also shown by analyzing the consumed active energy per unit of irrigation delivery. This energy indicator did not show high differences between both irrigation systems, with values between 0.31 kW m^3 (sprinkler irrigation systems) and 0.30 kW m^3 (drip irrigation systems). The similarity between these values is related to the filtering equipment required in drip irrigation systems, and the fact that in some of the drip irrigation networks analyzed the pumping system was oversized. The oversize was due to future enlargements of the irrigated area as a consequence of the increasing number of farmers interested in irrigating their plots. Moreover, the irrigation management of the analyzed drip irrigation networks was based on the demanded irrigation water, which explained the high-pressure head at the pumping station, increasing the energy costs in the analyzed areas.

In another study, Córcoles *et al.* (2016a) carried out an analysis of the performance of one irrigation district in Brazil, based on the use of performance indicators and energy audits. The study was carried out in the Baixo Acaraú irrigation district (DIBAU), located in the northeast of Brazil in the state of Ceara. The irrigation district has 8335 ha, with the Acaraú River the main source of water. Water is pumped from the river to a main canal, by a pumping station comprised of five pumps. Water is distributed along 40 km of canals and 110 km of pipes by gravity assisted by five lifting pumps. Water delivery is on-demand from small tanks at the entrance of the farm lots, where is pumped into the farm irrigation system. The irrigation system is pressurized, with drip and micro sprinkler irrigation systems the most commonly used. Regarding crop distribution, coconuts (38%), bananas (16%), orange trees (15%), and guavas (6%) are the most common crops.

The main system water delivery efficiency indicator reached values from 76% to 82%, which can be considered good enough for this irrigation district. This indicator is related to water distribution system losses and stored water evaporation, among others. The value of the consumed active energy per unit of irrigation water supply was very similar for all irrigation seasons (0.166 kW m^3 as average value). Regarding the installed power per unit of irrigated area, the results obtained showed a value close to 1.1 kW ha^{-1}, while installed power per unit of irrigation water supply was $7.31 \cdot 10^{-5}$ kW m^3.

According to the pumps measured, some problems were detected, such as malfunctioning of pumps. Only in the main pumping station, efficiency values close to 74% were obtained. Six of the pumps analyzed in the perimeter showed pumping efficiencies lower than 50%. In some cases, lower efficiencies were related to the presence of rubbish in the main channel, involving obstructions in pipe intakes. The energy costs were an important component of the total operation and maintenance costs representing 21%.

In general, switching from surface to pressurized irrigation systems results in a decrease in water use, mainly via reduced losses to percolation, but an increase in water (evapotranspiration) and energy consumption, together with maintenance and management costs. Thus, Jackson, Khan, and Hafeez (2010) reported that converting from surface irrigation to pressurized systems led to reductions in water application by 10% to 66% in a surface water supplied region in New South Wales, and a groundwater-dependent region in South Australia. However, in the surface water supplied region, energy consumption also increased by up to 163%. In the groundwater-dependent region, energy consumption was reduced by 12% to 44%, due to increased water use efficiency.

Jackson *et al.* (2011) examines water application and energy consumption relationships for different irrigation systems, and the ways in which the uncertainty of different parameters impacts on these relationships and associated emissions for actual farms in Australia. The risk and uncertainty analysis quantified the range of water and energy use that might be expected for a given irrigation method for each farm. Sensitivity analysis revealed the contribution of climatic (evapotranspiration and rainfall) and technical factors (irrigation system efficiency, pump efficiency, suction, and discharge head) impacted the uncertainty and the model output and water-energy system performance in general. The results indicate that where surface water is used, well-designed and managed flood irrigation systems will minimize the operating energy and carbon equivalent emissions. Where groundwater is the dominant use, the optimum system is a well-designed and managed pressurized system operating at the lowest discharge pressure possible that will still allow for efficient irrigation.

9.5.4 QUALITY OF ENERGY CONSUMED IN PUMPING STATIONS FOR IRRIGATION

In pressurized irrigation networks, especially those where the main water source is groundwater, energy cost is the main running cost. The use of variable speed drives in pumping stations involves a proper water flow and pressure that fits with the system demand curve.

Although preventive maintenance is necessary to reduce the possibility of breakdown in the electric facility, there are other causes of breakdown that cannot be controlled because of the high cost of these preventive measures. Among these causes, the most important is that generated by an inadequate power quality. This lack of quality is one of the main factors in breakdowns in pumping stations, mainly affecting the electronic converters that control pumps (Ferraci, 2004). In fact, low power quality generates several problems, such as load breakdown, decreases in efficiency of the pumping stations, and effects on the close electrical facilities. A perfect power supply should be always available considering voltage and frequency tolerances. Also, it should describe a sinusoidal wave shape. Low quality in the power supply can be due to low-quality energy supply from the electrical company or by the pumping station's electronic converters. Low-quality energy supply could be solved by electronic power management techniques. However, they are expensive techniques, therefore it is unfeasible to introduce these for improving power quality in pumping stations. Low-quality electronic converters could be improved by installing filters that reduce the harmonic disturbance introduced into the networks. This is one of the cheapest solutions. For example, in Spain, the installation of these kinds of filters is mandated by law. Components that generate harmonics are variable speed drives installed on some or all of the pumps. Although law mandates filter installation, some manufacturers include the filter as an extra that increases the price of the converter. Therefore, some pumping stations do not have them.

Many researchers (San Roman and Ubeda, 1998; Von Jouanne and Banerjee, 2001; Chang *et al.*, 2004; Djokic *et al.*, 2005) have analyzed the power supply quality, although it is not very common regarding to irrigated lands. Most reports focus on studying power supply quality for industrial applications, analyzing the presence of different quality indicators (interruptions, harmonics, and over-voltages, among others), and the effect on industrial devices.

Regarding this power supply quality, Córcoles *et al.* (2011) analyzed it in six pumping stations belonging to six WUAs in Castilla-La Mancha during three irrigation seasons (2006–2008). This study was focused on power quality supplied by the electric company and the disturbance generated by the electronic converters of the system.

In this study several electric quality parameters were analyzed divided in two groups (Pérez, Bravo, and Llorente, 2000): amplitude disruptions (interruptions, dips, swells) and wave shape disruptions (harmonics). These parameters were selected since they had a high influence on the performance of devices from WUAs. Moreover, these indicators were the most typical in determining power quality.

In all the pumping stations analyzed, several problems with power supply quality were detected. Some of them were caused by the electric company, such as interruptions, voltage dips, and swells. Other disruptions, such as harmonics, were due to loads included into the pumping stations. Despite the kind of disruption, none were higher than the limits established by European Standard (European Standard, 1999).

For wave disruptions, findings showed that variable frequency drives without proper filters generated a large proportion of the harmonics in the electrical network. This situation was reduced when fixed pumps were functioning because, in the overall current and voltage measurements, the fundamental wave had a higher impact than harmonics.

In the six WUAs analyzed and in most of the pumping stations in Castilla-La Mancha, only variable speed pumps were working for long periods of time. Therefore, it was very important to install proper filters which improve harmonic generation. In all WUAs analyzed, it was essential to carry out a proper equipment design, including protection and compensation systems which improve the working of the loads, reduce energy and maintenance costs, and extend the useful life of electrical facilities. Although in all WUAs the limits of power supply quality parameters were lower than European Standard EN 50160 established, they should be periodically evaluated in the near future to determine if power supply quality is declining.

9.6   RENEWABLE ENERGY IN IRRIGATION

The use of renewable energy resources in water distribution systems is becoming a new alternative for urban supply systems. For example, turbines are being installed which use the excess of energy where there are large differences in elevation. Hybrid systems that establish the optimal combination of several energy sources, such as solar, wind, and hydropower are being included (Ramos, Kenov, and Vieira, 2011). These measures contribute to decreased energy costs and perform a sustainable management of water distribution systems simultaneously.

In the agricultural sector, the implementation of renewable energy resources is increasingly common, such as the use of solar energy in the control of greenhouses or especially in pumping systems for irrigation water supply. It is necessary to intensify the analysis of the application of renewable energy, mainly wind and photovoltaic energy (PV), for applications in medium and large farms. A comprehensive analysis should be performed on the applicability of renewable energy in each country, which is highly influenced by the energy sector and distribution networks together with the seasonality of the irrigation energy demand (Figure 9.12).

The annual irrigation needs are concentrated in a few months. This very seasonal demand behavior is an obstacle to the incorporation of renewable generation. Power supply continuity and stability are needed, in contrast to the variability of solar and wind resources. These challenges make it necessary to study the economic feasibility of renewable energy integration by searching for systems specially adjusted to this kind of application.

Carroquino, Dufo-López, and Bernal (2015) addressed the economic feasibility of incorporating renewable energy systems into pumping for drip irrigation facilities in the Mediterranean area. They reported that the watering season matched the months of maximum solar radiation. In contrast, the seasonal behavior of the wind resources was the opposite, with minimums during the watering season. Therefore, wind turbines were not present in the best solutions. The optimal economic solutions to incorporate renewable energy were photovoltaic-diesel hybrid systems. In hybrid solutions, the Genset had only a few hours of operation per year, and the fraction of renewable energy was very high. The low investment cost of a diesel Genset made it the best solution to avoid oversizing the renewable generation and the associated budget overruns. Additionally, irrigation systems with proportionally smaller pumps and more pumping hours happened to be more cost-effective.

PV systems have been successfully applied to irrigate different types of crops in different climates and cropping conditions. The most widespread type of stand-alone photovoltaic irrigation systems is one that is designed to pump water from wells to an elevated storage reservoir. Water is then distributed to the plants by gravity or with a booster station. Another problem that must be faced is that irrigation application times are relatively short, while the production of energy is distributed during sunlight hours, so there is a gap between offer and demand. One possibility to cope with this gap could be to establish a net metering billing policy. Net metering is a billing mechanism that credits solar energy system owners for the electricity they do not use and add to the grid. The system owners can consume the energy they have in their accounts whenever they need within the applicable billing period. Although some countries have issued regulations regarding net metering billing policies, in other countries, like Spain, these regulations have not been implemented yet. In these cases, a stand-alone direct pumping photovoltaic irrigation system is necessary, using several irrigation sectors by plot. The system must include variable speed pumps in order to adjust the power consumed by the irrigation system to the power produced by the PV array, obtaining variable discharge. The use of reservoirs can help to adjust the variable discharge to a direct injection to the irrigation system or to the reservoir.

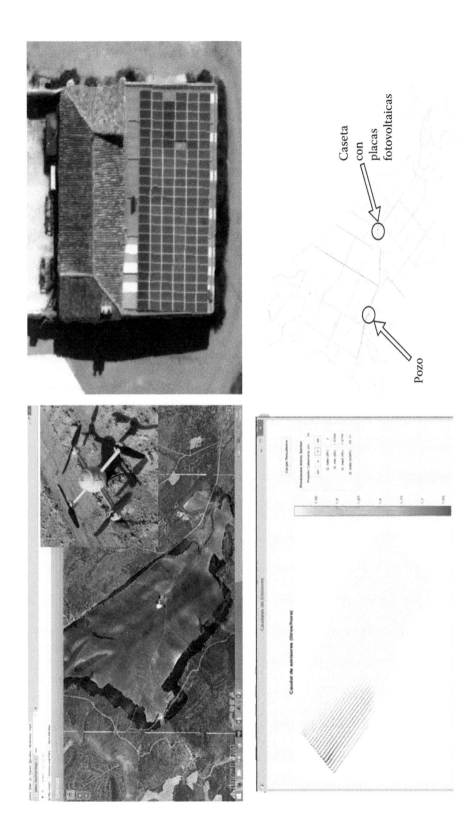

FIGURE 9.12 Examples of use of photovoltaic energy in drip irrigation.

Plot irrigation management under variable solar radiation conditions, crop water requirements, and irrigation system performance due to pressure variations, can contribute to an inefficient use of water with high variability of the irrigation system uniformity. Models to optimize the integration of photovoltaic power generation capacity and the pumping management are necessary.

López-Luque, Reca, and Martínez (2015) developed a model to assess the performance and reliability of standalone Photovoltaics Opportunity Irrigation (PVOI) systems for olive orchards under Mediterranean climate conditions, in order to obtain the economically optimal design of this system. The model includes different sub-models: the photovoltaic power generation capacity sub-model, the direct pumping management sub-model, and the sub-model that evaluates the economic and productive response of the crop to the application of water. The model simulates for increasing values of peak power of the photovoltaic system, the water and energy balance in the photovoltaic irrigation system, and the crop yield. It also calculates the investments and operational costs of the system and, finally, the net economic returns to the farmer. The operational strategy in the PVOI approach is to irrigate as long as enough power is supplied by the PV array. The results show that the total amount of water applied to the crop depends on the energy provided by the PV array, rather than on a strict fulfillment of the preestablished crop water irrigation requirements.

Carrillo-Cobo *et al.* (2014) analyzed the role of solar energy in the Bembézar left side irrigation district (Southern Spain) which has an irrigated area of 4000 ha. They combined sectoring as an energy saving strategy and a 2.15 MW photovoltaic (PV) system to supply energy to the sectors with higher energy consumption. Results showed that both measures together would reduce the energy costs by 71.7% and the greenhouse gas (GHG) emissions in 70.5%. The total investment was 2.8 M€, but with a payback period of only eight years. However, these results should be analyzed in-depth in the specific case of each country, considering the options of the proportion of energy consumed by the producer and injecting the surplus energy to the electrical grid, which can modify the results. Due to the increased costs of energy resources along with the reduction in unit costs of renewable energy resources, alternative energy sources are expected to play an important role in the future, which is another area where further research is needed.

9.7   CONCLUSION

Irrigation modernization, which is primarily focused on transforming open-channel gravity-based systems to sprinkler and drip irrigation, is essential in those countries with water shortages, but high production potential. Although the modernization process involves higher energy demand and an increase in investment, many advantages can be highlighted, such as increasing water efficiency and productivity, improving irrigation system operations and management, and improving farmers' working conditions. Calculating water conservation and energy consumption demonstrates the complexity of the energy and water efficiency balance. Hence, it is necessary to analyze the economic, social, and environmental viability of the irrigation modernization process for each case.

Optimization of irrigation systems design and management is essential for improving the water and energy efficiency of pressurized irrigation systems, reducing CO_2 emissions, contributing to the sustainable intensification of agriculture, and ensuring food production. Several decision support system tools have been developed for the design and management of irrigation infrastructure, which leads to further improvement in water and energy use, as well as other production inputs.

When groundwater is used, energy costs can constitute up to 70% of the total cost of water abstraction and application with a pressurized irrigation system (sprinkler or drip irrigation). Thus, optimizing pump efficiency and selection and adapting irrigation system design and management to different energy rate periods throughout the irrigation season is very important. Proper design and management of irrigation systems, dissemination of irrigation advisory services, web-GIS platform application, and utilities to transfer and share real-time information with farmers in a feedback process are some of the best tools for improving water consumption, energy, and the rest of production inputs.

More detailed studies are required to consider energy efficiency behavior along with useful pumping systems, while robust and efficient optimization algorithms that consider the optimization of all infrastructures as a whole are necessary for drawing a complete picture to understand farm productivity and profits related to water and energy efficiency.

Due to the increased costs of energy resources along with the reduction in unit costs of renewable energy resources, alternative energy sources are expected to play an important role in the future, which is another area where further research is needed.

NOTATION

The following symbols are used in this paper:

C_a	investment per unit of irrigated area ($€\,L^{-2}T^{-1}$)
C_A	water application cost ($€\,L^{-2}T^{-1}$)
C_{Aa}	investment cost in water application cost ($€\,L^{-2}T^{-1}$)
C_{Ae}	energy cost in water application cost ($€\,L^{-2}T^{-1}$)
C_e	operation cost per unit area ($€\,L^{-2}T^{-1}$)
C_m	maintenance cost per unit area ($€\,L^{-2}T^{-1}$)
C_T	water application cost per unit area ($€\,L^{-2}T^{-1}$)
CV_q	total coefficient of variation of flow rate (decimal)
C_w	water abstraction cost ($€\,L^{-2}T^{-1}$)
C_{wa}	investment water cost in water abstraction cost ($€\,L^{-2}T^{-1}$)
C_{we}	energy cost in water abstraction cost ($€\,L^{-2}T^{-1}$)
COPAM	Combined Optimization and Performance Analysis Model
DL	dynamic water level (m)
DOPIR	Design of Pressurized Irrigation
DOS	Diseño Optimo de Sondeos
DOT	daily operating time (T^{-1})
DRODIN	Design of Reservoirs of Regulation in On-Demand Irrigation Networks
DSS	decision support system
EU	emission uniformity (decimal)
GHG	greenhouse gases
GIS	geographic information system
GNSS	global navigation satellite systems
ICT	information and communication technologies
IAS	irrigation advisory services
L_m	water demand by the crops (L^3)
MAEEB	Model for Energy Analysis of Efficiency at Pumping Stations
N_{abs}	absorbed power (in kilowatts)
PRESUD	Pressurized Subunit Design
Q	design flow rate ($L^3\,T^{-1}$)
PV	Photovoltaics
PVOI	Photovoltaics Opportunity Irrigation
RDDC	Random Daily Demand Curves method
S	irrigated area (hectares)
SWT	static water table (meters)
Vs	total annual volume of irrigation water supply per unit irrigated area (L^3)
WEBSO	Water and Energy Based Sectoring Operation
WUA	Water User Association

GREEK SYMBOLS

Δh	difference in extreme pressure heads in the irrigation subunit (% of h_a)
Δq	difference in extreme emitter flow in the irrigation subunit (% of q_a)

REFERENCES

Abadia, R., Rocamora, C., Ruiz, A., and Puerto, H. (2008). Energy efficiency in irrigation distribution networks I: Theory. *Biosystems Engineering* 101 (1), 21–27.

Abadia, R., Rocamora, M. C., Corcoles, J. I., Ruiz-Canales, A., Martinez-Romero, A., and Moreno, M. A. (2010). Comparative analysis of energy efficiency in water user associations. *Spanish Journal of Agricultural Research* 8 (2), 134.

Al-Abed, N., Shudifat, E., and Amayreh, J. (n.d.). Modeling a rotation supply system in a pilot-pressurized irrigation network in the Jordan Valley, Jordan. *Irrigation and Drainage Systems* 17 (3), 163–77: Kluwer Academic Publishers.

Burt, C. M., and Plusquellec, H. L. (1990). Water delivery control. *Management of Farm Irrigation Systems*. American Society of Agricultural Engineers: St. Joseph, MI, 373–423.

Cakmak, B., Beyribey, M., Yildirim, Y. E., and Kodal, S. (2004). Benchmarking performance of irrigation schemes: A case study from Turkey. *Irrigation and Drainage* 53 (2), 155–63.

Calejo, M. J., Lamaddalena, N., Teixeira, J. L., and Pereira, L. S. 2008. Performance analysis of pressurized irrigation systems operating on-demand using flow-driven simulation models. *Agricultural Water Management* 95 (2), 154–62.

Carrillo Cobo, M. T., Rodríguez Díaz, J. A., Montesinos, P., López Luque, R., and Camacho Poyato, E. 2010. Low energy consumption seasonal calendar for sectoring operation in pressurized irrigation networks. *Irrigation Science* 29 (2), 157–69.

Carrillo-Cobo, M. T., Camacho-Poyato, E., Montesinos, P., and Rodríguez Díaz, J. A. (2014). Assessing the potential of solar energy in pressurized irrigation networks. The case of Bembézar MI Irrigation District (Spain). *Spanish Journal of Agricultural Research* 12 (3), 838.

Carrión, F., Tarjuelo, J. M., Hernández, D., and Moreno, M. A. (2013). Design of microirrigation sub-unit of minimum cost with proper operation. *Irrigation Science* 31, 1199–1211.

Carrión, F., Montero, J., Tarjuelo, J. M., and Moreno, M. A. (2014). Design of sprinkler irrigation subunit of minimum cost with proper operation: Application at a corn crop in spain. *Water Resources Management* 28 (14), 5073–89.

Carrión, F., Sanchez-Vizcaino, J., Corcoles, J. I., Tarjuelo, J. M., and Moreno, M. A. (2016). Optimization of groundwater abstraction system and distribution pipe in pressurized irrigation systems for minimum cost. *Irrigation Science* 34 (2), 145–59.

Carroquino, J., Dufo-López, R., and Bernal-Agustín, J. L. 2015. Sizing of off-grid renewable energy systems for drip irrigation in mediterranean crops. *Renewable Energy* 76, 566–74.

Chang, G., Hatziadoniu, C., Xu, W., Ribeiro, P., Burch, R., Grady, W. M., Halpin, M., *et al.* (2004). Modeling devices with nonlinear voltage-current characteristics for harmonic studies. *IEEE Transactions on Power Delivery* 19 (4), 1802–11.

Clément, R. (1960). Calcul des débits dans les réseaux d'irrigation fonctionnant "a la demande." *La Houille Blanche* 20 (5), 553–75.

Córcoles, J. I., Ortega, J. F., Carrión, P. A., De Juan, J. A., Tarjuelo, J. M., and Moreno, M. A. 2011. Power supply quality and harmonic generation in pumping stations for irrigation. *Applied Engineering in Agriculture* 27 (6), 1063–76: American Society of Agricultural and Biological Engineers.

Córcoles, J. I., De Juan, J. A., Ortega, J. F., Tarjuelo, J. M., and Moreno, M. A. (2012). Evaluation of irrigation systems by using benchmarking techniques. *Journal of Irrigation and Drainage Engineering* 138 (3), 225–34: American Society of Civil Engineers.

Córcoles, J. I., Tarjuelo, J. M., Carrión, P. A., and Moreno, M. A. (2015). Methodology to minimize energy costs in an on-demand irrigation network based on arranged opening of hydrants. *Water Resource Management* 29 (10), 3697–3710.

Córcoles, J. I., Frizzone, J. A., Lima, S. C. R. V., Mateos, L., Neale, C. M. U., Snyder, R. L., and Souza, F. (2016a). Irrigation advisory service and performance indicators in Baixo Acaraú Irrigation District, Brazil. *Irrigation and Drainage* 65 (1), 61–72.

Córcoles, J. I., Tarjuelo, J. M., and Moreno, M. A. (2016b). Pumping station regulation in on-demand irrigation networks using strategic control nodes. *Agricultural Water Management* 163, 48–56. http://www.sciencedirect.com/science/article/pii/S0378377415300937.

Daccache, A., Lamaddalena, N., and Fratino, U. (2009). On-demand pressurized water distribution system tmpacts on sprinkler network design and performance. *Irrigation Science* 28 (4), 331–39.

Djokic, S., Desmet, J., Vanalme, G., Milanovic, J. V., and Stockman, K. (2005). Sensitivity of personal computers to voltage sags and short interruptions. *IEEE Transactions on Power Delivery* 20 (1), 375–83.

Estrada, C., González, C., Aliod, R., and Paño, J. (2009). Improved pressurized pipe network hydraulic solver for applications in irrigation systems. *Journal of Irrigation and Drainage Engineering* 135 (4), 421–30: American Society of Civil Engineers.

European Standard. 1999. *Voltage Characteristics of Electricity Supplied by Public Distribution Systems. EN 50160.* Cenelec, Brussels, Belgium.

Fernández García, I., Rodríguez Díaz, J. A., Camacho Poyato, E., and Montesinos, P. (2013). Optimal operation of pressurized irrigation networks with several supply sources. *Water Resources Management* 27 (8), 2855–69.

Fernández García, I., Moreno, M. A., and Rodríguez Díaz, J. A. (2014). Optimum pumping station management for irrigation network sectoring: Case of Bembezar MI (Spain). *Agricultural Water Management* 144, 150–58.

Ferraci, P. (2004). La Calidad de La Energía Eléctrica. Centro de Formación Schneider. Cuadro Técnico No. 199. Barcelona, Spain.

Ghazalli, Mohd Azhari. (2004). Benchmarking of irrigation projects in malaysia: Initial implementation stages and preliminary results. *Irrigation and Drainage* 53 (2), 195–212.

Hassanli, A. M. and Dandy, G. C. (2000). Application of genetic algorithms for optimization of drip irrigation systems. *Iranian Journal of Science and Technology* 24 (1), 63–76.

Hirose, S. (1997). Determination of the capacity of a regulating pond in a pipeline irrigation system. *Rural and Environmental Engineering* 33, 67–78.

Izquiel, A., Carrión, P., Tarjuelo, J. M., and Moreno, M. A. (2015). Optimal reservoir capacity for centre pivot irrigation water supply: Maize cultivation in Spain. *Biosystems Engineering* 135, 61–72.

Jackson, T. M., Khan, S., and Hafeez, M. M. (2010). A comparative analysis of water application and energy consumption at the irrigated field level. *Agricultural Water Management* 97 (10), 1477–85.

Jackson, T. M., Hanjra, M. A., Khan, S., and Hafeez, M. M. (2011). Building a climate-resilient farm: A risk based approach for understanding water, energy and emissions in irrigated agriculture. *Agricultural Systems* 104 (9), 729–45.

Jayatillake, H. M. (2004). Application of performance assessment and benchmarking tool to help improve irrigation system performance in Sri Lanka. *Irrigation and Drainage* 53 (2), 185–93.

Jiménez-Bello, M. A., Martínez Alzamora, F., Bou Soler, V., and Bartolí Ayala, H. J. (2010.) Methodology for grouping intakes of pressurised irrigation networks into sectors to minimise energy consumption. *Biosystems Engineering* 105 (4), 429–38.

Jiménez-Bello, M. A., Royuela, A., Manzano, J., García Prats, A., and Martínez-Alzamora, F. (2015). Methodology to improve water and energy use by proper irrigation scheduling in pressurized networks. *Agricultural Water Management* 149, 91–101.

Kang, Y., Yuan, B., and Nishiyama, S. (1999). Design of micro-irrigation laterals at minimum cost. *Irrigation Science* 18 (3), 125–33.

Khadra, R. and Lamaddalena, N. (2006). A simulation model to generate the demand hydrographs in large-scale Irrigation Systems. *Biosystems Engineering* 93 (3), 335–46.

Kloezen, W. H., Kloezen, Wim H., and Garces-Restrepo, Carlos. (1998). *Assessing Irrigation Performance with Comparative Indicators: The Case of the Alto Rio Lerma Irrigation District, Mexico.*

Kumar, D., Heatwole, C. D., Ross, B. B., and Taylor, D. B. (1992). Cost Models for preliminary economic evaluation of sprinkler irrigation systems. *Journal of Irrigation and Drainage Engineering* 118 (5), 757–75: American Society of Civil Engineers.

Labye, Y. (1988). *Design and Optimization of Irrigation Distribution Networks.*

Lamaddalena, N., and Pereira, L. S. (2007a). Pressure-driven modeling for performance analysis of irrigation systems operating on demand. *Agricultural Water Management* 90 (1–2), 36–44.

Lamaddalena, N. and Pereira, L. S. (2007b). Assessing the impact of flow regulators with a pressure-driven performance analysis model. *Agricultural Water Management* 90 (1–2), 27–35.

Lamaddalena, N. and Sagardoy, J. A. (2000). *Performance Analysis of On-Demand Pressurized Irrigation Systems*. (Ed. Food and Agriculture Organization of the United Nations.)

Lamaddalena, N. and Khila, S. (2011). Energy saving with variable speed pumps in on-demand irrigation systems. *Irrigation Science* 30 (2), 157–66.

Lecina, S., Isidoro, D., Playán, E., and Aragüés, R. (2010). Irrigation modernization and water conservation in Spain: The case of Riegos Del Alto Aragón. *Agricultural Water Management* 97 (10), 1663–75.

López-Luque, R., Reca, J., and Martínez, J. (2015). Optimal design of a standalone direct pumping photovoltaic system for deficit irrigation of olive orchards. *Applied Energy* 149, 13–23.

Malano, H. M. and Burton, M. (2001). *Guidelines for Benchmarking Performance in the Irrigation and Drainage Sector*.

Malano, H., Burton, M., and Makin, I. (2004). Benchmarking performance in the irrigation and drainage sector: A tool for change. *Irrigation and Drainage* 53 (2), 119–33.

Mavropoulos, T. I. (1997). Sviluppo di una nuova formula per Il calcolo delle portate di punta nelle reti irrigue con esercizio alla domanda. *Rivista Di Irrigazione E Drenaggio* 44 (2), 27–35.

Molden, D., Sakthivadivel, R., and Habib, Z. (2001). *Basin-Level Use and Productivity of Water: Examples from South Asia*.

Montero, J., Martínez, A., Valiente, M., Moreno, M. A., and Tarjuelo, J. M. (2012). Analysis of water application costs with a centre pivot system for irrigation of crops in spain. *Irrigation Science* 31 (3), 507–21.

Moradi-Jalal, M., Mariño, M., and Afshar, A. (2003). Optimal design and operation of irrigation pumping stations. *Journal of Irrigation and Drainage Engineering* 129 (3), 149–54.

Moradi-Jalal, M., Rodin, S. I., and Mariño, M. A. (2004). Use of genetic algorithm in optimization of irrigation pumping stations. *Journal of Irrigation and Drainage Engineering* 130 (5), 357–65: American Society of Civil Engineers.

Moreno, M. A., Carrión, P. A., Planells, P., Ortega J. F., and Tarjuelo, J. M. (2007). Measurement and improvement of the energy efficiency at pumping stations. *Biosystems Engineering* 98, 479–86.

Moreno, M. A., Planells, P., Córcoles, J. I., Tarjuelo, J. M., and Carrión, P. A. (2009). Development of a new methodology to obtain the characteristic pump curves that minimize the total cost at pumping stations. *Biosystems Engineering* 102 (1), 95–105.

Moreno, M. A., Córcoles, J. I., Moraleda, D. A., Martinez, A., and Tarjuelo, J. M. (2010a). Optimization of underground water pumping. *Journal of Irrigation and Drainage Engineering* 136 (6), 414–20: American Society of Civil Engineers.

Moreno, M. A., Córcoles, J. I., Tarjuelo, J. M., and Ortega, J. F. (2010b). Energy efficiency of pressurised irrigation networks managed on-demand and under a rotation schedule. *Biosystems Engineering* 107 (4), 349–63.

Moreno, M. A., Ortega, J. F., Córcoles, J. I., Martínez, A., and Tarjuelo, J. M. (2010c). Energy analysis of water systems for irrigation: Proposal, monitoring, and evaluation of measures for improving the energy efficiency. *Irrigation Science* 28, 445–60.

Moreno, M. A., Medina, D., Ortega, J. F., and Tarjuelo, J. M. (2012). Optimal design of center pivot systems with water supplied from wells. *Agricultural Water Management* 107, 112–21.

Navarro Navajas, J. M., Montesinos, P., Camacho Poyato, E., and Rodríguez Díaz, J. A. (2012). Impacts of irrigation network sectoring as an energy saving measure on olive grove production. *Journal of Environmental Management* 111, 1–9.

Ortíz, J. N., De Juan, J. A., and Tarjuelo, J. M. (2010). Analysis of water application uniformity from a centre pivot irrigator and its effect on sugar beet (*Beta Vulgaris L.*) yield. *Biosystems Engineering* 105 (3), 367–79.

Pérez, Á., Bravo, N., and Llorente, M. (2000). *La Amenaza de Los Armónicos Y Sus Soluciones*: Thomson, Paraninfo, SA. Madrid, Spain.

Planells, P., Carrión, P., Ortega, J. F., Moreno, M. A., and Tarjuelo, J. M. (2005). Pumping selection and regulation for water-distribution networks. *Journal of Irrigation and Drainage Engineering* 131 (3), 273–81: American Society of Civil Engineers.

Planells, P., Ortega, J. F., and Tarjuelo, J. M. (2007). Optimization of irrigation water distribution networks, layout included. *Agricultural Water Management* 88 (1–3), 110–18.

Playán, E., Salvador, R., Faci, J. M., Zapata, N., Martínez-Cob, A., and Sánchez, I. (2005). Day and night wind drift and evaporation losses in sprinkler solid-sets and moving laterals. *Agricultural Water Management* 76 (3), 139–59.

Pulido-Calvo, I., Roldán, J., López-Luque, R., and Gutierrez-Estrada, J. C. (2003). Water delivery planning considering irrigation simultaneity. *Journal of Irrigation and Drainage Engineering* 129 (4), 247–55.

Pulido-Calvo, I., Gutiérrez-Estrada, J. C., López-Luque, R., and Roldán, J. (2006). Regulating reservoirs in pressurized irrigation water supply systems. *Journal of Water Supply: Research and Technology. AQUA* 55 (5), 367–81.

Ramos, H. M., Kenov, K. N., and Vieira, F. (2011). Environmentally friendly hybrid solutions to improve the energy and hydraulic efficiency in water supply systems. *Energy for Sustainable Development* 15 (4), 436–42.

Rodríguez-Díaz, J. A., Camacho-Poyato, E., López-Luque, R., and Pérez-Urrestarazu, L. (2008). Benchmarking and multivariate data analysis techniques for improving the efficiency of irrigation districts: An application in Spain. *Agricultural Systems* 96 (1–3), 250–59.

Rodríguez Díaz, J. A., López Luque, R., Carrillo Cobo, M. T., Montesinos, P., and Camacho Poyato, E. (2009). Exploring energy saving scenarios for on-demand pressurized irrigation networks. *Biosystems Engineering* 104 (4), 552–61.

Rodríguez Díaz, J. A., Montesinos, P., and Camacho Poyato, E. (2012a). Detecting critical points in on-demand irrigation pressurized networks—A new methodology. *Water Resources Management* 26 (6), 1693–713.

Rodríguez Díaz, J. A., Pérez Urrestarazu, L., Camacho Poyato, E., and Montesinos, P. (2012b). Modernizing water distribution Networks: Lessons from the Bembézar MD Irrigation District, Spain. *Outlook on Agriculture* 41 (4), 229–36: IP Publishing Ltd.

Rossman, L. A. (2000). EPANET 2 users manual, water supply and water resources division.

San Roman, T. and Ubeda, J. (1998). Power quality regulation in Argentina: Flicker and harmonics. *IEEE Transactions on Power Delivery* 13 (3), 895–901.

Van Der Gulik, T. (2003). Irrigation Equipment Cost: Canada.

Von Jouanne, A. and Banerjee, B. (2001). Assessment of voltage unbalance. *IEEE Transactions on Power Delivery* 16 (4), 782–90.

10 Improving Surface Irrigation

Malcolm H. Gillies, Joseph P. Foley,
and Alison C. McCarthy

CONTENTS

10.1 INTRODUCTION

Surface irrigation is the term used to describe the range of application techniques by which water is delivered to one or more points or edges of a field, and distributes over the field area under the influence of gravity. Surface irrigation is often referred to as flood irrigation, implying that the water distribution is uncontrolled, and, therefore, inherently inefficient. The term encompasses a large range of systems with varying degrees of complexity and performance. It is true that a significant proportion of surface irrigation systems do perform poorly, but, for the most part, much of this is due to design issues or incorrect management of the system rather than an inherent problem of the system itself.

Surface irrigation is the oldest form of irrigation and still remains the most common application technique across the globe. Irrigation was a necessary part of many of the world's ancient civilizations and evidence of various forms of surface irrigation has been found across Africa, the Middle East, Asia, and South America. Approximately 6000 years ago flood and furrow irrigation was practiced in the area surrounding the Tigris and Euphrates rivers in modern day Iraq (Postel, 1999). The Egyptians and Indians pioneered a system of "inundation canals" that are dug parallel to the river and rely on regular seasonal flooding (Cantor, 1970). In Egypt, these canals were used to supply water to basin irrigation layouts which resemble those still used in the region to this day. Evidence for similar primitive irrigation systems has also been found in other regions, such as China and Mexico.

Other forms of irrigation have only become a reality in the last 150 years with the development of pressurized sprinklers in the late nineteenth and early twentieth centuries, and the invention of drip irrigation in the mid-twentieth century. Pressurized irrigation, particularly large mobile machines and drip irrigation, is seen as a superior application technique. For millennia, farmers were reliant on a system which offered limited flexibility in terms of timeliness and rate of application. Now they have an engineered solution where perfect application is at least theoretically possible. Despite the perceived advantages of pressurized irrigation, much of the world's food and fiber is still grown using surface irrigation, and for good reason, as it has a number of advantages. Possibly the most significant of these is capital cost. Pressurized irrigation systems involve a higher capital investment. The highest cost option, drip irrigation, is typically restricted to higher-value horticultural and orchard crops, or in circumstances where great precision of irrigation applications is required. The major capital costs associated with surface irrigation is the grading of the land. This should be performed regardless of the choice of system to ensure adequate drainage of the field during rainfall events. In water-limited circumstances, farmers will only irrigate a portion of the total cropping area and then rotate to other fields in subsequent seasons. This practice, common in parts of Australia, is not compatible with high-cost infrastructure, as it will take much longer to recoup these costs if the field is only utilized every second or third season.

The energy requirements of irrigation are often neglected during farm planning, but as energy prices increase it is becoming an important concern. In many circumstances, surface irrigation requires no on-farm pumping, water flows from the scheme channels or on-farm reservoirs to the fields under the influence of gravity alone. Where pumping is required, pressurized sprinkler and drip systems will always involve considerably higher energy costs than the equivalent surface irrigation layouts. The requirement for reliable and consistent electrical power or petroleum supplies has proven to be a challenge in developing nations. Sprinkler and drip systems are designed to operate at a specific supply pressure and are reliant on access to this electrical power supply 24 hours per day. While not an issue in most developed nations, unreliable power supplies have forced designers to adopt high-system capacities with higher flowrates requiring higher capital costs. Many irrigation distribution schemes have been designed to supply high flow rates for short periods of time instead of lowering the continuing flow rates required for pressurized systems. Adoption of new systems either require on-farm storage or changes at the scheme level.

Surface irrigation can be managed with a low-skilled workforce, while other forms of irrigation require higher skill levels or specialist knowledge. Other forms of irrigation may also require more frequent maintenance and access to technical support. Neglect of these facts has resulted in many pressurized systems failing to achieve anticipated levels of performance (Burney and Naylor, 2012; Kulecho and Weatherhead, 2005; Maisiri *et al.*, 2005).

As previously stated, most of the irrigated area worldwide is under some form of surface irrigation, which, in most cases, is performing at a substandard level of efficiency. It is true that this performance can be increased through adoption of appropriate pressurized systems, but the truth is somewhat more complex. For example, promoters of drip irrigation would suggest that this single technology can save water and help save the water crisis, but as recent research has shown, there is little conclusive evidence that this is true (Van der Kooij *et al.*, 2013). Regardless of whether alternate systems offer any benefit, considering the cost and difficulty of the conversion, it is important to first ensure that the existing system is functioning optimally.

The proportion of irrigated land under surface irrigation has declined significantly over the past few decades in countries such as the United States (Koech *et al.*, 2015), but this is likely due to labor savings rather than water efficiency savings. Traditional surface irrigation is more labor-intensive than most other types of irrigation system; therefore, as wages rise in developing nations, it is reasonable to expect that the adoption of pressurized irrigation will increase. Rising labor costs are also the main driver behind the adoption of automation systems, as is described in Section 10.7.

This chapter draws from material written by the author which has been previously presented in a range of articles and theses (e.g., Gillies, 2008; Gillies and Smith, 2015; Koech *et al.*, 2014; Koech, 2012; Smith *et al.*, 2016).

10.2   SURFACE IRRIGATION THEORY

10.2.1   FORMS OF SURFACE IRRIGATION

The term surface irrigation is used to describe a range of different field layouts and irrigation water supply methods, and a broad classification is required to better understand the content of this chapter. The earliest irrigation systems, often referred to as "wild flooding," involved diverting flow from a river over cropping land in an uncontrolled fashion. This primitive technique was usually only possible during flood conditions, was wasteful, and suffered from poor uniformity and erosion issues. Most surface irrigation in the modern world can be termed "controlled flooding," where water is applied to individual fields or sections of fields in a controlled fashion. Some attempts have been made to classify modern surface irrigation into three broad categories: basin, border, and furrow irrigation.

Basin irrigation, or, more accurately level basin irrigation, describes the design by which the field is divided into sections called bays, each with a zero grade and surrounded with earth banks. The desired amount of water is applied rapidly, distributes itself over the soil surface, and is left to infiltrate. Basin irrigation is favored in soils with relatively low infiltration rates (Walker and Skogerboe, 1987). Fields are traditionally set up to follow the natural contours of the land, but the introduction of laser leveling and land grading has permitted the construction of large rectangular basins that are more appropriate for broadacre cropping. Traditionally, level basins are closed systems where all of the water applied to the individual basin is expected to infiltrate (Khanna and Malano, 2006). Alternatively, where land slope is sufficient, individual basins may be terraced so that the drainage from one basin is diverted into the next basin after an irrigation period is complete.

Furrow irrigation is the technique by which water is applied to a series of parallel furrows at the higher end of the field, and flows to the opposite end of the field in a controlled manner. The furrow itself is a small trench, typically between 200–800 mm wide, 100–300 mm deep, and normally orientated in the direction of the predominant slope. The water will flow down the length of the field in the direction of the furrow, which is usually straight, or in some cases, curved to minimize risks of erosion. The furrows are traditionally closely spaced with the plants located on the crest of the ridge between adjacent furrows. Alternatively, the field may be formed into beds, which are flat, elevated regions between the furrows and contain more than one row of plants. The actual distance between furrows is governed by a combination of the plant spacing, machinery track-width, irrigation management, and the soil's capacity for horizontal redistribution of water. Common furrow spacings are 0.75–1 m for cotton and maize, and 1.5–1.8 m for sugar cane. Shorter field and furrow lengths are commonly associated with a higher uniformity of application and an increased potential for runoff losses. However, high performance is also possible for longer furrows under the right soils and with sufficient flow rates. Furrow irrigation is the most common form of irrigation throughout the world (Burt, 1995), and is practiced in many row crops (such as cotton, legumes, and maize) and other summer grains, as well as sugarcane.

Border, border-check, or bay irrigation is the third common form of surface irrigation and is characterized by a series of rectangular bays separated by small earthen banks or borders. These bays are typically 200–500 m long and 20–50 m wide with no slope across the width and a constant slope in the longitudinal direction. Water is applied to the highest end of the bay, which then runs to the lower end of the bay under gravity. The interior of the borders is normally flat with no beds or furrows. Water is typically applied to the bay through a gate, pipe, or large siphon positioned in the middle of the highest end of the bay. These systems are common in the irrigated pasture industry.

Several alternative configurations have been proposed, but most can be considered as some combination of the three main types above.

10.2.2 THE PROCESS OF SURFACE IRRIGATION

The second major concept is that of the irrigation phases, terms given to describe key periods in the duration of the surface irrigation event. The following description applies to traditional furrow and border-check irrigation and may differ for alternative systems. Figure 10.1 provides a graphical representation. At the commencement of the irrigation, the entire field is dry. At time zero the water inflow is applied to the highest end of the furrow or border. The advance phase defines the time over which the resulting water front is moving down the field. The time at which water reaches the lowest end is occasionally referred to as the completion time. At this time, runoff commences on fields that are freely draining or accumulating if the tail end of the field is blocked. The time period from when the advancing water reaches the end of the field until the inflow ceases is termed the "storage phase." Once the inflow is stopped, it takes a short and often negligible time period for the water depth at the

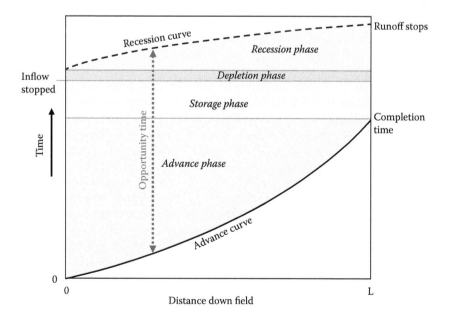

FIGURE 10.1   Representation of the phases of surface irrigation.

upstream end of the field to disappear, and this is termed the "depletion phase." The final surface irrigation phase occurs while water is receding towards the downstream end of the field, and is termed the "recession phase." The recession curve is somewhat theoretical, often in practice the recession does not proceed down the field in this clear manner.

In most cases, the inflow is stopped after the advance has reached the end of the field, as in Figure 10.1. In other situations, the inflow may be stopped well before the completion of advance, there will be no storage phase, and the advance and recession phases may occur simultaneously.

The opportunity, or ponding time as shown in Figure 10.1, determines the amount of infiltration at any location, and is the time difference between the arrival of the water advance front and the departure of the recession front and corresponding curves. The uniformity of the irrigation is directly linked to the difference in opportunity times over the length of the field. High irrigation uniformity is achieved when the advance and recession curves are as close to parallel as possible, and opportunity times are similar along the length of the field. The shape of the advance curve is dependent on the soil, field characteristics, and inflow hydrograph. For example, a soil with a higher steady intake rate will exhibit higher curvature in contrast to a low-permeability, heavy clay soil which will tend to have an advance curve which is closer to linear. The following section will focus on a hypothetical case with a fixed soil type and field design.

Figure 10.2 contains two irrigation events on the one field, one where the inflow rate is low, the advance is slow, and the water is stopped after a certain ponding time at the downstream end; and one where the inflow rate is high, the advance is rapid, and the water is cut off at a similar downstream ponding time. Long advance times with a steeper advance curve will result in high applications at the upstream end of the field and low applications at the tail end of the field (Figure 10.2a). Uniformity can be improved, in most cases, by increasing the speed of the water advance front, which will flatten the advance curve and provide opportunity times that are more uniform along the length of the field (Figure 10.2b). Systems with high uniformity should more easily allow the grower to achieve higher efficiency, as the irrigator does not need to overwater parts of the field in order to adequately irrigate areas that normally receive lower applications. However, this faster advance and higher uniformity irrigation will quickly lead to low application efficiencies if the inflow is not ceased in a timely fashion.

10.3   IRRIGATION PERFORMANCE

In this chapter, irrigation performance refers to the volumetric efficiency and uniformity rather than the water-use efficiency of the crop in terms of dry matter per unit of water. This term is often confounded by the plant variety, soil, climate, and crop management. The purpose of the irrigation is to replenish the soil moisture store, or part thereof. Considering the soil moisture balance at a single location, the soil moisture deficit will be zero when the root zone store is at field capacity, and then increase over time as the plant and soil lose this water through evapotranspiration. The depth of application required, assuming a zero loss to replenish this store back to the full point, is termed the soil moisture deficit, or Z_{req}.

10.3.1   VOLUMETRIC EFFICIENCY

A volume of water (Vol_{In}) is applied to the furrow or bay in any irrigation and at the completion of the event will be split into four components. The beneficial volume is added to the root zone ($Vol_{rootzone}$) and made available to the crop and the deep percolation and runoff and evaporative losses, so that:

$$Vol_{In} = Vol_{Rootzone} + Vol_{D.\ Perc} + Vol_{Runoff} + Vol_{Evap} \qquad (10.1)$$

Deep percolation occurs where any application depth is in excess of the Z_{req}, and the $Vol_{D.\ Perc}$ is a summation of the deep percolation depths on these areas of the field. While this water is lost to this irrigation event, this deep drainage can be beneficial in terms of salt leaching. On the other hand,

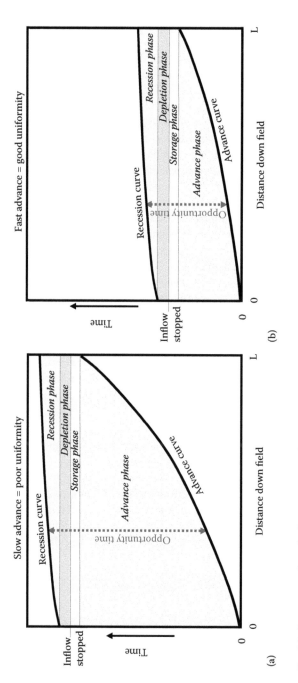

FIGURE 10.2 Graphical representation of advance curve and opportunity time for (a) an irrigation with poor uniformity and (b) irrigation with good uniformity.

runoff can be recovered for use elsewhere on the farm. The last term in Equation 10.1, the Vol_{Evap}, represents the water lost directly from the soil surface during the irrigation. While this is a loss for fallow ground, it is debatable whether this is a loss at all in normal in season irrigations. Where the crop has reached full canopy cover, this evaporation will simply offset some of the plant transpiration which is already higher than that of a free water surface.

The most simplistic expression of volumetric performance, the application efficiency (Ea) is defined as the proportion of water applied to the field which is stored in the root zone, and can be accessed by the plant:

$$Ea = \frac{Vol_{Rootzone}}{Vol_{In}} \times 100\% \tag{10.2}$$

Many modern surface irrigation systems contain runoff recycling or tailwater recovery drains that have the ability to recapture surface drainage from irrigations and rainfall. In these situations, it is appropriate to adjust the application efficiency to account for this reduction in the runoff loss, where R is the efficiency of the tailwater recycling system or the fraction of runoff which can be recovered:

$$Ea_R = \frac{Vol_{Rootzone}}{Vol_{In} - R \times Vol_{Runoff}} \times 100\% \tag{10.3}$$

The value of R is selected to reflect the expected losses to seepage and evaporation within the drainage channels. In reality, R would vary as there would be an initial loss due to dead storage and initial infiltration into dry drainage channels, and a higher loss rate when runoff volumes are large due to longer durations when seepage and evaporation can occur.

In most cases, particularly in non–water limited conditions, it is more important that the irrigation refills a significant portion of the soil moisture deficit over the field area. The requirement efficiency (Er), otherwise known as storage efficiency or effective water (Bali and Wallender, 1987), serves as a check of the irrigation adequacy, where Vol_{req} is the volume of infiltration required to satisfy the soil moisture deficit:

$$Er = \frac{Vol_{Rootzone}}{Vol_{req}} \times 100\% \tag{10.4}$$

Low values indicate that large portions of the field did not receive adequate water application. In most cases, farmers will endeavor to supply adequate moisture to the entire field ($Er \approx 100\%$) often at the expense of other performance indicators. Neither the Er or Ea give any indication of the proportion of the field area that receives the full application. It is possible to obtain high values (i.e., Er is greater than 90%), while parts of the field receive zero application. For this type of information, the efficiency value should be complemented by some indication of uniformity.

10.3.2 UNIFORMITY

For irrigation, uniformity describes the evenness of the applied water depths over the entire field. For pressurized systems, uniformity holds greater priority over efficiency due to the relative ease of improving efficiency through reduction of irrigation durations. Surface irrigation systems differ in that once the farmer has supplied water to the entire field area, application efficiency, Ea, demands the greatest priority. The distribution uniformity (DU), or, more accurately, low quarter distribution uniformity is the most common uniformity term employed for surface irrigation. It is expressed as the ratio of the average depth from that the quarter of the field with the lowest infiltrated depths to the average infiltrated depth ($\overline{D}$) over the entire field area (Kruse, 1978):

$$DU = \frac{\sum\limits_{i=0}^{B}(D(i))\Big/B}{\overline{D}} \times 100\% \quad B = \frac{N_D}{4} \tag{10.5}$$

where $D(i)$ are the infiltrated depths at each point i sorted in ascending order, N_D is the number of points where infiltrated depth is known, and the average infiltrated depth is given by:

$$\overline{D} = \frac{\sum\limits_{i=0}^{N}(D(i))}{N_D} \tag{10.6}$$

The absolute distribution uniformity (ADU) instead considers the ratio of the absolute minimum depth of infiltration (D_{min}) to the field average.

$$ADU = \frac{D_{min}}{\overline{D}} \times 100\% \tag{10.7}$$

The DU is usually preferred since if any part of the field receives a zero application the ADU cannot describe any further decrease in uniformity.

While uniformity is expressed using DU for surface irrigation, the more universally applied term is the Christiansen's Uniformity Coefficient (CU), usually applied to sprinkler systems but equally applicable in the context of surface irrigation.

$$CU = \left(1 - \frac{\sum\limits_{i=0}^{N_D}|D_i - \overline{D}|}{N_D \times \overline{D}}\right) \times 100\% \tag{10.8}$$

These are the traditional measures of uniformity. However, for surface irrigation where a significant part of the field can be over-irrigated, a more appropriate calculation may be the distribution uniformity of the root zone ($DURZ$) (Gillies and Smith, 2015) using the depth of infiltration in the root zone, D_{RZ}.

$$DURZ = \frac{\sum\limits_{i=0}^{B}(D_{RZ}(i))\Big/B}{\sum\limits_{i=0}^{N}(D_{RZ}(i))\Big/N_D} \times 100\% \quad B = \frac{N_D}{4} \tag{10.9}$$

The resulting expression closely resembles DU but will closely approach 100% in conditions of over-irrigation. The DURZ is similar to application efficiency of the low quarter (AELQ), as defined by Merriam *et al.* (1983), as proposed for sprinkler systems to combine the effects of uniformity and irrigation requirement (Kruse, 1978).

10.3.3 SUMMARY

All volumetric performance terms fall into one of three categories, efficiency (Ea), adequacy (Er), and uniformity (DU). Determining what constitutes the optimum irrigation is highly subjective and inevitably involves some compromise, as it is impossible to maximize all three simultaneously.

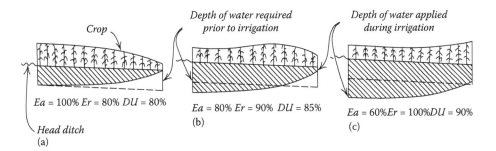

FIGURE 10.3 Infiltration profiles for three different surface irrigation events on the same field: (a) maximizing *Ea*, (b) with a balance between *Ea* and *Er*, and (c) maximizing *Er*.

The interaction between the three parameters was ideally illustrated by Israelsen and Hansen (1962), as shown in Figure 10.3. Here, the result of three different irrigation applications with varying inflow time and applied volumes on the same field are presented. The dashed line represents the soil moisture deficit (D_{req}) and the shaded region is the longitudinal profile of applied depths, assuming water is applied from the left side. Situation (a) is where the inflow was cut off, so that water only just reaches the end of the field, and, consequently, there is no runoff and no deep drainage, so the irrigation will be very efficient with an *Ea* of 100%. While there is no water waste, this irrigation fails to satisfy the crop over the downstream end of the field, giving it a low *Er*. Situation (c) represents the opposite, where the grower does not shut down the inflow rate until the D_{req} is satisfied over the entire length, resulting in perfect *Er* and high uniformity but poor *Ea*, as 40% of the applied water is lost to runoff or deep percolation. The compromise position (b) represents the irrigation management where the grower is willing to sacrifice some adequacy at the downstream end of the field to improve the *Ea*. The grower's decision will be influenced by the value of the crop, with (c) being more likely with high value crops and (a) where water is limited.

Figure 10.3 also displays the potential impact of waterlogging on crop yields. Most crops will have highest yields when the water application matches the required depth, and lower yields where *Er* is less than 100%. Waterlogging often coincides with deep drainage, and causes plant growth to slow or even stop until the excess water is drained. Waterlogging can also accelerate the loss of nitrates through volatilization. This negative impact of waterlogging on crops is often overlooked in irrigation management.

10.4 MEASUREMENT OF SURFACE IRRIGATION

Benchmarking current performance is essential before attempting to improve the performance of any irrigation system. Nowhere is this statement more correct than for surface irrigation, where the behavior of the system is governed by a combination of field design, water supply, soil characteristics, and hydraulic interactions.

The suite of measurements that are required for evaluating the performance of a surface-irrigated field include:

- Field geometry (slope, length, furrow shape, bay width)
- Soil moisture deficit (from crop water balance or infield measurement)
- Inflow rate and hydrograph
- Water advance data (at the correct number of locations for the infiltration estimation method)
- Runoff volumes (if possible)
- Moisture content of the soil after irrigation (if possible)

A typical setup for measurement of furrow irrigation is provided in Figure 10.4.

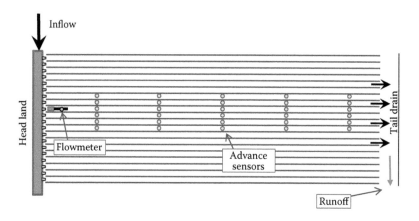

FIGURE 10.4   Field measurement of furrow irrigation.

The largest component of the water balance is inflow rate, which is, therefore, the most crucial measurement that must be collected. Flow information can be collected at a range of scales, from the farm scale down to within fieldpoint scale. It is recommended that records are taken over time to capture any changes to inflow rates over the duration of the irrigation. There are a range of devices which can be used to capture the inflow, including pipe flowmeters, ultrasonic Doppler velocity meters, or calibrated hydraulic structures, where flow can be inferred from water depth.

Water advance describes the movement of the water front from the water source across the basin or down the length of the field, as previously shown in Figure 10.1. The advance time refers to the time taken for water to reach particular points in the field and determines the opportunity time. As discussed in 10.5.3, advance is an easily measureable source of data which can be used to estimate soil intake properties. These inverse solutions each have different requirements in terms of number of points ranging from a single point for the simplistic methods; two points for the two-point method or greater than four points for numerical methods. The procedures described by (Dalton *et al.*, 2001) utilize five to six points approximately equally distributed over the field length, as shown in Figure 10.4. It is recommended that several positions be collected at each distance to overcome variability of infiltration, which can be significant (Gillies, 2008) and cause potential malfunctions in the measurement apparatus.

Runoff is a part of the volume balance which may be important or small compared to the inflow and infiltrated volumes. In many cases, it is difficult to measure the runoff volumes without altering the hydraulic behavior of the field, often the measurement device will tend to restrict drainage and cause waterlogging at the end of the field. For this reason, simulation models are commonly used to estimate runoff.

The distribution of soil moisture prior to and following an irrigation event is difficult to capture. Common practice is to assume uniform spatial properties prior to the irrigation and use simulation models to predict the distribution of moisture applications on completion of the event.

10.5   SIMULATION AND OPTIMIZATION

10.5.1   Background

Direct physical measurement of the data required for irrigation evaluations (e.g., the distribution of applied water and the volume of deep percolation) is expensive, labor-intensive, and difficult to perform. Thorough measurements of infiltrated depths across the field length are usually restricted to scientific trials. Simulation models, once calibrated, offer the potential to predict these elusive components of the water balance. In recent decades, a number of computer packages have been developed to simulate the process of surface irrigation. In general, they have three main uses where each package may perform one or more of the following, usually carried out in this sequence:

1. Identification of the infiltration curve and hydraulic characteristics
2. Evaluation of the current irrigation performance
3. Optimization of field design and management

The selection of models is diverse, where each has been developed with a specific purpose in mind, ranging from the simple direct solution of the volume balance to the complex Saint-Venant equations.

Simulation models offer an opportunity to test and evaluate potential alterations to irrigation design and management. Traditionally, optimization of irrigation management involved extensive field trials, which require substantial time to conduct and analyze. Alternatively, computer models provide the capability for the user to propose a change, simulate the irrigation and evaluate the likely performance, all within a matter of minutes. Prior to any optimization it is paramount to evaluate the current performance to provide a benchmark for any improvements.

10.5.2   THEORY

Surface water flow over an irrigated field can be described by the principles of conservation of mass (continuity) and conservation of momentum. The flow can be classified as unsteady open channel flow, where the water depths and velocities vary with distance and time, and are governed by a time-variant lateral outflow, the soil infiltration, i. The system can be simplified by assuming that the water velocity is constant at any cross section of the furrow or bay, thereby reducing the problem to one-dimensional flow. For a prismatic channel, the one-dimensional form of continuity and momentum (or, the Saint-Venant equations) can be expressed as (Walker and Skogerboe, 1987):

Conservation of mass (continuity):

$$\frac{\partial Q}{\partial x} + \frac{\partial A}{\partial t} + I = 0 \tag{10.10}$$

Conservation of momentum:

$$\frac{1}{Ag}\frac{\partial Q}{\partial t} + \frac{2Q}{A^2 g}\frac{\partial Q}{\partial x} + \left(1 - \frac{Q^2 T}{A^3 g}\right)\frac{\partial y}{\partial x} = S_0 - S_f \tag{10.11}$$

where Q is the discharge ($m^3\ s^{-1}$), A is the cross-sectional area of flow (m^2), x is the distance (m), t is time (s), I is the soil infiltration rate ($m^3\ m^{-1}\ s^{-1}$) per meter length of furrow, y is the depth of flow (m), g is the acceleration due to gravity ($m\ s^{-2}$), and S_0 and S_f are the bed slope and friction slope (dimensionless), respectively. This one-dimensional representation is also often preferred for basin irrigation whereas the theoretically correct, two-dimensional model encounters numerical stability issues during computations.

The soil-intake rate can be represented using any one of the mathematical models for soil infiltration such as the Green-Ampt (Warrick et al., 2005), Phillip (Tarboton and Wallender, 1989), Horton (Renault and Wallender, 1994), linear (Austin and Prendergast, 1997), and Kostiakov equations. However it is the modified Kostiakov or Kostiakov-Lewis equation that is most commonly applied and more ideally suited expression for describing infiltration during surface irrigation (Clemmens, 1983). The modified Kostiakov equation in terms of infiltrated depth Z or infiltration rate I is given by:

$$Z = k\tau^a + f_0 \tau \qquad OR \qquad I = ak\tau^{a-1} + f_0 \tag{10.12}$$

where τ is the opportunity or ponding time (s), and a, k, and f_0 are the empirical infiltration parameters, the values of which are determined by the units of depth and time. The parameter f_0 is often referred to as the final intake rate of the soil in $m^3\ m^{-2}\ min^{-1}$. For furrows the cumulative infiltration, Z represents infiltrated volume per unit length, has the units of $m^3\ m^{-1}$, and is related to the applied depth D defined earlier by dividing by the furrow spacing. For basins and borders, Z can be expressed as infiltrated volume per unit area $(m^3\ m^{-2})$. The infiltration rate I $(m^3\ m^{-2}\ min^{-1})$, is simply the derivative of the cumulative depth Z.

The most successful attempts to model the flow of water within a surface irrigation field are based on the one-dimensional form of the continuity (Equation 10.10) and some form of the momentum equation (Equation 10.11). The general terminology used to refer to these models in order of reducing complexity and accuracy are as follows:

- Full hydrodynamic: Solves the equations in their entirety
- Zero-inertia: Simplifies the equations by neglecting the time variant terms
- Kinematic wave: Reduces the momentum equation further to that of normal flow
- Volume balance: Neglects the momentum equation entirely

While it is generally accepted that the full hydrodynamic model is the most accurate approach, many have concluded that the complexity and accuracy of this model is not required, and the simplified models will suffice. The recent advances in microcomputer speed and memory capacity have greatly minimized the cost of this extra complexity and encouraged use of the more accurate approaches.

The most complex hydraulic model, the full hydrodynamic, is used to solve the continuity (Equation 10.10) and momentum (Equation 10.11) equations in their entirety, and are therefore considered to be the most accurate alternative. Historically, their use was restricted to scientific trials, or for validating the performance of the less complex models. Full hydrodynamic models are desired for field evaluation, since they can be used to describe water velocity and depth at any time and position during the event, and provide dependable predictions of runoff volumes and the final distribution of applied depths. Several models have been developed but one notable example is SIRMOD (Walker, 2005), which has been used extensively both in research (e.g., Horst et al., 2005; Raine et al., 1997; Smith et al., 2005) and commercial applications (Dalton et al., 2001).

The zero-inertia models are preferred by many researchers due to the ability to reduce computational requirements without sacrificing much of the accuracy of the complete equations (Ebrahimian and Liaghat, 2011). As the name suggests, this approach assumes that the inertia and acceleration terms within Equation 10.11 are insignificant and can be ignored which typically holds true in free-flowing conditions (Schwankl and Wallender, 1988). Doing this simplifies Equation 10.11 to

$$\frac{\partial y}{\partial x} = S_0 - S_f \tag{10.13}$$

WinSRFR (Bautista et al., 2009; Bautista et al., 2016) is possibly the best well known application of the zero-inertia model.

The kinematic wave model is a further simplification of the momentum equation by assuming that the friction slope is equal to the bed slope:

$$S_0 = S_f \tag{10.14}$$

This approach is so named because it describes the movement of a kinematic shock wave along a water surface (Walker and Skogerboe, 1987). It offers considerable advantages in computational speed but requires moderate downstream field slope (e.g, less than 0.1%) (Gharbi et al., 1993).

The simplicity of the kinematic wave model has allowed researchers to carry out modeling which requires large numbers of simulation runs. For example, Fonteh and Podmore (1994) applied it to describe the impact of spatially variable infiltration on irrigation performance. Rayej and Wallender

(1988) applied the kinematic wave to model spatially varying infiltration and wetted perimeters, work which was progressed further by Raghuwansi and Wallender (1996) and applied to study temporal variations in the field. The kinematic wave approximation is an option available within WinSRFR and SIRMOD III models.

The volume balance model is the most basic of the hydraulic models. It uses the same continuity equation (Equation 10.10) but completely neglects all components of the momentum equation. Without the momentum equation, the model cannot predict the water depths down the field, and therefore typically relies on an estimated or measured depth combined with an assumed surface water profile. The simplicity of the volume balance model lends itself to analytical expressions which can be solved directly without the need for computers. For this reason, the volume balance is often chosen for the inverse solution for infiltration.

Two-dimensional models for surface irrigation are not normally used in practice, but are slowly gathering attention with increasing computational power. These models are most relevant for irrigation layouts where water flow across the field is an important concern, such as for level basins and border-check irrigation with leveling imperfections. Clemmens (2003) developed a two-dimensional model of shallow water flow to evaluate the impact of field leveling precision. COBASIM, a two-dimensional, zero-inertia model was developed to model flow across level basins of irregular shape (Khanna et al., 2003), accounting for the hydraulic interaction between sequential basins. Others have developed models using the zero-inertia or full hydrodynamic models for describing the water flow within a single, closed basin. The general problem of these two-dimensional models is a lack of field measurement techniques and inverse solution procedures to enable practical application (Khanna and Malano, 2006).

10.5.3 INVERSE SOLUTIONS FOR INFILTRATION

Soil infiltration is second only in magnitude and importance to inflow rate in the water balance for surface irrigation. Therefore, it is no surprise that knowledge of the soil intake rate is essential for design and management of these systems. Simulation models can be used in a rough investigative manner with generic soil intake parameters, but application to a specific event or field requires specific values of these parameters for the field of interest.

Infiltration rates are the direct result of the physical and chemical soil properties, which suggests that the best way to quantify their value is through direct measurement. The single ring and double ring infiltrometers consist of a cylinder typically 250–600 mm in diameter, inserted vertically in the soil, and filled with water. Infiltration rates can be obtained by recording the rate that the water level drops (falling head) or amount of water which has to be added (constant head) to the ring. These devices are most commonly used to establish saturated hydraulic conductivity or final intake rates (infiltration after long ponding durations), but do not adequately describe the infiltration process which occurs during surface irrigation.

Infiltration variability poses a significant problem for this simple measurement process. Infiltration rates vary across all scales, from the paddock scale down to the sub-meter scale. Field measurements have demonstrated that 100% of the variation expected across a typical field can be observed within the sample size of 1 m^2 (Van Es et al., 1999). Some have attempted to define the minimum number of samples or ring area (McKenzie and Cresswell, 2002) to overcome small-scale variability, but wider variation across the field remains. It is generally accepted that ring infiltrometers do not provide reliable results for soils with high clay contents (vertic soils) that experience appreciable shrinking and swelling after wetting (McKenzie and Cresswell, 2002). Unfortunately, these high clay content soils are commonly surface irrigated.

Standard infiltrometers rely on still, ponded water and cannot properly account for the impacts of water velocity, depth, wetted perimeters, and temperature on the infiltration rate. Furrow infiltrometers have been used in an attempt to account for the impact of furrow shape and water depth (Bautista and Wallender, 1985; Childs et al., 1993) and recirculating furrow infiltrometers have been used to include the impact of flowing water (Lentz and Bjorneberg, 2001), but both are limited to a short section of a single furrow.

It is generally accepted that field scale measurements of soil infiltration provide data which is more representative of the known soil properties (Selle *et al.*, 2011) and more relevant to understanding the behavior of surface irrigation. For this reason, the most successful attempts at quantifying infiltration during irrigation events use the approach of transforming the entire furrow, bay, or field into an infiltrometer. Unlike the standard infiltrometer, parts of the measurement area may be dry or ponded at any particular point in time and, therefore, this approach normally requires use of a simulation model to predict water movement. These techniques are referred to here as an "inverse solution" because they involve solving for the infiltration input parameters though forcing the model of choice to match measured field data. There have been numerous published approaches, but most rely on measurements of field geometry and inflow combined with advance times or velocities.

The volume balance hydraulic model is the most simplistic and historically the most common choice for inverse solutions of infiltration. The two-point method (Elliott and Walker, 1982) uses the volume balance model to provide a direct analytical solution to the values of two of the Kostiakov infiltration parameters, a and k, from measured advance times at two distances down the field. McClymont and Smith (1996) developed INFILT, which was capable of estimating all three Kostiakov parameters from three or more advance points. This was further improved in the development of IPARM to utilize runoff data, in addition to the advance (Gillies and Smith, 2005) and variable inflow hydrographs. The volume balance model was also been utilized to inversely solve for other infiltration functions, such as the Horton Equation in RAIEOPT (Mailhol, 2003), the Phillip equation (Shepard *et al.*, 1993), and the SCS intake function (Valiantzas *et al.*, 2001).

The volume balance is too simplistic to be applicable in all situations and, therefore, various researchers (Clemmens, 1991; Katopodes *et al.*, 1990; Yinong *et al.*, 2006) have employed the zero-inertia model to solve this inverse solution problem. The inverse problem has also been applied to the full hydrodynamic model with the major advantage being that the process can then utilize data collected during all phases of the irrigation. Two notable examples of this are the MultiLevel calibration in SIRMOD III (Walker, 2005) which assumes that each of the parameters is determined by the data from a particular phase of the irrigation, and the SISCO calibration (Gillies and Smith, 2015) which can fit the three parameters of the Kostiakov infiltration and the Manning roughness to any combination of advance, runoff, recession, and water depths.

The final result of the inverse solution process is to provide the missing inputs for the simulation model so that it can be calibrated to the field of interest. Once the model has been calibrated it can be used to optimize the field design.

10.5.4 Optimization

Simulation models serve as a tool to identify and evaluate improved irrigation practices for a given field. The use of models such as SIRMOD and WinSRFR for this process has been well-documented in the scientific literature (Pereira *et al.*, 2007; Horst *et al.*, 2005; Hornbuckle, 1999) but there have been few examples of follow up studies to evaluate the effectiveness of this optimization process.

A good case study can be observed with the Australian cotton industry, where in the late 1990s, application efficiencies for individual irrigations varied from 17% to 100% (Dalton *et al.*, 2001; Smith *et al.*, 2005), with deep drainage below the root-zone identified as a major loss. Through computer modeling and optimization with SIRMOD, Smith *et al.* (2005) demonstrated that increasing the furrow inflow rates above normal practice to 6 L/s, and reducing the cutoff time (T_{co}) accordingly would increase the potential average *Ea* to 75%. Later, measurements collected in that industry over the subsequent years and analyzed with SISCO indicated that growers have adopted those practices and were achieving that increased level of efficiency (Roth *et al.*, 2014; Gillies, 2012). This example demonstrates the great potential for improvements in water use efficiency through the measurement and modeling process. Table 10.1 contains the summary results of

TABLE 10.1

Summary of Optimization of 614 Irrigation Events in the Australian Cotton Industry Collected from 1998 to 2012

	Measured (Average)	Optimized (Average)
Flow rate (L/s per 2 m width)	4.4	5.7
Run time (hours)	12.6	8.1
Total Water applied (ML/ha)	1.312	1.086
Application efficiency (%)	64.2	72.7
Application efficiency with tailwater recycling* (%)	75.7	84.7
Infiltration (mm)	105.5	91.5
Deep percolation per irrigation (mm)	27.9	13.9
Runoff (mm)	25.6	17.0
Potential water saving based on total applied (mm)		22.6
Potential water saving with tail water recycling (mm)		15.5

Source: Gillies, M., 2012. Benchmarking Furrow Irrigation Efficiency in the Australian Cotton Industry—Final Report. Cotton Catchment Communities, CRC.

* Assuming that 85% of the tail water is recovered.

614 individual measured irrigation events collected between 1998 and 2012, and the optimized performance of those events using practical changes to inflow rates and cutoff times. Here, through measurement and optimization, it was possible to raise the average application efficiency, with tailwater recycling from 75.7% to 84.7%, with a halving in the deep percolation and a reduction in runoff volume. It is important to realize that this data represents that of an industry that already believes they are efficient users of water, and the potential benefits may be much greater in other industries.

General rules for specific combinations of soil type and designs can be derived from scientific studies, but the best guidelines can only be achieved through optimizations based on models that have been calibrated with representative field data. Ideally, that field data should be collected from the fields under investigation.

Most modeling techniques focus on the measurement and simulation of a single furrow or bay, or perhaps a small number of furrows. It is up to the user to determine if variability is significant, and then to decide on how the measurements and simulations may be carried out to account for this variability. Ito *et al.* (2005) investigated the optimal sample size in terms of sampling cost and expected impact on profit with a recommended three samples for fields with less than 400 furrows, and nine samples for over 500 furrows. These numbers can be expected to change with the magnitude of the variability and the costs. Both the previously-mentioned SIRMOD and WinSRFR models can only simulate individual furrows. SISCO (Gillies and Smith, 2015), on the other hand, models the flows independently and combines the results to facilitate whole field assessment and optimization.

10.6 IMPROVED PERFORMANCE BY DESIGN

10.6.1 ALTERNATIVE FLOW REGIMES

The flow rate and its duration play a significant role in determining irrigation performance, and so it is no surprise that alternative flow regimes have been proposed to improve efficiency and uniformity. In traditional furrow irrigation, the inflow rate is conceptualized as remaining constant throughout the entire irrigation event. A constant flow during the event and across the season simplifies the management and has lower demands in terms of infrastructure. A range of alternative inflow hydrographs have been described in the literature and are summarized in Figure 10.5.

Surge irrigation is a technique where the flow that is typically higher than the constant flow used on that field is pulsed on and off, as shown in Figure 10.5a. In suitable soils, the technique can accelerate the advance phase while reducing the volume of applied water (Walker and Skogerboe, 1987) and increasing the uniformities of applied depths (Purkey and Wallender, 1989). The primary mechanisms responsible for this decline are a consolidation of the soil surface, filling of cracks with water and/or sediment load, surface sealing during the recession of each inflow cycle, and accelerated disintegration of soil aggregates as the result of rapid wetting (Kemper et al., 1988). Decreased infiltration rates translate to faster water advance rates so that the advance phase can be completed more quickly with less water (Horst et al., 2007), better uniformity of opportunity times (Turral et al., 1992), and a higher application efficiency over the field length due to less tailwater and deep percolation. As a consequence, surge irrigation has been shown to improve both the DU and Ea (El-Dine and Hosny, 2000; Trout and Kemper, 1983; Purkey and Wallender, 1989).

The effectiveness of surge irrigation is largely dependent on the soil texture and stability. Sandy loam soils are best suited to the technique, because of the large reduction in infiltration rates during the recession after each surge (Walker and Skogerboe, 1987). The impact of surge is greatest at the start of the season, where the soil has low stability (Kemper et al., 1988), but the effect diminishes later when the soil surface is already smooth and consolidated (Horst et al., 2007). Many clay soils experience a rapid decline in intake rates under traditional inflow, and therefore surging has limited benefit (Walker and Skogerboe, 1987). Field evaluations on two Australian clay soils with different clay contents (25% and 74% to 82%) experienced a slight reduction in infiltration with surging, but this was small relative to the inherent variation in infiltration properties (Smith et al., 1992). Simulation of the surging phenomenon is challenging, as it is difficult to predict or measure the changes in the soil due to the wetting and drying cycles (Fekersillassie and Eisenhauer, 2000).

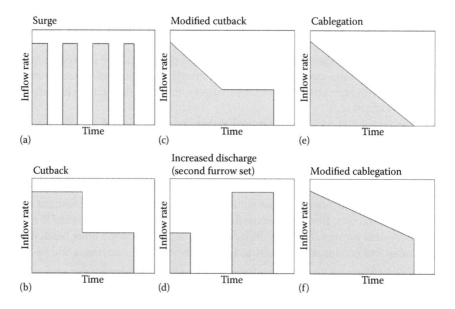

FIGURE 10.5 Sample inflow hydrographs for alternative inflow regimes. (From Gillies, M. H., 2008. Managing the Effect of Infiltration Variability on Surface Irrigation. *Faculty of Engineering and Surveying.* Toowoomba, University of Southern Queensland.)

Increasing inflow rates is a simple way to accelerate the advance phase, and therefore improve the uniformity of applied depths, but may result in high runoff losses, particularly if a long opportunity time is required. One approach to reduce runoff in these situations is to commence the irrigation with a high inflow rate, but then reduce the flow when the water has reached the end of the field. In this case, inflow rates are typically "cutback," often by as much as half, at a time approximately equal to or immediately after completion of the advance phase. A 50% reduction in discharge is often selected (Figure 10.5b) due to the ease of implementation. For example, in Spain the common practice uses a flexible pipe to supply water to each furrow through two holes, one of which is blocked at the completion of the advance time (Mateos and Oyonarte, 2005). Izadi *et al.* (1991) demonstrated an increase in *Ea* by 5% for cutbacks, while still satisfying 100% *Er*. Modified cutback irrigation, Figure 10.5c, is a variant where the inflow is slowly reduced during the initial phase and remains constant after water has reached the end of the field. Through modeling it was demonstrated that cutbacks and modified cutbacks have higher potential *Ea* than conventional irrigation, but are difficult to implement in practice (Alazba, 1999).

A combination of the surge and cutback methods, the increased discharge flow method (Figure 10.5d), or inverse cutback (Vazquez-Fernandez *et al.*, 2005) is another, less common alternative to the constant inflow regime. Water is applied to the furrows at a low flow rate until the advance reaches the end of the field. At this point, the flow is diverted to only half of the furrows and sustained until the desired volume is applied. The flow is then diverted to the remaining furrows, which are irrigated in the same manner. The primary advantage of this technique is that the reduction in soil erosion from the upper reaches of the furrow, as erosion is reduced when water advances over wet rather than dry soil. To improve uniformity, Vazquez-Fernandez (2006) proposed that the inflow should be increased earlier, when the advance has reached one quarter of the field length. This technique of increasing discharge has performed favorably on blocked furrows, resulting in *DU* values of 90.5% compared to 83.2% (Vazquez-Fernandez *et al.*, 2005), and 83% compared to 69% (Vazquez-Fernandez, 2006) against conventional constant inflow.

The major drawback of cutback irrigation is the difficulty in reducing the flow rates during the irrigation. Cablegation, an automated system using rigid-walled gated pipe (Kemper *et al.*, 1987; Shahidian and Serralheiro, 2012) provides a means of achieving a flow which starts at a high rate and decreases gradually with time, as shown in Figure 10.5e and f.

Despite the research in this area and the demonstration of potential benefits, the vast majority of surface irrigation systems are designed for a constant inflow rate. Success of the surge and cutback methods requires a high level of understanding of the soil infiltration rates that vary between events. With the exception of surge irrigation, it is unlikely that most of these strategies will see significant adoption outside of scientific studies.

10.6.2 ALTERNATE AND WIDE SPACED FURROW

In conventional furrow irrigation, the field is designed with furrows located between each plant row or bed, and each furrow receives the same inflow rate. One method to speed up the advance rate and potentially conserve water is to irrigate every second furrow or use a wider bed spacing (Stone *et al.*, 1979). Wide-spaced irrigation was shown to improve *Ea* by 15%, while offering a higher return per unit water than for every furrow irrigation (Stone and Nofziger, 1993). This practice is widespread in the Australian cotton industry, where it is commonplace to irrigate every second furrow (two meters) in a crop planted on a one-meter spacing. Host *et al.* (2005) found that alternate furrow irrigation could reduce the seasonal water use by 200–300 mm for long furrows. Alternate furrows also encourages the development of deep root systems, with greater biomass and higher uniformity across the row width. Hence, the plants can utilize a greater proportion of the applied water which may be reflected in reduced deep percolation.

There are two potential issues with alternate row irrigation, lateral infiltration problems and increased significance of variability. In soils with poor lateral soakage, alternate furrow irrigation may fail to fill the profile, which is further exacerbated for the initial irrigations of the season where the roots of the crop are underdeveloped. When every furrow is irrigated, the crop may extract water from either neighboring furrows, hence reducing the significance of between furrow variability. Trout and Mackey (1988) found that variability between furrows overstates the actual plant water variability between 25% and 50% due to this integration effect. In contrast, as irrigated furrow spacing is widened, the plant may only be able to draw water from one furrow, increasing the impact of non-uniformity. It might be expected that widely-spaced furrows have a negative impact on variability of advance between furrows, but this has not been observed in the field (Hodges *et al.*, 1989).

Intermittent furrow irrigation, a variant of alternate furrow, is carried out by applying water to every second furrow in one irrigation, and then to the alternate dry furrows during the following event, switching between each consecutive irrigation. Field trials suggest that this technique has the ability to reduce water use and improve root development, while retaining comparable yields to the conventional practice (Kang *et al.*, 2000), but studies on the use of this technique are limited. This approach may potentially allow growers to employ partial root zone drying (PRD), a practice used in drip irrigation to induce a stomatal response to reduce transpiration and induce beneficial plant stresses. Studies by White and Raine (2009) and McKeering (2004) indicated that it was not possible to induce a PRD effect in commercial cropping enterprises on cracking clay soils.

A newer approach used to assist farmers to cope with limited water situations is "skip-row" planting. Any conventional crop is grown with plants on a regular spacing, but in skip-row some of the regular rows are missed during planting. This planting technique, practiced by some dryland farmers, provides each plant with a larger soil volume for root exploration, maximizes the ability to capture rainfall during the season (Dugdale *et al.*, 2008), and improves the ability of the plant to cope with short-term drought (Payero *et al.*, 2012). Skip-row cropping will result in lower yields per unit area when water is plentiful, and may present challenges in terms of weed control (Gwathmey *et al.*, 2008). Some of the configurations used by Australian irrigated cotton farmers on regular, one-meter spacings are "single skip," where every third furrow is not planted; "double skip," with two planted rows followed by two skipped rows; and a "super single," where only every third row is planted. Experiments on cotton demonstrated that while skip plantings have lower yields per unit area, they offered greater irrigation water use efficiency in water limited conditions with yields of 4.66 and 2.99 bales/megaliter, compared to 2.21 bales/megaliter for conventional plantings (Dugdale *et al.*, 2008).

10.6.3 TAILWATER RECYCLING

As previously mentioned, capture and reuse of tailwater can greatly improve the efficiency of surface irrigation systems. Drainage recovery systems can serve as a means to reduce the loss of water during irrigations, but also provide an opportunity to harvest surface runoff during heavy rainfall events. Irrigation runoff also contains levels of contaminants, such as applied nutrients and pesticide residues, both of which are potentially harmful to nearby natural waterways. Hence, tailwater recycling has significant benefits for the farming enterprise and the surrounding environment. Adoption is dependent on the value of water and legislative pressures. For example, in the Australian cotton industry, it is estimated that 95% of surface irrigation systems employ tailwater recycling (Roth *et al.*, 2014), which is being driven by low water security and regulations regarding pesticide runoff. The drawbacks of tailwater recycling are a loss of land for construction of drainage channels, the necessity for water storages, and the pumping cost of lifting recycled water back to the storage or supply channel. Anecdotal evidence indicates that approximately 85% of tailwater can be recovered and used for the next irrigation with the appropriate drainage recycling system. Field measurements across 631 measured irrigation events showed that the Ea increased from 64.6% to 76.1% when accounting for tailwater recycling (Gillies, 2012).

10.6.4 IMPORTANCE OF LAND LEVELING

Field slope is often perceived as a major contributor to the performance of surface irrigation. Increasing the slope and lowering the surface roughness of the field does, in theory, increase the speed of water advance, and therefore improving uniformity and potential efficiency, but the impact compared to other factors (such as flow rate) is minimal (Renault and Wallender, 1996). Moreover, altering the field slope involves a significant earthmoving cost, and poses agronomic problems where the topsoil is shallow. Slope tends to have a greater impact on the rate of recession at the end of the irrigation and the ability to drain surface water following rainfall events. This statement is supported by modeling studies such as the work of Zapata and Playan (2000), who found that accounting for variation in soil elevation within a level basin slightly improved the fit to measured advance, but gave a greater improvement in the fit to the measured recession. Therefore, land leveling is more of a drainage design concern than an irrigation design concern.

In furrow irrigation, the flow into each furrow is a function of the water head relative to the control point, i.e., the downstream end of the siphon, the level of the gated pipe outlet, or the water level at the start of the furrow. Furrows are usually managed and controlled as a set, therefore the flows across all furrows in that set need to be as uniform as possible. Uniformity of inflows is only achievable with a tight control over the levels of the flow control point, which usually means that particular care must be taken with land leveling at the upstream end of the field.

Surface irrigation is reliant on gravity to distribute water across the field, therefore field slope and uniformity of that slope is important. While the magnitude of the slope has minimal effect, any areas of the field with zero or reverse slope could result in ponding or waterlogging, and produce a negative impact on the crop. For level basin irrigation or fields with close to zero slope (i.e., less than 0.01% slope) any imperfections in elevation or microtopography within the field can have a negative impact on the performance. The limited studies that have been conducted have struggled to separate the influence of field elevations from infiltration variability (Playan *et al.*, 1996). Widespread use of laser leveling has greatly improved the uniformity of field levels, but there are many instances where leveled fields may have imperfections due to the natural movement and settlement of the soil, machinery compaction, or cultivation. For example, Grabham (2012) conducted a detailed survey in a reverse grade furrow field and found that wheel-trafficked furrows were 16–18 mm lower and had a 12% larger cross section than non-trafficked furrows. In this field, the 17-mm elevation difference was equivalent to 170 m of normal design slope. The impact of microtopography on the irrigation uniformity is magnified if the field inflow is lower and the advance is slower.

In conclusion, it can be stated that while the magnitude of field slope has minimal impact on performance, surface irrigation systems should be designed and constructed so that the spatial distribution of elevations within furrows and/or bays permit water to be rapidly applied to the entire field and be drained quickly upon completion of the irrigation, or following rainfall.

10.6.5 OPTIMIZATION OF FIELD TOPOGRAPHY

As a natural extension of realizing the importance of field slope, a number of researchers have proposed moving away from the traditional constant grade used in furrow and border-check irrigation. Improvements to leveling equipment, including widespread use of GPS-controlled graders and scrapers, have opened up the possibility of implementing the desired elevation profile across the length of the furrow or the entire field. From a cost perspective, varying the slope over the field area might be desired so that the design can be achieved through minimal earthworks. These designs are relatively new, and limited data is available on their performance.

Some researchers have applied simulation models to optimize field topography to improve irrigation performance. For example, Gonzalez-Cebollada *et al.* (2016) devised a computer model which starts with a constant grade, simulates the irrigation, raises the elevation of areas with high infiltrated depths, and lowers the elevation of areas with lower depths. This process is repeated until the model converges on the optimum curved profile for the chosen one-dimensional or two-dimensional field. They were able to show that it is theoretically possible to achieve perfect uniformity through the optimal topography (González-Cebollada *et al.*, 2016). Morris *et al.* (2015) applied the ANUGA model to identify bay designs and microtopography that would decrease the pondage time following border-check irrigation of pasture. This is an interesting area of study, which may lead to practical solutions once this work moves to field trials.

Others, such as Playan *et al.* (2004), have identified benefits associated with level slope fields, but, as discussed, anything above a level field requires careful control over the field topography and/or a rapid water advance.

10.6.6 ALTERATION OF SOIL PROPERTIES

The soil infiltration rate is a major driver in the performance of surface irrigation. The soil is often seen as the uncontrollable factor, and optimal irrigation performance is achieved by matching the field design (e.g., field length and slope) and management (inflow rate and duration) to suit the soil conditions.

It is well-established that traffic-induced soil compaction leads to a reduction in the hydraulic conductivity of the soil. The magnitude of this change is difficult to predict, with its dependence on loading weight, machine speed, tire slippage, and tire pressure, among other factors (Antille *et al.*, 2016). The standard machine and implement widths used in most cropping industries results in a pattern of trafficked and non-trafficked furrows which causes a significant problem for managing furrow irrigation (Gillies, 2008). Conversely, cultivation will normally increase the infiltration rate of the soil but the effects are often temporary and may diminish after a small number of wetting and drying cycles.

Some have taken the approach that infiltration can be controlled through purposely compacting or smoothing furrows in a controlled fashion. For example, Hunsaker *et al.* (1999) dragged a weighted, torpedo-shaped object down every furrow to reduce roughness and infiltration. The benefits of furrow smoothing or shaping are minimal, and there has been limited use outside of field trials.

Many soils suffer from low infiltration or poor lateral water movement from the furrow into the plant row. The resulting irrigations will fail to satisfy the crop requirements, but may have excessive runoff losses. From limited experimental trials and grower investigations, it appears that mulch on the soil surface can improve infiltration rates. This behavior has been observed in sugarcane, where the mulch load resulting from a green trash blanket field (not burnt before harvest) appears to increase infiltration rates. The mechanisms for this behavior and the magnitude of the impact have yet to be established, but this phenomenon has been observed several times by the author and has been documented previously by Raine and Bakker (1996).

10.6.7 NEW FIELD LAYOUTS

With the continuous push to improve water, labor, and energy efficiencies, growers and cropping industries have been looking into modified layouts to improve the performance of surface irrigation. Most are touted as offering improved performance, but have not been adequately proven when compared to the traditional forms of surface irrigation. This section will touch on two examples of alternative surface systems; bankless irrigation and automated short furrow.

Bankless furrow irrigation systems have gained great interest in the past ten years in Australia in the drive to reduce labor costs associated with furrow irrigation. The name bankless emphases a wide variety of layouts, but the basic premise of bankless irrigation is that water flows from the supply channel straight into the field, without the need for any siphons or gated pipe to distribute that water (North, 2008). The field is divided into a series of bays which correspond to the number of furrows which can be irrigated simultaneously in one set. The supply channel runs along one side or both sides of the bays, and the bays are terraced so that the bays can be flooded consecutively, one at a time. The only control required is to open and close gate structures in the supply channel at the edge of each of the bays. The slope in the field may be close to zero, where water is applied from both ends of the furrow, and will be reverse grade where water is applied to the bottom end of the furrow and drains off the same end, or normal-sloped furrow where water is applied to the top end of the field. Figure 10.6 provides a schematic of the reverse grade, bankless layout. Another popular bankless layout, the "roll-over" design consists of a series of terraced bays on zero-slope, where water flows into those furrows from both ends of the field. Few of these layouts have been evaluated and those that have suggest great sensitivity to uniformity of field leveling and wheel traffic, elevation drop between bays, and bay to bay flow interactions (Grabham, 2012). Further research is required in order to benchmark these systems and provide guidance on how to optimize performance.

Automated short furrow irrigation (ASFI) is variant of furrow irrigation, where water is applied at several locations down the field length rather than just at the top of the field, effectively splitting a conventional furrow field into a series of short fields (Lecler *et al.*, 2008; Lecler *et al.*, 2011). ASFI can be envisaged as a hybrid between drip irrigation and surface irrigation. The fact that water is only applied to short sections of the field length overcomes much of the problems associated with the variation in opportunity times between furrows and over the length of the field. The main pipe runs along the side of the field and a length of small diameter (i.e., 10 mm), polyethylene pipe is connected to the mainline approximately every 30 m running across the furrows. Emitters are positioned along this lateral according to the furrow spacing. Water is applied to the first 30 m section for the required time, and then to each subsequent section in turn. A 30 m spacing was shown to be optimal for the soil at the trial site, but spacings of up to 200 m may be appropriate for clay soils. Measurements

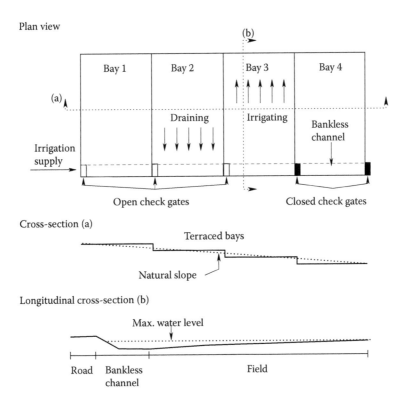

FIGURE 10.6 Schematic of a reverse grade bankless system. (From Grabham, M. K., 2012. *Performance Evaluation and Improvement of Bankless Channel Surface Irrigation Systems*. University of Southern Queensland.)

demonstrated that ASFI achieved high uniformity, even when only applying depths of 10 mm (Lecler *et al.*, 2008), something that would be impossible under normal furrow irrigation in that field.

The above two case studies serve only as examples of the approaches that are being adopted by farmers in order to reduce labor and/or improve water use efficiency. Any new irrigation technique poses new challenges and new questions in terms of measurement, potential performance, and optimal management strategies that should be measured and evaluated.

10.7 REMOTE CONTROL AND AUTOMATION

10.7.1 RATIONALE FOR AUTOMATION

In most cases, high performance is only possible through careful and timely management of the irrigation system. The farmers may know exactly how to achieve high efficiency, but usually this involves a higher degree of monitoring and control than existing management levels. As labor costs increase, this high level of control is no longer financially or practically possible. Automation of simple mundane tasks, such as opening and closing valves or gates, means that irrigation control decisions are implemented at the correct time or at least at the time desired by the irrigation manager. High labor costs and low labor availability is driving many irrigators to innovate and try new techniques, such as bankless irrigation in Australia (Grabham, 2012), despite the absence of any evidence of improved irrigation performance. Automation systems also provide a platform for system monitoring, which, in some cases, may be of greater value to the irrigator than the control aspects themselves. The ability to remotely observe system flowrates and pressures, channel levels, or soil moisture status all ultimately result in the potential for improved irrigation management. Automation does not guarantee optimum performance, but it does provide a practical way for farmers to achieve good performance while keeping their labor costs to a minimum.

10.7.2 AUTOMATED FLOW CONTROL FOR FURROW IRRIGATION

In furrow irrigation, water is either delivered to the field using pressurized pipelines or open channels. Where channels are used, the water is normally applied to the field using small siphon tubes positioned on each furrow (Figure 10.7a), or larger pipes buried in the channel wall (called "pipes though the bank" or PTBs), supplying water to a group of furrows (Figure 10.7b). For a pressurized supply to furrow fields, it is common practice to use a gated pipe with outlets positioned

(a)   (b)

FIGURE 10.7   Examples of (a) siphons and (b) large PTB pipes for furrow irrigation.

on each furrow, or valve risers (also known as alfalfa values) supplying a group of furrows. With the exception of pressurized pipe systems, none of these techniques easily lend themselves to mechanical actuation. The siphons are particularly troublesome, and involve a high labor cost, as each siphon must be started and stopped individually by hand.

There are a number of options in which a siphon head, ditch-supplied field can be modified in order to facilitate automated control. All of these options can be implemented without altering the on-farm, open-channel delivery system.

  a. Replacing the siphons with large PTB pipes spaced every 10 to 30 furrows (Figure 10.7b); the result is a system which requires modifications at the top of the field to ensure even spreading of the water. The problem is that this system requires control units at every PTB (Figure 10.8).
  b. Installation of gated pipe to distribute water evenly from the PTB's to the furrows (Figure 10.9); this option has the advantage of being able to irrigate a larger number of furrows with each PTB, but the use of gated pipe requires some additional water head compared to the other systems, which may not be achievable through large, open-channel irrigation systems.

FIGURE 10.8   PTB's automated using Aquator at Moree, NSW, Australia.

FIGURE 10.9   Gated pipe and PTB controlled with Rubicon radio unit at Narrabri, NSW, Australia.

   c. Conversion of the field into bankless irrigation by removing the channel wall on the field side and relevelling the field to the required configuration and grade; this option is costly due to the large amount of earthmoving required, and will create a system which is more difficult to manage.

   d. Building a secondary head ditch and set of small pipes through the bank (sPTBs) or "permanent siphons"; here a third head ditch bank is constructed parallel to the existing head ditch with a length corresponding to the number of furrows that will be irrigated simultaneously. A drop gate structure is installed to regulate flow from the normal head ditch into this secondary blind head ditch. This configuration involves loss of 6–8 m of crop area at the top of the field and careful placement of the sPTBs during construction.

For any control system to be cost effective it must be designed such that the control node is able to regulate flows over the largest possible area. The cost of automating the systems above is highest for option (a), and decreases in given order, with (c) and (d) being the lower cost alternatives due to the area controlled by each unit. For example, in trials conducted in the Australian cotton industry, option (a) had a control area of 3.3 ha per radio control unit, whereas option (d) has been trialed with a control area of 18 ha per radio control unit, meaning an approximate fivefold reduction in automation costs per unit area.

The systems shown in Figures 10.9 and 10.10 represent systems that have been recently installed on Cotton farms in Northern NSW, and are currently being evaluated by the Cotton Research and Development Corporation.

(a)

(b)   (c)

FIGURE 10.10 Secondary head ditch and permanent siphon layout: (a) Wee Waa, NSW, Australia, Rubicon gate between primary and secondary channel, (b) and (c) sPTB pipes in bank of secondary head channel.

Gated pipelines provide a furrow irrigation system which can be controlled and automated, compared to siphons. The pipe traverses along the top end of the field with an outlet positioned according to each irrigated furrow. The outlets may be individually or manually controlled, but it is more commonplace to control the flow to all outlets through a single valve at the upstream end of the pipeline. Gated pipes may be either rigid PVC; aluminum, as in Figure 10.11a; or flexible walled plastic, as shown in Figure 10.11b, to enable easy transport and storage. Surge irrigation, an early form of automated irrigation, involves switching the flow between two sides of a T-piece pipe fitting, as shown in Figure 10.11a, several times during the advance.

Cablegation (Figure 10.12) is another means to automate gated pipe, where a piston is pulled through the inside of the gated pipeline slowly, and allows water to flow out of each outlet and a

(a) (b)

FIGURE 10.11 Examples of gated pipe (a) rigid aluminium pipe and surge valve, (b) lay-flat gated pipe.

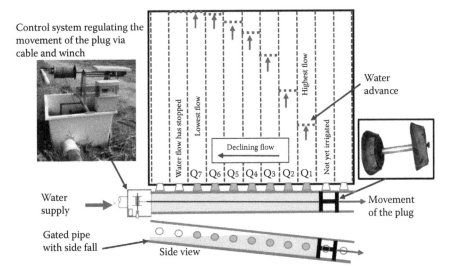

FIGURE 10.12 Schematic explaining the process of cablegation.

(a)   (b)

FIGURE 10.13 Automation of butterfly valves in gated pipe systems (controlled by WiSA actuators).

number immediately upstream as the piston passes that outlet (Kemper *et al.*, 1987). Shahidian and Serralheiro (2012) describe a system where the movement of the plug is adjusted according to the infiltration rate of the soil as detected by real-time field measurements.

Control of gated pipe systems merely require actuation of the valves at the hydrant before the water enters the pipe. Surge valves (Figure 10.11a) and linear actuators on butterfly valves (Figure 10.13) are two ways to achieve this. Surge valves provide the ability to switch the flow between two branches of the T-piece, and must be combined with a secondary upstream valve if the system is to be fully automated. The valves shown in Figure 10.13 represent automation systems which are currently being evaluated by Sugar Research Australia (SRA) where valve actuators can be operated and scheduled remotely from a smartphone via an infield radio network (Gillies *et al.*, 2017).

10.7.3 Flow Control for Basin and Border Check Irrigation

For level basin and border check irrigation systems, water is typically delivered to the field using open channels, and then is diverted into the bay or between bays using a small number of gated control structures. For border check irrigation, there is normally only one inflow point per bay. In Australia, there are a number of technology providers that sell these actuated gates to farmers (e.g., Figure 10.14a and b). Automation of these systems involves replacing the cable winch drop gate with equipment that can be opened or closed with electric linear or rotary actuators of some form, as shown in Figure 10.14a and b. Automation is also possible for low pressure on-farm pipe distribution systems such as the pipe and riser outlet in Figure 10.14c.

The most common options available for automated control of flow into surface irrigation fields only involve starting and stopping the flow, and there is typically no ability to regulate the flowrate. One way to achieve flow regulation is to alter water levels or heads upstream of the control structure. This level of control would require accurate monitoring of the incoming flow rate or water levels.

10.7.4 Sensors for Monitoring and Control

A major benefit of automated systems is the ability to continuously and remotely monitor the status of the irrigation system. These measurements may be used to inform the automation process directly, to allow the farm manager to check the system is running properly, or serve as sources of data for record keeping.

(a)   (b)   (c)

FIGURE 10.14 Automated border check outlets (a) Rubicon drop gate, (b) Automated Padman stop, and (c) AWMA pipe and riser outlet.

In surface irrigation, the deceleration of the water front advance is an integrator of the soil infiltration, water inflow, and hydraulic behavior occurring in the field. Measurement of water advance is a routine part of the field evaluation and modeling process, as mentioned previously, but it can also be a useful indicator of the best time to stop the irrigation. A number of the commercial surface automation systems include some form of advanced detection probes, which can be either programmed to send the irrigation manager a notification (i.e., a cellular phone text message), cause the system to switch to the next irrigation set, or as the trigger for real time control (Section 10.7.7). The water advance may be measured using a contact sensor (Figure 10.15a and b), a water depth sensor (Figure 10.15c), or a buried soil moisture sensor (Figure 10.15d) to avoid damage from machinery or grazing animals. The position of the sensor(s) along the length of the field is determined by the required time period before notice of the control decision, and the hydraulic behavior of the system and would ideally be informed by an irrigation assessment at the site. Research indicates that a single sensor may be sufficient (Smith *et al.*, 2016).

Pressure and flow information may be of general interest to the irrigation manager, but can also serve as a simple way to detect system problems before any damage is caused. For example, if flow is zero and the pressure is high, there is a possibility that a valve may be jammed. If the pressure is low and flow is zero, then perhaps the water level at the pump intake has dropped, and the pump needs to

(a)   (b)   (c)   (d)

FIGURE 10.15 (a) G&M Poly water monitor (from G&MPoly, 2015, Aquator. http://gmpoly.com.au /product_range/aquator.), (b) SMS Chatterbox Bay Monitor (Padman, 2015), (c) Rubicon field depth sensor, (d) Decagon soil moisture probe.

(a)   (b)   (c)

FIGURE 10.16 Water level measurement using (a) WIKA PST for pressurized systems, (b) submersible PST inside vertical cylinder, (c) Rubicon ultrasonic level sensor.

be shut down to avoid damage. Measurements of flow collected over the season provide a continuous record of the system performance and permit the farmer to identify fields or sections of fields with an irrigation performance issue.

Most surface irrigation systems rely on gravity flow channels, and the irrigation system is reliant on water levels in these channels being maintained at appropriate levels without overtopping. Over recent years, systems such as Total Channel Control™ (Mareels *et al.*, 2005; Nayar and Aughton, 2007) have become widespread in scheme level control, but are yet to be implemented at the farm level. Measurements of channel water levels are the first stage of this channel control system. Two practical approaches to measure water level are low range pressure sensing transducers (PST) (Figure 10.16a and b) or ultrasonic level sensors (Figure 10.16c). The pressures and water levels in these systems are generally low, therefore the sensors should be capable of sub-centimeter accuracy or better.

Generally, management is simplified if the system is able to maintain channels at a consistent elevation during the event and for subsequent events on that same field. High performance will be difficult to maintain if the channel levels upstream of the field are unpredictable. Water levels measured at key locations can be used to infer system flowrates, and often at a lower cost and greater simplicity than direct flow measuring devices.

Other sensing devices may be beneficial to the surface irrigation system, in addition to those listed above, for direct feedback of the irrigation while it is underway. Multi-depth soil moisture sensors can be used to assist with irrigation scheduling. Tipping bucket rain gauges are another useful device that can be connected to some of the automation systems and allow a more complete field water balance to be developed.

10.7.5 Real Time Control

Typically, the process of surface irrigation optimization involves collecting measurements from an irrigation event, calibration of a simulation model, use of that model to identify appropriate adjustments, and implementation of those adjustments for the next irrigation. The idea of real-time control requires that the measurement, optimization, and control are integrated into an automated system. This offers the possibility to almost completely remove the problem of temporal variability in infiltration rates that occurs between irrigation events. True real-time control systems are rare, but many researchers have used simulation models to investigate the possible performance under such a system. For example, Raine *et al.* (1997) found that irrigation application efficiencies for a sugar cane field in Queensland, Australia measured over a season ranged between 27% to 55%, with an average of 41% and an Er of 98%. Optimizing each irrigation individually using SIRMOD modeling indicated that with real-time control based on field measurements, the Ea could be increased to 93% while maintaining a high Er 90% (Raine et al., 1997). Similarly, Reddell and Latimer (1987) concluded that a real-time control system would be able to improve the Ea to 90%, with distribution uniformities of 85%.

Infield implementation of real-time control systems is rare. The high requirement of these systems for automated sensing and water control requires a great degree of technical skill and a higher cost compared to traditional surface irrigation. Some notable examples of real time control systems include:

a. The Advanced Rate Furrow Irrigation System (ARFIS), described by Reddell and Latimer (1987), which was comprised of a water advance sensor, telemetry system, microcomputer, and two-position solenoid valve. Constant inflow is supplied to the furrow for the duration of the advance phase. When water is detected, the system calculates the volume required and pulses the flow in a surge-like technique for a duration long enough to satisfy that volume.

b. The RIOS system, where a series of up to 14 detectors along the length of the field are connected via cable to a computer unit which predicts the infiltration rate and determines when the irrigation is to be shut off.

c. The real time optimization system, described by Koech et al. (2014), based around a computer connected to a sensor and control valve via a radio network. The flow rate through siphons and PTBs was inferred from logged channel levels, and when a single advance sensor in the field detected water, it launched the full hydrodynamic model to self-calibrate based on the measurements, and optimized the irrigation time to maximize performance.

d. The PLC controlled cablegation system described by Shahidian and Serralheiro (2012), where advance sensors positioned at the midpoint and end of the furrow are used to determine the speed at which the plug would be moved through the gated pipe.

Real-time control systems must be able to collect advanced measurements at different locations in the field, determine flowrates into furrows, and transfer this data to a central processing unit that will optimize and implement the control.

10.7.6 THE CONCEPT OF PRECISION IRRIGATION

Precision irrigation, once a term used to describe drip irrigation, is defined as applying precise amounts of water to crops at precise locations at precise times which might imply uniform application over the field (Smith and Baillie, 2009). A better definition was given by Smith et al. (2010) who conceptualized that a precision irrigation system is one that can:

1. Determine the timing, magnitude, and spatial pattern of applications for the next irrigation to give the best chance of meeting the seasonal objective (i.e., maximization of yield, water use efficiency, or profitability)

2. Be controlled to apply exactly (or as close as possible to) what is required

3. Through simulation or direct measurement, know the magnitude and spatial pattern of the actual irrigation applications, and the soil and crop responses to those applications

4. Utilize these responses to best plan the next irrigation

The components, scale, and form of this system may vary depending on the irrigation application system type, but there is no reason why this concept does not apply to all forms of irrigation.

With the advent of variable rate center-pivot technology, irrigation applications can now be tailored precisely to match the heterogeneity of the soil (Haghverdi et al., 2016). One robust approach is to base this irrigation prescription on the soil available water holding capacity of the root zone, which can be captured through use of apparent electrical conductivity processed with the appropriate geostatic methods (Haghverdi *et al.*, 2015). Hedley and Yule (2009) provide an example of how this spatial EC mapping can be used to predict the daily water status of the soil.

The main difference between surface irrigation and more precise application techniques is the size of the management unit. For example, variable rate center pivot irrigators may be able to vary water application on the scale of several meters, but for surface irrigation, the minimum control size may be in the order of hectares. Conventional irrigation management, including surface irrigation, strives for uniform application. A key concept of precision irrigation is that the irrigation system should be able to spatially vary the application to achieve the best result in terms of crop yield, quality, or profit.

10.7.7   FROM AUTOMATED TO SMART AUTONOMOUS SYSTEMS

The ultimate aim of the irrigation system is to apply water where and when it is needed by the crop. Automated systems provide a tool which facilitates this aim, and offers the potential for management decisions to be made autonomously. A smart autonomous irrigation system is defined here as precision irrigation which is being achieved through use of an irrigation automation system. The concept of smart autonomous irrigation is still in its infancy, with limited commercial applications but a great amount of scientific interest.

The best integrator of irrigation history and field conditions is the crop growth. It is unsurprising that most efforts to achieve precision irrigation outcomes focus on crop measurements or crop modeling. The SOFIP model (Mailhol *et al.*, 2005) is one good example of using crop modeling to inform surface irrigation management. SOFIP is comprised of a hydraulic model (RAIEOPT) to predict water application, a crop model (PILOTE) to quantify the soil water use and predict crop yield, and a parameter generator (PG) to predict soil infiltration parameters required to run the RAIEOPT model. While the model developed by Mailhol *et al.* (2005) did not use crop measurements, it was unique in the way it accounted for furrow to furrow variability in the optimization process.

McCarthy *et al.* (2010a) developed the VARIwise software, which applies control strategy theory to provide site specific irrigation of crops. Control systems require the following three components: sensors that measure the status of the system; control strategy, that processes these measurements to determine the control outcome that achieves performance objective (optimization of productivity or efficiency); and actuation, that implements control strategy decision. Typical control strategies use a model that represents the interactions in the system to perform mathematical optimization of inputs to achieve a set of performance objectives. VARIwise applies this control theory to agriculture, where crop production models are used for iteratives executed with different irrigation application dates and volumes, until the combination that achieves the performance objective is reached.

The basic architecture of the VARIwise system showing major inputs and outputs is given in Figure 10.17. VARIwise first divides the field into a large number of management cells (which may be as small as 1 m^2). VARIwise simulates crop response to in each cell using a crop production model, and uses spatial soil, plant, irrigation, and weather data to calibrate the crop model for each cell in the field on a sub-day and time-scale. The resulting simulations are able to predict daily crop growth and soil water use for every cell in the field. Infield observations of key plant parameters are used to adjust predicted plant growth back to reality. Promising novel approaches such as described by McCarthy *et al.* (2010b) are using non-destructive machine vision to assess crop growth. The complete VARIwise system can then be used to formulate optimal irrigation strategies to achieve the desired crop outcome. Assessments completed in cotton have shown that VARIwise can provide improved yields and water use efficiencies in both homogenous and spatially varied fields for surface irrigation systems (McCarthy *et al.*, 2012) and overhead irrigation systems (McCarthy *et al.*, 2014a, McCarthy *et al.*, 2014b). Work is continuing to determine the most appropriate control strategy and improve the way in which infield data is collected.

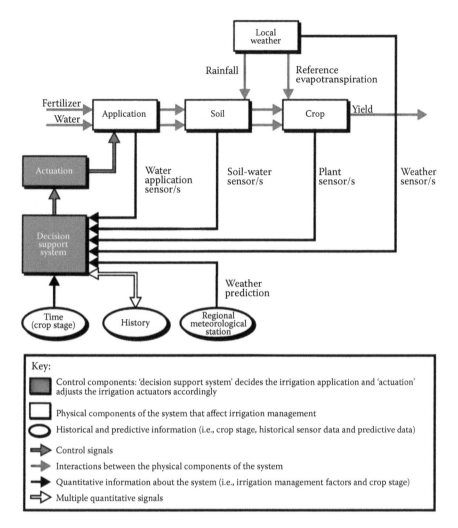

FIGURE 10.17   Conceptual adaptive control system for variable-rate irrigation—the basis of the simulation model VARIwise.

Other forms of irrigation, such as drip and center pivot machines have enjoyed far more attention in terms of automated scheduling systems. One notable source of data is the canopy temperature, which provides a simple way to assess plant water status over large areas (González-Dugo *et al.*, 2006). Canopy temperature has been used to inform the timing aspects of irrigation scheduling for drip irrigation (Wanjura *et al.*, 1992) and to determine precision application patterns for variable rate center pivot irrigators (Manuel *et al.*, 2015). It will be only a matter of time until these same technologies are implemented in surface irrigation systems now that automation is becoming more widespread.

10.8   CONCLUSION

Surface irrigation has and will continue to play an important role in agricultural production across the globe. While remaining unchanged for thousands of years, surface irrigation, like many aspects of agricultural production, is now benefiting from the range of technological advances available to our modern society. Over the past 50 years, there has been a body of research leading to measurement tools, modeling approaches, new field designs, and water application techniques. Much of this research has been implemented with promising results. More recently, there has been a push for automated systems which aid farmers in implementing these improved practices. Looking toward the future, automated systems combined with crop and environmental monitoring will no doubt lead to further improvements in water use efficiency and farm productivity.

NOTATION

τ	the opportunity or ponding time (s or minutes)
A	cross sectional area of flow (m^2)
a	empirical Kostiakov-Lewis exponent term
ADU	absolute distribution uniformity (%)
AELQ	application efficiency of low quarter (%)
DU	low quarter distribution uniformity (%)
DURZ	low quarter distribution uniformity of the depths added to the root zone (%)
Ea	application efficiency (%)
Ea_R	application efficiency taking into account tailwater recycling (%)
Er	requirement efficiency (%)
f_0	empirical Kostiakov-Lewis steady intake term
g	acceleration due to gravity (m s^{-2})
I	soil infiltration rate per meter length of furrow (m^3 m^{-1} s^{-1}).
k	empirical Kostiakov-Lewis coefficient
N_D	number of positions with known infiltrated depth
PRD	partial root zone drying
Q	Discharge (m^3 s^{-1})
R	efficiency of the tail water recycling system (fraction)
S_0	bed slope as a fraction
S_f	friction slope
T	top width of flow (m)
t	time (s or minutes)
T_{CO}	cutoff time (minutes)
$Vol_{D.Drainage}$	volume of water lost to deep drainage below the root zone (m^3)
Vol_{In}	volume of water applied to the field (m^3)
Vol_{req}	volume of water required to replenish the soil moisture deficit over the field (m^3)
Vol_{Runoff}	volume of water lost through surface drainage or runoff during the irrigation (m^3)
x	distance from the upstream end of the field (m)
y	depth of flow (m)
$D(i)$	infiltrated depth at a specified position i (m)
$\overline{D}$	average depth infiltrated across the field length (m)
D_{req}	soil moisture deficit (m)
$D_{RZ}(i)$	depth of infiltration stored in the root zone at a specified position i (m)

REFERENCES

Alazba, A. A. (1999). Simulating furrow irrigation with different inflow patterns. *Journal of Irrigation and Drainage Engineering* 125, 12–18.

Antille, D. L., Bennett, J. M. and Jensen, T. A. (2016). Soil compaction and controlled traffic considerations in Australian cotton-farming systems. *Crop and Pasture Science* 67, 1–28.

Austin, N. R. and Prendergast, J. B. (1997). Use of kinematic wave theory to model irrigation on cracking soil. *Irrigation Science* 18, 1–10.

Bali, K. and Wallender, W. W. (1987). Water application under varying soil and intake opportunity time. *Transactions of the ASAE* 30, 442–8.

Bautista, E., Clemmens, A. J., Strelkoff, T. S., and Schlegel, J. (2009). Modern analysis of surface irrigation systems with WinSRFR. *Agricultural Water Management* 96, 1146–54.

Bautista, E., Schlegel, J. L., and Clemmens, A. J. (2016). The SRFR 5 Modeling System for Surface Irrigation. *Journal of Irrigation and Drainage Engineering* 142.

Bautista, E. and Wallender, W. W. (1985). Spatial variability of infiltration in furrows. *Transactions of the ASAE* 28, 1846–51.

Burney, J. A. and Naylor, R. L. (2012). Smallholder irrigation as a poverty alleviation tool in sub-Saharan Africa. *World Development* 40, 110–23.

Burt, C. (1995). *The Surface Irrigation Manual: A Comprehensive Guide to Design and Operation of Surface Irrigation Systems*. Exeter: Waterman Industries.

Cantor, L. M. (1970). *A World Geography of Irrigation*. New York: Praeger Publishers Inc.

Childs, J. L., Wallender, W. W., and Hopmans, J. W. (1993). Spatial and seasonal variation of furrow infiltration. *Journal of Irrigation and Drainage Engineering* 119, 74–90.

Clemmens, A. J. (1983). Infiltration equations for border irrigation models. *Advances in Infiltration*, Proc. of the National Conference, Chicago, IL. ASAE, St. Joseph, Michigan.

Clemmens, A. J. (1991). Direct solution to surface irrigation advance inverse problem. *Journal of Irrigation and Drainage Engineering* 117, 578–93.

Clemmens, A. J. (2003). Field Verification of a Two-Dimensional Surface Irrigation Model. *Journal of Irrigation and Drainage Engineering* 129, 402–11.

Dalton, P., Raine, S. R., and Broadfoot, K. (2001). Best management practices for maximizing whole farm irrigation efficiency in the Australian cotton industry. Final report to the Cotton Research and Development Corporation. National Center for Engineering in Agriculture Report. 179707/2. USQ, Toowoomba.

Dugdale, H., Harris, G., Neilsen, J., Richards, D., Roth, G., Williams, D., and Wigginton, D. (2008). WATERpak—A guide for irrigation management in cotton and grain farming systems. *D. C. S.* Third Ed. (Ed. David Wigginton.) Cotton Research and Development Corporation and Cotton Catchment Communities, CRC, Narrabri, NSW.

Ebrahimian, H. and Liaghat, A. (2011). Field evaluation of various mathematical models for furrow and border irrigation systems. *Soil and Water Research* 6, 91–101.

El-Dine, T. G. and Hosny, M. M. (2000). Field evaluation of surge and continuous flows in furrow irrigation systems. *Water Resources Management* 14, 77–87.

Elliott, R. L. and Walker, W. R. (1982). Field evaluation of furrow infiltration and advance functions. *Transactions of the ASAE* 25, 396–400.

Fekersillassie, D. and Eisenhauer, D. E. (2000). Feedback-controlled surge irrigation: I. Model development. *Transactions of the American Society of Agricultural Engineers* 43, 1621–30.

Fonteh, M. and Podmore, T. (1994) Furrow irrigation with physically-based, spatially-varying infiltration. *Journal of Agricultural Engineering Research* 57, 229–36.

G&MPoly. (2015). Aquator. http://gmpoly.com.au/product_range/aquator.

Gharbi, A., Daghari, H. and Cherif, K. (1993). Effect of flow fluctuations on free draining, sloping furrow, and border irrigation systems. *Agricultural Water Management* 24, 299–319.

Gillies, M. (2012). Benchmarking furrow irrigation efficiency in the Australian Cotton Industry—Final Report. Cotton Catchment Communities, CRC.

Gillies, M., Attard, S., Jaramillo, A., Davis, M., and Foley, J. (2017). Smart Automation of Furrow Irrigation in the Sugar Industry. Australian Society of Sugar Cane Technologists. Cairns.

Gillies, M. H. (2008). Managing the effect of infiltration variability on surface irrigation. *Faculty of Engineering and Surveying*. Toowoomba, University of Southern Queensland.

Gillies, M. H. and Smith, R. J. (2005). Infiltration parameters from surface irrigation advance and runoff data. *Irrigation Science* 24, 25–35.

Gillies, M. H. and Smith, R. J. (2015). SISCO—Surface Irrigation Simulation Calibration and Optimisation. *Irrigation Science*.

González-Cebollada, C., Moret-Fernández, D., Buil-Moure, I., and Martínez-Chueca, V. (2016). Optimization of field topography in surface irrigation. *Journal of Irrigation and Drainage Engineering* 142.

González-Dugo, M. P., Moran, M. S., Mateos, L., and Bryant, R. (2006). Canopy temperature variability as an indicator of crop water stress severity. *Irrigation Science* 24, 233.

Grabham, M. K. (2012). Performance evaluation and improvement of bankless channel surface irrigation systems. University of Southern Queensland.

Gwathmey, C. O., Steckel, L. E., and Larson, J. A. (2008). Solid and skip-row spacings for irrigated and nonirrigated upland cotton. *Agronomy Journal* 100, 672–80.

Haghverdi, A., Leib, B. G., Washington-Allen, R. A., Ayers, P. D., and Buschermohle, M. J. (2015). High-resolution prediction of soil available water content within the crop root zone. *Journal of Hydrology* 530, 167–79.

Haghverdi, A., Leib, B. G., Washington-Allen, R. A., Buschermohle, M. J., and Ayers, P. D. (2016). Studying uniform and variable rate center pivot irrigation strategies with the aid of site-specific water production functions. *Computers and Electronics in Agriculture* 123, 327–40.

Hedley, C. B. and Yule, I. J. (2009). A method for spatial prediction of daily soil water status for precise irrigation scheduling. *Agricultural Water Management* 96, 1737–45.

Hodges, M. E., Stone, J. F., Garton, J. E., and Weeks, D. L. (1989). Variance of water advance in wide-spaced furrow irrigation. *Agricultural Water Management* 16, 5–13.

Hornbuckle, J. W. (1999). Modelling furrow irrigation on heavy clays, high water tables and tiled drained soils using SIRMOD in the Murrumbidgee irrigation area. Armidale, NSW, Australia, University of New England.

Horst, M. G., Shamutalov, S. S., Goncalves, J. M., and Pereira, L. S. (2007). Assessing impacts of surge-flow irrigation on water saving and productivity of cotton. *Agricultural Water Management* 87, 115–27.

Horst, M. G., Shamutalov, S. S., Pereira, L. S., and Goncalves, J. M. (2005). Field assessment of the water-saving potential with furrow irrigation in Fergana, Aral Sea Basin. *Agricultural Water Management* 77, 210–31.

Hunsaker, D. J., Clemmens, A. J., and Fangmeier, D. D. (1999). Cultural and irrigation management effects on infiltration, soil roughness, and advance in furrowed level basins. *Transactions of the ASAE* 42, 1753–64.

Israelsen, O. W. and Hansen, V. E. (1962). *Irrigation Principles and Practices*. New York: John Wiley and Sons.

Ito, H., Wallender, W. W., and Raghuwanshi, N. S. (2005). Optimal sample size for furrow irrigation design. *Biosystems Engineering* 91, 229–37.

Izadi, B., Studer, D., and McCann, I. (1991). Maximizing set-wide furrow irrigation application efficiency under full irrigation strategy. *Transactions of the ASAE* 34, 2006–14.

Kang, S. Z., Shi, P., Pan, Y. H., Liang, Z. S., Hu, X. T., and Zhang, J. (2000). Soil water distribution, uniformity and water-use efficiency under alternate furrow irrigation in arid areas. *Irrigation Science* 19, 181–90.

Katopodes, N. D., Tang, J. H., and Clemmens, A. J. (1990). Estimation of surface irrigation parameters. *Journal of Irrigation and Drainage Engineering* 116, 676–96.

Kemper, W. D., Trout, T. J., Humpherys, A. S., and Bullock, M. S. (1988). Mechanisms by which surge irrigation reduces furrow infiltration rates in a silty loam soil. *Transactions of the ASAE* 31, 821–9.

Kemper, W. D., Trout, T. J., and Kincaid, D. C. (1987). Cablegation: Automated supply for surface irrigation. (Ed. D. Hillel) *Advances in Irrigation*. London: Academic Press Inc.

Khanna, M. and Malano, H. M. (2006). Modelling of basin irrigation systems: A review. *Agricultural Water Management* 83, 87–99.

Khanna, M., Malano, H. M., Fenton, J. D., and Turral, H. (2003). Two-dimensional simulation model for contour basin layouts in Southeast Australia. II: Irregular shape and multiple basins. *Journal of Irrigation and Drainage Engineering* 129, 317–25.

Koech, R. (2012). Automated real-time optimisation for control of furrow irrigation. *Faculty of Engineering and Surveying*. Toowoomba, University of Southern Queensland.

Koech, R. K., Smith, R. J., and Gillies, M. H. (2014). A real-time optimisation system for automation of furrow irrigation. *Irrigation Science*, 1–9.

Koech, R., Smith, R., and Gillies, M. (2015). Trends in the use of surface irrigation in Australian agriculture. *Australian Water Association Water Journal* 42, 11.

Kruse, E. G. (1978). Describing irrigation efficiency and uniformity. *Journal of Irrigation and Drainage Engineering* 104, 35–41.

Kulecho, I. K. and Weatherhead, E. K. (2005). Reasons for smallholder farmers discontinuing with low-cost microirrigation: A case study from Kenya. *Irrigation and Drainage Systems* 19, 179–88.

Lecler, N., Mills, D., and Smithers, J. (2008). *Automated Short Furrow: A System for Precision Irrigation*. Sixth Australian Controlled Traffic Farming Conference. Dubbo, NSW, ACTFA.

Lecler, N., Mills, D., and Smithers, J. (2011). Design and evaluation of an automated short furrow irrigation system. 2011 Society for Engineering in Agriculture Conference: *Diverse Challenges, Innovative Solutions*. Surfers Paradise, QLD, Australia, Engineers Australia.

Lentz, R. D. and Bjorneberg, D. L. (2001). Influence of irrigation water properties on furrow infiltration: Temperature effects. Retrieved from http://topsoil.nserl.purdue.edu/fpadmin/isco99/pdf/ISCOdisc/SustainingTheGlobalFarm/P196-Lentz.pdf.

Mailhol, J. C. (2003). Validation of a predictive form of Horton infiltration for simulating furrow irrigation. *Journal of Irrigation and Drainage Engineering* 129, 412–21.

Mailhol, J. C., Ruelle, P., and Popova, Z. (2005). Simulation of furrow irrigation practices (SOFIP): A field-scale modelling of water management and crop yield for furrow irrigation. *Irrigation Science*, 1–12.

Maisiri, N., Senzanje, A., Rockstrom, J., and Twomlow, S. J. (2005). On-farm evaluation of the effect of low-cost drip irrigation on water and crop productivity compared to conventional surface irrigation systems. *Physics and Chemistry of the Earth, Parts A/B/C*, 30, 783–91.

Manuel, A. A., Susan, A. O. S., and Steven, R. E. (2015). Advances in irrigation management tools: The development of ARSmartPivot. 2015 ASABE/IA Irrigation Symposium: Emerging Technologies for Sustainable Irrigation—A Tribute to the Career of Terry Howell. C.OMMAS.R.X.X.X., Proc.

Mareels, I., Ooi, S. K., Aughton, D., and Oakes, T. (2005). Total Channel Control™—The value of automation in irrigation distribution systems. *SCADA and Related Technologies for Irrigation District Modernization*, 11.

Mateos, L. and Oyonarte, N. A. (2005). A spreadsheet model to evaluate sloping furrow irrigation accounting for infiltration variability. *Agricultural Water Management* 76, 62–75.

McCarthy, A. C., Hancock, N. H., and Raine, S. R. (2010a). VARIwise: A general-purpose adaptive control simulation framework for spatially and temporally varied irrigation at sub-field scale. *Computers and Electronics in Agriculture* 70, 117–28.

McCarthy, C. L., Hancock, N. H., and Raine, S. R. (2010b). Applied machine vision of plants: A review with implications for field deployment in automated farming operations. *Intelligent Service Robotics* 3, 209–17.

McCarthy, A. C., Hancock, N. H., and Raine, S. R. (2014a). Development and simulation of sensor-based irrigation control strategies for cotton using the VARIwise simulation framework. *Computers and Electronics in Agriculture* 101, 148–62.

McCarthy, A. C., Hancock, N. H., and Raine, S. R. (2014b). Simulation of irrigation control strategies for cotton using model predictive control within the VARIwise simulation framework. *Computers and Electronics in Agriculture*, 101, 135–147.

McCarthy, A., Smith, R., and Hancock, N. (2012). Real-time adaptive control of furrow irrigation: Preliminary results of cotton field trial. Proc. of the Irrigation Australia Conference (IAL 2012). Irrigation Australia Ltd.

McClymont, D. J., and Smith, R. J. (1996). Infiltration parameters from optimization on furrow irrigation advance data. *Irrigation Science* 17, 15–22.

McKeering, L. M. (2004). Evaluating the potential to impose partial root zone drying (PRD) on clay soils in vommercial cotton production systems. *University of Southern Queensland Faculty of Engineering and Surveying*.

McKenzie, N. J., and Cresswell, H. P. (2002). Selecting a method for hydraulic conductivity. (Ed. N. J. McKenzie, K. J. Coughlan, and H. P. Cresswell.) *Soil Physical Measurement and Interpretation for Land Evaluation*. Collingwood: CSIRO Publishing.

Merriam, J. L., Shearer, M. N., and Burt, C. M. (1983). Evaluating irrigation systems and practices. (Ed. M. E. Jensen.) *Design and operation of farm irrigation systems: Irrigation and Drainage Notes*. Michigan: American Society of Agricultural Engineers.

Morris, M., Githui, F., and Hussain, A. (2015). Application of Anuga as a 2D surface irrigation model. MODSIM2015, Twenty-First International Congress on Modelling and Simulation. Broadbeach, Queensland, Australia. Modelling and Simulation Society of Australia and New Zealand.

Nayar, M. and Aughton, D. (2007). Canal automation and cost recovery—Australian experience using rubicon total channel control™. *Agriculture and Rural Development Discussion Paper 33—The Role of Technology and Institutions in the Cost Recovery of Irrigation and Drainage Projects*. The World Bank.

North, S. (2008). A review of basin (contour) irrigation systems I: Current design and management practices in the Southern Murray-Darling Basin, Australia. *Irrigation Matters Series*. CRC for Irrigation Futures.

Padman. (2015). Padman Stops—Innovative Irrigation Solutions. Retrieved from http://www.padmanstops.com.au/.

Payero, J., Robinson, G., Harris, G., and Singh, D. (2012). Water extraction of solid and skip-row cotton. Sixteenth Agronomy Conference 2012—Capturing Opportunities and Overcoming Obstacles in Australian Agronomy. University of New England, Armidale, NSW.

Pereira, L. S., Gonçalves, J. M., Dong, B., Mao, Z., and Fang, S. X. (2007). Assessing basin irrigation and scheduling strategies for saving irrigation water and controlling salinity in the upper Yellow River Basin, China. *Agricultural Water Management* 93, 109–22.

Playán, E., Faci, J. M., and Serreta, A. (1996). Characterizing microtopographical effects on level-basin irrigation performance. *Agricultural Water Management* 29, 129–45.

Playán, E., Rodríguez, J., and García-Navarro, P. (2004). Simulation model for level furrows. I: Analysis of field experiments. *Journal of Irrigation and Drainage Engineering* 130, 106–12.

Postel, S. (1999). *Pillar of Sand: Can the irrigation miracle last?* New York; London: Norton.

Purkey, D. R. and Wallender, W. W. (1989). Surge flow infiltration variability. *Transactions of the ASAE* 32, 894–900.

Raghuwanshi, N. S. and Wallender, W. W. (1996). Modeling seasonal furrow irrigation. *Journal of Irrigation and Drainage Engineering* 122, 235–42.

Raine, S. R. and Bakker, D. M. (1996). Increased furrow irrigation efficiency through better design and management of cane fields. *Australian Society of Sugar Can Technologists*. Mackay.

Raine, S. R., McClymont, D. J., and Smith, R. J. (1997). The development of guidelines for surface irrigation in areas with variable infiltration. (Ed. B. T. P. O. Egan.) *Australian Society of Sugar Cane Technologists*. Cairns, Australia: Watson Ferguson.

Rayej, M. and Wallender, W. W. (1988). Time solution of kinematic-wave model with stochastic infiltration. *Journal of Irrigation and Drainage Engineering* 114, 605–21.

Reddell, D. L. and Latimer, E. A. (1987). Field evaluation of an advance rate feedback irrigation system. Irrigation Systems for the Twenty- First Century, Proc. 1987 Irrigation and Drainage Division Specialty Conference. Portland, OR. ASCE, New York, NY.

Renault, D. and Wallender, W. W. (1994). Furrow advance-rate solution for stochastic infiltration properties. *Journal of Irrigation and Drainage Engineering* 120, 617–33.

Renault, D. and Wallender, W. W. (1996). Initial inflow variation impacts on furrow irrigation evaluation. *Journal of Irrigation and Drainage Engineering* 122, 7–14.

Roth, G., Harris, G., Gillies, M., Montgomery, J., and Wigginton, D. (2014). Water-use efficiency and productivity trends in Australian irrigated cotton: A review. *Crop and Pasture Science* 64, 1033–48.

Schwankl, L. J. and Wallender, W. W. (1988). Zero-inertia furrow modeling with variable infiltration and hydraulic characteristics. *Transactions of the ASAE* 31, 1470–5.

Selle, B., Wang, Q. J., and Mehta, B. (2011). Relationship between hydraulic and basic properties for irrigated soils in southeast Australia. *Journal of Plant Nutrition and Soil Science* 174, 81–92.

Shahidian, S. and Serralheiro, R. P. (2012). Development of an adaptive surface irrigation system. *Irrigation Science* 30, 69–81.

Shepard, J. S., Wallender, W. W., and Hopmans, J. W. (1993). One-point method for estimating furrow infiltration. *Transactions of the ASAE* 36, 395–404.

Smith, R. and Baillie, J. (2009). Defining precision irrigation: A new approach to irrigation management. Irrigation Australia 2009: Irrigation Australia Irrigation and Drainage Conference Proc. Irrigation Australia Ltd.

Smith, R., Baillie, J., McCarthy, A., Raine, S., and Baillie, C. (2010). Review of precision irrigation technologies and their application. USQ, Toowoomba, National Centre for Engineering in Agriculture.

Smith, R. J., Raine, S. R., and Minkevich, J. (2005). Irrigation application efficiency and deep drainage potential under surface irrigated cotton. *Agricultural Water Management* 71, 117–30.

Smith, R. J., Uddin, J. M., Gillies, M. H., Moller, P., and Clurey, K. (2016). Evaluating the performance of automated bay irrigation. *Irrigation Science* 34, 175–85.

Smith, R. J., Walton, R. S., and Loxton, T. (1992). Infiltration and surge irrigation on two clay soils in Queensland. Eighth Conference on Engineering in Agriculture—Quality Soils, Quality Food, Quality Environment. Albury, Australia.

Stone, J. F., Garton, J. E., Webb, B. B., Reeves, H. E., and Keflemariam, J. (1979). Irrigation water conservation using wide-spaced furrows. *Soil Science Society of America Journal* 43, 407–11.

Stone, J. F. and Nofziger, D. L. (1993). Water use and yields of cotton grown under wide-spaced furrow irrigation. *Agricultural Water Management* 24, 27–38.

Tarboton, K. C. and Wallender, W. W. (1989). Field-wide furrow infiltration variability. *Transactions of the ASAE* 32, 913–8.

Trout, T. J. and Kemper, W. D. (1983). Factors which affect furrow intake rates. Advances in Infiltration, Proc. of the National Conference. Chicago, IL. ASAE, St. Joseph, Michigan.

Trout, T. J. and Mackey, B. E. (1988). Furrow inflow and infiltration variability. *Transactions of the ASAE* 31, 531–7.

Turral, H. N., Malano, H. M., and McMahon, T. A. (1992). Evaluation of surge flow in border irrigation. Eighth Conference on Engineering in Agriculture—Quality Soils, Quality Food, Quality Environment. Albury, Australia.

Valiantzas, J. D., Aggelides, S., and Sassalou, A. (2001). Furrow infiltration estimation from time to a single advance point. *Agricultural Water Management* 52, 17–32.

Van der Kooij, S., Zwarteveen, M., Boesveld, H., and Kuper, M. (2013). The efficiency of drip irrigation unpacked. *Agricultural Water Management* 123, 103–10.

Van Es, H. M., Ogden, C. B., Hill, R. L., Schindelbeck, R. R., and Tsegaye, T. (1999). Integrated assessment of space, time, and management-related variability of soil hydraulic properties. *Soil Science Society of America Journal* 63, 1599–608.

Vazquez-Fernandez, E. (2006). Comparison between continuous-flow and increased-discharge irrigations in blocked-end furrows using a mathematical model. *Applied Engineering in Agriculture* 22, 375–80.

Vazquez-Fernandez, E., Lopez-Tellez, P., and Chagoya-Amador, B. (2005). Comparison of water distribution uniformities between increased-discharge and continuous-flow irrigations in blocked-end furrows. *Journal of Irrigation and Drainage Engineering* 131, 379–82.

Walker, W. R. (2005). Multilevel calibration of furrow infiltration and roughness. *Journal of Irrigation and Drainage Engineering* 131, 129–36.

Walker, W. R. and Skogerboe, G. V. (1987). *Surface Irrigation: Theory and Practice*. Englewood Cliffs: Prentice-Hall.

Wanjura, D. F., Upchurch, D. R., and Mahan, J. R. (1992). Automated Irrigation Based on Threshold Canopy Temperature 35.

Warrick, A. W., Zerihun, D., Sanchez, C. A., and Furman, A. (2005). Infiltration under variable ponding depths of water. *Journal of Irrigation and Drainage Engineering* 131, 358–63.

White, S. C. and Raine, S. R. (2009). Physiological response of cotton to a root zone soil moisture gradient: Implications for partial root zone drying irrigation. *Journal of Cotton Science* 13, 67–74.

Yinong, L., Shaohui, Z., Di, X., and Meijian, B. (2006). An optimized inverse model used to estimate soil infiltration parameters and the manning's roughness coefficient under surface irrigation conditions. *CIGR World Congress—Agricultural Engineering for a Better World*. Bonn, Germany.

Zapata, N. and Playan, E. (2000). Simulating elevation and infiltration in level-basin irrigation. *Journal of Irrigation and Drainage Engineering* 126, 78–84.

11 Advanced Tools for Irrigation Scheduling

Susan A. O'Shaughnessy and Ruixiu Sui

CONTENTS

11.1 INTRODUCTION

Water resources for agriculture are becoming increasingly constrained due to their limitation, drought conditions, and competition for quality water from other economic sectors. Yet, irrigated agriculture is critical to meet increasing global demands for food, feed, and fiber. Compared with dryland farming, irrigated agriculture produces at least double the yield of forage and grain crops, as discussed by Colaizzi *et al.* (2009) and shown in field studies by O'Shaughnessy *et al.* (2014). Irrigated agriculture also helps to stabilize yields. During times of drought conditions or sporadic rainfall, supplemental irrigation (especially during critical growth stages such as tasseling and silking for corn, or anthesis and grain filling for other grain) aids to overcome yield loss (Mastrorilli *et al.*, 1995; Jain *et al.*, 2007; Steduto, 2012). Increased cereal production has resulted mainly from greater inputs of fertilizer, water, and integrated pest management (Tilman *et al.*, 2002), as well as improved crop genetics (Pingali, 2012). Accordingly, increasing food production while maintaining aquatic ecosystems will depend on increasing crop water use efficiency (WUE) (Pretty *et al.*, 2006). Therefore, scientific irrigation scheduling is critical to meet the increasing global demand for agricultural products with limited water and land resources.

Although water for agriculture is becoming limited in regions of North America, the United States is well-positioned to help meet globalized agricultural demands (Macdonald *et al.*, 2015). Currently, at least 65% of irrigated land in the United States is watered with pressurized irrigation systems, while 35% is irrigated with gravity flow systems (Figure 11.1) (NASS, 2013). These percentages are almost a complete reversal from the amount of irrigated area watered by pressurized and gravity flow systems in the 1980s (Howell, 2001). Accompanied by the increase in pressurization was a decrease in average water withdrawals (207 to 159 billion m^3) for irrigation (ERS, 2013). The transition from gravity flow irrigation methods to pressurized systems not only helped to increase water application efficiency, but has also benefited farmers by enabling better irrigation control and reducing labor expenses; perhaps the best example is the center pivot irrigation system.

In Europe, approximately 25% of total freshwater abstracted is for agricultural use. This annual amount represents 45.5 billion m^3. Of the water applied in this area, 36.7% is by gravity-fed methods, while sprinklers and drop irrigation (irrigating plants by placing the water low drop by drop or with micro sprinklers) represent 30.1% and 33.2% of the methods, respectively (Figure 11.2). However, pressurized sprinkler irrigation systems are becoming more prevalent in the North and South of Europe (Eurostat, 2010).

Surface or gravity-fed irrigation also dominates in Asia, which is the method used over 96.8% of irrigated hectares, while only 2.8% is by sprinkler and less than 1% is accomplished with drip irrigation as reported by Aquastat (2012), using 2009 data.

In the late 1990s, Howell (1996) reported that few fundamentals governing irrigation scheduling have changed since the 1970s. The fundamentals of irrigation scheduling include determining the amount of water to replenish crop water use or evapotranspiration, i.e., water that has evaporated from the soil and plant surfaces and water transpired through the crop; and deciding when to irrigate so as to minimize yield loss due to water stress. Crop water needs vary depending on atmospheric demand, the type of crop, and the growth stage of the crop. The timing of the irrigation events and the amount of water to apply are critical. The main methods of scientific irrigation scheduling can be categorized as soil-based, plant-based, and weather-based (Henggeler *et al.*, 2011). Early methods of soil water sensing began as early as the 1930s (Smith-Rose, 1933), as reported by Or and Wraith (2002). Weather-based irrigation scheduling began in the 1950s (Penman, 1952; Bayer, 1954;

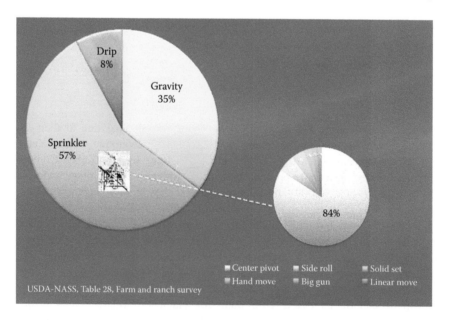

FIGURE 11.1   The increase in pressurized irrigation systems provides a growing inventory for automation and control for irrigation scheduling, Farm and Ranch Irrigation Survey, 2013.

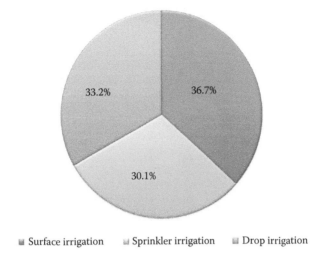

FIGURE 11.2 Percent of hectares irrigated by method in Europe. (Data from Eurostat Statistics Explained, 2010. http://ec.europa.eu/eurostat/statistics-explained/index.php/Agri-environmental_indicator_-_irrigation #Further_Eurostat_information. Accessed November 15, 2016.)

Pruitt and Jensen, 1955). Decades later, the same can be said of irrigation scheduling fundamentals, but changes in the delivery of the scheduling methods have occurred with the development of more robust plant and soil water sensors, the introduction of radio frequency (RF) telemetry for outdoor wireless sensor network systems, and advancements in software applications and algorithms used for models and to formulate decision support for irrigation management.

Sensor-based irrigation scheduling uses weather, plant, and soil water sensors to help determine the optimal timing of irrigation and amount of water to apply for a single event. The possible benefits of sensor-based irrigation scheduling include water conservation without significant penalty to crop yields, saving time for growers who manage multiple fields, and preventing groundwater pollution and environmental harm caused by deep percolation and runoff (Sadler et al., 2005).

With today's challenges from water restrictions, climate variability, and environmental regulations, it is not enough for farmers to adopt efficient sprinkler irrigation systems to sustain profitability. Today, farmers must ensure that their sprinkler design, application methods, and farming practices are as efficient as possible to comply with water regulations and cope with water restrictions. Scientific irrigation scheduling is meant to help a farmer maintain crop water productivity (economic yield per unit of water used by a crop), manage goals of maximizing yields and returns, or use water efficiently in the situation of limited water resources. Other benefits from scientific irrigation management include controlling soil water depletion at the root zone, reducing evapotranspiration (ET) during non-critical growth stages, water conservation, preventing non-point source pollution, eliminating water wastage, decreasing deep percolation, and preventing nutrient leaching. Sprinkler design, agronomic practices, irrigation application methods, tillage practices, and inputs of water, fertilizer, and chemicals should be optimized to maximize return and minimize negative impacts to the environment (Howell, 1996) (Figure 11.3).

Growers must continue to find ways to manage water more efficiently. For those who already use low-pressurized sprinkler irrigation systems, this can be quite challenging. However, opportunities for furthering water conservation do exist, and include improving irrigation scheduling methods, optimizing inputs, and preventing water losses from deep percolation and runoff to help protect the environment. Much of what is needed today is real-time information concerning crop and soil water status, weather data, and information in the form of decision support. This chapter highlights today's advanced tools for scientific irrigation scheduling, which includes innovations in soil water and plant sensors, the integration of RF telemetry to take advantage of sensor network systems, and the use of software-based applications for data acquisition, management, and decision making.

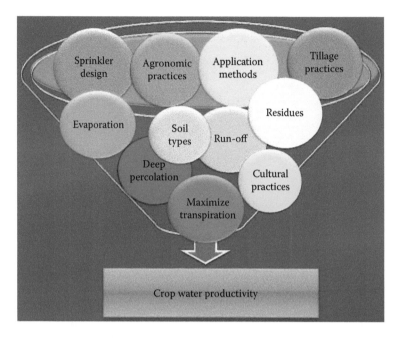

FIGURE 11.3 Maximizing crop water productivity depends on proper irrigation system design, appropriate farming practices and application methods, system maintenance, and responsible management of resources and inputs.

11.2 RECENT NOVEL TECHNOLOGIES

11.2.1 Soil Water Sensing

Soil water sensors are used to characterize soil water content on either a mass or volume basis. There are many types of soil water sensors available today, all of which work by measuring a surrogate property that is empirically or theoretically related to the soil water content (Evett *et al.*, 2012). For each type of sensor, some type of calibration in the soil of interest is required (Evett *et al.*, 2008). Today's soil water sensors are capable of continuously recording measurements, storing data in logging units, or transmitting data to a base station or the cloud using cellular communication. Soil water sensors can help farmers and land managers fine-tune site-specific soil nutrient and water management (Viscarra Russel and Bouma, 2016).

11.2.1.1 Time Domain and Water Content Reflectometry

Soil water sensors based on time domain reflectometry (TDR) and water content reflectometry are comprised of waveguides (stainless steel probes or wires) and a pulse generator. Standalone sensors contain circuitry in the head of the sensor that propagates an electromagnetic signal at frequencies of MHz to GHz, which travel along the probe's waveguides. The internal circuitry calculates the time it takes the reflected wave to return, and converts the information into an apparent dielectric constant and the soil water content of the soil surrounding the waveguide (Or and Wraith, 2002). Acclima* (Meridian, Idaho) has recently developed small TDR sensors (TDR-315) with three stainless steel waveguides (20 cm long) and circuitry within the head of the sensor that generates an electromagnetic pulse. The CS655 water content reflectometry sensor (developed by Campbell Scientific in Logan, Utah) is an example of another standalone electromagnetic soil water sensor that makes indirect measurements of soil water content by using an oscillator circuit located in the head of the

---

* The mention of trade names, commercial products, or companies in this publication is solely for the purpose of providing specific information and does not imply recommendation or endorsement by the U.S. Department of Agriculture.

sensor. The oscillation frequency of the generated electromagnetic wave changes as soil water content changes along the waveguide.

Both of these standalone sensors continuously record measurements while wired to a data logging unit. Because these sensors support the serial digital interface communication standard (SDI-12), a number of sensors can be wired into a single data recorder. Multiple sensors placed at a single location but at different depths produce a "sensing node" with radio frequency telemetry (RF) integrated into the data recorder, these sensing nodes become distributed networks of soil water sensors and can be located in various locations throughout a field to assess soil moisture variability at the field scale (Vellidis et al., 2008; Hedley and Yule, 2009; Kim et al., 2009).

Time domain and water content reflectometry sensors provide accurate volumetric soil water content measurements within 1% or 2% of volumetric water content, and can be easily established as an automated monitoring system to continuously measure soil water content (Skierucha et al., 2012). Due to their smaller length, these two types of sensors are suitable for high-resolution exploration of subsurface flows with minimal disturbance to soil. The order and information outputted from these sensors depends on the manufacturer, and can include soil water content, soil temperature, bulk relative permittivity, and soil electrical conductivity. The downside of using these sensors is that installation to greater depths can be labor intensive, difficult, and cause soil disturbance (Evett et al., 2012). Also, the use of multiple sensors within a single field can become expensive. If TDR methods can be made easier to install with less soil disturbance, this method may eventually replace soil water measurements with the neutron probe. Casanova et al. (2013) designed a multisegmented waveguide on an access tube (WOAT). The intention of this prototype down-hole TDR sensor was to ease the assembly and installation of an accurate soil water sensor in the vertical direction to a desired depth. The digital sensor provided quality TDR waveforms and longtime reflection coefficients without the trouble of cables. A patent for the WOAT was recently issued and prototypes for this sensor continue to be tested in the field (Patent No. 8,947, 102 B1, 2015).

In addition to TDR and WCR sensors, other electromagnetic (EM) sensors such as time domain transmissometry (TDT), capacitance sensors, and resistance sensors have been rapidly developed and adopted for soil moisture measurement as well (Blonquist et al., 2005; Vellidis et al., 2008; Sui et al., 2012). The TDT sensors provide circuitry in the head of the probe that generate and sample pulses. These sensors are less accurate than TDR and WCR sensors for measuring soil water content (Evett et al., 2009). However, they are inexpensive, easy to install and maintain, and capable of monitoring soil moisture status for irrigation scheduling after being well-calibrated (Blonquist et al., 2006; Yoder et al., 1997; Leib et al., 2003; Kizito et al., 2008).

Another effective method for remotely monitoring soil water content in the top layers of the soil profile is with radiometers that operate in the microwave range (30 cm to 1 mm). Vulfson et al. (2013) developed remote sensing system comprised of a scatterometer and a gamma-ray radiometer to determine soil water content at the three depth intervals—0 to 5, 0 to 30, and 5 to 30 cm. The scatterometer in this study transmitted electromagnetic pulses in the P-band ($\lambda = 68$ cm), while the gamma-ray radiometer detects radiation in the range of 50–3000 keV. Radiation in these wavelengths have a low sensitivity to surface roughness and vegetation. Ground-truthing of soil water content was achieved by gravimetric measurements of wet and dry soil samples, and with in situ time domain reflectometry sensors. Direct surface roughness measurements were made with an automated laser system. This study reported that the values of soil water content determined by the scattorometer and gamma-ray radiometer showed the best correlation, with ground samples taken at the depth of 0–5 cm, with correlation coefficients of upwards of 0.9 for sand, sandy loam, and loam soils.

A recent novel method for estimating surface and near-surface soil moisture on a large scale is the Cosmic-Ray Soil Moisture Observing System (COSMOS). This stationary probe, when situated above the ground (approximately 1.5 m), measures cosmic ray neutrons above the soil surface (Zreda et al., 2008). The intensity of fast neutrons at this level is inversely correlated to soil moisture content to a depth of a few centimeters (Cosh et al., 2016). The radius of the sensing footprint ranges from 120 to 240 m, while the depth of influence is limited to 83 cm in dry soils and 15 cm in wet soils (Köhli *et al.*, 2015). Fast neutron intensity is corrected for temporal changes in the incoming neutron density, atmospheric pressure, water vapor, and surrounding vegetation (Zreda *et al.*, 2012). The instrument is calibrated against the area-average soil moisture obtained from a large number of soil moisture readings from *in situ* soil water sensors located within the same footprint as the range of influence of the COSMOS. It is important to realize that the probe is influenced by near surface moisture such as ponds, biomass, and atmospheric water vapor (Rosolem *et al.*, 2013). However, if the surrounding sources of water are quantified, the COSMOS readings can be acceptably adjusted. Coopersmith *et al.* (2014) reported that corrections based on measured leaf area indices (LAI) of crops surrounding a COSMOS probe provided estimates of soil-water content with a RMSE less than $0.03 \text{ m}^3 \text{ m}^{-3}$ of the weighted average of soil water content of an *in situ* network of soil water sensors within the same footprint as that influenced by the COSMOS probe.

Because of their large footprint, it is not likely that these sensors will be used to schedule irrigations at the field scale. However, the COSMOS sensor is currently being used to improve the scaling of soil moisture content in the top layer of the soil profile by enabling the calibration and validation of a variety of *in situ* sensors located near the surface. A network of COSMOS sensors may help to overcome variability in measurement of soil water at the surface, which is inherent from multiple point measurements with *in situ* sensors (Desilets *et al.*, 2010). Additionally, a network of COSMOS sensors may support weather or climate forecasting (Rosolem *et al.*, 2013), and provide data assimilation for hydrological models (Köhli *et al.*, 2015). In a separate work, Dong *et al.* (2014) calibrated a COSMOS rover with a GPS receiver over an *in situ* test bed of impedance probes (ML2x, Theta Probe, Delta-T Devices). Results from the study demonstrated that a COSMOS rover can be used to effectively map soil moisture to a depth of 0–5 cm and could be used for determining the average soil moisture at spatial scales for purposes of satellite calibration.

11.2.2   ADVANCED TECHNOLOGIES WITH PLANT SENSORS

Plant sensors for irrigation scheduling can be categorized into two main groups; invasive, which make contact with the plant; and noninvasive, which do not contact the plant, but provide proximal sensing measurements (Jones, 2004). Invasive plant sensors are used especially for irrigation management of high value crops such as grapes, fruit orchards, and in some cultures, for cotton irrigation scheduling.

11.2.2.1   Invasive Plant Sensors

Water stress is a major yield-reducing factor. Both invasive and noninvasive types of plant sensors have been in development for at least the past 30 years as a tool for irrigation scheduling. Plant-based sensors have been adopted to some extent by researchers and the industry to monitor plant water stress and control irrigation scheduling. Examples of invasive plant sensors include pressure bombs to measure leaf water potential (Kite and Hanson, 1984; Cohen *et al.*, 2005), dendrometers (Klepper *et al.*, 1971; Livellara *et al.*, 2011), dendrographs (Impens and Schalck, 1965), and sap flow gauges (Cohen *et al.*, 1981; Herzog *et al.*, 1995; Fernández and Cuevas, 2010). These sensors require some level of manual intervention, destructive sampling, or the sensors to be clamped onto the plant or inserted into tissue. Albeit cumbersome, invasive-type plant sensors continue to be used in vineyards (Centeno *et al.*, 2010), cotton fields (Grimes *et al.*, 1987), and in various kinds of orchards for irrigation scheduling (Huguet *et al.*, 1992; Fernández *et al.*, 2008).

11.2.2.2   Noninvasive Plant Sensors

Most of the advancements in plant-based irrigation scheduling encompass the use of noninvasive plant sensors. Sensors of this type include radiometers, RGB cameras, infrared thermometers, thermal imagery, and satellite imagery. An obvious reason for their advancement is that non-invasive monitoring provides greater crop coverage, and decision-making is based on data measurements that are averaged rather than point measurements. Point measurements may or may not be representative of whole-field crop status. Early detection of plant stress (abiotic or biotic) using canopy temperature is possible since changes in leaf temperature in non-thermogenic plants mainly results from alterations of transpiration in response to stress (Chaerle and Van der Straeten, 2000).

11.2.2.3 Infrared Thermometers

Thermal energy is electromagnetic radiation with wavelengths in the region of 3–14 μm. A healthy plant under well-watered conditions will transpire at its full potential, allowing the canopy surface to cool. However, when available water for plants becomes limited, the crop reduces its transpiration through stomatal closure, which leads to a reduction in latent heat flux, causing canopy surface temperature to rise. Since the peak thermal radiation for plants is in the infrared region at the spectral range near 9.8 μm, infrared thermometer sensors that filter radiation in the range of 8–14 μm are useful instruments to measure crop canopy temperature. The amount of thermal radiant energy emitted from a canopy surface (W) (W m^{-2}) is represented by Equation 11.1 Stefan–Boltzmann law (Kaplan, 2007):

$$W = \varepsilon\sigma T^4 \tag{11.1}$$

where ε is the emissivity of the canopy surface, σ is Stefan-Boltzmann's constant (5.6703×10^{-8} W m^{-2} K^{-4}); and T is the absolute surface temperature of the target (K). Emissivity quantifies an object's ability to emit radiation and varies depending on surface roughness. Emissivity for plant leaves has been established to be between 0.93 and 0.98 (Tanner, 1963; Idso et al., 1969; López et al., 2012) for different species, while effective emissivity of a closed canopy is often represented as 0.98 to 0.99 (Jones et al., 2003; Campbell and Norman, 1998).

Algorithms use spot measurements of canopy temperature made by infrared thermometers and canopy temperature information from pixels in thermal imagery to develop thermal stress indices and characterize areas of water-stressed crops. Wanjura et al. (1995) and Peters and Evett (2008) demonstrated that infrared thermometry could be used to schedule irrigations automatically using a thermal stress index derived from canopy temperature.

Although single measurements are made with infrared thermometers, continuous geo-referenced measurements from infrared thermometers provide ample information to map areas of crop water stress temporally and spatially (Peters and Evett, 2007). Within-field maps of crop water stress can be made by converting georeferenced canopy temperature into a stress index and using kriging (Figure 11.4a) (O'Shaughnessy et al., 2011). In addition, average seasonal thermal stress indices can be used to predict yields (Figure 11.4b) (O'Shaughnessy and Evett, 2010).

Thermal stress indices are developed to characterize the level of water stress that a crop is experiencing. The most well-known crop water stress indices are the empirical and theoretical crop water stress indices developed by Idso et al. (1981) and Jackson et al. (1981), respectively. Over the years, there have been a number of different permutations of CWSI developed, which are described in a review paper by Maes and Steppe (2012).

11.2.2.4 Infrared Thermometry and the TSEB Model

The two-source energy balance (TSEB) model estimates latent and sensible heat fluxes of the vegetative and soil components of cropped areas, which can be converted into estimates of ET (Colaizzi et al., 2014). The model was initially developed in the mid-1990s by Norman et al. (1995). With estimates of ET in hand, the correct amount of water to apply can be determined. Ben-Asher et al. (2013) calculated instantaneous ET from an array of infrared thermometers (IRTs) mounted on a moving irrigation system and reported that these estimates compared well with a one-dimensional soil water balance model. Colaizzi et al. (2014) are using the model to improve water management for site-specific irrigation methods by incorporating the TSEB model into a plant feedback irrigation scheduling supervisory control and data acquisition (ISSCADA) system.

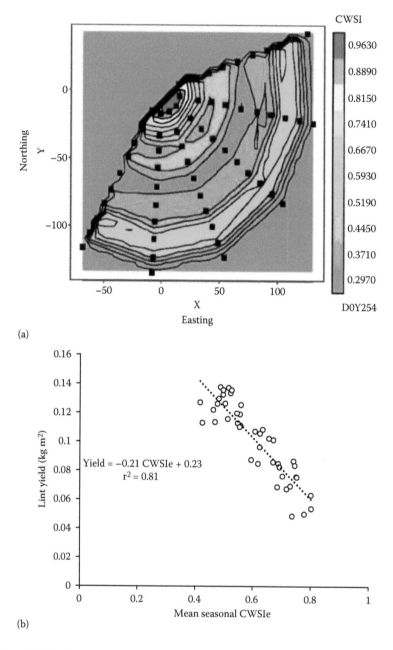

(a)

(b)

FIGURE 11.4 (a) Map of crop water stress index (CWSI) near the end of the irrigation season (2007) for cotton grown under a center pivot field where irrigation treatment levels (100%, 67%, 33%, and 0%) were applied in a concentric pattern. The color scale represents the empirical CWSI calculated for the last day of irrigation scheduling, September 11, 2007. The purple and blue colors representing less water-stressed cotton and areas in green and yellow representing more water-stressed cotton. (b) Graph showing negative linear relationship between cotton lint yield and the mean seasonal empirical crop water stress for cotton grown in 2007 under different irrigation treatment levels of 100%, 67%, 33%, and 0%.

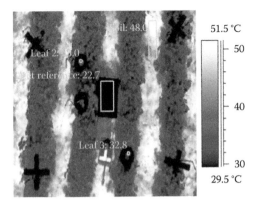

FIGURE 11.5 Thermal image from a forward looking infrared (FLIR) camera taken from lift located approximately 21 meters above a center pivot field overlooking a plot that was irrigated at 33% replenishment of soil water depletion to field capacity.

11.2.2.5 Thermal Imagery

While infrared thermometry is economically feasible compared with thermal imagery, the latter has the advantage of providing thermal data at the pixel level in a spatial context. The image resolution or number of pixels for an imaging instrument varies per manufacturer and model. For example, a small inexpensive thermal camera may have a resolution of 160 by 120 pixels, while a top grade industrial model will typically have a resolution of 1024 by 768 pixels (Kaplan, 2007). Shaded and sunlit components of a target become more obvious. Cotton canopy planted in circular rows in a center pivot field appear as relatively cool surfaces (32.8–33°C) compared with dry soil (48.0°C) shown in furrows between the plant rows (Figure 11.5). Methods of acquisition include boom and aerial platforms with sensors mounted onboard. Often the manufacturer of the commercial thermal imaging camera provides software with the capability to digital analysis and temperature extraction at the pixel level.

Researchers have used thermal images for site-specific assessment of crop water status in olive orchards (Ben-Gal *et al.*, 2009), vineyards (Möller *et al.*, 2007), and in cotton fields (Padhi *et al.*, 2012). Cohen *et al.* (2005) converted aerial thermal imagery of crop canopy into maps of leaf water potential (LWP) because farmers in Israel are more comfortable with leaf water potential measurements and use LWP thresholds for irrigation scheduling (Alchanatis *et al.*, 2009). The maps clearly showed the different levels of irrigation treatments. Converting thermal imagery into information that describes areas of deficit soil water levels within a field could be used for mapping crop water stress and triggering irrigations using preestablished thresholds if the frequency of newly generated maps is timely to meet crop water demands. Disadvantages of thermal imagery are the greater cost of the sensor, and processing the image requires a more complex method (analysis by a separate software, or stitching images together in the case of aerial imagery platforms) when compared with the evaluation of a spot measurement.

11.2.2.6 Ground-Based Spectral Reflectance Sensors

Spectral reflectance sensors are a type of radiometer that contain optical detectors and filters to measure and pass a specific wavelength of radiation emitted from a surface. Ground-based radiometers provide a single measurement in time that represents the average signal produced from a specific target (one pixel). The size of the pixel depends on the field-of-view of the instrument and the height of the instrument above the target. For example, a 20-degree field-of-view resolves a 1.0 m spot at a working distance of 2.8 m. Commercial spectral radiometers (Crop Circle, Holland Scientific, Lincoln, Nebraska; Green Seeker, Trimble, Sunnyvale, California), and CCD (charge-couple device) cameras have been used in agricultural applications for phenotyping (Thorp *et al.*, 2015), estimating plant canopy cover and yields (Rodriguez-Moreno, 2016), disease detection (Kobaysahi *et al.*, 2000), and assessing and controlling nitrogen application levels in plants (Li *et al.*, 2014). Radiometers can be active sensors using an artificial modulated light source, or they can be passive sensors, using the sun as a light source. Measurements from active sensors are less dependent on changing sun angles and cloud cover. Either type of radiometer used for monitoring crops should be maintained at distances above the canopy that range from 10 to 200 cm (Kipp *et al.*, 2014). Multispectral sensors have been used for irrigation scheduling by estimating basal crop coefficients, K_{cb}, from the normalized difference vegetative index (NDVI) calculated from active spectral reflectance sensors, and multiplied by reference evapotranspiration ET_o, calculated from local weather data (Hunsaker *et al.*, 2003; Hunsaker *et al.*, 2005; Hunsaker *et al.*, 2015). This method was compared with traditional irrigation methods for the area and found to substantially reduce irrigation water for surface-irrigated cotton. Stone *et al.* (2016) used remotely sensed NDVI measurements to schedule irrigations for a variable rate irrigation (VRI) system and compared the method to Irrigator Pro for Corn and soil water measurements with soil water potentiometers. All three methods produced similar outcomes in corn production.

Another useful multiband optical sensor incorporates an ultrasonic distance sensor and a global positioning system (GPS) receiver to be used as a ground-based remote sensing system (Sui *et al.*, 2005; Sui and Thomasson, 2006). The multiband optical sensor is capable of measuring crop canopy reflectance in four wavebands, including blue band (400–500 nm), green band (520–570 nm), red band (610–710 nm), and near-infrared band (750–1100 nm). The infrared temperature sensor detects plant canopy temperature. Information from the ultrasonic sensor is converted to plant height. The GPS receiver is used to locate the position where the measurements are taken. An intelligent data acquisition (DAQ) device in this system collects the data from these sensors and the spatial information from the GPS receiver at an interval of one second. This system can be implemented on a moving platform. When the platform moves across the field, the sensors automatically scan the plant canopy. One measurement of canopy spectral reflectance at the four bands, canopy temperature, and plant height is taken in each second and the data along with the spatial coordinates of each measurement position are stored into a memory card in the DAQ. Plant health conditions can be diagnosed using the information collected, and plant stress indices such as water stress index and nutrient stress index can be developed. This sensing system has been used to determine spatial nitrogen status and variability of crop height and canopy spectral reflectance within a cropped field. Crop height and canopy reflectance characteristics are indicators of plant vigor and yield potential, and could be used to schedule irrigations (Figure 11.6). More work is needed in this area to provide decision support algorithms.

In a recent study, O'Shaughnessy *et al.* (2015) used passive multiband sensors to qualify surface radiometric temperature data taken from IRTs on a moving center pivot lateral. Canopy temperature was determined to be from soil or from vegetation using NDVI, calculated simultaneously with the acquisition of the thermal data. Temperatures contributed mainly by soil were not used to calculate a thermal stress index. The spectral and thermal sensors were contained within the same cylindrical housing (Figure 11.7) and viewed the same target.

Sui and Baggard (2015) developed a wireless data acquisition (WDAQ) system to measure plant canopy temperature and plant height in real-time in situ. Each WDAQ unit consisted of a GPS receiver, programmable data logger, infrared temperature sensor, ultrasonic distance sensor, solar power supply, and wireless data transmitter/receiver. Multiple WDAQ units are installed on a lateral of center pivot. As the pivot moves around the field, the WDAQ system makes continuous and simultaneous measurements of plant canopy temperature and plant height, and records spatial coordinates at each measurement location (Figure 11.8). Data collected are wirelessly transferred to a receiver for download. As mentioned previously, plant canopy temperature is correlated with plant water stress. Plant height can be an indicator of plant health and light use efficiency (Li *et al.*, 2016). Therefore, site-specific measurements of plant canopy temperature and plant height can possibly be used for plant-based irrigation scheduling and automatic creation of prescriptions for variable rate irrigation.

FIGURE 11.6 Plant phenomics system contains three main categories of sensors: optical sensors, ultrasound, and global positioning system. The sensor system can be mounted on a moving tractor or a moving irrigation system and data collected as the moving platform travels across the field.

FIGURE 11.7 Wireless multiband sensor containing three bands in the visible range (red, blue, and green), one band in the near infrared (NIR) range and a thermal band. Eighteen of these sensors were deployed on a moving center pivot lateral for a variable rate irrigation system to qualify and measure crop canopy temperature.

FIGURE 11.8 Center-pivot-mounted WDAQ system scans corn plant canopy to make real-time measurements of plant canopy temperature and plant height coupled with the spatial information as the center pivot moves around the field. The infrared temperature sensor and ultrasonic distance sensor were installed facing down toward plant canopy. The data logger was housed in a weather-proof box.

11.2.2.7 Computer Vision Devices

Multiband sensors can have varied utility (i.e., detection of disease, nutrient deficiency, water stress, and multiple sensors within a system), which tends to drive up the cost. With the advent of systems on a chip, miniature imaging sensors, and reduced upfront costs, researchers have turned to computer vision devices as tools for crop and irrigation management (Figure 11.9). Computer vision or machine vision devices are systems that acquire, process, and analyze images. The analysis provides

FIGURE 11.9 Computer vision thermal device prototype using a single-board computer, a digital RGB camera, and a thermal sensor to calculate percent canopy cover and detect crop water stress or biotic stress.

information for decision support, such as disease detection (Casanova *et al.*, 2014; Lee and Ehsani, 2015), crop nitrogen content (Mao *et al.*, 2015), crop water stress (Mangus *et al.*, 2016), and fruit and weed detection (Lee *et al.*, 2010; Burgos-Artizzu *et al.*, 2011). The computer vision device developed by Escarabajal-Henarejos *et al.* (2015), calculates the percent of crop canopy cover, converts that value into a crop coefficient, and uses this information in conjunction with a water balance equation (similar to FAO 56) to indicate when and how much irrigation to apply. The information provided to a user concerning disease and weed detection could be used to determine if irrigations should be withheld or reduced as in the case of an area that contains more weeds than crop or an area within a field where the majority of the crop is diseased.

11.2.2.8 Surface Renewal and Eddy Covariance Methods

Estimating crop evapotranspiration using methods of FAO 56 with meteorological instrumentation has played a key role in irrigation scheduling (Allen *et al.*, 1998). The method incorporates the use of meteorological instrumentation to determine evapotranspiration levels of a crop canopy and uses the energy balance equation as its basis for ET calculations, i.e., R_n- G- λ ET-H = 0, where R_n is net radiation, G is soil heat flux, λ ET is latent heat flux, and H is sensible heat flux (all units are W m^{-2}).

Other researchers use the eddy covariance (EC) method to estimate crop water use at the field-scale level. EC method is capable of measuring exchanges of carbon dioxide, water vapor, methane, and energy between the surface of the earth and the atmosphere. EC systems have been used for monitoring agroecosystems and estimating crop ET for irrigation scheduling. A typical EC system can consist of a variety of gas analyzers, including a CO_2/H_2O Analyzer and a three-dimensional (3D) sonic anemometer (Figure 11.10). In addition to these basic instruments, biomet (biological and meteorological) sensors (including a net radiometer, soil heat flux plates, soil temperature sensors, and a precipitation gauge) may be used in an EC system to collect ancillary data for filling

FIGURE 11.10   An eddy covariance system for estimating crop ET and monitoring agroecosystems.

measurement gaps and interpreting flux results. The CO_2/H_2O Analyzer is capable of making simultaneous CO_2 and H_2O flux measurements in the free atmosphere. The sonic anemometer is a three-dimensional sonic sensing device that measures three orthogonal wind components and the speed of sound. In eddy covariance systems, a 3D sonic anemometer measures the turbulent fluctuations of horizontal and vertical wind, which are then used to calculate momentum flux and friction velocity. Data from the CO_2/H_2O Analyzer and 3D sonic anemometer are used to calculate the crop ET and other variables.

One drawback of this method is that instrumentation consists of costly sensors and data analysis is complex and intensive. Another is that EC systems do not fully close the surface energy budget (Alfieri et al., 2012; Leuning et al., 2012). Anderson and Dong (2014) report closure when annual data was paired and analyzed from two EC systems located in irrigated sugarcane fields in Hawaii. One tower was situated in a location where the wind created continuous turbulence, while the second was located in a field where there was much less turbulence.

Recently, efforts have concentrated on a simpler approach, termed surface renewal, to measure ET using high-frequency measurements of air temperature near the crop canopy (McElrone et al., 2013). The method analyzes the time series of temperature data using structure function to determine temperature ramp characteristics to estimate sensible heat flux density using only thin single-wire thermocouples (76 μm diameter) (Spano et al., 2000). Rosa and Tanny (2015) determined that surface renewal provided a low-cost and feasible method for estimating ET over a cotton crop. Castellví et al. (2008) demonstrated that the surface renewal method performed better than the eddy covariance method in closing the energy balance. Suvočarev et al. (2014) determined that surface renewal techniques to estimate ET for peach orchards was as reliable as eddy covariance measurements. Actual irrigation scheduling studies and reports on crop yields and WUE with this technique will help identify its usefulness to the agricultural industry.

11.2.3 Wireless Sensor Network Systems

11.2.3.1 Aboveground Wireless Sensor Network Systems

In recent years, engineering advances in infrared instrumentation led to self-powered infrared thermometers (IRTs) that allowed for easier direct recording with various electronic data loggers and, thereby, continuous crop surface temperature monitoring. Similar technologies were utilized in satellites to remotely monitor the earth's surface temperature for much larger land areas. More recently, the integration of wireless communication with remote thermal-based plant measurements is a critical advancement for production agriculture. In the United States and other developed countries, farm size is increasing (Van Vliet et al., 2015). With larger production fields, it becomes more of a challenge for farmers to inspect their fields for disease, pest infestation, damage of crops from farm equipment, and lack of maintenance on equipment.

Wireless sensors and wireless sensor networks are preferable to wired because they are capable of operating throughout the duration of an irrigation season, providing continuous canopy temperature monitoring without interfering with farm equipment or farm operations. Similar to wired sensor systems, data from wireless sensor networks may be stored to a logging-unit or host computer in the field or to a remote location accessed at a specific website or cloud-based. However, advantages of wireless network systems over wired systems are the ease of deployment and network maintenance, since wires and cables are eliminated and wireless units may be replaced easily and will be recognized by the network automatically. Scalability of wireless IRTS coupled with locating them on a moving irrigation system allows for greater spatial coverage for crop monitoring than what could be achieved with in situ sensors. Figure 11.11 shows a wireless network of 18 canopy temperature sensors spaced over a length of 317 m (960 ft) over wheat infected with a virus. Companies that produce wireless IRTs for outdoor use include Apogee, Everest Scientific, Dynamax Inc., SmartField, and Omega Engineering.

FIGURE 11.11 Array of 18 wireless infrared thermometers (SAP-IRTs, Dynamax Inc., Houston, Texas) located in a wheat field infected with Wheat Streak Mosaic Virus. A trap strip along the west edge of the plot was inoculated with WSMV. Sensors were placed stationary in the field to help detect the spread of disease which was predicted to be in the direction of the prevailing wind.

Wireless networks also allow flexibility in relocating sensors to different areas of interest within a field throughout the growing season or temporarily removing sensors with minimal effort so as not to interfere with farm equipment. Importantly, wireless sensor networks also offer continuous measurements in challenging environments (e.g., along one-fourth or half-mile moving lateral pipelines of irrigation systems). The increased density of observations allows for improved validation of models and hypotheses (Burgess *et al.*, 2010), and daily characterizations of crop water stress using canopy temperature allow for more in-depth sensitivity analysis between plant response and environmental fluxes (Blonquist Jr. *et al.*, 2009). The most popular communication protocols used in agriculture are WiFi, Bluetooth, and Zigbee (Figure 11.12). These protocols operate in the different bandwidths, with the lower bandwidths providing greater range at the expense of power consumption.

11.2.3.2 Underground Wireless Sensor Networks

Because soil water sensors are installed *in situ* and maintained in a stationary location, it would be beneficial to maintain the RF telemetry components below the soil surface as well to avoid obstructing farm operations. However, there are extant challenges including affordability, durability, and signal propagation through rock or different textured soil and water content levels (Parameswaran *et al.*, 2013); operating frequency of the RF telemetry; and burial depth of soil water sensors (Akyildiz *et al.*, 2009). The lossy medium of soil requires underground radios to have greater transmission power than aboveground networks. Akyildiz and Stuntebeck (2006) describe the effect of soil properties on signal transmission and the appropriate communication architecture, wireless sensor network (WSN) topology, and antenna design to help overcome this difficulty. Bogena *et al.* (2009) describes a hybrid WSN system, which uses the low-cost Zigbee communication protocol and hybrid topology with a mixture of ground end devices wired to several soil sensors and aboveground

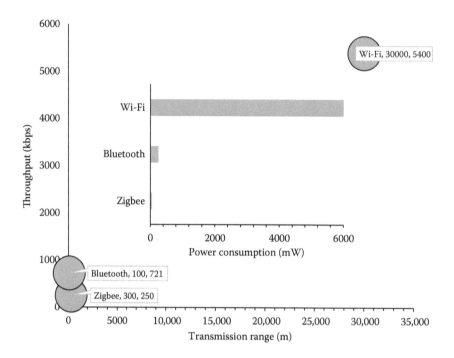

FIGURE 11.12 Graphs showing relationship between data throughput with transmission range and power consumption of the three common communication protocols (Zigbee, Bluetooth, and Wi-Fi) for wireless sensors used in agricultural applications.

router devices. Research continues to focus on the challenges associated with wireless underground transmission to make these networks practical and cost-effective.

11.2.4 Advances in Irrigation Scheduling Methods

Advanced irrigation scheduling methods are comprised of a combination of instruments that provide direct or indirect measurements of weather parameters, soil water content and/or plant status, and modeling programs that indicate when to irrigate and how much water to apply. Often the software programs require specific information relative to the crop and soil type, planting date, and other agronomic practices. The purpose of such advanced systems is to give agricultural producers the confidence to conserve water without significantly reducing yields or penalizing the quality.

Direct contact sensors continue to be used today to determine the irrigation timing of cotton and high value crops including table grapes and fruit trees. Examples of these direct contact sensors include pressure bomb chambers that measure stem leaf water potential and linear variable displacement transducers (LVDT) (Mirás-Avalos *et al.*, 2016), which provide continuous measurements of trunk or stem diameter fluctuations. Trunks and stems shrink during the day and increase at night. Diurnal fluctuations are the result of trunks and stems losing more water from transpiration than the plant can take up through its root system during daylight hours (Link *et al.*, 1998). Relationships developed between the fluctuations and measured environmental parameters, such as VPD (Goldhamer *et al.*, 2003) and reference evapotranspiration (Doltra *et al.*, 2007) are used to provide an indication of water-stress. Conesa *et al.* (2016) demonstrated that wireless sensor networks of LVDTs combined with stem leaf water potential measurements provided a means to measure maximum daily tree shrinkage and determining irrigation timing for table grapes, while De LaRosa *et al.* (2016) used LVDTs to regulate deficit irrigation of nectarine trees. The LVDT sensors are now available as wireless mesh networks allow, scaling up the number of sensors to make measurements on multiple plants simultaneously and to provide the flexibility in selecting different plants over time.

Remote plant sensors are better suited for production agriculture, especially grain production, where field sizes are relatively large, greater than 30 ha, and producers typically farm several fields. Sensor-based irrigation management has the potential to reduce water losses due to inefficient application practices, e.g., over-irrigation, which can lead to waterlogging, and water or nutrient losses from deep percolation or runoff. Pumping less water also reduces energy costs. The use of sensor-based methods for irrigation scheduling requires management of data streams from sensor network systems that are weather, plant, or soil water–based or a combination of any of the three. Feedback systems require that thresholds be established to determine when and how much to irrigate (Jones, 2004).

An irrigation scheduling sensing system, combines plant or soil water sensing with algorithms for to provide support for the timing and amounts of water to apply. One established plant feedback system for irrigation scheduling supervisory control and data acquisition is the ARSmartPivot (Andrade *et al.*, 2015). The system uses a client/server software and wireless sensor networks of infrared thermometers and weather sensors to detect crop water stress and build prescription maps to trigger irrigations based on thermal stress indices. Water is applied in variable amounts depending on the level of the stress index. Prescription maps are built throughout the growing season to meet variable crop water needs. However, the maps are interactive, allowing the user to make changes or accept the water recommendations. Osroosh *et al.* (2015) describe a second plant-feedback system for an automatic drip irrigation scheduling scheme for apple trees using a dynamic CWSI threshold. The system was comprised of stationary infrared thermometers located over apple tree canopies and the acquisition of local meteorological data. Stem leaf water potential measurements and soil water content measurements with the neutron probe were used to assess the effectiveness of the dynamic threshold concept. Results demonstrated that the new algorithm resulted in fewer false irrigation signals on cloudy, humid, or cool days.

Sui and Baggard (2015) described a sensing system for irrigation scheduling, using soil water sensing feedback and meteorological data. Meteorological data, air temperature, relative humidity, solar irradiance, and wind speed were collected by a CR1000 (Campbell Scientific, Logan, Utah). The soil-sensing feedback system provided soil water information used EC-5 and 5TM soil water sensors (Decagon Devices, Inc., Pullman, Washington) for irrigation scheduling of corn, cotton, and soybean. Soil water sensors were installed in 16 different locations under a center pivot field. At each location, three sensors were wired to a data logger (models Em50R or Em50G) and installed at depths of 15, 30, and 61 cm below the ground surface. Each sensor automatically reported soil water measurements at a time interval of one hour during the growing season. Each data logger sent soil water information hourly to a single data station (ECH2O, Decagon Devices, Inc., Pullman, Washington). The CR1000 was also connected to the data station through a serial cable. The data station then transferred all field data to the internet using a cellular modem. Soil water content measured at the three depths were interpreted using a weighted average method to better reflect the status of soil water in the plant root zone. Irrigations were triggered as soil moisture content was reduced to approximately 50% of available soil water holding capacity.

Vellidis *et al.* (2008) also deployed a wireless network of soil water sensors in a field to manage irrigation for cotton. Each soil-sensing node contained three Watermark soil moisture sensors installed at different depths. Because of the relatively low cost of these types of sensors, a number of soil-sensing nodes were installed across the field, providing a good representation of available soil water variability in the field. Information from such an extensive network could be used to build dynamic prescription maps, if soil water availability between the nodes could be interpolated to provide adequate information to meet the resolution of the sprinkler irrigation system.

11.2.4.1 Software-Based Irrigation Scheduling Tools

The use of computers has become increasingly common in the agricultural industry. A number of producers in the United States use web-based servers to access information or remotely control the operation of their sprinkler irrigation systems (Kranz *et al.*, 2012). Also available via the Internet are software tools to aid in decision support of irrigation scheduling. Many of these scheduling software programs incorporate a type of checkbook type of water balance scheduling method and local agrometeorological data. A central website that lists a number of irrigation-scheduling tools (Agricultural Water Conservation Clearinghouse, 2016) and includes such examples as AZSCHED (University of Arizona), KanSched2 (Kansas State University), and North Dakota Agricultural Weather Network (NDAWN). Steele *et al.* (2010) described the algorithms used for the NDAWN scheduler in detail, which is similar to the other scheduling tools listed. All of these programs require specific input from producers relative to the agronomic practices (e.g., crop type, planting date, soil type, and previous crop cultivated), and the continuous input of rainfall and irrigation amounts. These scheduling methods could help give confidence to farmers where climate variability is high and seasonal rainfall amounts are unpredictable. However, it is difficult to assess the cost-effectiveness of such tools. Stone *et al.* (2016) conducted a comparison between three irrigation scheduling methods over corn using a VRI center pivot system. One method was use of the Irrigator Pro for Corn (USDA-ARS National Peanut Research Laboratory), a software program where management zones were established over areas of different soil types. A second method was included with management based on the normalized difference vegetative index (NDVI), remotely determined using a spectral radiometer mounted on a tractor, and combining the information with a seven-day soil water balance equation. The third method was based on soil water potential measurements and a threshold of 50% maximum allowable depletion. Results showed that all three methods produced corn grain yields and water use efficiencies that were similar during a dry and wet growing season. Similar results between methods should instill confidence, so that a producer could choose the method most comfortable to them to manage the irrigation of crops.

Other web-based irrigation scheduling tools available to farmers, crop consultants, and resource managers allow access not only to repositories of weather data but offer data from satellite imagery to develop water budgets for irrigation scheduling at the field-scale level. Two examples are the California Irrigation Management Information System (CIMIS), which is comprised of weather stations located throughout California and linked to a central server that reports reference evapotranspiration (Hart *et al.*, 2009). The Satellite Irrigation Management Support project (SIMS) combines data from NASA's Terrestrial Observation and Prediction System (TOPS), Landsat and MODIS satellite imagery, and reference evapotranspiration from several weather station networks (Trout *et al.*, 2014).

11.2.4.2 Software Applications

The popularity of mobile devices has fostered the change in the delivery method for decision support tools. According to Rose *et al.* (2016), effective design and delivery of these decision support systems are critical for producer adoption. Many are now available for use with mobile computing devices and are in demand by agricultural producers (Bartlett *et al.*, 2014; Migliaccio *et al.*, 2015; Migliaccio *et al.*, 2016; Vellidis *et al.*, 2016). An example of an app that was designed for producers who manage large-sized fields is the Water Irrigation Scheduling for Efficiency Application developed by Colorado State University and agricultural stake-holders (Bartlett *et al.*, 2015; Andales *et al.*, 2014). Similar to most applications, this application uses a cloud-based platform that allows acquisition of different types of data (soil physical properties, weather, and reference evapotranspiration) and displays the information using a dashboard of a specific field; soil type characteristics, such as field capacity and permanent wilting, and available water in the profile. A user (farmer, irrigator, or crop consultant) must input at least the type of crop planted, the planting and emergence dates, and an estimate of the initial soil water content. Irrigation amounts for each irrigation event must be added and daily rainfall events must be downloaded. This app provides a water-balance method of determining when to irrigate based on the concept of maximum allowable depletion (MAD), while the suggested amount to irrigate is based on a percentage of replenishing soil water depletion to field capacity. Essentially this is a modern version of the checkbook method of irrigation scheduling in a format that performs the calculations for the user, and provides the information in real-time with a graphic display. Migliaccio *et al.* (2016) also describes Smart Irrigation apps based on ET or water balance algorithms developed for real-time irrigation scheduling of avocadoes, citrus fruits, cotton, peanuts, strawberries, and vegetables. Vellidis *et al.* (2016) developed a smartphone irrigation-scheduling app for cotton that uses meteorological data, soil characteristics, crop phenology, crop coefficients, and irrigation applications to determine when the root zone soil water deficit was greater than 40%. The three-year study reported that the application achieved results similar to other irrigation scheduling methods in Georgia and Florida.

11.2.4.3 Modeling Programs for Irrigation Scheduling

Although farmer adoption of models is limited (Levidow *et al.*, 2014), hydraulic models can be used to provide water managers with prognostic outcomes regarding crop ET requirements for irrigation scheduling and policy makers with information concerning agroecological and environmental impacts from irrigation. Optimal irrigation scheduling using hydraulic modeling is typically based on optimum crop yield or crop water use. There are a number of extant models that integrate soil hydraulic properties, including HYDRUS-2D (Šimůnek *et al.*, 2008; Šimůnek *et al.*, 2016), a transient numerical model; the Soil Water Atmosphere Plant (SWAP) Model that simulates water, heat, and solute transport (Van Dam *et al.*, 1997; Rallo *et al.*, 2010); and the Root Zone Water Quality Model, RZWQM2 (Ahuja *et al.*, 2000). Some models are tied to a particular type of irrigation method, for example, WinSRFR a new generation software for hydraulic analysis (Bautista *et al.*, 2009) and FIS-DSS (flexible irrigation scheduling decision support system; Yang *et al.*, 2017) are used for surface irrigation scheduling. The latter was developed for irrigation districts in China and can be adapted to different scenarios. Another advantage of using hydraulic models is that these models can be used to estimate plant response to deficit irrigation levels (RZWQM2-Ma *et al.*, 2012; SWAP-Ma *et al.*, 2010), which is a strategy that is becoming more important with limited water supplies. Models that represent water status around an irrigation system can also be used to determine the optimal location for placing soil water sensors. Using numerical modeling, Soulis *et al.* (2015) reported that the optimal location of TDT soil water sensors for irrigation scheduling of a drip system is 10 cm below the soil surface and 11 cm from the dripline. Dabach *et al.* (2015) stated that the HYDRUS 2D/3D model indicated that tensiometers located close to drippers within a geotextile drip interface improved the control and water application efficiency of high frequency drip irrigation systems. The strength of models is their ability to simulate thousands of outcomes in a short period, while sparing the expense and length of time required from field experiments.

11.3 CONCLUSION

Irrigated agriculture is needed to meet demands for agricultural products, but farmers are challenged with limited quality water supplies, environmental and regulatory policies, climate variability, and competition for water from other sectors. Scientific irrigation scheduling could help mitigate these challenges by providing management practices that enable the application of the correct amount of water at the appropriate time. In the United States and Europe, more than 63% of irrigated land is accomplished with pressurized irrigation systems. These systems are inherently more efficient than gravity-flow methods, yet, prudent use of water supplies for agriculture remains critical. In Asia, the vast majority of irrigation is with gravity-flow methods (greater than 96%); even then, scientific irrigation scheduling to improve water conservation is key to sustaining irrigated agriculture. Advanced irrigation scheduling tools such as TDT, TDR, and microwave-based soil water sensors, wireless invasive plant sensors, and radiometric-based noninvasive plant sensors for remote and proximal sensing are available as irrigation scheduling tools today and work well with pressurized systems. In addition to sensors and sensor network systems, current advancements in irrigation scheduling also include software programs integrated with moving irrigation systems that direct dynamic irrigation scheduling and control. More pervasive are the web-based programs and electronic applications for tablets and smartphones that use real-time local data to conveniently provide weather information and decision support for irrigation scheduling at the farm scale. Decision support tools for surface irrigation are also important, and exist mainly in the form of software programs that allow custom inputs and provide simulated outputs for optimizing irrigation scheduling. These resources provide farmers and crop consultants with expedient accessibility to information. Future work is required to refine decision support tools for irrigation scheduling to ensure maximum crop yield, and quality as well as maximum crop WUE, where water for irrigation is limited.

Disclaimer: The U.S. Department of Agriculture (USDA) prohibits discrimination in all its programs and activities on the basis of race, color, national origin, age, disability, and where applicable, sex, marital status, familial status, parental status, religion, sexual orientation, genetic information, political beliefs, reprisal, or because all or part of an individual's income is derived from any public assistance program. (Not all prohibited bases apply to all programs.) Persons with disabilities who require alternative means for communication of program information (Braille, large print, audiotape, etc.) should contact USDA's TARGET Center at (202) 720-2600 (voice and TDD). To file a complaint of discrimination, write to USDA, Director, Office of Civil Rights, 1400 Independence Avenue, S.W., Washington, DC 20250-9410, or call (800) 795-3272 (voice) or (202) 720-6382 (TDD). USDA is an equal opportunity provider and employer.

REFERENCES

Agricultural Water Conservation Clearinghouse. (2016). http://agwaterconservation.colostate.edu/Tools Calculators.aspx.

Ahuja, L. R., Rojas, K. W., Hanson, J. D., Shaffer, M. J., and Ma, L. (2000). Root zone water quality model, modeling management effects on water quality and crop production. *Water Resources*.

Akyildiz, I. A. and Stuntebeck, E. P. (2006). Wireless underground sensor networks: Research challenges. *Ad Hoc Networks* 4, 669–86.

Akyildiz, I. F., Sun, Z., and Vuran, M. C. (2009). Signal propagation techniques for wireless underground communication networks. *Physical Communication* 2, 167–83.

Alchanatis, V., Cohen, Y., Cohen, S., Möeller, M., Sprinstin, M., Meron, M., Tsipris, J., Saranga, Y., and Sela, E. (2010). Evaluation of different approaches for estimating and mapping crop water status in cotton with thermal imaging. *Precision Agric.* 11 (1), 27–41.

Alfieri, J. G., Kustas, W. P., Prueger, J. H., Hipps, L. E., Evett, S. R., Basara, J. B., Neale, C. M. U., French, A. N., Colaizzi, P., Agam, N., Cosh, M. H., Chavez, J. L., and Howell, T. A. (2012). On the discrepancy between eddy covariance and lysimetry-based surface flux measurements under strongly advective conditions. *Advances in Water Res.* 50, 62–78.

Allen, R. G., Pereira, L. S., Raes, D., Smith, M. (1998). Crop evapotranspiration: Guidelines for computing crop water requirements. FAO Irrigation and Drainage, paper 56. Rome, Italy: United Nations FAO.

Andales, A. A., Bauder, T. A., and Arabi, M. (2014). A mobile irrigation water management system using a collaborative GIS and weather station networks. (Ed. L. R. Ahuja, L. Ma, and R. Lascano.) *Practical Applications of Agricultural System Models to Optimize the Use of Limited Water, Advances in Agricultural System Models to Optimize the Use of Limited Water*. Advances in Agricultural Systems Modeling. ASA, CSSA, and SSSA, Madison, Wisconsin, 53–84.

Anderson, R. G. and Dong, W. (2014). Energy budget closure observed in paired eddy covariance towers with increased and continuous daily turbulence. *Agric. Forest Meteor.* 184, 204–9.

Andrade, M. A., O'Shaughnessy, S. A., and Evett, S. R. (2015). ARSmartPivotv.1: Sensor-based management software for center pivot irrigation systems. ASABE Paper No. 152188736. ASABE, St. Joseph, Michigan.

AQUASTAT. (2012). Irrigation in Southern and Eastern Asia in Figures. AQUASTAT Survey (Ed. K. Frenken.) FAO. Rome.

Bayer, L. D. (1954). The meteorological approach to irrigation control. *Hawaiian Planter's Record* 54, 291–8.

Bartlett, A. C., Andales, A. A., Arabi, M., and Bauder, T. A. (2015). A smartphone app to extend use of a cloud-based irrigation scheduling tool. *Computers and Electronics in Agric.* 111, 127–30.

Bautista, E., Clemmens, A. J., Strelkoff, T. S., and Schlegel, J. (2009). Modern analysis of surface irrigation systems with WinSRFR. *Agric. Water Manage.* 96, 1146–54.

Ben-Asher, J., Yosef, B. B., and Volinsky, R. (2013). Ground-based remote sensing system for irrigation scheduling. *Biosyst. Eng.* 114, 444–53.

Ben-Gal, A., Agam, N., Alchanatis, V., Cohen, Y., Yermiyahu, U., Zipori, I., Presnov, E., Sprintsin, M., and Dag, A. (2009). Evaluating water stress in irrigated olives: Correlation of soil water status, tree water status, and thermal imagery. *Irrig. Sci.* 27, 367–76.

Blonquist, J. M., Jones, S. B., Robinson, D. A. (2005). Standardizing characterization of electromagnetic water content sensors: Part II. Evaluation of seven sensing systems. *Vadose Zone J.* 4, 1059–69.

Blonquist, J. J. M., Jones, S. B., and Robinson, D. A. (2006). Precis irrigation scheduling for turfgrass using a subsurface electromagnetic soil moisture sensor. *Agric. Water Manage.* 84, 153–65.

Burgos-Artizzu, X. P., Ribeiro, A., Guijarro, M., and Pajares, G. (2011). Real-time image processing for crop/weed discrimination in maize fields. *Comput. Electron. Agric.* 75, 337–46.

Casanova, J. J., Schwartz, R. C., and Evett, S. R. (2012). Design and field tests of a directly coupled waveguide on access tube soil water sensor. *Appl. Engr. Agric.* 30 (1), 105–12.

Casanova, J. J., O'Shaughnessy, S. A., Evett, S. R., and Rush, C. M. (2014). Development of a wireless computer vision instrument to detect biotic stress in wheat. *Sensors.* 14:17753–69.

Castellví, F., Snyder, R. L., and Baldocchi, D. D. (2015). Surface energy-balance closure over rangeland grass using the eddy covariance method and surface renewal analysis. *Agric. Forest Meteor.* 148, 1147–60.

Centeno, A., Baeza, P., and Lissarrague, J. R. (2010). Relationship between soil and plant water status in wine grapes under various water deficit regimes. *HorTech.* 20 (3), 585–93.

Chaerle L. and Van der Straeten, D. (2000). Imaging techniques and the early detection plant stress. *Trends in Plant Sci.* 5 (11), 495–501.

Cohen, Y., Fuchs, M., and Green, G. C. (1981). Improvement of the heat pulse method for determining sap flow in trees. *Plant, Cell and Environ.* 4, 391–7.

Cohen Y., Alchanatis V., Meron M., Saranga Y., and Tsipris J. (2005). Estimation of leaf water potential by thermal imagery and spatial analysis. *J. Exp. Botany* 56 (417), 1843–52.

Colaizzi, P. D., Gowda, P. H., Marek, T. H., and Porter, D. O. (2009). Irrigation in the Texas High Plains: A brief history and potential reduction in demand. *Irrig. Drain.* 58, 257–74.

Colaizzi, P. D., Agam, N., Tolk, J. A., Evett, S. R., Howell, T. A., Gowda, P. H., O'Shaughnessy, S. A., Kustas, W. P., and Anderson, M. C. (2014). Two-source energy balance model to calculate E, T, and ET: Comparison of Priestley-Taylor and Penman-Monteith formulations and two-time scaling methods. *Trans. ASABE* 57, 479–98.

Conesa, M. R., Torres, R., Domingo, R., Navarro, H., Soto, F., and Perez-Pastor, A. (2016). Maximum daily trunk shrinkage and stem water potential reference equations for irrigation scheduling in table grapes. *Agric. Water Manage.* 172, 51–61.

Coopersmith, E. J., Cosh, M. H., and Daughtry, C. S. T. (2014). Field-scale moisture estimates using COSMOS sensors: A validation study with temporary soil water networks and leaf area indices. *J. Hydro.* 519, 637–43.

Cosh, M. H., Ochsner, T. E., McKee, L. Dong, J., Basara, J. B., Evett, S. R., Hatch, C. E., Small, E. E., Steele-Dunn, S. C., Zreda, M., and Sayde, C. (2016). The soil moisture active passive sensor test bed (SMAP-MOISST): Testbed design and evaluation of *in situ* sensors. *Vadose Zone J.* 15 (4), 1–11.

Dabach, S., Shani, U., Lazarovitch, N. (2015). Optimal tensiometer placement for high-frequency subsurface drip irrigation management in heterogeneous soils. *Agric. Water Manage.* 152, 91–8.

Doltra, J., Oncins, J. A., Bonany, J., and Cohen, M. (2007). Evaluation of plant-based water status indicators in mature apple trees under field conditions. *Irrig Sci* 25, 351. doi:10.1007/s00271-006-0051-y.

Dong, J., Ochsner, T. E., Zreda, M., Cosh, M. H., and Zou, C. B. (2014). Calibration and validation of the COSMOS rover for surface soil moisture measurement. *Vadose Zone J.*, 1–8. doi:10.2136/vzj2013.08.0148.

Escarabajal-Henarejos, D., Molina-Martínez, J. M., Fernández-Pacheco, D. G., Cavas-Martínez, F., and García-Mateos, G. (2015). Digital photography applied to irrigation management of Little Gem lettuce. *Agric. Water Manage.* 151, 148–57.

Eurostat Statistics Explained. (2010). http://ec.europa.eu/eurostat/statistics-explained/index.php/Agri-environmental_indicator_-_irrigation#Further_Eurostat_information. Accessed November 15, 2016.

Evett S. R., Heng, L. K., Moutonnet, P., and Nguyen, M. L. (Eds.) (2008). Field Estimation of Soil Water Content: A Practical Guide to Methods, Instrumentation, and Sensor Technology. IAEA-TCS-30. International Atomic Energy Agency, Vienna, Austria. ISSN 1018–5518.

Evett, S. R., Schwartz, R. C., Tolk, J. A., and Howell, T. A. (2009). Soil profile water content determination: Spatiotemporal variability of electromagnetic and neutron probe sensors in access tubes. *Vadose Zone J.* 8, 926–41.

Evett, S. R., Schwartz, R. C., Casanova, J. J., and Heng, L. K. (2012). Soil water sensing for water balance, ET and WUE. *Agric. Water Manage.* 104, 1–9.

Fernández, J. E., Green, S. R., Caspari H. W., Diaz-Espejo, A., and Cuevas, M. V. (2008). The use of sap flow measurement for scheduling irrigation in olive, apple, and asian pear trees, and in grapevines. *Plant and Soil.* 305 (1), 91–104.

Fernández, J. E. and Cuevas, M. V. (2010). Irrigation scheduling from stem diameter variations: A review. *Agric. Forest Meteor.* 150, 135–51.

Girona, J., Mata, M., Del Campo, J., Arbonés, A., Bartra, E., Marsal, J. (2006). The use of midday leaf water potential for scheduling deficit irrigation in vineyards. *Irrig. Sci.* 24, 115–27.

Goldhamer, D., Fereres, E., and Salinas, M. (2003). Can almond trees directly dictate their irrigation needs? *Calif. Agr.* 57 (4), 138–44. doi: 10.3733/ca.v057n04p138.

Grimes, D. W., Yamada, H., and Hughes, S. W. (1987). Climate-normalized cotton leaf water potentials for irrigation scheduling. *Agric. Water Manage.* 12, 293–304.

Hart, Q. J., Brugnach, M., Temesgen, B., Rueda, C., Ustin, S. L., and Frame, K. (2009). Daily reference evapotranspiration for California using satellite imagery and weather station measurement interpolation. *Civil Engr. Environ. Sys.* 26 (1), 19–33.

Hedley, C. B. and Yule., I. J. (2009). Soil water status mapping and two variable-rate irrigation scenarios. *Precision Agric.* 10 (4), 342–55.

Henggeler, J. C., Dukes, M. D., Mecham, B. Q. (2011). Irrigation scheduling. (Ed. L.E. Stetson and B.Q. Mecham.) *Irrigation.* Sixth Ed. Falls, Church, VA: Irrigation Association, 491–564.

Howell, T. A. (1996). Irrigation scheduling research and its impact on water use. Proc. of the International Conference, San Antonio, TX. ASAE. St. Joseph, Michigan.

Huguet, J. G., Li, S. H., Lorendeau, J. Y., and Pelloux, G. (1992). Specific micro morphometric reactions of fruit trees to water stress and irrigation scheduling automation. *J. Horticultural Sci.* 67 (5), 631–40.

Hunsaker, D. J., Pinter Jr., P. J., Barnes, E. M., and Kimball, B. A. (2003). Estimating cotton evapotranspiration with multispectral vegetation index. *Irri. Sci.* 22, 95–104.

Hunsaker, D. J., Barnes, E. M., Clarke, T. R., Fitzgerald, G. J., and Pinter Jr., P. J. (2005). Cotton irrigation scheduling using remotely sensed and FAO-56 basal crop coefficients. *Trans. ASAE* 48 (4), 1395–407.

Hunsaker, D. J., French, A. N., Waller, P. M., Bautista, E., Thorp, K. R., Bronson, K. F., and Andrade-Sanchez, P. (2015). Comparison of traditional and ET-based irrigation scheduling of surface-irrigate cotton in the arid southwestern United States. *Agric. Water Manage.* 159, 209–44.

Idso, S. B., Reginato, R. J., Reicosky, D. C., and Hatfield, J. L. (1981). Determining soil-induced plant water potential depressions in alfalfa by means of infrared thermometry. *Agron J.* 73 (5), 826–30.

Impens, I. L. and Schalck, J. M. (1965). A very sensitive electronic dendrograph for recording radial changes of a tree. *Ecology.* 46, 183–4.

Jackson, R. D., Idso, S. B., Reginato, R. J., and Pinter, P. J. (1981). Canopy temperature as a crop water stress indicator. *Water Resources. Res.* 17, 1133–8.

Jain, M., Prasad, P. V., Boote, K. V., Hartwell, A. L., and Chourey, P. S. (2007). Effects of season-long high-temperature growth conditions on sugar-to-starch metabolism in developing microspores of grain sorghum (*Sorghum Bicolor L. Moench*). *Planta.* 227 (1), 67–9.

Jones, H. G. (2004). Irrigation scheduling: advantages and pitfalls of plant-based methods. *J. Exp. Botany* 55 (407), 2427–36.

Kaplan, H. (2007). *Practical Applications of Infrared Thermal Sensing and Imaging Equipment.* 49–60. Spie Press: Bellingham, Washington.

Kranz, W. L., Evans, R. G., Lamm, F. R., O'Shaughnessy, S. A., and Peters, R. T. (2012). A review of mechanical move sprinkler irrigation control and automation technologies. *Appl. Engr. Agric.* 28 (3), 389–97.

Kim, Y., Evans, R. G., and Iversen, W. M. (2009). Evaluation of closed-loop site-specific irrigation with wireless sensor network. *Journal of Irrigation and Drainage Engineering* 135 (1), 25–31.

Kipp, S., Mistel, B., and Schmidhalter U. (2014). The performance of active spectral reflectance sensor as influenced by measuring distance, device temperature, and light intensity. *Comput. Electron. Agric.* 100, 24–33.

Kizito, F., Campbell, C. S., Campbell, G. S., Cobos, D. R., Teare, B. L., Carter, B., Hopmans, J. W. (2008). Frequency, electrical conductivity and temperature analysis of a low-cost capacitance soil moisture sensor. *Journal of Hydrology* 352 (3–4), 367–78.

Kobayashi, R., Kanda, E., Kitada, K., Ishiguro, K., and Torigoe, T. (2000). Detection of rice panicle blast with multispectral radiometer and the potential of using airborne multispectral scanners.

Köhhli, M., Schrön, M., Zreda, M. Schmidt, U., Dietrich, P., and Zacharias, S. (2015). Footprint characteristics revised for field-scale soil moisture monitoring with cosmic-ray neutrons. *Water Res. Res.* 51 (7), 5772–90.

Klepper, B., Browning, V. D., and Taylor, H. M. (1971). Stem diameter in relation to plant water status. *Plant Physiology* 48, 683–5.

Lee, W. S., Alchanatis, V., Yang, C., Hirafuji, M., Moshou D., and Li, C. (2010). Sensing technologies for precision specialty crop production. *Comput. Electron. Agric.* 74, 2–33.

Lee, W. S. and Ehsani, R. (2015). Sensing systems for precision agriculture. *Florida. Comput. Electron. Agric.* 112, 2–9.

Leib, B. G., Jabro, J. D., and Matthews, G. R. (2003). Field evaluation and performance comparison of soil moisture sensors. *Soil Science* 168 (6), 396–408.

Leuning, R., Van Gorsel, E., Massman, W. J., and Isaac, R. R. (2012). Reflections on the surface energy imbalance problem. *Agric. Forest Meteor.* 156, 65–74.

Levellara, N., Saavedra, F., and Salgado, E. (2011). Plant-based indicators for irrigation scheduling in young cherry trees. *Agric. Water Manage.* 98, 684–90.

Levidow, L., Zaccaria, D., Maia, R., Vivas, E., Todorovic, M., and Scardigno, A. (2014). Improving water-efficient irrigation: Prospects and difficulties of innovative practices. *Agric. Water Manage.* 146, 84–94.

Li, F., Miao, Y., Feng, M., Yuan, F., Yue, S., Gao, X., Liu, Y., Liu, B., Ustin, S. L., and Chen, X. (2014). Improving estimation of summer maize nitrogen status with red edge-based spectral vegetation indices. *Field Crops Res.* 157, 111–23.

Li, W., Niu, Z., Chen, H., Li, D., Wu, M., and Zhao, W. (2016). Remote estimation of canopy height and aboveground biomass of maize using high-resolution stereo images from a low-cost unmanned aerial vehicle system. *Ecological Indicators.* 67, 637–48.

Link, S. O., Thiede, M. E., and Van Bavel, M. G. (1998). An improved strain-gauge device for continuous field measurement of stem and fruit diameter. *J. Experimental Botany* 49 (326), 1583–7.

Ma, L., Trout, T. J., Ahuja, L. R., Bausch, W. C., Saseendran, S. A., Malone, R. W., and Nielsen, D. C. (2012). Calibrating RZWQM2 model for maize responses to deficit irrigation. *Agric. Water Manage.* 103, 140–9.

Ma, Y., Feng, S., Huo, Z., and Song, X. (2010). Application of the SWAP model to simulate the field water cycle under deficit irrigation in Beijing, China. *Mathematical and Computer Modeling* 54, 1044–52.

MacDonald, G. K., Brauman, K. A., Sun, S., Carlson, K. M., Cassidy, E. S., Gerber, J. S., West, P. C. (2015). Rethinking agricultural trade relationships in an era of globalization. *BioScience* 65 (3), 275–89.

Maes, W. H. and Steppe, K. (2012). Estimating evapotranspiration and drought stress with ground-based thermal remote sensing in agriculture: A review. *J. Exp. Botany.* 63 (13), 4671–712.

Mangus, D. L., Sharda, A., and Zhang, N. (2016). Development and evaluation of thermal Infrared imaging system for high spatial and temporal resolution of crop water stress monitoring of corn within a greenhouse. *Comput. Electron. Agric.* 121, 149–59.

Mao, H., Gao, H., Zhang, X., and Kumi, F. (2015). Nondestructive measurement of total nitrogen in lettuce by integrating spectroscopy and computer vision. *Scientia Horticulturae.* 184, 1–7.

Mastrorilli, M., Katerji, N., and Rana, G. (1995). Water efficiency and stress on grain sorghum at different reproductive states. *Agric. Water Manage.* 28 (1), 23–4.

McElrone, A. J., Shapland, T. M., Calderon, A., Fitzmaurice, L., Paw, U., and Snyder, R. L. (2013). Surface renewal: An advanced micrometeorological method for measuring and processing field-scale energy flux density data. *J. Vis. Exp.* (82). e50666, doi:10.3791/50666.

Migliaccio, K. W., Morgan, K. T., Fraisse, C., Vellidis, G., and Andreis, J. H. (2015). Performance evaluation of urban turf irrigation smartphone app. *Comput. Electron. Agric.* 118, 136–42.

Migliaccio, K. W., Morgan, K. T., Vellidis, G., Zotarelli, L., Fraisse, C., Zurweller, A. A., Andreis, J. H., Crane, J. H., and Rowland, D. L. (2016). Smartphone apps for irrigation scheduling. *Trans. ASABE.* 59 (1), 291–301.

Mirás-Avalos, J. M., Pérez-Sarmiento, F., Alcobendas, R., Alarcón, J. J., Mounzer, O., and Nicolás, E. (2016). Reference values of maximum daily trunk shrinkage for irrigation scheduling in mid-late maturing peaches. *Agric. Water Manage.* 171, 31–9.

Möller, M., Alchanatis, V., Cohen, Y., Meron, M., Tsipris, J., Naor, A., Ostrovsky, V., Sprinstsin, M., and Cohen, S. (2007). Use of thermal and visible imagery for estimating crop water status of irrigated grapevine. *J. Exp. Botany* 58 (4), 827–38.

Norman, J. M., Kustas, W. P., and Humes, K. S. (1995). A two-source approach for estimating soil and vegetation energy fluxes form observations of directional radiometric surface temperature. *Agric. Forest Meteorol.* 77, 263–93.

Or, D. and Wraith, J. M. (2002). Soil water content and water potential relationships in Soil Physics Companion. (Ed. A. W. Warrick.) CRC Press: Boca Raton, Florida. 49–84.

O'Shaughnessy, S. A. and Evett, S. R. (2010). Canopy temperature based system effectively schedules and controls center pivot irrigation of cotton. *Agric. Water Manage.* 97, 1310–6.

O'Shaughnessy, S. A., Evett, S. R., Colaizzi, P. D., and Howell, T. A. (2011). Using radiation thermography and thermometry to evaluate crop water stress in soybean and cotton. *Agric. Water Manage.* 98, 1523–35.

O'Shaughnessy, S. A., Evett, S. R., Colaizzi, P. D., Tolk, J. A., and Howell, T. A. (2014). Early and Late maturing grain sorghum under variable climatic conditions in the Texas High Plains. *Trans. ASABE.* 57 (6), 1583–94.

O'Shaughnessy, S. A., Evett, S. R., and Colaizzi, P. D. (2015). Dynamic prescription maps for site-specific variable-rate irrigation of cotton. *Agric. Water Manage.* 159, 123–38.

Osroosh, Y., Peters, R. T., Campbell, C. S., and Zhang, Q. (2015). Automatic irrigation scheduling of apple trees using theoretical crop water stress index with an innovative dynamic threshold. *Comput. Electron. Agric.* 118, 193–203.

Padhi, J., Misra, R. K., and Payero, J. O. (2012). Estimation of soil water deficit in an irrigated cotton field with infrared thermography. *Field Crops Res.* 126, 45–55.

Parameswaran, V., Zhou, H., and Zhang, A. (2013). Wireless underground sensor network design for irrigation control: Simulation of RFID deployment. Seventh International Conference on Sensing Technology, 842–9, IEEE.

Patented Soil Water Censor (No. 8,947,102 B1, "Soil Water and Conductivity Sensing System," issued February 3, 2015). CRADA has resulted in commercial introduction of a low-cost, low-power, very accurate soil water content and bulk electrical conductivity sensor (Acclima, Inc., TDR-315).

Penman, H. L. (1952). The physical basis of irrigation control. *Proc. Intl. Hort. Congr.* 13: 913–24, London.

Peters, T. R. and Evett, S. R. (2007). Spatial and temporal analysis of crop stress using multiple canopy temperature maps created with an array of center pivot mounted infrared thermometers. *Trans. ASABE.* 50 (3), 919–27.

Peters, R. T. and Evett, S. R. (2008). Automation of a center pivot using the temperature-time threshold method of irrigation scheduling. *J. Irrig. Drain. Engr.* 134 (3), 286–91.

Pingali, P. L. (2012). Green revolution: Impacts, limits, and the path ahead. Proc. National Academy of Sciences, USA. 109 (31), 12302–8.

Pretty, J. N., Noble, A. D., Bossio, D., Dixosn, J., Hine, R. E., Penning de Vries, F. W. T., and Morison, J. I. L. (2006). Resource-conserving agriculture increases yields in developing countries. *Environ. Sci. and Tech.* 40 (4), 114–9.

Pruitt, W. O. and Jensen, M. C. (1955). Determining when to irrigate. *Agric. Engr.* 36, 389–93.

Rallo, G., Agnese, C., Blanda, F., Ninacapilli, M., and Provenzano, G. (2010). Agrohydrolocial models to schedule irrigation of mediterranean tree crops. *Italian J. Agrometeorology* 11–21.

Rodriguez-Moreno, F., Zemek, F., Kren, J., Pikl, M., Lukas, V., and Novak, J. (2016). Spectral monitoring of wheat canopy under uncontrolled conditions for decision-making purposes. *Comput. Electron. Agric.* 125, 81–8.

Rosa, R. and Tanny, J. (2015). Surface renewal and eddy covariance measurements of sensible and latent heat fluxes of cotton during two growing seasons. *Biosys. Engr.* 136, 149–61.

Rose, D. C., Sutherland, W. J., Parker, C., Lobley, M., Winter, M., Morris, C., Twinning, S., Foulkes, C., Amano, T., and Dicks, L. V. (2016). Decision support tools for agriculture: Towards effective design and delivery. *Agric. Sys.* 149, 165–74.

Rosolem, R., Shuttleworth, W. J., Zreda, M., Franz, T. E., Zeng, X., and Kurc, S. A. (2013). The effect of atmospheric water vapor on neutron count in the cosmic-ray soil moisture observing system. *J. Hydro. Meteor.* 14 (5), 1659–71.

Sadler, E. J., Evans, R. G., Stone, K. C., and Camp, C. R. (2005). Opportunities for conservation with precision irrigation. *J. Soil Water Cons.* 60 (6), 371–9.

Šimůnek, J., Van Genuchten, M. T., and Šejna, M. (2008). Development and applications of the HYDRUS and STANMOD software packages and related codes. *Vadose Zone J.* 7, 587–600.

Šimůnek, J., Van Genuchten, M. T., and Šejna, M. (2016). Recent developments and applications of the HYDRUS computer software packages. *Vadose Zone J.* 2–25.

Skierucha, W., Wilczek, A., Szyplowska, A., Slawiński, C., and Lamorski, K. (2012). A TDR-based soil moisture monitoring system with simultaneous measurement of soil temperature and electrical conductivity. *Sensors*. 12, 13545–66.

Smith-Rose, R. L. (1933). The electrical properties of soils for alternating currents at radio frequencies. *In Proc. Roy. Soc.* 140, 359. London.

Soulis, K. X., Emaloglou, S., and Dercas, N. (2015). Investigating the effects of soil moisture sensors positioning and accuracy on soil moisture based drip irrigation scheduling systems. *Agric. Water Manage.* 258–68.

Spano, D., Snyder, R. L., Duce, P., and Paw, U. K. T. (2000). Estimating sensible and latent heat flux densities form grapevine canopies using surface renewal. *Argic. Forest Meteor.* 104 (23), 171–83.

Steduto, P., Hsiao, T. C., Fereres, E., and Raes, D. (2012). Crop yield response to water. FAO irrigation and drainage paper 66. Rome, Italy: United Nations FAO.

Steele, D. D., Scherer, T. F., Hopkins, D. G., Tuscherer, S. R., and Wright, J. (2010). Spreadsheet implementation of irrigation scheduling by the checkbook method for North Dakota and Minnesota. *App. Engr. Agric.* 26 (6), 983–95.

Stone, K. C., Bauer, P. J., and Sigua, G. C. (2016). Irrigation management using an expert system, soil water potentials, and vegetative indices for spatial applications. *Trans. ASABE* 59 (3), 941–8.

Stone, K. C., Bauer, P., Busscher, W., Millen, J., Evans, D., and Strickland, E. (2010). Variable-rate irrigation management for peanut in the Eastern Coastal Plain. ASABE Paper No. 1IRR10-8977. ASABE: St. Joseph, Michigan.

Sui, R., Wilkerson, J. B., Hart, W. E., Wilhelm, L. R., and Howard, D. D. (2005). Multispectral sensor for detection of nitrogen status in cotton. *Appl. Engr. Agric.* 21 (2), 167–72.

Sui, R. and Thomasson, J. A. (2006). Ground-based sensing system for cotton nitrogen status determination. *Trans. ASABE* 49 (6), 1983–91.

Sui, R., Fisher, D. K., and Barnes, E. M. (2012). Soil moisture and plant canopy temperature sensing for irrigation application in cotton. *Journal of Agricultural Science* 4 (12), 93–105.

Sui, R. and Baggard, J. (2015). Wireless sensor network for monitoring soil moisture and weather conditions. *Appl. Engr. Agric.* 31 (2), 193–200.

Suvočarev, K., Shapland, T. M., Snyder, R. L., and Martínez-Cob, A. (2014). Surface renewal performance to independently estimate sensible and latent heat fluxes in heterogeneous crop surfaces. *J. Hydrology*.

Tilman, D., Cassman, K. G., Matson, P. A., Naylor, R., and Polasky, S. (2002). Agricultural sustainability and intensive production practices. *Nature*. 418, 671–7.

Thorp, K. R., Gore, M. A., Andrade-Sanchez, P., Carmo-Silva, A. E., Welch, S. M., White, J. W., and French, A. N. (2015). Proximal hyperspectral sensing and data analysis approaches for field-based plant phonemics. *Comput. Electron. Agric.* 118, 225–36.

Trout, T., Melton, F., and Johnson, L. (2014). A web-based tool that combines satellite and weather station observations to support irrigation scheduling. Proc. Twenty-Sixth Annual Central Plains Irrigation Conference. Burlington, CO.

USDA-ARS, Irrigator Pro for Corn. (2016). https://www.ars.usda.gov/research/software/download/?softwareid =248. Accessed on November 17, 2016.

Van Dam, J. C., Huygen, J., Wesseling, J. G., Feddes, R. A., Kabat, P. Van Walsum, P. E. V., Groenendijk, P., and Van Diepen, C. A. (1997). Theory of SWAP Version 2.0. Simulation of Water Flow, Solute Transport and Plant Growth in the Soil–Water–Atmosphere–Plant Environment. Report 71. Department Water Resources, Wageningen Agricultural University.

Van Vliet, J. A., Schut, A. G. T., Reidsama, P., Descheemaker, K., Slingerland, M., Van de Ven, G. W. J., and Giller, K. E. (2015). De-mystifying family farming: Features, diversity and trends across the globe. *Global Food Security*. 5, 11–8.

Vellidis, G., Tucker, M., Perry, C., Kvien, C., and Bednarz, C. (2008). A real-time wireless smart sensor array for scheduling irrigation. *Comput. Electron. Agric.* 61, 44–50.

Vellidis, G., Liakos, V., Andreis, J. H., Perry, C. D., Porter, W. M., Barnes, E. M., Morgan, K. T., Fraisse, C., and Migliaccio, K. W. (2016). Development and assessment of a smartphone application for irrigation scheduling in cotton. *Comput. Electron. Agric.* 127, 249–59.

Viscarra Russel, R. A. and Bouma, J. (2016). Soil sensing: A new paradigm for agriculture. *Agric. Sys.* 148, 71–4.

Vulfson, L., Genis, A., Blumberg, D. G., Kotlyar, A., Freilikher V., and Ben-Asher, J. (2013). Remote sensing in microwave and gamma ranges for the monitoring of soil water content of the root zone. *International Journal of Remote Sensing* 34 (17), 6182–201. doi: 10.1080/01431161.2013.793863.

Wanjura, D. F., Upchurch, D. R., Mahan, J. R. (1995). Control of irrigation scheduling using temperature-time thresholds. *Trans ASAE* 38 (2), 403–9.

Yang, G., Liu, L., Guo, P., and Li, M. (2017). A flexible decision support system for irrigation scheduling in an irrigation district in China. *Agric. Water Manage.* 179, 378–89.

Yoder, R. E., Johnson, D. L., Wilkerson, J. B., and Yoder, D. C. (1997). Soil water sensor performance. *Applied Engineering in Agriculture* 14 (2), 121–33.

Zreda, M., Desilets, D., Ferré, T. P. A., and Scott, R. L. (2008). Measuring soil moisture content noninvasively at intermediate spatial scale using cosmic-ray neutrons. *Geophys. Res. Lett.* 35, L21402. doi:10.1029 /2008GL035655.

Zreda, M., Shuttleworth, W. J., Zeng, X., Zweek, C., Desilets, D., Franz, T., and Rosolem, R. (2012). COSMOS; the Cosmic-ray soil moisture observing system. *Hydrol. Earth Syst. Sci.* 16, 4079–99.

12 Minimizing Evaporation Loss from Irrigation Storages

Pam Pittaway, Nigel Hancock, Michael Scobie, and Ian Craig

CONTENTS

12.1 INTRODUCTION

12.1.1 WATER LOSS ASSESSMENT AND IMPORTANCE OF DAM MONITORING

Viewed via satellite or aircraft, a landscape feature of most warm countries around the world is the considerable number of glistening storages used for stock watering, and particularly, irrigation. Such storages are almost always shallow, these being defined as having a water depth of less than 5 m when full. Evaporation rates for shallow open water in Australia are generally within the range 2–5 m annually, i.e., a major proportion of the stored volume. Estimates of the total fresh water lost due to evaporation vary greatly due to differing climatic conditions, but are generally assumed to be 10% to 30% of the total water utilized for agricultural purposes (Gallego-Elvira *et al.*, 2011). Since 60% to 80% of available fresh water is consumed for the purposes of irrigation, this amounts to a very great deal of water globally. The challenge is on for engineers as well as farmers to meet water and food security requirements for a predicted world population of 10 billion people by 2050. Minimizing loss of water from irrigation storages therefore remains an issue of obvious priority.

Critical to this process is the adoption by the agricultural irrigation industry of a standardized approach for adequately estimating open water storage evaporation, and this estimation needs to be independent of water depth measurement because water depth is also influenced by seepage through the base of the dam. There are two major options: these are the use of an adjacent "evaporation pan" as a miniature physical simulation of the dam; and the calculation of evaporation from local meteorological variables, specifically solar radiation, humidity, wind speed, and temperature. Each has major difficulties and disadvantages, and comparison in the context of agricultural water management is set out in Allen *et al.* (1998). These arise not just from the practicalities of instrumentation and maintenance, but also because the location of a relatively small area of open water surface involves complex physical processes.

Nevertheless, it is generally agreed that the preferable meteorologically-based approach is that of Penman-Monteith (PM), and its use for irrigated cropping is recommended by the Food and Agriculture Organization (FAO) of the United Nations as "FAO56" (Allen *et al.*, 1998). This is widely accepted as "best practice." FAO56-based estimates of open water evaporation are now available from websites such as the Australian Bureau of Meteorology (BOM), for most regions in Australia. If regional data is unavailable, or if greater accuracy is needed appropriate to local conditions, FAO56 evaporation can be calculated by automatic weather stations (AWS) located near to the dam. Armed with such data, farmers can now directly assess how their dam is performing in terms of evaporative loss, and therefore also in terms of additional losses due to dam seepage.

There is a range of methods for measuring water storage losses. Most simply, the rate of loss of water depth in millimeters per day can be using a depth ruler mounted on a fixed post. For convenient and automated daily measurement, the position of a float on the surface may be measured using a liquid level sensing instrumentation (traditionally involving a cable, counterweight, and rotary shaft encoder; but preferably using magnetostrictive technology which is less prone to mechanical friction or stiction errors). However, successful deployment of such instrumentation requires a stilling well, which usually implies the cost of a permanent installation. A simpler and usually more convenient approach is to use a submerged pressure sensitive transducer (PST) suspended near the base of the storage (Craig *et al.*, 2006). A PST produces an electrical measurement of water pressure, corresponding directly with depth at the transducer which may then be conveniently recorded (logged).

If the observed loss generally agrees with the FAO56 estimate, the farmer can reasonably assume that their dam is not leaking. The location of particular storage, e.g., with respect to natural shelter from trees or buildings, can result in enhanced or diminished evaporation; plus day-to-day variations between observed loss and daily FAO56 evaporation estimates may be expected, due to changes in the thermal storage of the water body in (delayed) response to sudden changes in local weather, but these will average out over a period of a few days. But if the observed loss is *significantly greater* that the PM-FAO56 estimate on a continuing basis, this additional loss can be attributed to seepage from the storage.

Seepage losses can either be through the earthen liner at the base of the dam, or leakage through cracks in the walls of the dam. Seepage is also a major cause of agricultural water loss in Australia, and because its magnitude tends to increase slightly each year as the liner and walls gradually degrade, it can go unnoticed by farmers for years (Craig *et al.*, 2007a). All Australian farmers are strongly encouraged to regularly monitor the depth of water in their storages, and compare them to the FAO56 estimate of evaporative loss. This way, regular maintenance of dam liners and walls can be scheduled. The economic benefit of saving water in this way should more than cover the costs of an AWS for monitoring temperature, relative humidity, solar radiation, and wind speed.

12.1.2 EVAPORATION THEORY APPLIED TO OPEN WATER STORAGES

The Penman 1948 equation, upon which PM-FAO56 is based, is commonly presented (in energy flux units, watts per meters) as:

$$\lambda E_0 = \frac{\Delta(R_n - G) + \Delta\rho C_p(e_s - e)/r_a}{\Delta + \gamma} \tag{12.1}$$

Here, λ is latent heat of vaporization of water, E_0 is the evaporation rate, R_n is the net radiation, G is heat flux to water and underlying sediment, Δ is the slope of the saturation vapor pressure temperature curve, ρ is the mean air density at constant pressure, C_p is the specific heat of air, $(e_s - e)$ is the vapor pressure deficit (VPD) of the airflow (which increases as humidity decreases), γ is the psychrometric constant, and r_a is aerodynamic resistance (which decreases as windspeed increases). Using instantaneous values of each variable (which, commonly, will vary greatly throughout the day) this equation is repeatedly evaluated to calculate E_0 in units of millimeters per second. These are then summed, usually in the AWS logger, to produce the daily total, i.e., E_0 in millimeters per day. Equation 12.1 is simplified from the full Penman-Monteith (PM) version which is required for evaporation from vegetated surfaces (Allen *et al.*, 1998; Craig *et al.*, 2007b).

Despite being well-accepted as providing useful estimates of evaporative loss, the approach of Equation 12.1, automated using an AWS, has some shortcomings. Firstly, the heat flux G into (by day) and out of (by night) is difficult to estimate. Secondly, the water body has significant heat storage capacity, and thermal stratification and other effects can occur in the water column beneath the surface. Ideally, the net radiation flux R_n should be measured over the water surface. However, this is usually impractical, and hence the values measured adjacent to the dam will be in error as regards the albedo (reflection factor) for incoming solar radiation, and the albedo of a water surface varies greatly with sun angle. The outgoing terrestrial radiation will also be in error due to the water-land temperature difference but this is less significant, except near dawn and dusk. Finally, to further complicate matters, an extensive open water surface sets up its own microclimate, influenced by the dam walls, and consequently humidity and wind speed will usually be somewhat different to that measured over a nearby land surface.

Given these shortcomings it is commonly assumed that the evaporation performance of a dam can be physically simulated by an evaporation pan, of which the "Class A pan" (25 cm deep and with a 120.7 cm diameter) is most commonly used. However, the atmospheric interactions and the energy storage of such a small water body are very different from those of a dam of multi-megaliter capacity. Despite extensive research, the choice of an appropriate "pan factor" to relate pan evaporation to actual evaporation for a particular dam has proven very difficult and the pan evaporation losses recorded can be 1.3 to 2.1 times the FAO56 reported amount (Allen *et al.*, 1998).

In relatively hot countries, at least, the dominant driver of daily evaporative loss is the daytime incoming solar net radiation R_n, and this is the principal reason that PM-FAO56 provides a reliable estimate. However, evaporation occurring during the night (i.e., non-incoming, R_n dependent) can be 0% to 20% of the daily total, as illustrated by changes in dam height recorded with a PST unit (Craig, 2006). Both FAO56 and energy balance methods successfully predict that the rate of evaporative loss per day for a typical agricultural water storage in Southeast Queensland, Australia, would be approximately 5 mm per day during summer, with a typical range of 2–7 mm, depending upon meteorological conditions (Craig *et al.*, 2007b; McJannet *et al.*, 2011).

Other parameters such as the precise temperature at the air-water interface, known as the water skin temperature, is also important, as is air velocity and humidity measured immediately above the water. These parameters are both inconvenient and difficult to measure, and are likely to only loosely correlate with AWS parameters recorded at a 2 m height over land. The reason why the PM–FAO56 approach is still reliable, for hot weather at least, is the strongly dominating effect of daytime solar radiation, especially during summertime conditions. Also, important for improved overall accuracy of analysis is the complete thermal history of the water column and underlying sediment. This depends not only upon energy received on the day of measurement, but also during the previous days and weeks. Thorough determination of the complete energy balance evolution with time is therefore required for more accurate estimates of evaporation loss on regional catchment scales, and accurate evaporation estimates throughout the year, especially in winter.

Finally, at the research level, sophisticated techniques such as infrared scintillometry and, independently, barge-mounted eddy covariance (ECV) atmospheric measurement, may be used to assess dam evaporation. Using these techniques, the long-term energy balance has recently been determined at a medium-sized (17 ha) agricultural storage located in Southeast Queensland, Australia (McJannet et al., 2008, 2011, 2013; McGloin et al., 2014a, 2014b). The water surface energy balance was determined from:

$$R_n - S_w - S_a + Q_r - Q_p - Q_s - H - E = 0 \tag{12.2}$$

Here, R_n is the net radiation (measured over the water surface), S_w is the change in heat stored within the water column, S_a is the change in heat stored in the air column below the net radiation measurement height, Q_r is the energy added through rainfall, Q_p is energy addition or removal via inflows or outflows, Q_s is energy transfer to sediments beneath the water, H is the sensible heat flux, and E is the latent heat flux. Gaining a fuller understanding of the entire energy balance has enabled researchers to more accurately correlate information across the various different approaches to quantify the evaporation from farm storages. Further work in this area should involve the development of a robust combination of theoretical science and practical application to provide an accurate measurement of daily water loss from open water bodies. This is a critical step towards the ability to fairly assess the potential of Evaporation Mitigation Technologies (EMTs).

12.1.3   PERFORMANCE OF VARIOUS DAM EVAPORATION MITIGATION TECHNOLOGIES (EMTs)

As introduced in Section 12.1.2, there are significant losses from water storages in Australian agriculture. These water losses represent a real cost for farmers in terms of reduced irrigation (and therefore production) potential. As a result, there have been a range of EMTs developed in attempt to slow or halt the evaporation loss.

The current methods recognized for evaporation mitigation from agricultural storages are physical covers which are suspended above the water surface; physical covers which float upon the water surface; floating modules, which are not continuous covers, but are discrete modular units; and chemical methods, as summarized in Table 12.1.

TABLE 12.1

Summary of Current Knowledge of Evaporation Reduction Techniques

EMT Type	Evaporation Reduction (%)	Advantages	Problems	References
Suspended covers—e.g., shadecloth	70% to 90%	Once installed, good for 30 plus years	Expensive to install (due to extensive human labor cost)	Craig et al., 2005; Martínez-Alvarez, 2006; Finn and Barnes, 2007; Maestre-Valero et al., 2008, 2009, 2011; Gallego-Elvira et al., 2011
Continuous covers—e.g., bubble wrap	50% to 80%	Few	Very expensive, difficult to install, manage, and maintain	Craig et al., 2008
Floating modules	30% to 60%	Simple to implement, capital investment can be phased	Expensive unless recycled plastics used (although could be combined with solar panels)	Hassan et al., 2015; Segal and Burstein, 2010; Simon et al., 2016; Ferrer-Gisbert et al., 2013; Santafe et al., 2014
Chemical monolayer	5% to 30%	Low cost, only apply when there is water in storage	Highly variable field performance, physical/chemical degradation issues	Hancock et al., 2011; Gallego-Elvira et al., 2013; Pittaway et al., 2015

What is not known in precise terms is the economic potential, and consequences, of the various methods available for reducing evaporative loss of stored agricultural water in Australia.

12.2   REDUCING EVAPORATION LOSSES USING PHYSICAL COVERS

12.2.1   Shadecloth

Experimental trials carried out by the National Centre for Engineering in Agriculture (NCEA) at the University of Southern Queensland (USQ) demonstrated that shadecloth (Figure 12.1) shows great promise for the protection and security of high value water. The results in terms of evaporation saving were consistently high (70% to 90%), during the 2000–2005 monitoring, which was undertaken both at research plot, and at farm scale. However, the cost of installation (mainly manual labor) becomes significant for storages greater than 4 ha in size. Obviously, to reduce cost per megaliter stored, storages should be excavated as deeply as possible.

An additional advantage of shadecloth installation is the reduction in light intensity reduces the potential for algal blooms, improving water quality. Ingress of wind-borne dust is also reduced, reducing water turbidity and salinity, and contamination from pesticide spray drift. The technology is particularly suited to supplying high quality water for drip irrigation systems, deployed for high value horticultural crops (Maestre-Valero *et al.*, 2011). Once installed, shadecloth is almost entirely maintenance free. Suspending the shadecloth above the water reduces the need for cleaning, as the fabric dries rapidly after rain, minimizing the potential for epiphytic growth. The polyethylene cloth has a life expectancy of approximately 30 years and will eventually become brittle, but will continue to function satisfactorily until a storm or hail causes damage. Hail, on reasonably new cloth, just bounces and causes little damage, and collected hail is discharged through hail chutes to the water body.

In summary, the key limitation with the shadecloth option is the capital investment cost, which is mainly human labor at installation. The challenge remains to reduce this cost using innovative engineering technologies. Growing fish beneath the cloth can help to reduce costs, but even taking this into account, it is still difficult to demonstrate economic viability, even with the highest value crops in Australia.

FIGURE 12.1   Suspended shadecloth structure installed at a fruit farm in Stanthorpe, Queensland, Australia.

12.2.2 Continuous Floating Covers

Continuous floating covers are routinely used in swimming pool applications to retain heat and reduce evaporation. These covers provide an impervious physical barrier and are therefore very effective at reducing evaporation losses. They also block sunlight penetration which will reduce algal growth. At less than a few hundred square meters in size (swimming pool covers), they can be easily deployed and retracted. However, when the cover is several hectares in size, as required in agricultural installations, a range of operational difficulties may be encountered.

Often agricultural storages are filled and emptied on a seasonal basis. This requires the continuous floating cover to be installed to allow for both full and empty storage scenarios. The covers need to be anchored to the walls or crest of the storage to ensure that it stays in place during windy conditions. However, the covers do not need to cover the entire water surface, and can be floated over a portion of the storage. This will reduce the actual evaporation savings but will allow for increasing and decreasing water depth.

Experimental trials carried out during the period 2000–2005 (Craig *et al.*, 2008) highlighted some of the operational difficulties associated with large scale (4 ha) continuous floating covers (Figure 12.2). These difficulties included stretching and development of slack areas; sinking and surface ponding (which heat up and evaporate water very quickly); and windblown dust accumulation and surface weed growth (Figure 12.2). The latter difficulty requires the surface of the cover to be cleaned, which is difficult in practice and expensive to undertake, such that the cost-effective deployment of this strategy was limited to storages of less than 1 ha (Craig *et al.*, 2008).

While the physical ability of continuous covers to reduce the evaporation losses is very high (70% to 90%), the practicalities of this technology on large agriculture storages may reduce its overall evaporation saving (Craig *et al.*, 2006).

FIGURE 12.2 Continuous floating cover, installed at a trial site located near St. George, Queensland. Weed growth and ponding is apparent at the water's edge.

12.2.3   FLOATING MODULES

Floating modules potentially overcome the operational difficulties of sinking and surface ponding, associated with continuous floating covers. The design and construction of modules includes purpose-built, plastic modules (Segal and Burstein, 2010), spherical plastic balls originally deployed as bird deterrents (Figure 12.3), recycled empty or partially full plastic bottles (Hassan *et al.*, 2015; Simon *et al.*, 2016), and natural palm fronds (Al Hassoun *et al.*, 2011). Theoretically, floating covers reduce evaporative loss by reducing the velocity of wind close to the water surface, and by reducing the transmission of solar radiation into the water body (Segal and Burstein, 2010). A key feature of floating modules is the ability to float to the windward shore, where rates of evaporative loss can be relatively high due to the shallow depth (Craig *et al.*, 2006). However, surface coverage and the stability of the modules to wind and wave turbulence has a large impact on performance. In Australian tank trials, the effectiveness of floating covers in retarding evaporative loss almost halved (from 38% to 20%) when surface coverage was reduced from 78% to 42% (Hassan *et al.*, 2015).

The most commonly trialed floating modules (constructed from inert, hydrophobic plastics) should not be directly toxic to the aquatic environment. However, the color and reflectivity of floating modules may affect their performance in reducing evaporative loss, and their impact on water quality. Nonreflective modules are likely to transmit heat into the water body, which may counteract any reduction in solar transmittance associated with shading. From a water quality perspective, the water-calming associated with floating modules may increase the likelihood of thermal stratification, conditions known to favor the development of algal blooms (Borman *et al.*, 2004).

A novel feature of floating modules is the potential to substitute relatively expensive, purpose-built modules (e.g., Segal and Burstein, 2010) with readily available, recycled modules (e.g., Simon *et al.*, 2016). Estimates for deploying purpose-built modules vary from US\$8/m^3 to US\$30/m^3, whereas the deployment of recycled, polyethylene terephthalate bottles was estimated to cost

FIGURE 12.3   Bird-ball floating modules. The economic success claimed for this technology is based upon the simplicity of production and deployment of the plastic sphere (poured into the water body directly from a tip-truck). (Reproduced from http://www.dailymail.co.uk/sciencetech/article-3194098/Could-plastic-balls-bring-relief-drought-stricken-California-Los-Angeles-releases-96-million-spheres-protect-reservoir-water.html)

FIGURE 12.4 Floating solar PV array experimental facility based in Adelaide. (Reproduced from http://www
.suntrix.com.au/featured/floating-solar-solution-for-water-industry-a-winner)

US$0.09/m^3. Floating modules may also be deployed to achieve more than one objective. Black
plastic balls (often called "bird balls") originally deployed to deter water fowl from colonizing water
storages (Figure 12.3) is one example, and the deployment of floating solar arrays (Figure 12.4) for
power generation is another (Ferrer-Gisbert *et al.*, 2013; Santafe *et al.*, 2014). The installation costs
of this evaporation mitigation strategy would be offset by the solar power produced. In one Spanish
case study (Ferrer-Gisbert *et al.*, 2013), the infrastructure, transport, and labor costs of deploying the
floating the platforms, foundations, and elastic joints, inverters, and photovoltaic cells was €0.2m.
However, this 100 kW floating system cost 30% more than a conventional grid-connected photovoltaic installation.

Research and development of new designs of float modular covers continues around the world.
The ability to mass produce and deploy in batches to cover only part of the storage means that the
capital costs of the technology can be phased over a number of years. This may prove a significant
benefit to the success of this EMT.

12.3   REDUCING EVAPORATION LOSSES WITH CHEMICAL COVERS

12.3.1   Existing Knowledge: Monolayers, Capillary Waves, and Evaporation

The potential for an artificial monolayer to reduce evaporative loss was demonstrated under laboratory conditions in 1925 (Rideal, 1925). The study of surface-active compounds (surfactants) in the
1950s and 1960s lead to the development of commercial detergents, and to trials of artificial
monolayers deployed at the field scale. Surfactants are molecules that contain a water-loving
(hydrophilic) head, and a water-repelling (hydrophobic) tail (Barnes, 2008). These amphiphilic
(hydrophilic and hydrophobic) molecules interact with water molecules at the solid/liquid and liquid/
air interface, to reduce surface tension. A reduction in surface tension increases the surface area,
exposing the surface, which in practice is how a detergent improves the effectiveness of "cleaning."

In the absence of a surfactant, the greater energy available to water molecules at the surface
induces a spontaneous contraction (surface tension; Figure 12.5). The temperature of the surface of a

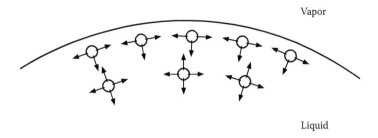

FIGURE 12.5 Direction of attractive forces between water molecules at the air/water interface, and within the bulk phase. Molecules at the surface have a higher free energy that induces the spontaneous contraction of the surface area, referred to as surface tension. (Reproduced from Davies, J., and Rideal, E., 1963, *Interfacial Phenomena*. Second Edition. London, UK: Academic Press.)

body of water is often colder than the immediate subsurface, due to the loss of heat as water molecules evaporate (latent heat loss or evaporative cooling). This colder water is heavier than the less dense, warmer water below, with the force of surface tension inducing regular descending microcurrents. Warmer, less dense water rises, with this cyclic motion generating capillary waves. Capillary waves increase the roughness of the water surface, reducing the resistance to evaporation.

Evaporation is the result of water molecules at the surface gaining enough energy to overcome the adhesive forces of subsurface water molecules escaping into the gaseous boundary layer (Figure 12.6). Wind turbulence and low humidity reduce any resistance within the gaseous boundary layer, increasing the rate of evaporative loss from the water body. The physical covers discussed in the first section of this chapter reduce evaporative loss at the macro scale, by protecting the water surface

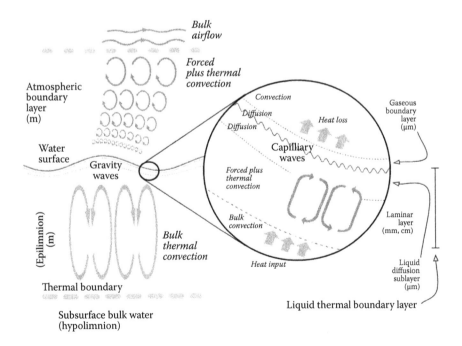

FIGURE 12.6 Scales at which physical floating and suspended covers and artificial monolayers retard evaporative loss. Floating and suspending covers protect the water surface from forced and thermal convection. An artificial monolayer calms capillary waves and increases the resistance of the gaseous and liquid thermal boundary layers to evaporation. (Reproduced from Hancock, N. et al., 2011, *Australian Journal of Multidisciplinary Engineering* 8, 1–10.)

from both forced (wind-driven) and thermal (heat-driven) convection. Artificial monolayers work at the micro scale, by calming capillary waves and by increasing the resistance imposed by the gaseous boundary layer and the water layer immediately below the air-water interface (the liquid thermal boundary layer; Figure 12.6).

Monolayers are films one molecule thick, formed at the air-water interface by amphiphilic compounds (Barnes, 2008). Some monolayers are capable of spontaneously spreading to form a surface film that reduces evaporative loss. The long-chain, linear (unbranched), fully saturated fatty alcohols cetyl and stearyl alcohol (hexadecanol and octadecanol) pack together to produce a condensed phase that maintains the high surface pressure required to retard evaporative loss. These compounds have been studied most commonly, as they are biodegradable, and are considered sufficiently benign for application in the food and cosmetics industries. Under laboratory conditions, monolayers of cetyl alcohol and stearyl alcohol reduce evaporation by up to 50%. However, results from field trials are extremely variable, ranging from 0% to 40% (McJannet *et al.*, 2008). Poor monolayer performance has most commonly been attributed to wind speed (Barnes, 2008), with turbulence breaking up and beaching the film. Some air movement is certainly helpful, and perhaps necessary, to ensure full monolayer coverage of the water surface, but winds greater than 2–4 m.s^{-1} progressively destroy the film (Frenkiel, 1965). Automatic dispensing overcame this problem, with monolayer applied only during periods of low wind speed, conducive to film formation (Figure 12.7).

In early laboratory studies, increasing the carbon chain length of fatty alcohols from 16 to 18 improved the resilience of the artificial monolayer to wind, as has replacing the hydrophilic alcohol head-group with an ether head-group (glycol monoalkyl ether; Barnes, 2008). However, long-chain compounds must be mixed with a suitable non-toxic solvent to assist in spreading. Early studies also showed monolayers formed with the 16-carbon chain hexadecanol were readily degraded by common

FIGURE 12.7   Shore application of an experimental monolayer formulation. The calm, reflective surface indicates the extent of surface coverage of the monolayer. The natural microlayer (green surface scum compressing the monolayer from the lower left of the picture) on this water storage was most concentrated at the windward and lee shores, developing sufficient "surface pressure" (Barnes, 2008) to inhibit the lateral spreading of the monolayer.

TABLE 12.2

Artificial Monolayer Compounds (Barnes, 2008) and Equivalent Naturally Occurring Compounds Extracted from Waxy Coatings on Sand in Australia (Franco *et al.*, 2000)

Carbon Chain Length	Artificial Monolayer Compounds	Natural Plant Waxes Isolated from Water-Repellent Sand Grains
C12	Lauryl alcohol, lauric acid	Dodecanoic acid
C14	Myristyl alcohol Myristic acid	Tetradecanoic acid, tetradecanoic acid methylester, tetradecanoic acid 12-methyl
C16	Cetyl alcohol Palmitic acid	Hexadecanoic acid, hexadecanoic acid methylethylester
C18	Stearyl alcohol Stearic acid	Octanoic acid, octadecadienoic acid methylester, octadecanoic acid
C20	Arachidic alcohol arachidic acid, vinyl stearate	Eicosane, eicosane cyclohexane, eicosanoic acid methyl ester, eicosylester
C22	Behenyl alcohol, behenic acid	Docosanol, docosanoic acid

aquatic bacteria (Chang *et al.*, 1962). A diversity of fatty alcohol compounds occur naturally, as components of plant waxes (Table 12.2). Decaying leaf litter and bark is transported in overland flow to lakes and rivers during storm events. Waxy, amphiphilic compounds derived from the litter concentrate at the air-water interface of lakes and streams, forming natural microlayers (Pittaway and Van den Ancker, 2010). Fatty acid compounds with a carbon chain length of 16 carbons are also synthesized as storage compounds by aquatic bacteria, explaining why hexadecanol monolayers are so readily degraded. Monolayers must be biodegradable to minimize any adverse environmental impact, but must also persist for a minimum period to provide cost-effective evaporative reduction savings (Barnes, 2008). Little information previously existed on how the maintenance of an artificial monolayer affected natural microlayer processes.

12.3.2 KNOWLEDGE GAPS: THE MICROLAYER, MICROMETEOROLOGY, AND MONOLAYERS

The existence of natural microlayers has been known since the 1940s, evident as a fine film collecting pollen and dust at the water's edge of most lakes, rivers, estuaries, and on the ocean (Norkrans, 1980). The concentration of organic compounds in the microlayer enhances microbial activity well above subsurface water activity. Microlayer chemistry varies from monolayer-like long-chain, linear, non-saturated alcohols and acids to complex aromatic ring compounds, referred to as aquatic humic substances. The ultraviolet (UV) light absorbed by these brown or black colored substances releases energy which chemically degrades carbonyl and other covalent bonds (direct photodegradation or photolysis; Figure 12.8), producing smaller organic compounds and highly reactive chemical species (Vahatalo, 2009). These reactive chemical species further degrade other non-UV absorbing microlayer compounds (indirect photodegradation), naturally "cleansing" the water.

The impact of these natural microlayer processes on artificial monolayers has not previously been studied. The artificial monolayer compounds hexadecanol, octadecanol, and glycol monoalkyl ether lack the chemical structures responsible for absorbing UV light, and are therefore not susceptible to direct photodegradation (Pittaway and Van den Ancker, 2010). However, until recently, no published information on the susceptibility of these compounds to indirect photodegradation was available. Many ecologically important physical processes also occur across the microlayer. Ecologists are concerned repeat monolayer application may disrupt these processes, reducing the transfer of sparingly soluble gases including oxygen, and may increase the frequency of blooms of toxin-producing

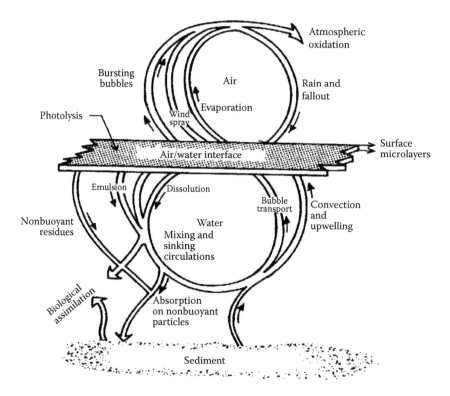

FIGURE 12.8 Natural physical, chemical, and biological processes that occur within the microlayer of water storages, rivers, estuaries, and the ocean. (Reproduced from Norkrans, B. 1980, *Advances in Microbial Ecology* 4, 51–85.)

blue-green algae (McJannet *et al.*, 2008). Blue-green algae (cyanobacteria) grow best in warm water, and can exploit thermally stratified conditions by using gas vacuoles to rise into the warmer, still layer (Bormans *et al.*, 2004). Until recently, most research on the impact of artificial monolayers on physical microlayer processes was at the macro scale (the atmospheric boundary layer in Figure 12.6), not the micro scale (i.e., the gaseous and liquid thermal boundary layers).

12.3.3 Current Research Findings: Micrometeorology, Monolayers, and Microlayers

Research at the laboratory scale typically restricted water depths to the centimeter scale (Barnes, 2008), too shallow for natural convection currents to operate. Recent glass-house research using deeper water and fans to control wind speed has provided insights into how micro-meteorological conditions affect the efficacy of three condensed monolayers in reducing evaporative loss (Gallego-Elvira *et al.*, 2013). The greater wind resilience of the glycol monoalkyl ether monolayer was confirmed, but at zero wind and with wind at 3 m.s^{-1}, the efficacy of all three monolayers in retarding evaporation was not or only marginally different from the clean water control. These results highlight the need to apply monolayer product only during periods of low wind speed, i.e., greater than 0 but less than 3 m.s^{-1}, when the impact of the monolayer will be greatest (Brink *et al.*, 2011).

The Gallego-Elvira *et al.* (2013) research also highlights the evaporation retarding property of a condensed monolayer does not incrementally increase water temperature. In the absence of a condensed monolayer, the surface roughness of a cold surface film reduces the resistance of the gaseous and liquid thermal boundary layers to molecular diffusion (Figure 12.9). At the micro-level within

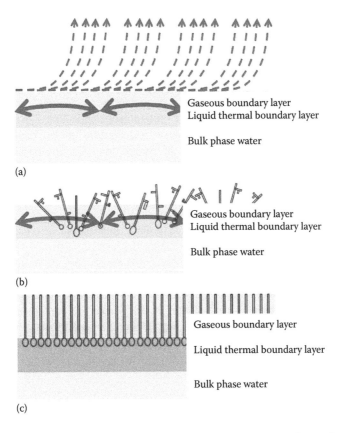

FIGURE 12.9 Microscale evaporation processes at the surface, and the impact of a condensed monolayer on microlayer processes. Capillary waves induced by surface tension and a cold surface film i.e., the curved blue arrows in (a) and (b) reduce the resistance imposed by the gaseous (pink layer) and liquid thermal (dark blue layer) boundary layers, increasing the rate of evaporation depicted by the red dotted arrows (a). Amphiphilic microlayer molecules (b) lack the packing nature of artificial microlayer molecules (c) and do not calm capillary waves or increase resistance. A condensed monolayer (c) calms capillary waves, increasing the resistance to evaporative loss (c).

these interfacial boundary layers, the transport of water molecules and heat is by diffusion (Gladyshev, 2002). Diurnal changes in the transfer of heat across the microlayer are modified by the presence of a condensed monolayer (Pittaway *et al.*, 2015a). During the day when the convective heat transfer is downward, a condensed monolayer increases the resistance of the gaseous and liquid thermal boundary layers to molecular diffusion (Figure 12.9c), "insulating" the microlayer. During the night when convective heat transfer is upward, a condensed monolayer also insulates the microlayer from heat loss.

Natural microlayer molecules have zero or minimal impact on evaporative loss, as they lack the chemical uniformity required to pack together to increase resistance to molecular diffusion (Figure 12.9b). However, the UV light–absorbing properties of aquatic humic substances within the microlayer have a major impact on the environmental resilience of monolayers (Pittaway *et al.*, 2015b). In the absence of aquatic humic substances (on clean water), the susceptibility of hexadecanol to microbial degradation was confirmed, with the addition of two carbons (octadecanol) and the substitution of the alcohol with an ether head-group (glycol monoalkyl ether) increasing resilience. In the presence of aquatic humic substances (naturally brown water), the microbially resilient glycol monoalkyl ether monolayer is highly susceptible to indirect photodegradation. These results indicate water quality may also be contributing to the highly variable field performance of artificial monolayers in reducing evaporative loss. On clear water (low concentration of aquatic humic substances), performance under light wind may be very good, but on brown water, the monolayer may not persist for the duration required for cost-effective savings. Many farm dams in Australia are turbid and brown, indicating appropriate monolayer selection will be required to ensure the persistence of the product for the five days to be considered cost-effective as a water conservation strategy.

These results highlight the need for a "smart" monolayer application system, for two major reasons. First, the system should dispense monolayer formulations only during periods when the micrometeorological conditions favor cost-effective evaporation reduction (Brink *et al.*, 2011). A decision support system is essential to ensure the environmental resilience of the monolayer is matched to the water quality of the target storage: monolayer formulations based on hexadecanol, octadecanol, and glycol monoalkyl ether are suitable for clean, clear water storages, but only octadecanol will resist microbial and photodegradation on brown water storages. And second, the smart monolayer application system is required to provide intermittent or continuous application appropriate to the prevailing meteorological conditions (Brink *et al.*, 2011).

Figure 12.10 shows the development of a practical floating "Smart Monolayer Application System" (SMAS) developed by NCEA, USQ (Brink *et al.*, 2011). The on-board GPS-based system receives information from nearby automatic weather stations and also other ancillary equipment including Seepage and Evaporation Meters (SEMs). A WiFi-linked coordinator unit adjacent to the storage processes these data to optimize application of the chemical monolayer across the water storage and commands the SMAS units to commence, continue, and cease the application. The software continuously receives a data stream comprising wind speed, wind direction, air temperature, solar radiation, relative humidity and rainfall; and, from these data, the decision software evaluates the principal questions:

i. Does the evaporation rate warrant monolayer application? (Recognizing the value of the remaining water *at that* time, e.g., to irrigate a near-to-harvest crop, and that monolayer product is expensive)

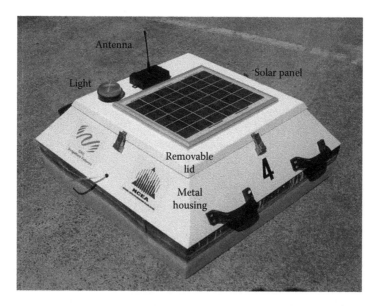

FIGURE 12.10   GPS based propelled floating Smart Monolayer Application System (SMAS) developed by NCEA USQ, for monolayer application direct to water surfaces. The system optimizes the delivery, according to weather information received from an Automatic Weather Station (AWS) located on the bank of the water storage.

ii. Is it cool, overcast, or raining? (i.e., Application not required)

iii. Is there low/moderate wind speed? (i.e., Representing optimal conditions for layer survivability and spreading performance)

Providing the environmental variables are favorable (i.e., (i) = yes; (ii) = no; (iii) = yes) the control system can then calculate the dosage rate required and deploy one or several SMAS units to minimize water loss due to evaporation.

12.3.4 FUTURE RESEARCH

Our understanding of the impact of artificial monolayers on microlayer processes has improved substantially over the last decade. However, further research on the impact of natural micrometeorological processes on evaporative loss is required to improve the efficacy of smart monolayer application systems. Two key areas are in improving smart monolayer application systems (Brink *et al.*, 2011), and in understanding if the resistance to molecular diffusion imposed by a warm surface film is greater than the resistance imposed by a condensed monolayer (Figure 12.11).

Rates of evaporation are greatest with wind, when a cold surface film reduces the resistance to molecular diffusion imposed by the gaseous and liquid thermal boundary layers (Figure 12.11a). However, under calm, warm conditions, a less dense, warm surface film may develop (Figure 12.11b). Warm surface films are thermally stable, damping capillary wave formation and increasing the resistance of the liquid thermal boundary layer. Under these conditions, the resistance imposed by an artificial monolayer may be smaller than the existing resistance imposed by the warm surface film.

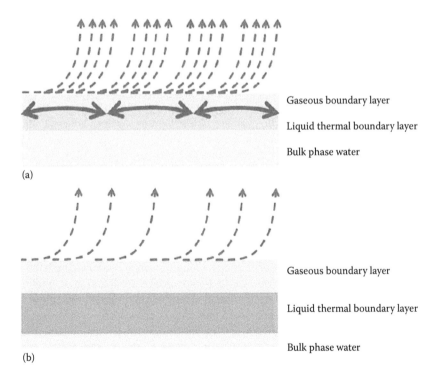

FIGURE 12.11 Impact of a cold surface film (a) and a warm surface film (b) on the resistance to molecular diffusion imposed by the liquid thermal boundary layer (dark blue layer). Capillary waves, i.e., the blue arrows (a) formed under a cold surface film reduce the resistance of the gaseous and liquid boundary layers. The thermally stable warm surface film calms capillary waves.

Knowing if and when these conditions prevail, and only applying monolayer when a cold surface film prevails, will improve the cost-effectiveness of monolayer technology. Future research is also required to produce new monolayer products with improved resistance to wind, microbial, and photo-degradation (Prime *et al.*, 2012, 2013).

More research also needs to be carried out on chemical monolayers, to see if new candidate compounds can be formulated, and made less susceptible to physical, chemical, and biological degradation compared to the products which are currently commercially available.

12.4   CONCLUSION

The need to actively minimize water loss from irrigation storages is well recognized, and a range of techniques are available to achieve this. However, the choice, and particularly the cost-effectiveness, of each technique will vary greatly with circumstances. It follows that continuous monitoring of all aspects of irrigation water usage provides essential baseline information from which to make informed decisions on the choice and deployment of mitigation techniques for any particular storage.

While physical covers offer a potentially long-term solution, the structure and materials have limited lifetimes, and of course, are only effective when the storage is actually holding water! In contrast, and at much less initial capital cost, molecular monolayer films can be deployed only as required and can provide significant reduction in the rate of evaporation *under certain circumstances*. The physics of evaporation at the open water surface is complex, but detailed analysis reveals the major changes in microscale conditions which occur during the diurnal cycle, and hence explains why highly variable mitigation performance has often been observed over successive days—there are certain conditions during which the applied surface film is simply unable to provide any evaporation reduction. But it has now been demonstrated that with autonomous robotic technology the diurnal changes can be accommodated and the deployment of material limited to only those periods (hours) for which the material can provide cost-effective reduction in evaporation. During other periods, e.g., high wind speed or inappropriate surface thermal conditions, it must be accepted that monolayer application will not be effective.

The evaluation of mitigation techniques remains challenging. Although simple daily measurement can produce data on millimeter depth changes, unless there is a pair of adjacent, similar-sized storages on which only one has monolayer material applied, the megaliters saved due to the monolayer application cannot be directly determined. Furthermore, any assessment of monolayer utility must be interpreted with respect to prevailing weather conditions in relation to long-term climates for the particular storage site, and also with respect to the agronomic usage of the irrigation water (timing: crop stage and value). All of these practical considerations may be expected to vary from season to season.

The evaluation of the efficacy of monolayer films is further complicated by the fact that the measurement of simple daily totals alone is not adequate. Rather, such measurements must be taken at sub-diurnal intervals, preferably hourly or shorter, and interpreted in relation to both the prevailing conditions at the microscale of the water surface (hot/cold surface film), and to the thermal storage changes and microclimate of the water body measured at similar sub-diurnal timescales. To achieve this, the use of relatively complex automatic instrumentation is unavoidable. Obviously, only when the circumstances are favorable for the monolayer (to inhibit evaporation) can the performance of the material be fairly evaluated. The authors hold the view that while the technology available for mitigation by physical covers appears mature and offers a range of mitigation solutions, recent research in water surface science—biophysics, material development (monolayers), and sophisticated deployment technology—indicates that the use of monolayer-type materials can now be optimized, and their performance reliably evaluated.

ACKNOWLEDGMENTS

The authors gratefully acknowledge Emeritus Professor Geoff Barnes, University of Queensland, for many years of discussions and advice. The authors also gratefully acknowledge the Queensland Government CRC for Irrigation Futures (CRC-IF), and the National Center for Engineering in Agriculture (NCEA), and University of Southern Queensland (USQ) for their continuous support.

REFERENCES

Al Hassoun, S. A., Mohsen, A., Al Shaikh, A., Al Rehaili, A., Misbahuddin, M. (2014). Effectiveness of using palm fronds in reducing water evaporation. *Canadian Journal of Civil Engineering* 38, 1170–4.

Allen, R. G., Pereira, L. S., Raes, D., and Smith, M. (1998). Crop evapotranspiration—Guidelines for computing crop water requirements. FAO Technical Paper 56. Food and Agriculture Organization of the United Nations, Rome.

Barnes, G. (2008). The potential for monolayers to reduce the evaporation of water from large water storages. *Agricultural Water Management* 95, 339–53.

Bormans, M., Ford, P., Fabbro, L., and Hancock, G. (2004). Onset and persistence of cyanobacterial blooms in a large, impounded tropical river, Australia. *Marine and Fresh Water Research* 55, 1–15.

Brink, G., Symes, T., and Hancock, N. (2011). Development of a smart monolayer application system for reducing evaporation from farm dams: Introductory paper. *Australian Journal of Multidisciplinary Engineering* 8, 121–9.

Chang, S., McClanahan, M., and Kabler, P. (1962). Effect of bacterial decomposition of hexadecanol and octadecanol in monolayer films on the suppression of evaporation loss of water. *Retardation of Evaporation by Monolayers: Transport Processes*. (Ed. V. Le Mer.) London, UK: Academic Press.

Craig, I., Green, A., Scobie, M., and Schmidt, E. (2005). *Controlling evaporation loss from water storages*. NCEA Publication No. 1000580/1, Queensland, 207.

Craig, I. P. (2006). Comparison of precise water depth measurements on agricultural storages with open-water evaporation estimates. *Agricultural Water Management* 85, 193–200.

Craig, I. P., Mossad, R., and Hancock, N. (2006). Development of a CFD-based dam evaporation model. International Symposium on Environmental Health Climate Change and Sustainability, 8. Queensland University of Technology, Kelvin Grove Campus, Brisbane, Queensland, Australia.

Craig, I., Aravinthan, V., Baillie, C., Beswick, A., Barnes, G., Bradbury, R., Connell, L. *et al.* (2007a). Evaporation, seepage, and water quality management in storage dams: A review of research methods. *Environ Health: Climate Change Special* Issue 7 (3), 84–97.

Craig, I., Schmidt, E., and Hancock, N. (2007b). EvapCalc software for the determination of dam evaporation and seepage. IEAust Society for Engineering in Agriculture (SEAg) Conference, South Terrace, Adelaide.

Craig, I. P. (2008). Loss of storage water through evaporation with particular reference to arid zone pastoralism in Australia. NCEA 1001858/1. DKCRC Working Paper 19, The WaterSmart™ Literature Reviews, Desert Knowledge CRC, Alice Springs.

Davies, J., and Rideal, E. (1963). *Interfacial Phenomena*. Second Edition. London, UK: Academic Press.

Ferrer-Gisbert, C., Ferran-Gozalves, J., Redon Santafe, M., Ferrer-Gisbert, P., Sanchez-Romero, F., and Torregrosa-Soler, J. (2013). A new photovoltaic floating cover system for water reservoirs. *Renewable Energy* 60, 63–70.

Finn, N. and Barnes, S. (2007). The benefits of shade-cloth covers for potable water storages, *CSIRO Textile & Fiber Technology*. CSIRO Gale Pacific, 42.

Franco, C., Clarke, P., Tate, M., and Oades, J. (2000). Hydrophobic properties and chemical characterization of natural water-repellent materials in Australian sands. *Journal of Hydrology* 231–2, 47–58.

Frenkiel, J. (1965). Evaporation reduction: Physical and chemical principles and review of experiments. *Arid Zone Research* 27, Evaporation Reduction, UNESCO, Paris.

Gallego-Elvira, B., Martínez-Alvarez, V., Pittaway, P., Brink, G., Martín-Gorriz, B. (2013). Impact of micrometeorological conditions on the efficiency of artificial monolayers in reducing evaporation. *Water Resources Management* 27, 2251–66.

Gallego-Elvira, B., Baille, A., Martín-Górriz, B., Maestre-Valero, J. F., and Martínez-Alvarez, V. (2011). Energy balance and evaporation loss of an irrigation reservoir equipped with a suspended cover in a semi-arid climate (southeastern Spain). *Hydrological Processes* 25, 1694–703.

Gladyshev, M. (2002). *Biophysics of the Surface Microlayer of Aquatic Ecosystems*. London: IWA Publishing.

Hassan, M. M., Peirson, W. L., Neyland, B. M. and Fiddis, N. M. (2015). Evaporation mitigation using floating modular devices. *Journal of Hydrology* 530, 742–50.

Hancock, N., Pittaway, P., Symes, T. (2011). Towards a biophysical understanding of observed performance of evaporation suppressant films applied to agricultural water storages—First analyses. *Australian Journal of Multidisciplinary Engineering* 8, 1–10.

Maestre-Valero, J., Martinez-Alvarez, V., Gallego-Elvira, B., and Pittaway, P. (2011). Effects of a suspended shadecloth cover on water quality of an agricultural reservoir for irrigation. *Agricultural Water Management* 100, 70–5.

Martínez-Alvarez, V., Baille, A., Molina-Martínez, J. M., and González-Real, M. M. (2006). Efficiency of shading materials in reducing evaporation from free water surfaces. *Agricultural Water Management* 84, 229–39.

Martínez-Alvarez, V., Calatrava-Leyva, J., Maestre-Valero, J. F., and Martín-Górriz, B. (2009). Economic assessment of shadecloth covers for agricultural irrigation reservoirs in a semi-arid climate. *Agricultural Water Management* 96, 1351–9.

Martínez-Alvarez, V., González-Real, M. M., Baille, A., Maestre-Valero, J. F., and Gallego-Elvira, B. (2008). Regional assessment of evaporation from agricultural irrigation reservoirs in a semi-arid climate. *Agricultural Water Management* 95, 1056–66.

McGloin, R., McGowan, H., McJannet, D. (2014a). Effects of diurnal, intraseasonal and seasonal climate variability on the energy balance of a small subtropical reservoir. *Int. J. Climatol.* doi 10.1002/joc.4147.

McGloin, R., McGowan, H., McJannet, D., and Burn, S. (2014b). Modelling sub-daily latent heat fluxes from a small reservoir. *Journal of Hydrology* 519, 2301–11.

McJannet, D., Cook, F., Knight, J., Burn, S. (2008). Evaporation reduction by monolayers: Overview, modeling and effectiveness. Report series ISSN: 1835-095X, CSIRO Water for a Healthy Country Flagship, Brisbane, Queensland.

McJannet, D. L., Cook, F. J., McGloin, R. P., McGowan, H. A., and Burn, S. (2011). Estimation of evaporation and sensible heat flux from open water using a large-aperture scintillometer. Water Resources Research 47 (W05545), 14.

McJannet, D. L., Cook, F. J., McGloin, R. P., McGowan, H. A., Burn, S., and Sherman, B. S. (2013). Long-term energy flux measurements over an irrigation water storage using scintillometry. *Agric. For. Meteorol.* 168, 93–107.

Norkrans, B. (1980). Surface microlayers in aquatic environments. *Advances in Microbial Ecology* 4, 51–85.

Pittaway, P., Herzig, M., Stuckey, N., and Larsen, K. (2015b). Biodegradation of artificial monolayers applied to water storages to reduce evaporative loss. *Water Science and Technology* 72, 1334–40.

Pittaway, P., Martinez-Alvarez, V., and Hancock, N. (2015a). Contrasting covers reveal the impact of an artificial monolayer on heat transfer processes at the air-water interface. *Water Science and Technology* (in press). doi:10.2166/wst.2015.379.

Pittaway, P. and Van den Ancker, T. (2010). Properties of natural microlayers on Australian freshwater storages and their potential to interact with artificial monolayers. *Marine and Freshwater Research* 61, 1083–91.

Pittaway, P., Martínez-Alvarez, V., Hancock, N., and Gallego-Elvira, B. (2015). Impact of artificial monolayer application on stored water quality at the air-water interface. *Water Science & Technology* 72 (7), 1250–6.

Prime, E., Tran, D., Plazzer, M., Sunartio, D., Leung, A., Yiapinis, G., Baoukina, S., Yarovsky, I., Qiao, G., and Solomon, D. (2012). Rational design of monolayers for improved water evaporation mitigation. *Colloids and Surfaces A: Physicochemical and Engineering Aspects* 415, 47–58.

Prime, E. L., Tran, D. N. H., Leung, A. H. M., Sunartio, D., Qiao, G. C., and Solomon, D. H. (2013). Formation of dynamic duolayer systems at the air-water interface by using non-ionic hydrophilic polymers. *Aust. J. Chem.* 66, 807–13.

Rideal, E. K. (1925). On the influence of thin surface films on the evaporation of water. *The Journal of Physical Chemistry* 29 (12), 1585–8.

Santafe, M., Ferrer-Gisbert, P., Sanchez-Romero, F., Torregrosa-Soler, J., Ferran-Gozalves, P., and Ferrer-Gisbert, C. (2014). Implementation of a photovoltaic floating cover for irrigation reservoirs. *Journal of Cleaner Production* 66, 568–70.

Segal, L. and Burstein, L. (2010). Retardation of water evaporation by a protective float. *Water Resources Management* 24, 129–37.

Simon, K., Shanbhag, R., and Slocum, A. (2016). Reducing evaporative water losses from irrigation ponds through the reuse of polyethylene terephthalate bottles. *Journal of Irrigation and Drainage Engineering* 142.

Vahatalo, A. (2009). Light, photoreactivity, and chemical products. *Encyclopaedia of Inland Waters*. (Ed. G. Likens.) Amsterdam: Elsevierdirect, 761–73.

Wetzel, R. (2001). *Limnology: Lake and River Ecosystems*. Third Edition. San Diego: Academic Press.

Section III

Harvesting and Post-Harvest Technology

13 Advances in Silage Harvest Operations

C. Amiama, J. Bueno, and J. M. Pereira

CONTENTS

13.1   INTRODUCTION

The sustainability of forage-based animal production systems is vital to supply quality, safe, and affordable food for the ever growing human population. Silage is a preservation process that allows high quality forage, which is produced seasonally, to be available all year round. Producing high quality silage can be an expensive operation. A poor decision in the choice of harvesting systems can increase the production costs and affect profitability significantly. The choice of silage harvesting systems depends on many factors, whether it is a self-propelled forage harvester, or tractors with different equipment, such as a self-loading forage wagon system, rotary mowers, conditioners, and so forth. The final decision will depend on particular circumstances, such as the number of fields to harvest, plot size, accessibility, time available to perform the operations, and the purchasing cost of equipment (some small farms are not able to invest in the newest technology). In order to make a right decision, relevant information has to be available, so in this topic about "advances in silage harvest operations" we will explain the techniques, technologies, and the new tendencies for improving the use of silage harvest equipment. We will start the chapter characterizing the silage process and their variants, depending on the crop and the silage procedure. In Section 13.4, equipment for silage operations will be described (rotary mowers, conditioners, rakes, balers, forage wagons, and the self-propelled forage harvester). Finally, we will analyze the new tendencies for improving the use of the equipment, which is mainly linked to the improvement of the organization of the work.

13.2   SILAGE PROCESS CHARACTERIZATION

The balanced feeding of livestock throughout the whole year involves the need to collect the surplus forage production, which is produced seasonally, and to store it by means of a preservation process that allows high quality forage to be available all year round. Fresh forage from various crops (like alfalfa, sunflower, corn, and grass, among others) can be preserved by means of silage.

Silage is a preservation process by which using different techniques and technologies, fresh forage is chopped, loaded, transported, unloaded, stored, compacted, and packed in a manner that eliminates and excludes as much oxygen as possible. This process produces a spontaneous microbial fermentation in the absence of oxygen, the stabilization of the forage, and its long term preservation.

The quality of the product obtained depends fundamentally on the following factors:

1. Establishment of the optimal moment for forage collection
2. Size and uniformity of the chopped forage
3. Moisture content of the plant
4. Method and speed of filling
5. The quality of silage storage (packing)

The fermentation and stabilization process can be divided into four stages:

Phase 1: Aerobic phase. The atmospheric oxygen present in the plant material decreases rapidly due to respiration of the plant matter and aerobic microorganisms raising the interior temperature of the silage. While filling, it is very important to proceed quickly and eliminate as much air as possible by compacting in order to avoid the respiration of the forage and the consumption of carbohydrates and highly digestible energy contained within it.

Phase 2: Fermentation phase. This phase begins to occur in an anaerobic environment. It can last from several days to several weeks, depending on the characteristics of the silaged material and the conditions at the time of silage. If the fermentation is successful, acetic acid is produced first, which would cause a decrease in pH, and simultaneously, lactic acid is produced by bacteria that begin to multiply increasing the acidity of the silage, lowering the pH to values between 3.8 and 5.0. The faster the fermentation is completed, the lower the consumption of soluble carbohydrates and more nutrients are preserved in the silage.

Phase 3: Stable phase. While the environment is kept without air, few changes occur. The presence of the majority of the microorganisms from Phase 2 is slowly reduced. Some acidophilic microorganisms survive this period of inactiveness; others, like clostridia and bacillus, survive as spores. The pH of the silage stabilizes reaching a final value which depends on the production and storage conditions. When working with pastures with a high protein content, the final pH can reach values approximating 4.5, while with corn that value can reach 3.8 to 4.0 (Bragachini *et al.*, 2008).

Phase 4: Aerobic deterioration phase. This phase begins on opening the silage and exposing it to air, although it can also occur before extraction because of accidental damage to the covering caused by various factors (e.g., rodents or birds). The deterioration period can be divided into two stages. The first is due to the degradation of the organic acids that preserve the silage, by way of yeasts and occasionally bacteria that produce acetic acid. This prompts an increase in pH, which permits the beginning of a second stage of deterioration, in which the temperature increases and also the activity of microorganisms that deteriorate the silage. The aerobic deterioration occurs in almost all silages when opened and exposed to air, nonetheless, the rate of this deterioration depends on the concentration and activity of the organisms that cause it.

13.3   PROCESS VARIANTS

The silage can be made using any crop or mixture containing soluble carbohydrates (barley, wheat, oats, alfalfa, clover, corn, and grass). In the process that leads to the production of the silage, we can

FIGURE 13.1   The machines used are conditioned by the crop type and the silo.

encounter different ways of addressing the task depending on the type of crop, the machinery employed in the silage operations, or the silage storage systems (tower silos, bunker silos, silage piles, silo bags, and balage) (Figure 13.1). In corn silage, self-propelled forage harvesters collect and chop the plant material, and deposit it into trucks or wagons. In grass, silage a wider range of machinery with different possibilities is usually used.

13.4   MACHINES FOR SILAGE HARVESTING OPERATIONS

13.4.1   ROTARY MOWERS

The first step in silage production is cutting the plants. The traditional sickle cutter bar machinery, very adequate for their clean cuts and not producing tears in the stalks allowing for quick re-growth of the pasture, has given way to rotary mowers, with far superior work capacities when compared to sickle cutter bar. The rotary movement cutting systems are based on the spinning of blades with a free end that permits them to remain perpendicular to the axis of rotation due to the effect of centrifugal force. The blades spin at a high peripheral speed (60–90 m.s^{-1}), producing the cutting of the plant stalks on impact. They don't need to be sharpened, just periodically replaced when the cutting edge has been worn away, so they require little maintenance time. On the market, we find two types of rotary mowers: disc mowers and drum mowers.

Disc mowers are those rotary mowers that are made up of one or several cutter bars that support, on its underside, five to nine rotors that are 40–50 cm in diameter (discs) (Figure 13.2). The disc transmission is carried out underneath with a disc rotation speed of between 2500 and

FIGURE 13.2   A disk mower transmission. (Adapted from Pottinger, http://www.poettinger.at/en_in [January 2015], own photograph.)

3000 revolutions per minute. The cutter bar that contains the transmission slides along the ground supported by steel skids, which prevent the accumulation of soil and intercept any impacts. The cutting height is regulated by inclining the body of the mower or by skids.

Drum mowers are rotary mowers that are made up of one or several cutter bars that support, on its upper side, two or four rotors of over 60 cm in diameter (drums) (Figure 13.3). The drum transmission is carried out in the upper part at a drum rotation speed of between 1500 and 2500 rpm, in such a way as to reach the necessary peripheral rotation speed for cutting ($60-90$ m.s^{-1}), according to the diameter of the drum. Given that the transmission is carried out further from the ground and needs lower rotation speeds than disc mowers, this type of machine is better adapted to tougher conditions in which high working temperatures can be reached. The cutting height can be adjusted between $30-60$ mm by means of a regulating mechanism found in the upper part of the drum.

Rotary mowers stand out for being able to be hooked up in frontal, lateral, or rear positions, they can work at speeds of $10-12$ km/hr with working widths that range from 2 to 4 m in simple machines, and $8-10$ m in combined ones. The work performances depend on the working speed and the width of the machine, oscillating between $2-10$ ha.h^{-1}. The adoption of disc machines will allow producers to harvest forage at a faster rate when weather conditions are ideal, resulting in higher overall forage quality.

The latest advances in forage harvesting involve the use of combined machines with conditioning mechanisms incorporated into the mower, cross conveyer belts, floating systems, and control terminals, which are easy to use and compatible with the tractor's ISOBUS terminal.

The use of combined mowers has permitted a considerable increase in the working width of each pass, even exceeding 10 m (Figure 13.4). In addition, the use of modern floating systems that improve the adaptation to the terrain (Figure 13.5), minimizing losses and increasing precision at high working speeds, has allowed for considerably increased work performance.

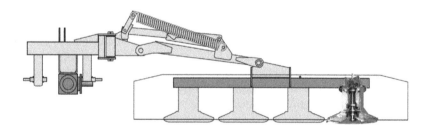

FIGURE 13.3 A drum mower transmission. (Adapted from Pottinger, http://www.poettinger.at/en_in [January 2015].)

FIGURE 13.4 Disc mower with three combined mowing modules.

FIGURE 13.5 Machine to ground adaptation (alpha-motion). (Adapted from Pottinger, http://www .poettinger.at/en_in [January 2015].)

FIGURE 13.6 Forage swathing by means of a conditioned mower with a cross conveyer belt.

The combination of a rotary mower with other mechanisms, like conditioners and cross conveyer belts, even though this can limit advancing speeds in high production conditions, this allows the combination of three operations in just one pass (mowing, pre-drying, and swathing; Figure 13.6), considerably reducing forage collection costs.

Some mowers include the options of gathering and distributing, which increases their versatility. The wide distribution option permits the acceleration of the drying process. The windrower forms parallel swaths and the grouper joins them, creating large sized rows and reducing the number of passes made by the forage wagon or the forage harvester.

13.4.2 CONDITIONERS

Once the plant has been cut, it continues breathing, with a rate related with forage temperature and moisture, causing loss of nutrients and dry matter. The objective of the forage conditioners is to accelerate the loss of water contained within the plant tissue favoring its evaporation, thereby increasing the concentration of nutrients after mowing, and the quality of the silage is improved by favoring the rapid reduction in pH. This can be achieved by equipment associated with the mower (conditioners), or by means of systems which are independent from the mowing process (rakes).

13.4.3 IMPELLER CONDITIONERS OR FINGERS

The job of the fingered conditioners consists in removing the cuticle, which is present on the surface of leaves and stalks, in order to accelerate the release of water. The forage is accelerated with a horizontal axis rotor with several rows of fingers, and a plate or brake (Figure 13.7), which favors the lacerating or scraping of the stalks and leaves. The intensity of the conditioning can be adjusted with a lever, automatically in more advanced equipment, altering the separation of the deflecting plate with respect to the rotor, or tailoring the number of revolutions to the conditioner. The higher the revolutions, the higher the work rate.

This type of conditioning is recommended when working with short growing plants (under 1.2 m) and those that don't have meaty stalks. With meaty or thicker stalks the drying of the leaves

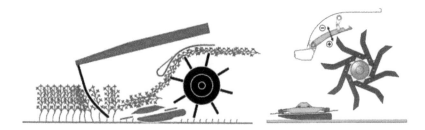

FIGURE 13.7 Fingered conditioner or impeller. (Adapted from John Deere https://www.deere.com/es [December 2015]; adapted from Krone https://www.krone.de [December 2015].)

and stalks is not even. Also, when the plants are taller, the fingers begin to grind the leaves, resulting in loss of quality and quantity of forage.

13.4.4 ROLLER CONDITIONERS

Roller conditioners work by applying intermittent pressure on the plants by passing the forage between two rollers with studs or a grooved profile (Figure 13.8), the rollers spin in opposite directions at a speed in the order of three to four times that of the ground speed of the machine. They crush and shatter the stalks, opening pathways to release the water content of the plant tissue. By regulating the approximation and the pressure of the rollers, the level of conditioning of the rollers can be chosen and differentiated from that of the leaves, obtaining forage with a larger amount of leaves and therefore greater digestibility and protein value.

Roller conditioners have no limitations with respect to forage height and volume, as this doesn't limit the passage through them, for which they adapt to any plant height. Their use is recommended with forages rich in leaves like legumes (alfalfa or clover) and grass with meaty stalks, as well as having a lower power requirement when operating with large volumes of pasture.

13.4.5 RAKES

When handling high yielding forage with very humid environmental and ground conditions, and using mowers with no conditioning or grouping tools, it's possible to achieve the conditioning and swathing with rakes.

There are several different types of rakes on the market, such as parallel-bar rakes, wheel rakes, and rotary rakes. Although rotary rakes are the most expensive system, they have been gradually gaining popularity, and are now the most widely-used system, especially when handling fresh forage, as they are the best at handling heavy and moist forage, achieving well-formed swaths, which allow for good air circulation accelerating loss of water in the forage. The operation of the most advanced

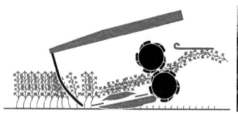

FIGURE 13.8 Roller conditioner. (Adapted from John Deere https://www.deere.com/es [December 2015]; adapted from Class http://www.claasofamerica.com [January 2016].)

systems is based on a series of spinning arms (6 to 15), usually with four double teeth on each arm mounted on a rotor. The teeth spin on an axis perpendicular to the ground where the height of their insertion point is constant, but the teeth oscillate, going from a vertical position at the attacking front where they sweep the forage transversely to the advancing direction, to an almost horizontal position at the rear, releasing the forage and thus facilitating the formation of the row (Figure 13.9). These rows have the sufficient density to allow its subsequent collection by means of balers, forage wagons, or self-propelled forage harvesters.

The working width is variable according to the number of rotors (one, two, three, four, or six rotors), ranging from 3 to 4 m with single rotary rakes, to almost 20 m that can be reached by six rotors (Figure 13.10). The latest advances in electronic systems compatible with any ISOBUS terminal allow for automatic control over the depth, working width, and swath width; programming

FIGURE 13.9   Windrowing using rakes. (Adapted from Claas. http://www.claas.es/ [January 2016].)

FIGURE 13.10   Six-rotor windrower. (Adapted from Krone. https://www.krone.de [December 2015].)

the values comfortably from the tractor, facilitating management, and achieving a greater precision in the work.

When using windrowers, special importance must be given to carrying out the work cleanly, avoiding contamination from soil which could provoke undesirable fermentations in the process of preserving the forage, and thus losses in the silage. For this, the teeth should be adequately regulated in order for them to span the whole working width at a uniform height with respect to the ground. It is also desirable that they be supported by pairs of wheels joined in tandem, as this limits the effect of irregularities in the terrain reducing contamination of the forage. Another important aspect is the rotation speed of the rotor, especially in forages which are rich in leaves like clover or alfalfa, for which special attention should be paid to finding the adequate combination between the advancing speed of the tractor and the rotation speed of the rotor in order to avoid the detachment of the leaves from the forage.

13.4.6 BALERS

Balers were initially designed to condition forage dried on the ground, however, developments in silage films and machinery capable of working with moist forage has allowed bale silage to be an interesting option in some situations, especially in smaller-sized farms, but it is also employed as a secondary system (often associated with facilitating grazing management during mid-summer) on many dairy and larger sized farms.

Bale silage consists of taking forage with 40% to 60% moisture, baling it in a round baler, and wrapping it with a stretch film. It does not require structures for storing silage, takes less fuel and less dry matter losses compared to silage chopping, and can be placed in locations around the farm or any other location. Nonetheless, in order to prevent damages, a high standard of bale wrapping, handling, and storage is required. The objective with the baled silage system must therefore be to rapidly achieve adequately anaerobic conditions and maintain them thereafter. Since each bale of silage typically has six to eight times the surface area in contact with plastic compared to clamp silage, it is clearly important that the plastic film is of good quality, is properly applied, and is not subsequently damaged (O'Kiely et al., 2007). Gentle handling of bales before and after wrapping is essential in order to maintain the shape of the bales and the integrity of the seal provided by the plastic film.

Traditionally, this process was carried out by two differentiated machines: a round baler and a wrapping machine. The forage is collected with a pick-up, at the front, it is introduced into a chamber where it is subjected to compression forming a cylindrical bale 1 to 2 m in diameter times 1.2 m in width. There are three ways of carrying out the compression: in a fixed chamber, in a variable chamber, or in a semi-variable fixed chamber. The fixed chambers use rollers to apply pressure on the forage from the exterior edge of the cylinder, whereby the central part of the bale is less compressed, which may cause fermentation problems if air is left in the center. The variable chambers use belts, and manage to compress the whole forage material to the same pressure although their complexity makes them more prone to breakdowns. The semi-variable fixed chambers are the latest system to be introduced on the market, in an attempt to combine the robustness of the fixed systems with the quality of compression in the variables. The first part of the filling is carried out in a small fixed chamber which, once full, increases in diameter like a variable chamber. Once the bale is formed, a tying mechanism using string or net wraps the bale to maintain its cohesion when it is deposited on the ground.

Subsequently and in the briefest time possible, the bale should be isolated from the exterior with a film wrapping. The first rotating platform wrappers have given way to double rotating arm wrappers, in which the bale is wrapped by means of two rolls of polyethylene mounted on a rotating structure, and the horizontal rotation of the bale is carried out by rollers on which its lateral surface is rested (Figure 13.11).

FIGURE 13.11   First rotating platform wrappers have given way to rotating arm wrappers.

The convenience of simplifying the process, reducing on the one hand the dependence on labor, and on the other getting the bale wrapped in the least time possible, has conditioned the development of the new machines. The most advanced systems, baler-wrapper combinations, integrate two operations into one machine, eliminating the use of a second tractor and operator (Figure 13.12). With the baling and wrapping modules coming from the same manufacturer, all machine functions and cycles are automatically controlled and perfectly sequenced to provide an absolutely smooth performance, eliminating the risk of crop contamination, as bales are not placed on the ground before film wrapping. As soon as the baling chamber is filled or the preset density is reached, the system signals the operator to stop. The net is fed into the baling chamber and net wrapping starts. Then, the tailgate opens and the bale is transferred onto the wrapping table. As the baler resumes baling, the wrapper starts wrapping and stops automatically when the preset number of wraps has been applied. The next time the combination stops, because the current baling cycle is completed, the wrapping table tips to the rear to drop the bale onto a rubber mat, discharging the bale. With the ISOBUS unit, all baling and wrapping functions are sequence-controlled, at the same time, the system gives audible and visual alarms to update the operator on all operations.

FIGURE 13.12   Baler wrapper combination integrates two operations into one machine. (Adapted from Krone, https://www.krone.de [December 2015], own photographs.)

13.4.7 FORAGE WAGONS

Once the grass is cut and gathered into swaths it should be chopped, collected, and transported to the silo. One of the most common procedures is centered on the use of forage wagons operated by a tractor. The concept consists of a number of systems that work in tandem, including pick-up, loading assembly, and an inclined scraper floor sloped toward the loading bay. The pick-up, located on the front part of the loader wagon, collects the forage by means of an elevator with several rows (five to nine) of dual tines and a height adjustable baffle plate that deliver an uninterrupted crop flow that is introduced into a canal in which a loading rotor together with a variable number of knives chop it and force its passage into the wagon with a scraper floor. The crop is transported toward the rear of the wagon as it gets filled (Figure 13.13).

The modern forage wagons are characterized for possessing a high number of knives, some models can have over 40, which permit chopping lengths of under 40 mm, although it is less precise than in self-propelled forage harvesters. Some wagons have an automatic sharpening system enabled using the operator terminal on the tractor. The complete set of knives are fully sharpened directly on the loader wagon, depending on the wear of the knives. This considerably reduces maintenance expenses, and at the same time, guarantees long-lasting, optimal cutting quality with lower energy consumption and increased output.

The loading rotor is driven via a gear equipped with a load sensing bolt (or a load cell), which measures the drive torque and can be adjusted to the type of crop for controlling the automatic loading function, switching the scraper floor on and off accordingly. In some systems, a strain gauge measures the load on the crossbeam that braces the lower part of the front wall. As soon as the gauge senses a load and the crop density in the wagon reaches the limit, the chain and slat floor starts moving automatically, filling the machine uniformly. Additionally, a volume sensor on the front flap at the top senses the degree of filling. As soon as the flap moves beyond a preset angle, the system activates the chain and slat floor. The operator can use either the load sensing system, the volume sensing system, or both systems together. The inclined scraper floor delivers the crop directly to the loading bay after chopping, with no need for a steep and narrow loading channel above the rotor. This reduces the power required for loading and fuel consumption. The wedge shape made by the inclined scraper floor allows the harvested material to be compacted, it remains in an almost upright position, and is pushed towards the rear of the wagon as a whole. Finally, a sensor on the tailgate measures the level inside the loading chamber indicating to the driver as a value on the control terminal, allowing the full load capacity to be used efficiently.

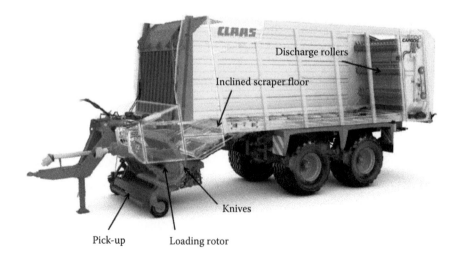

FIGURE 13.13   Silage loader wagon. (Adapted from Claas, http://www.claas.es/ [December 2015].)

FIGURE 13.14 Intelligent tractor–silage trailer combination. (Adapted from John Deere and Pöttinger, https://www.poettinger.at/landtechnik/download/ladewagenkombination_en.pdf [December 2015].)

The dimensions and load capacity of the forage wagons has tended to increase over the last few years with the aim of reducing transport costs. Some models can exceed 60 m^3. Also, these forage wagons serve two purposes equally well: either running alongside or behind the forage harvester, or loading crops by themselves using their pick-up system. By removing the loading assembly and discharge rollers, some models can be transformed from a loader wagon to chopper transport wagon in a few minutes; freeing up an additional load capacity, and protecting components not required in transporting the chopped material.

The current state of technology based on ISOBUS principles has allowed for the development of intelligent control systems for tractor-loader wagon combination. The system allows the speed to be automatically adjusted during use to keep the mass flow constant according to the swath volume and optimum operating point, thus decreasing the risk of blockage in the wagon. The most advanced equipment decisions about the loading process performance take place by data readings of the several sensors, swath geometry, travelling speed, position of the pickup, rotor torque, scraper floor condition, and fill level (Figure 13.14).

13.4.8 SELF-PROPELLED FORAGE HARVESTERS

The increase in farm size and the rapid rise in popularity of corn cultivation for silage in the 1960s called for ever more powerful harvesting systems. Attached and trailed implements were gradually replaced by self-propelled forage harvesters (SPFH). Self-propelled forage harvesters have three to four times the capacity of a pull-type harvester, so the shift to self-propelled machines has dramatically reduced the total unit sales (Bernardes and Chizzotti, 2012). At present, contracted operations with self-propelled forage harvesters have become common practice for silage harvesting all over the world (Figure 13.15).

The power range in the newest series of SPFH from the leading machinery manufacturers usually goes from 300 to 800 kW (400–1000 HP), with diesel engines from 6 to 12 cylinders and from 9 to 24 liters of displacement. The commercial models have a base unit unloaded weight (without header) from 11 to 18 t, as shown in Figure 13.16.

New Holland SP 818-1961   John Deere 5200-1972   Claas Jaguar 60 SF-1973

FIGURE 13.15   First self-propelled forage harvester models. (Courtesy of New Holland, John Deere, and Claas.)

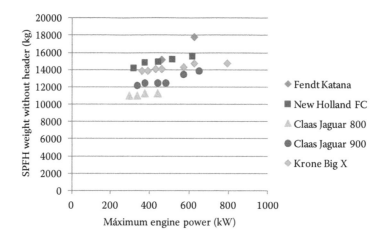

FIGURE 13.16   SPFH base unit weight without header (kilograms) and maximum engine power (kilowatts).

A SPFH is a machine for universal use with a range of headers with which it can be attached. Grass pickup, corn rotary header, whole-crop header, and corn picker for earlage are the most common four headers for a SPFH (Figure 13.17).

The headers pick up the crop and convey it to the chopping unit that is formed by four elements: feeder, drum, corn cracker, and discharge accelerator. During the periods in which corn is not chopped, the corn cracker is removed from the chopping unit. The feeder has four or six feed rollers, with a transmission that allows settings for different cutting lengths. A metal detection system is placed in the feed rollers and, in some models, a stone detector. Since 2005, some manufacturers have been selling mass flow and yield detection systems based on the measurement of the displacement of the feeder rolls (volume flow system) as options for their self-propelled forage choppers (Schmidhalter *et al.*, 2008). At present, other brands include this same yield metering system as an option.

The drum or cutting cylinder chops the forage with the combined action of knives and a shear bar. These cutterheads mount from 20 to 64 knives, with diameters from 630 to 720 mm, and a width of 630 to 850 mm. There are two main options for arranging the knives in the cutting cylinder: in a V-shape or chevron-style, or the multi-knife cylinder. It is essential to keep the knives on the forager chopping drum sharp, and to adjust the counter blade appropriately. Manufacturers include automatic sharpening systems and sensors for adjusting the gap between knives and shear bar. One of the last innovations is to measure the gap with an inductive method and two special sensors instead of the help of vibration sensors (Agritechnica, 2015) (Figure 13.18).

FIGURE 13.17   SPFH header range. (Courtesy of New Holland.)

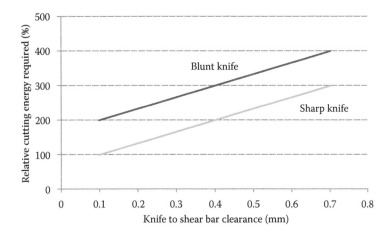

FIGURE 13.18   Effect of knife sharpness and clearance from the shear bar on energy requirements for self-propelled forage harvesters. (Adapted from McClure, J. R., 1990, New Holland Model 1915/2115 Forage Harvesters. *American Society of Agricultural Engineers*, Paper 90, 1517.)

The corn cracker or crop processor usually consists of two serrated steel rollers, or two rollers with beveled discs, turning at different speeds. Roller gap can be set from the cab and changed on the go. The kernels are crushed and broken by the rollers, allowing for better treatment of their contents. When the corn cracker is not in use, it is moved from the shaft to the parking position. For extended periods of work in grass, the corn conditioner is removed with the help of an integrated crane or hoist included in the SPFH. It is important that all work relating to installation and removal can be carried out quickly and easily, with sufficient space and good accessibility of corn cracker and drive belts.

The discharge accelerator is a blower that ensures efficient crop transfer to the truck or trailer. The gap between the blower and the crop processor can be adjusted for adapting the throwing capacity. In some SPFH models, there is a system which enables the operator to alter the position of the blower depending on the crop being harvested. The system features different settings. For instance, in grass mode the blower is situated 20 cm closer to the cutter head, and offers savings of up to 40 HP to enhance overall machine efficiency (Figure 13.19).

The chopped crop is expelled through the discharge spout to the trailer. The discharge spout has a modular design in order to adjust to different working widths. Spout extensions are available to get the optimum discharge distance for harvesting with large headers. The angle of rotation of the spout

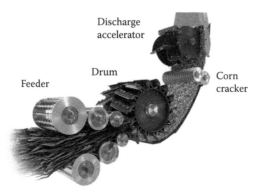

FIGURE 13.19 Elements of the chopping unit in a SPFH. (Adapted from Fendt https://www.fendt.com /[December 2015].)

is usually 210°, but with optional equipment some manufacturers offer a rotation angle up to 225°. The spout rotation enables trailers to be filled both on the right and left sides of the harvester. The discharge tube can be equipped with systems for controlling the spout movement with 3D cameras or laser scanner that automatically detects the trailer edge and monitor filling (Figure 13.20). With these systems, operators can offer maximum focus on crop flow and field progress, filling the trailers perfectly without spillages when working by day or night.

When the forage harvester has silage additive tanks, the liquid dosing nozzles are located in the spout or in the crop accelerator. Applying silage additives while chopping has become a standard service offered by professional contractors. At present, all SPFH are provided with a big tank for standard acid additives, with a capacity of 275 to 400 liters. Some brands also provide a secondary small tank for concentrated lactic acid bacteria solutions, with a capacity of 20 to 30 l. The dosage

FIGURE 13.20 Laser scanner system for automatic control of spout movement. (Adapted from Krone, https:// www.krone.de [December 2015].)

is controlled in the display and fixed or variable dosing rates based on moisture, with the possibility of throughput or constituent readings.

Sensors for measuring crop moisture are also located in the spout. Resistive or near infrared (NIR) moisture sensing systems provide the operator with a real-time and an average moisture reading on the cab monitor. This moisture reading can be used for automatically calibrating the chop length and for the precise application of additives, improving the silage packing. The near-infrared technology sensor mounted in the spout of some SPFH brands also measures crop constituents in real-time. Acid detergent fiber, neutral density fiber, starch, protein, and sugar readings are possible in alfalfa, barley, corn earlage, grass, maize, wheat, and whole crop. Investigations carried out by Kormann and Auernhammer (2002) have clearly shown that the NIR moisture measurement systems are applicable for use on the self-propelled forage harvester (Figure 13.21).

Hydrostatic transmission is the type of ground drive in the SPFH with field speeds of up to 25 km.h^{-1} and road speeds of up to 40 km.h^{-1}. Tire pressure adjustment of the front tires from the cab is offered in some brands. It allows reduction of the tire pressure from 2.0 bar to 1.2 bar, increase in traction, and to decrease fuel consumption during field work.

Guidance systems are available for operating with a self-propelled forage harvester. GPS guidance systems are at present widely used in agricultural machines. Automatic steering with GPS reduces the stress on the operator significantly, and enables the working width to be used effectively. When harvesting grass with the pickup header, systems with 3D cameras detect the swath as a three-dimensional image. Correction signals are transmitted to the steering mechanism in the event of deviations in the swath shape or direction. The steering axle then responds to these steering commands. In corn headers, mechanical touch sensors can be mounted for sensing row position. The signals generated by these sensors are translated into corrective steering impulses.

Claas   New Holland   Krone   John Deere

FIGURE 13.21 Sensors for measuring crop moisture and crop constituents in the spout of SPFH. (Courtesy of Claas, New Holland, Krone, and John Deere).

FIGURE 13.22 Self-propelled bunker forage harvester. (Courtesy of Duran agricultural machinery.)

Self-propelled forage harvesters usually work by discharging the chopped crop into trailers pulled by tractors or in trucks. But the combination of forage harvester with its own hopper is also possible (Figure 13.22). This enables very large transport vehicles to be used, for instance, for long-distance haulage to the clamp silo. A bunker forage harvester is able to work with only one truck, as the chopped crop is discharged in its own hopper during the truck trip.

13.4.9 FORAGE HANDLING IN THE SILO

Equipment at the silo must be sized accordingly in order to adequately handle the rate of forage received. This is especially critical in bunker or stack silos, where high densities are dependent on good packing practices (Bernardes and Chizzotti, 2012). There has been less technological development in this area than in other ensiling phases. The conventional machine used for spreading and packing the forage in horizontal silos is the agricultural tractor with four-wheel drive (Weinberg and Ashbell, 2003). The pushing tractors can commonly use blades or rollers as tools for spreading and compaction of the silo. Some producers have experimented with alternative packing vehicles, such as industrial wheel or trucked loaders. Industrial wheel loaders are heavier than agricultural tractors on an equivalent power basis (Holmes and Bolsen, 2009) (Figure 13.23).

Filling should be as fast as possible in order to exclude air from the crop quickly and to minimize the losses that result from plant respiration and the activity of aerobic microorganisms (Weinberg and Ashbell, 2003). Harvesting forage requires perfect coordination between several different pieces of equipment (forage harvesters, transport vehicles, and packers) in order to avoid bottlenecks. Several issues affect the process of filling the silo: width of the silo, fill level, and gradient of the access ramp. For a correct management of the process, it is necessary to carefully coordinate the packing equipment, as well as avoid excessive slopes on the access ramp (Cascudo et al., 2014).

FIGURE 13.23 Spreading and packing the forage in bunker silos with agricultural tractors and industrial wheel loaders.

13.5   IMPROVING THE EFFICIENCY OF THE HARVESTING OPERATION

Over the last few years, progressive advances in remote sensing, guidance systems, fleet management, and information and communication technologies (ICT) has demanded a change in planning, scheduling, and posterior operating of harvest operations. As a result, the traditional agricultural operations and planning methods must be supplemented with new planning features (Botchis, Sorensen, and Busato, 2014).

In Section 13.5, we will analyze the new tendencies for improving the use of the equipment, which is mainly linked to the improvement of the organization of the work.

In order to carry out a correct planning, it is essential to determine the influence that the parameters of the field and crop have on the performance of the self-propelled forage harvesters (SPFH). In studies carried out by Amiama, Bueno, and Álvarez (2008a) and Amiama, Bueno, and Pereira (2010) it has been concluded that "crop yield" is the variable that best accounts for the effective field capacity of the harvester. "Field area" is the variable with the second-best correlation values. The highest effective field capacities are associated with the largest fields. Under these conditions, it seems advisable to implement techniques that enable an increase in the average area of the fields, such as land consolidation or transborder farming. The model constructed by the authors is valid for the purposes of machinery management planning.

With the information provided by the model, prior to the start of the harvesting season, it is necessary to perform strategic planning in order to determine the right combination of resources according to the fields being harvested. The logistic process includes multiple harvest, transport, and storage operations. The harvest operation involves SPFH, trucks for transport, and machinery for silage packing. All of these need to be coordinated, and the number of equipment components needs to be adjusted according to the field's capacity—all under a tight schedule, with a large number of fields for harvesting. Bottlenecks within transport or unloading operations can reduce the capacity of harvest operations.

For this reason, using simulation models to analyze and optimize these complex systems can be useful (Amiama *et al.*, 2015a). A simulation demonstrated the usefulness of systems analysis in predicting the amount and cost of biomass supply in appropriate resource allocation to minimize bottlenecks. The equipment setup that provides lower costs varies depending on the SPFH considered. As an example, in Figure 13.24, an analysis is shown of the costs obtained in Northern Spain from the simulation of the process, with the aim of determining the number of transport vehicles that incur the least costs (Figure 13.24).

The use of simulation tools allows for:

* Comparing the performance of the system with different SPFH header widths
* Evaluating different resource combinations, in order to minimize the total system costs
* Determining which elements have a greater impact on the behavior of the system, via a sensitivity analysis

On an operational level, it is necessary to develop decision support tools which facilitate the management of the fleet of vehicles participating in the harvesting process. Along these lines, Amiama *et al.* (2015b) have developed a tool that serves to improve the SPFH performance, reducing the traveling distance between fields, and to design an efficient planning for transport vehicles. The focus is on searching the routes that provide reduced traveling distances for the SPFH by prioritizing the harvest starting date for each farmer, and matching the SPFH and number of trucks in order to minimize the total cost of the maize silage harvesting cycle. In field tests, this tool has showed on average savings of 879.80 € per harvester in the maize harvesting season. In Figure 13.25, an example of a route generated by the decision support system is shown.

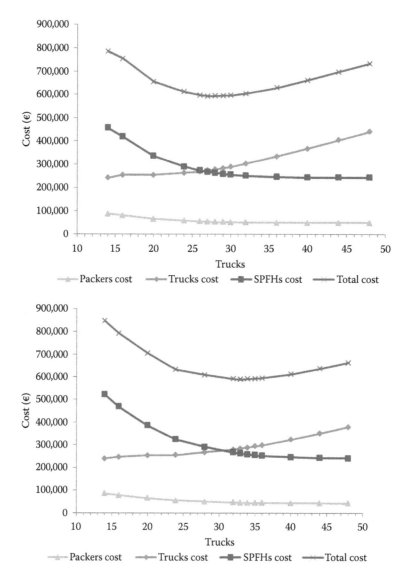

FIGURE 13.24 Cost analysis of harvesting season with six-row SPFH (top) and eight-row SPFH (bottom), with 3.35 t.min^{-1} of packing capacity.

At this operational level, other authors have focused, in order to improve the effective field capacity, on analyzing the most adequate trajectories that should be followed by the harvesters (Bakhtiari *et al.*, 2013). These methods were difficult to implement until relatively few years ago, due to the operator's difficulty in distinguishing the next track to be followed, according to the optimal plan, as the patterns designed aren't often repeated. Nonetheless, the gradual appearance of auto steering systems supports the execution of this kind of patterns. These technologies allow optimization of the trajectories based on different optimization criteria; such as total, or non-working traveling distance; total, or non-productive operational time; or risk of soil compaction. These systems are a prelude to the appearance, in a not too distant future, of the first prototypes of autonomous harvesting vehicles. Current tests with this technology are restricted to the simpler harvesting processes, which require less supervision such as transport operations (Brown, 2013).

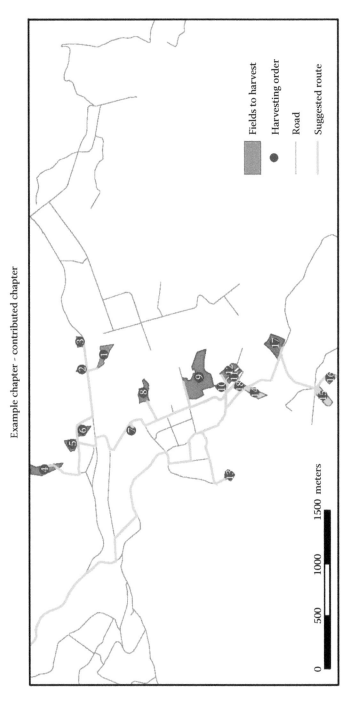

FIGURE 13.25 Route provided by HSDT (different colors in fields to harvest denote different owners).

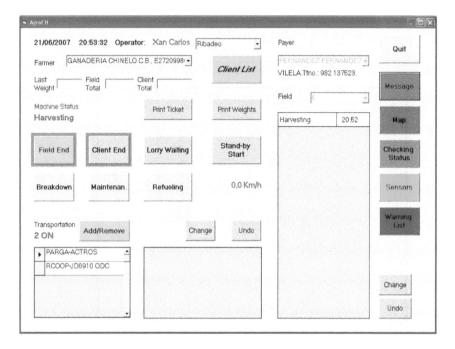

FIGURE 13.26 Data entry interface.

Finally, assessment of the machinery's performance is required in order to compare the planned operation and the actual executed operation. For this purpose, Amiama *et al.* (2008b) developed an online information and documentation system for the performance data of a forage harvester (Figure 13.26).

The following online information was recorded: performance data (operation speed, location, harvested area, and quantity harvested), machine settings (knife, drum, rotation, and speed), machine warnings (oil levels, oil pressure, and oil temperature), and a display map. In designing the software, it is very important to select the amount of information given to the driver. An excess amount of information could lead the driver to stop paying attention to the screen. A correct management of the information provided for the driver would result in better equipment operation, and therefore a lower cost. In addition, if the driver is notified when a threshold value in one of the parameters has been exceeded, they can act immediately and avoid major damages. The development of systems for the detection of anomalies for agricultural machines is also needed for the automated evaluation and real-time, re-planning of tasks. Craessaerts, De Baerdemaeker, and Saeys (2010) have also developed systems along these lines.

In summary, the improvement in the efficiency of the process demands changes in how the harvest is tackled. Efforts should go further than making individual improvements in the equipment, considering all the vehicles participating in the process as a whole. Failure to do so would mean that the improvements achieved in a particular task may not imply an improvement in the overall harvesting process.

13.6 CONCLUSION

The silage process involves a wide and variable range of equipment, depending on the silage harvesting system chosen.

The latest advances in forage harvesting entail the use of combined machines (mowers and conditioners, baler wrapper combinations) and to increase the working width with the goal of achieving higher effective field capacities. Also, there has been progress in connectivity of tractor equipment with the use of ISOBUS technology.

The current trend is to increase the SPFH power and versatility, exchanging headers, setting cutting lengths, and setting the blower position discharge spout, with the aim to adapt the machine to the specific needs of each crop. Additionally, SPFH are provided with systems that allow to apply silage additives or other solutions while chopping, in order to save time and to get a better distribution of the solution.

Advances are also aimed to determine mass flow and yield at harvest time, as well as other parameters of interest (crop moisture, acid detergent fiber, neutral density fiber, starch, and protein). In the ergonomics field, great strides are being made with the guidance systems.

The improvement of the efficiency at harvest operation requires strategic planning and to implement tools that allow for the right management and control at the operational level. Using simulation models, DSS for route management, technologies for trajectories optimization, and onboard telemetric systems, are examples for a better management of the harvest process. The key is to considerer all equipment involved in the harvest as a whole.

REFERENCES

Agritechnica 2015. Agritechnica Innovations (2015). Gold Medals. Hanover, Germany.

Amiama, C., Bueno, J., and Álvarez., C. J. (2008a). Influence of the physical parameters of fields and of crop yield on the effective field capacity of a self-propelled forage harvester. *Biosystems Engineering* 100, 198–205.

Amiama, C., Bueno, J., Álvarez, C. J., and Pereira, J. M. (2008b). Design and field test of an automatic data acquisition system in a self-propelled forage harvester. *Computers & Electronics in Agriculture* 61 (2), 192–200.

Amiama, C., Bueno, J., and Pereira, J. M. (2010). Prediction of effective field capacity in forage harvesting and disk harrowing operations. *Transactions of ASAE* 53 (6), 1739–45.

Amiama, C., Pereira, J. M., Castro, J., and Bueno, J. (2015a). Modelling corn silage harvest logistics for a cost optimization approach. *Computers and Electronics in Agriculture* 118, 56–65.

Amiama, C., Cascudo, N., Carpente, L., and Cerdeira-Pena, A. (2015b). A decision tool for maize silage harvest operations. *Biosystems Engineering* 134, 94–104.

Bakhtiari, A., Navid, H., Mehri, J., Berruto, R. and Bochtis, D. D. (2013). Operations planning for agricultural harvesters using ant colony optimization. *Spanish Journal of Agricultural Research* 11 (3), 652–60.

Bernardes, T. F. and Chizzotti, F. H. M. (2012). Technological innovations in silage production and utilization. *Revista Brasileira de Saúde e Produçao Animal* 13 (3), 629–41.

Bochtis, D. D., Sorensen, C. G. C and Busato, P. (2014). Advances in agricultural machinery management: A review. *Biosystems Engineering* 126, 69–81.

Bragachini, M., Cattani, P., Gallardo, M. and Peiretti, J. (2008). Forrajes conservados de alta calidad y aspectos relacionados al manejo nutricional (High quality conserved forage and aspects related to nutritional management). Córdoba, Argentina: INTA E.E.A.

Brown, H. (2013). From precision farming to autonomous farming: How commodity technologies enable revolutionary impact. Working paper, *Robohub*. http://www.robohub.org/from-precision-farming-to -autonomous-farming-how-commodity-technologies-enable-revolutionary-impact/ (Accessed November 5, 2015).

Cascudo, N., Campa, Y., Sousa, S., and Amiama, C. (2014). Estudio de la incidencia de la geometríadel silo sobre la capacidad de extendido y compactado (Effect of the geometry of the silo in the filling). *Spanish Journal of Rural Development* (3), 9–20.

Claas. (2015). Forage harvesters. *Jaguar 900 series. Claas UK Brochure*, 85.

Holmes, B. J. and Bolsen, K. K. (2009). What's new in silage management? In *Proceedings 15th International Silage Conference*, 2009. University of Wisconsin, Madison, WI, 61–76.

Craessaerts, G., De Baerdemaeker, J., and Saeys, W. (2010). Fault diagnostic systems for agricultural machinery. *Biosystems Engineering* 106, 26–36.

Kormann, G. and Auernhammer, H. (2002). Continuous moisture measurements in self-propelled forage harvesters. *Landtechnik* 5 (2002), 264–5.

McClure, J. R. (1990). New Holland Model 1915/2115 Forage Harvesters. *American Society of Agricultural Engineers*, Paper 90, 1517.

O'Kiely, P., Forristal, D. P., O'Brien, M., McEniry J., and Laffin C. (2007). Technologies for restricting mold growth on baled silage. *Beef Production Series* (81) Teagasc, Grange Beef Research Center, Ireland.

Schmidhalter, U., Maidl, F. X., Heuwinkel, H., Demmel, M., Auernhammer, H., Noack, P. and Rothmund, M. (2008). Precision Farming—Adaptation of Land Use Management to Small Scale Heterogeneity. *Perspectives for Agroecosystem Management*. (Ed. P. Schröder, J. Pfadenhauer, and J.C. Munch), 121–199. Elsevier B.V.

Weinberg, Z. G. and Ashbell, G. (2003). Engineering aspects of ensiling. *Biochemical Engineering Journal* (13), 181–188.

14 Drying of Agricultural Crops

D. M. C. C. Gunathilake, D. P. Senanayaka,
G. Adiletta, and Wiji Senadeera

CONTENTS

14.1   INTRODUCTION

The drying of foods and crops is a major operation in the food industry, consuming large quantities of energy. Dried foods are stable under ambient conditions, easy to handle, possess extended storage life, and can be easily incorporated during food formulation and preparation. The drying operation is used either as a primary process for preservation, or a secondary process in certain product manufacturing operations.

Drying operations alone account for 10% to 25% of the total energy in the food processing industry worldwide (Mujumdar and Passos, 2000). The processed food and beverage industry is Australia's largest manufacturing industry, accounting for 21.3% of total manufacturing turnover and the addition of 18.7% of industry value. It is one of only two manufacturing sectors that are net exporters (the other being metal product manufacturing) (DAFF, 2005). One of the main goals in designing and optimizing industrial drying processes is to reduce moisture at minimum costs. Presently, climate energy conservation plays a major role to make the process sustainable. To design a more efficient process, energy use and quality changes, as well as heat and mass transfer during processing, must be investigated. There are many drying methods available to achieve this task

(Akipinar *et al.*, 2006). The effect of hydrodynamics conditions existing in the drying bed influences moisture removal and the energy efficiency of the process (Senadeera, 2009).

Loss of moisture during drying of biological materials is accompanied by physical changes and internal structural changes (Rahman, 2005) which affect the final quality attributes of the dried product. Therefore, it may also affect the water absorption rate while temporary exposure to high humidity conditions which will, in turn, have an implication on the handling susceptibility of the product (Odilio *et al.*, 2006; Odilio *et al.*, 2010).

Most engineers do not realize that drying is a problem of preservation or structural transformations rather than the removal of water (Aquilera, 2003). During the past decade, many advances in technology and methods have emerged with the goal of minimizing degradation of various quality attributes of food products during drying. Therefore, careful selection of drying techniques and the optimization of drying conditions play a significant role. Physical changes (such as deformation of shape and size, as well as color changes and microstructural changes) have a direct impact on consumer decisions in buying a product.

Agricultural crops are important for the human diet, depending upon their nature, including vitamins, mineral, and fibers. Some crops are highly seasonal and are usually available in plenty. In the peak seasons, the selling price of crops becomes too low, leading to heavy losses for the grower. Furthermore, this leads to an unnecessary stock in the market, resulting in the spoilage of large quantities. Due to its seasonal nature, a need was felt to preserve crops over a period of time for use during off-seasons. Preservation of crops can prevent the huge wastage and make them available in the off-season at remunerative prices. Additionally, the seasonal nature of availability has led to efforts to extend the shelf life of crops by dehydration. Various drying methods have been practiced for dehydrating crops. These technologies need sophisticated equipment and skill training, the adoption of which appears difficult at the field or rural level.

There are many drying techniques are available. The most common technique is via air, in which heat is applied by convection, which carries away the vapor as humidity from the product. Examples of this include sun drying and artificial drying. Other drying techniques are vacuum drying and fluidized bed drying, where agricultural products are kept in vacuum conditions and water is used to evaporate and fluidize the material. These methods are suitable for heat sensible crops. Drum drying is another method, where a heated surface is used to provide energy; and spray drying that atomizes the liquid particles to remove moisture, like in milk powders. Special drying and curing techniques are used for preservation of big onion crops. This shows that drying is one of the most important preservation techniques or methods for food crops. In this chapter, theories of agricultural crop drying, methods, different dryers, and recent development in drying technology are presented. The methods used to dry grain or legumes, parboiled paddy, perishable crops, and fruits and vegetables are also discussed.

14.2 THE FUNDAMENTALS OF DRYING OF AGRICULTURAL CROPS

A crop is any cultivated plant that is harvested for food, livestock fodder, biofuel, medicine, or other uses. Agricultural crops can be broadly classified as grains, legumes, fruits, vegetables, and agricultural crops are broadly classified as durable and perishable crops. The history of agriculture dates back thousands of years, and its development has been driven and defined by different climates, cultures, and technologies. In the civilized world, industrial agriculture based on large-scale monoculture farming has become the dominant agricultural methodology. Durable crops (such as grains and legumes) contain relatively low amounts of water, and perishable crops (such as fruits and vegetables) contain high amounts of water during harvesting. Therefore, durable crops can be stored for extensive periods of time in comparison to perishable crops. However, durable crops need to be dried with sufficiently low moisture content before sending to stores for safe, long-term storage. Drying is an important operation in terms of improving and extending the shelf-life of durable crops.

As explained above, durable crops, such as grains and pulses, should be dried up to their equilibrium moisture content with warehouse atmospheric conditions (the most preferable warehouse atmospheric condition for safe storage of a durable crop is between 25–30°C, with a relative humidity of 65% to 75%). Thus, it can be said that drying is the procedure used to remove excess moisture from the grain to reduce moisture levels to a level which is acceptable for safe storage.

Drying results in a reduction of losses during storage from causes such as:

- Premature and unseasonable germination of the grain
- Development of molds
- Proliferation of insects

14.2.1   GRAIN/LEGUME MOISTURE CONTENT

The moisture content of a product is a numerical value expressed as a percentage. This is determined by the relationship between the weight of the water contained in a given sample of grain and the total weight of that sample. There are two ways to measure moisture content i.e., wet basis and dry basis. However, wet basis moisture percentage levels are commonly used to explain the moisture levels of grains. Method suggested by the Association of Official Analytical Chemists (AOAC, 2000) can be used to measure the moisture content of grains i.e., grain samples dried for 24 hours at 120°C. Grain sample weight should be measured before and after oven drying, and the moisture percentage is calculated by using the following equation:

$$M_{wb} = \frac{W_w}{W_w + W_{dm}} \times 100$$

Furthermore, the following equation can be used for converting moisture percentage from wet basis to dry basis moisture percentage:

$$M_{db} = \frac{100 \times M_{wb}}{100 - M_{wb}}$$

Where,

M_{wb} = moisture content (wet basis), M_{db} = moisture content (dry basis),
W_w = weight of water,
W_{dm} = weight of dry matter.

14.2.2   RELATIVE HUMIDITY (RH)

Knowledge about relative humidity is required for understanding most grain drying theories and practical situations. Grains are "hydroscopic," meaning that in ambient air they can either give off or absorb water in the form of vapor. However, at a given temperature the air cannot absorb unlimited quantities of water vapor. The air is considered to be "saturated" when it is unable to absorb water vapor at a given temperature and has a relative humidity of 100%.

The relative humidity of the air, expressed by percentage, is defined as the relationship between the weight of the water vapor contained in 1 kg of air and the weight of water vapor contained in 1 kg of saturated air at a given temperature. Table 14.1 shows the maximal weights of water vapor contained in 1 kg of air, where RH% is the relative humidity of the air (in percent):

$$RH\% = \frac{\text{Weight of water vapor in 1 kg of air}}{\text{Weight of water vapor in 1 kg of saturated air}} \times 100$$

TABLE 14.1
Maximal Weights of Water Vapor Contained in 1 kg of Air

Air temperature	0°C	10°C	20°C	30°C	40°C
Maximal water vapor weight (in grams)	3.9	7.9	15.2	28.1	50.6
Air temperature	50°C	60°C	70°C	80°C	90°C
Maximal water vapor weight (in grams)	89.5	158.5	289.7	580.0	1559

Air containing a given amount of water vapor tends to become saturated if its temperature is lowered. On the contrary, if the "drying power" of air is required to be increased (its capacity for absorbing more water vapor), it is necessary to heat it. For example, air saturated at 15.2 g of water vapor per kilogram of air at a temperature of 20°C has a 100% relative humidity. When temperature is increased, a kilogram of air can hold more water vapor (28.1 g of water vapor is required to saturate a kilogram of air at 30°C). Therefore, RH is decreased to 15.2 to 28.1, or 54%.

14.2.3   AIR-GRAIN EQUILIBRIUM

For a certain category of products and for a given temperature, equilibrium in the exchange of water vapor between the grain and the air is represented by the curve called "hydroscopic equilibrium curve." This curve shows, at a given temperature, the equilibrium between the moisture content of the grain (MC%) and the relative humidity of the air (RH%). Figure 14.1 gives an example of the equilibrium curve at four temperatures (0, 30, 60, and 120°C).

14.2.4   EQUILIBRIUM MOISTURE CONTENT

Any hygroscopic material (including grain) has its own characteristic balance (or equilibrium) between the moisture it contains and the water vapor in the air to which it is in contact with. This is known as the equilibrium moisture content (EMC). When food grains containing a certain amount of moisture are exposed to air, moisture moves from the grain to the air, or vice versa, until there is a balance between the vapor pressure of grain and ambient air. Each food grain has a characteristic moisture balance curve, which is obtained by plotting a graph of moisture content against the relative humidity and temperature of the air. Curves for some common food grains are given in Figure 14.2.

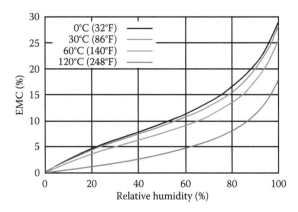

FIGURE 14.1 Equilibrium moisture equilibrium curves at four temperatures. (From https://commons .wikimedia.org/wiki/File:Hailwood-Horrobin_EMC_graph.svg.)

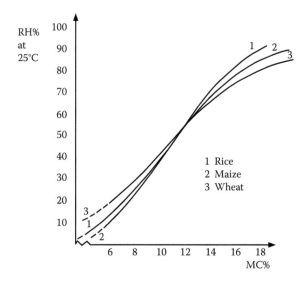

FIGURE 14.2 Equilibrium moisture content for different grains.

These values must be considered only as a guide, as the equilibrium values of different types and varieties of grain vary. The EMC will also vary slightly with temperature. For most cereals, it will decrease by approximately 0.5% for every 10°C temperature increase. At the same percentage, the relative humidity of the air containing a certain amount of moisture is exposed to air, and moisture moves from the grain to the air, or vice versa, until there is a balance between the moisture in the grain and in the air.

As sacks are porous and allow air to circulate readily through the crop, it is generally acceptable to allow the grain to be stored at a moisture content that is 1% to 2% higher than in bins or containers with nonporous walls. In addition to temperature and moisture content, the storage of grains can also be affected by the atmosphere. If damp grain is held in a sealed container, the respiration of grain and insects will consume the available oxygen. As the oxygen is depleted, it is replaced with carbon dioxide, which inhibits the activity of the insects and fungal problems. This will decrease to the point that it virtually ceases. However, storage in this manner can cause tainting of the grain, which renders it less acceptable for human consumption. Storage of seed grain requires conditions that will not only maintain peak viability but will avoid the possibility of germination while in storage. Low oxygen during storage may decrease viability and high moisture content, resulting in the reduction in viability of seeds. Therefore, this should be avoided for seed storage. At the same time, to avoid any danger of germination or fungal and insect damage while in storage, seeds should be dried 1% to 2% percent more than for human consumption. Additionally, it is important to keep the temperature of the seed as low as possible.

14.2.5 Drying Process

Drying of products can thus be obtained by circulating air at varying degrees of heat through a mass of grains. As it moves, the air imparts heat to the grain, while absorbing the humidity of the outermost layers. In terms of physics, the exchange of heat and humidity between the air and the product to be dried is seen in the following phenomena:

- Heating of the grain, accompanied by a cooling of the drying air
- Reduction in the moisture content of the grain, accompanied by an increase in the relative humidity of the drying air. However, this process does not take place uniformly.

The water present in the outer layers of grain evaporates much faster and more easily than that of the internal layers.

Thus, it is much harder to lower the moisture content of a product from 25% to 15% than from 35% to 25%. It would be a mistake to think that this difficulty can be overcome by rapid drying at high temperature. In fact, such drying conditions create internal tensions, producing tiny cracks that can lead to the rupture of the grain during subsequent treatments.

For drying grain, essentially two methods are used:

- Natural drying
- Artificial drying

Both methods have advantages and disadvantages, and no ideal method exists that permits all needs to be met.

14.2.6 NATURAL DRYING

The natural drying method consists essentially of exposing the high moist grain or legumes to the air (in sun or shade). To obtain the desired moisture content, the grain or legumes are spread in thin layers on a drying floor, where it is exposed to the air (in sun or shade) for a maximum of 10 to 15 days. To encourage uniform drying, the grain or legumes must be stirred frequently, especially if it is in direct sunlight.

Furthermore, for drying to be effective, the relative humidity of the ambient air must not be higher than 70%. For that reason, grain must not be exposed at night. In fact, by bringing about an increase in the relative humidity of the air, the cold of the night fosters re-humidification of the grain. For the same reasons, this method should not be used in humid regions or during the rainy season.

It must be remembered that insufficient or excessively slow drying can bring about severe losses of product during storage from the self-generated heat of "green" grain. Finally, prolonged exposure of grain to atmospheric factors, and thus to pest attack (insects, rodents, and birds) and microorganisms (molds), can also cause losses of product.

Despite these drawbacks, natural drying is advantageous in the following situations:

- When atmospheric conditions favor a reduction in moisture content over a reasonably short time-span
- When the quantities of grain to be dried are modest
- When production organization and socioeconomic conditions do not justify the cost of installing artificial drying equipment

14.2.7 ARTIFICIAL DRYING

The introduction of high-yielding crop varieties and the progressive mechanization of agriculture now make it possible to harvest large quantities of grain with high moisture content in a short amount of time. In humid tropical and subtropical zones, due to unfavorable weather conditions at harvest time, it is often difficult to safeguard the quality of products. In order to satisfy the need for an increasing agricultural production, it is therefore necessary to dry the products in relatively brief periods of time, whatever the ambient conditions. Consequently, it is necessary to resort to artificial drying. This method consists of exposing the grain to a forced ventilation of air that is heated to a certain degree in special appliances called "dryers."

14.2.8 TEMPERATURE AND PSYCHOMETRICS OF ARTIFICIAL DRYING

Grain stored in bins or sacks may have temperatures or moisture content, which are too high. If ambient temperatures are low, then air alone may cool the stored grain enough to prevent mold and insect damage while the moisture content is slowly reduced to a safe level. If the air temperature is too high (over 10°C), drying may be hastened by heating, as this further increases its capacity to absorb moisture.

Figure 14.3 shows an example of the effect of heating air, increasing its capacity to absorb moisture. In this case, the ambient air at 25°C and 70% RH is heated to 45°C and 24% RH. When passing through the grain, it gains enough moisture to again reach 70% RH while the temperature drops to 30.1°C. Each kilogram of air would then have removed $(0.023 - 0.0167) = 0.0063$ kg of moisture. Generally, whether the air returns to 70% RH or to some other level will depend on the initial moisture content and air velocity through the grain.

14.2.9 DRYING AND DEHYDRATION

One of the prime goals of food processing or preservation is to convert foods into stabilized products that can be stored for extended periods of time to reduce their post-harvest losses. Processing extends the availability of seasonal commodities, retaining their nutritional values, and adds convenience to the products. In particular, it has expanded the markets of products and ready-to-serve convenience foods all over the world. Several process technologies have been employed on an industrial scale to preserve fruits and vegetables; the major ones are canning, freezing, and dehydration. Among these,

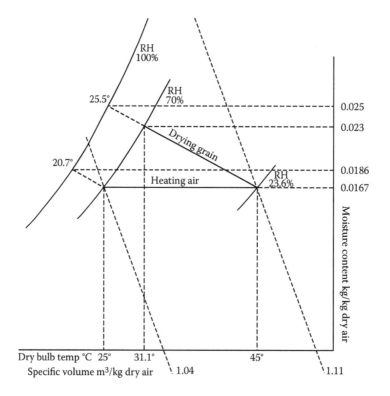

FIGURE 14.3   Effect of heating air for drying.

dehydration is especially suited for developing countries with poorly-established low temperatures and thermal processing facilities. It offers a highly effective and practical means of preservation to reduce postharvest losses and offset the shortages in supply.

The technique of drying is probably the oldest method of food preservation practiced by mankind. The removal of moisture from the produce prevents the growth of spoilage causing microorganisms and minimizes many of the moisture mediated deterioration reactions. It brings about substantial reduction in weight and volume and hence, minimizes packing, storage, and transportation costs, and also enables storability of the product under ambient temperatures. Dried agricultural crops have to be packed using a material which is a barrier to moisture. For this purpose, at the rural level, transparent high- or low-density polyethylene or film can be used, and the bags are heat sealed.

On the other hand, drying is a suitable alternative for post-harvest management especially in countries where poorly established low temperature distribution and handling facilities exist. It is noted that over 20% of the world's perishable crops are dried to increase shelf-life and promote food security (Grabowski *et al.*, 2003). The preservation of agricultural crops through drying dates back many centuries and is based on sun and solar drying techniques.

14.2.10 DRYING/DEHYDRATION FUNDAMENTALS

Dehydration involves the simultaneous application of heat and removal of moisture from foods. Thermal drying is one of the most widely used methods of drying foods. In this process, heat is mainly used to remove water from the fruits and vegetables. The mechanisms of moisture transfer depend mainly on the types or physicochemical state of food materials and the drying process. Fruits and vegetable materials can be classified as homogeneous gels, porous materials with interconnecting pores or capillaries, and materials having an outer skin that is the main barrier to moisture flow. The type or structure of foods always plays an important role in the drying process. In terms of transport phenomenon, it is considered as both heat and mass transport processes inside and outside of the food materials. Hence there are two resistances: heat transfer and mass transfer.

The drying curve usually plots the drying rate versus drying time or moisture contents. Three major stages of drying can be observed in the drying curve:

1. Transient early stage, during which the product is heating up (transient period)
2. Constant or first period, in which moisture is comparatively easy to remove (constant rate period)
3. Falling or second period, in which moisture is bound or held within the solid matrix (falling rate period)

14.2.10.1 Constant-Rate Period

When food is placed into a drier, there is a short initial settling down period as the surface heats up to the wet-bulb temperature. Drying then commences and, provided water moves from the interior of the food at the same rate as it evaporates from the surface, the surface remains wet. This is known as the "constant-rate period" and continues until a certain critical moisture content is reached. The surface temperature of the food remains close to the wet-bulb temperature of the drying air until the end of the constant-rate period, due to the cooling effect of the evaporating water. In practice, different areas of the food surface dry out at different rates and, overall, the rate of drying declines gradually towards the end of the constant-rate period.

14.2.10.2 Falling-Rate Period

When the moisture content of the food falls below the critical moisture content, the rate of drying slowly decreases until it approaches zero at the equilibrium moisture content (that is, the food comes into equilibrium with the drying air). This is known as the "falling-rate period." Non-hygroscopic foods have a single falling-rate period, whereas hygroscopic foods have two or more periods. In the first period, the plane of evaporation moves from the surface to inside the food, and water vapor diffuses through the dry solids to the drying air. The second period occurs when the partial pressure of water vapor is below the saturated vapor pressure, and drying is by desorption. During the falling-rate period(s), the rate of water movement from the interior to the surface falls below the rate at which water evaporates to the surrounding air, and the surface therefore dries out (assuming that the temperature, humidity, and air velocity are unchanged). If the same amount of heat is supplied by the air, the surface temperature rises until it reaches the dry-bulb temperature of the drying air. Most heat damage to food can therefore occur in the falling-rate period, and the air temperature is controlled to balance the rate of drying and the extent of heat damage. Most heat transfers are accomplished by convection from the drying air to the surface of the food, but there may also be heat transfer by radiation. If the food is dried in solid trays, there will also be conduction through the tray to the food.

The falling-rate period is usually the longest part of a drying operation and, in some foods (for example, grain drying), the initial moisture content is below the critical moisture content and the falling-rate period is the only part of the drying curve to be observed. During the falling-rate period, the factors that control the rate of drying change. Initially the important factors are similar to those in the constant-rate period, but gradually the rate of water movement (mass transfer) becomes the controlling factor. Water moves from the interior of the food to the surface by the following mechanisms:

- Liquid movement by capillary forces, particularly in porous foods
- Diffusion of liquids, caused by differences in the concentration of solutes at the surface and in the interior of the food
- Diffusion of liquids which are adsorbed in layers at the surfaces of solid components of the food
- Water vapor diffusion in air spaces within the food, caused by vapor pressure gradients or capillary condensation
- Vapor liquid flow due to differences in total pressure
- Gravity induced liquid moisture flow

During drying, one or more of the above mechanisms may be taking place and their relative importance can change as drying proceeds. For example, in the first part of the falling-rate period, liquid diffusion may be the main mechanism, whereas in later parts, vapor diffusion may be more important. It is therefore sometimes difficult to predict drying times in the falling-rate period. The mechanisms that operate depend mostly on the temperature of the air and the size of the food pieces. The size of the food pieces has an important effect on the drying rate in both the constant-rate and falling-rate periods. In the constant-rate period, smaller pieces have a larger surface area available for evaporation, whereas in the falling-rate period, smaller pieces have a shorter distance for moisture to travel through the food. Calculation of drying rates is further complicated if foods shrink during the falling-rate period. Other factors which influence the rate of drying include:

- The composition and structure of the food that has an influence on the mechanism of moisture removal; similarly, moisture is removed more easily from intercellular spaces than from within cells. Rupturing cells by blanching or size reduction increases the rate of drying but may adversely affect the texture of the rehydrated product. Additionally, high concentrations of solutes such as sugars, salts, gums, starches, etc., increase the viscosity and lowers the water activity, and thus reduce the rate of moisture movement.
- The amount of food placed into a drier in relation to its capacity (in a given drier, faster drying is achieved with smaller quantities of food). In practice, the rate at which foods dry may differ from the idealized drying curves.

The absence of a constant rate period indicates that the drying is controlled from the beginning by internal mass transfer resistance. The moisture content at the point when the drying period changes from a constant to a falling rate can be considered as the critical moisture content. The critical moisture content depends on the characteristics of the food and the drying conditions. The critical moisture content varies from 0.78 to 0.83 (kg/kg, wet basis) for vegetables and 0.85 to 0.89 (kg/kg, wet basis) for fruits. At high moisture contents, liquid flows due to capillary forces. At a decreasing moisture content, the amount of liquid in the pores also decreases and a gas phase is built up, causing a decrease in liquid permeability. Gradually, the mass transfer is taken over by vapor diffusion in a porous structure. At the saturation point, liquid is no longer available in the pores and mass transfer is taken over completely by vapor diffusion.

The first falling rate period is postulated to depend on both internal and external mass transfer rates; while the second period, during which drying is much slower, is postulated to depend entirely on internal mass transfer resistance. The slower rate may be due to the solid–water interaction or glass–rubber transition. The drying behavior of food materials depends on the porosity, homogeneity, and hygroscopic properties.

The moisture is transferred from the solid materials by diffusion or a capillary mechanism. During the diffusion mechanism, the concentration gradient is the driving force. Water diffusion can be in the form of liquid or vapor. In the case of liquid diffusion, osmotic pressure could be the driving force for water movement. In the capillary mechanism, the moisture moves due to surface tension, and does not conform to the laws of diffusion. A porous material contains a complicated network of interconnecting pores and channels extending to the exterior surface. As water is removed, a meniscus is formed across each pore, which sets up capillary forces by the interfacial tension between the water and the solid. Capillary forces act in a direction perpendicular to the surface of the solid.

14.2.11 Mathematical Models for Grains and Legume Drying

Several mathematical models have been developed based on mass balance in moisture transfer. The following models are the commonly used models for grain and legumes drying.

14.2.11.1 Modeling of Hot Air Drying of Bulk Grain

The heat balance for bulk grain in a fluidized bed dryer is defined in the following equations. Fluidized bed drying has been one of the most successful techniques of grain and legume drying. During fluidized bed drying, solid particles are suspended in a gas stream, and high rates of heat and mass transfer take place between gas and solid phases. Compared with other drying techniques, the entire bed is dried homogenously. The absolute humidity of drying air (W_{out}) passing through the bed (control volume) was calculated by the following equation:

$$\rho_p(1 - \varepsilon)C_p\left(\frac{d\theta}{dt}\right) = \rho_p(1 - \varepsilon)hh_{fg}\left(\frac{dM}{dt}\right)$$
$$+ \rho_a V_a C_a(T_{in} - T_{out})$$

(14.1)

In Equation 14.1, θ is the grain temperature (in degrees Celsius), T is the air temperature (Celsius), and M is the moisture content of the grain (kilogram per kilogram). The moisture transfer between the grain and drying air is described by the following relationship:

14.2.11.2 Mass Balance Equations

The absolute humidity of drying air (W_{out}) passing through the bed (control volume) was calculated by the Equations 14.2 and 14.3, where W is the absolute humidity of drying air in dry basis (kilogram/kilogram):

$$\rho_a V_a (W_{out} - W_{in}) = -\rho_p (1 - \varepsilon) h \left(\frac{dM}{dt} \right) \qquad (14.2)$$

$$\rho_a v_a A (W_{out} - W_{in}) = -\rho_s (1 - \varepsilon) V \left(\frac{dM}{dt} \right) \qquad (14.3)$$

Here ρ_a and V_a are air density (kilogram per meters cubed) and air velocity (meters per second), respectively, and A is the area of bed section (meters squared). The drying rate, (decimeters per displacement ton), was calculated as follows:

In this model, an individual grain is assumed to have the role of an element for describing the heat and mass transfer to and from the bulk of grain during drying. As they have mentioned in Equation 14.4:

$$\left(\frac{dM}{dt} \right) = \left(\frac{dMR}{dt} \right) (M_0 - M_e) \qquad (14.4)$$

Where MR is the moisture ratio, and M_e and M_0 are the equilibrium and initial moisture content of a grain on a dry basis (kilogram per kilogram), respectively. The thermodynamics parameters such as ρ_a, W, and C_a were calculated using the equations presented in Table 14.2.

14.2.11.3 Thin Layer Drying Equation for Soybean During Hot Air Drying

A thin-layer drying equation was used to predict the changes in the moisture content of grain under several air conditions. Equation 14.5 can be presented for single-kernel drying. To describe the changes in the moisture content of a single kernel during drying, an empirical thin-layer drying equation for soybean was used based on Page's equation.

$$MR = \frac{M - M_e}{M_0 - M_e}$$

$$= exp \left(-(0.0116 T^{0.2413}) \left(\frac{t}{60} \right)^{\left(0.6997 - 0.0046T + 0.000061T^2 \right)} \right) \qquad (14.5)$$

TABLE 14.2
Thermodynamics of Air–Vapor System

Property	Expression
Air density (kg/m^3)	$\rho_a = \dfrac{101.325}{0.287\, T_{abs}}$
Specific heat capacity of air (J/kg K)	$C_a = 1009.26 - 0.0040403\, T + 0.00061759\, T^2 - 0.0000004097\, T^3$
Latent heat of vaporization of water (J/kg)	$h_{fg} = 2503000 - 2386(T_{abs} - 273.16),\ 273.16 \leq T_{abs} \leq 533.16$
Saturated vapor pressure (kPa)	$P_{vs} = 0.1\, exp \left(27.0214 - \dfrac{6887}{T_{abs}} - 5.31\, \ln \left(\dfrac{T_{abs}}{273.16} \right) \right)$
Relative humidity (Pa/Pa)	$RH = \dfrac{101.3\, W}{0.62189\, P_{vs} + W\, P_{vs}}$

14.3 GRAIN/LEGUMES DRYERS

In its construction, the basic elements of a dryer are:

- The body of the dryer, which contains the grain or legumes to be dried
- The hot air generator, which permits heating of the drying air
- The ventilator, which permits circulation of the drying air through the mass of grain

Two main types of dryers are used:

- Static or discontinuous dryers
- Continuous type dryers

The formers are inexpensive and can treat only modest quantities of grain; thus, they are better adapted to the needs of small and medium scale centers for the collection and processing of products. As for the latter, these are high-flow dryers that need a more complex infrastructure, complementary equipment and, above all, special planning and organization. They are therefore more appropriate for big centers, silos, or warehouses, where very large quantities of product are treated.

14.3.1 STATIC OR DISCONTINUOUS DRYERS

A current of hot air moves from the bottom to top through a thick layer of grain. Drying of the mass of grain does not take place in a uniform fashion. As it moves from the bottom to the top, the drying air imparts heat to the grain and absorbs moisture, losing its "drying power" in the process. The lower layers will therefore dry more rapidly than the upper ones. Most batch dryers (except the Louisiana State University dryer, where grains are continuously circulated inside during whole operation) have this problem because most of the dryer's drying area is manufactured as a tank.

During the drying process, the mass of grain is thus found to be divided into three areas:

- Area of dry grain
- Drying area
- Area of humid grain

The imaginary line separating the area of dry grain from the area undergoing drying is called the "drying front."

14.3.1.1 Batch Dryers for Small-Scale Grain Drying

This type of batch dryer is used to dry grain and parboiled paddy. Figure 14.4 shows a small-scale dryer containing husk-fired furnaces with heat exchangers to supply the heat required for heating the drying air. The dryer can be fabricated in medium scale workshops. It should be noted, if direct oil or kerosene-fired burners are used for this type of dryers, the operating cost is significantly increased. Therefore, paddy husk was used as fuel for heating of air.

14.3.1.2 Batch Type Continuous Flow Dryers for Medium- and Large-Scale Grain Drying

Basically, the Louisiana State University (LSU) dryer is a continuous flow, medium batch type (static) and large-scale dryer for drying raw grain (Figure 14.5). Drying efficiency is comparatively high in these dryers because cocurrent heated air and grain flow is practiced in these dryers. Most of these dryers have husk-fired furnaces with heat exchangers and some have steam heat exchangers to supply the heat required for heating the air. The drying air temperature is in the range of 110–120°C.

FIGURE 14.4   Small-scale husk fired indirectly heated grain dryer.

For example, Paddy is recirculated continuously within the dryer until it was dried to the desired moisture content in about 8 to 12 hours. Then the blower is turned off and the dried paddy was emptied and transferred to the mill. It was claimed that paddy dried by this procedure gave dark colored rice and some kernels were heat damaged. It is recommended that the use of the LSU dryer as a batch in two stages. In the first stage, it was recommended that paddy should be dried from 35% to about 16% moisture content using a drying air temperature of 90°C at an air flow rate of 60 m^3/min per ton of paddy for about three hours.

Then, the batch of paddy should be transferred into a tempering bin for a period of at least six hours. Paddy should be tempered for six to eight hours in a tempering bin to equalize the moisture within the grain. In the second stage, the paddy should be dried after tempering at a temperature of 70–75°C for about an hour by recirculation, so that the moisture content of the paddy drops to 14%. The last half hour of the second stage drying should be used for cooling the paddy. It is claimed that paddy produced in the above manner yielded good quality parboiled rice with very little breakage (2.5%) and the efficiency of drying and utilization of heat was also much better.

14.3.1.3   Batch Drying in Fluidized Bed Dryer

This dryer has a husk-fired furnace with heat exchangers and some have steam heat exchangers to supply the heat required for heating the drying air. The dryer is provided for the drying of materials in the fluidized bed. Figure 14.6 shows the fluidized batch type fluidized bed dryer. Normally the

FIGURE 14.5   Medium- and large-scale batch type LSU grain dryer.

FIGURE 14.6 Grain fluidized bed dryer.

velocity of drying air is 8 m^3/min, the drying air temperature of 50°C was used, and it takes one hour to dry 500 kg per day.

Removal of dried material has to be done manually. At any stage of the drying operation, the velocity of air entering the drying chamber is maintained in such a way that the portion of grains, which are dried always remains separated and floating over the bed, while in the paddy bed there is a mixture of wet, partially dried, and dried fractions. This dryer is also comparatively efficient.

14.3.1.4 Batch Type Maize (Corn) Dryer

Maize is dried by spreading it directly on ground or on mats and allowing the sun to dry it. Drying time is dependent on the weather conditions. The main disadvantage of this method is contamination of the product with dust, dirt, and other airborne impurities. Furthermore, contamination occurs due to stray animals. During cloudy weather, drying cannot be accomplished within one day and has to be prolonged for several days, thus leading to microbial and chemical spoilage, resulting in the deterioration of product quality. This dryer (Figure 14.7) was developed for drying maize cobs, which would preserve its quality and substantially increase farmers' incomes. It was observed that the developed dryer can be used to dry 1000 kg of maize cobs containing moisture of 26% to 33.7% to a moisture content of 13% that is fit to store within 15.5 to 26.6 hours. The average drying air temperature was 46°C. Firewood was used as a fuel source and the requirement of fire wood was 12 kg/hr.

14.3.1.5 Batch Type Small-Scale Fresh Paddy Dryer

Sun drying is the most common paddy drying practice of them and under tropical climatic conditions, high rainfall seriously effects sun drying. It becomes extremely difficult for drying of freshly harvested paddy and parboiled paddy during rainy weather. Mechanical dryers are used only at large-scale rice millers. Hence, this dryer (Figure 14.8) was developed especially for small-scale fresh paddy drying.

It was observed that the developed dryer can be used to dry 500–700 kg of freshly harvested paddy, containing 19.8% to 20.9% of moisture to a moisture content of 13% to 14% that are fit to store within 3.1 to 4.25 hours; also, 400–600 kg of parboiled paddy containing 35.6% to 35.8% of moisture can be dried at a moisture content of 13% to 14% that can be milled within 8.22 to

FIGURE 14.7   Batch type maize dryer.

9.88 hours. The drying air temperature was varied within 55–72.5°C. Optimum batch size for better quality final product was 600 kg per batch in both cases of drying freshly harvested paddy and parboiled paddy. Relevant drying times were 3.4 hours and 8.87 hours, respectively. The cost of drying per kilogram of freshly harvested paddy and parboiled paddy were 1.56 SLR and 2.38 SLR, respectively.

14.3.2   Continuous Type Dryers

14.3.2.1   Continuous Fluidized Bed Dryer

The dryer (Figure 14.9) is provided for drying of materials in fluidized bed which comprises of a drying bed having an air inlet for injecting hot air into the bed, the arrangement for feeding the grain or legumes into the drying chamber column, and a single or more discharge outlets for the automatic and continuous removal of dried material in a suspended state. The air inlet at the bottom of the drying chamber is connected through a proper diffuser is supplied with air at a preset temperature, with relative humidity and velocity. The bed of the grain or legumes charged into the drying chamber gets itself classified in respect of dried and non-dried fractions. The moist grains move down, while

FIGURE 14.8   Batch type small scale fresh paddy dryer.

FIGURE 14.9 Continuous fluidized bed dryer.

the progressively dried grains move up during the passage of hot air through the column. The classified grains soon reach a point when the top layer of dried grains with the desired moisture content will leave the surface and float up at the preset air properties. The dried grain discharge ports provided at the upper section of the chamber then collect the grains, take them away from the drying chamber, and immediately feed them into the cooling unit for cooling the grains.

14.4 DRYING OF PARBOILED PADDY

Paddy after parboiling contains about 35% moisture and it has to be dried to about 14%, which is the moisture level for proper milling and safe storage. The manner in which this excess moisture is removed is of considerable importance. If the moisture is removed at a very slow rate, the micro-organisms will grow and spoil the parboiled paddy partially or fully. On the other hand, if drying is done rapidly and continuously, cracks may develop and rice will break during milling. However, if parboiled paddy is uniformly dried by any means (shade, sun, or by hot air) practically no breakage of rice will occur. However improper drying conditions may result in breakage as high as 100%.

14.4.1 Shade Drying of Parboiled Paddy

Parboiled paddy could be dried under shade, but it would take a long time and is therefore not usually practiced. Due to the slow drying rate, the milling results of paddy dried under shade are best. Proper shade drying of raw paddy can yield excellent milling quality with less than 1% breakage.

14.4.2 Sun Drying of Parboiled Paddy

Figure 14.10 shows the sun drying and tempering of parboiled paddy. Sun drying of parboiled paddy is the most common practice, which is followed by big and small rice mills and individuals. The cost of sun drying per ton of paddy is the minimum. The main drawback in this method is that it is depend upon weather conditions. It cannot be carried out throughout the year and also the drying rate is too slow. Sun drying in a single stage causes considerable damage to the milling quality of paddy. The milling quality of paddy is not affected until the moisture content reaches 16%. Suspension of drying and tempering of paddy for about two hours by heaping the paddy and covering it with gunny, then drying in the second stage improves the milling quality considerably without adding much to the cost of drying. Sometimes the first stage of drying can be carried out in the day time and the second stage (with overnight tempering) on the next day. The cost of parboiling and drying is a very important consideration for popularizing a method of parboiling. While calculating the cost of parboiling and drying per unit quantity of paddy, many factors should be carefully considered. Cost of the

FIGURE 14.10   Sun drying and tempering of parboiled paddy.

equipment for any method should be considered as per present market prices. For example, in the case of sun drying, the yard might have been built 50 years back and it must have cost less at that time. However, for better evaluation, today's cost of the land and yard must be considered. If there are losses in any method, the losses should be included in the cost. The choice of parboiling units for evaluating the cost should be of the same capacity as far as possible. Cost per unit quantity of paddy processed would be a better evaluation than the cost per hour or day.

The above-explained artificial dryers can be used for drying of parboiled paddy. The LSU dryer can be used for efficiently drying of parboiled paddy. Most rice mills use LSU dryers for drying of parboiled paddy.

14.5   DRYING OF PERISHABLE CROPS

Commerce in fruits and vegetables arise from the importance of this commodity in human diet. Humans have kept fruits and vegetables in their diet to provide variety, taste, aesthetic appeal, and to meet certain essential nutritional requirements. Furthermore, some fruits and vegetables can be important supplementary source of carbohydrates, minerals, and protein. The possible beneficial effects of dietary fiber derived from fruits and vegetables are currently under scrutiny as part of a reexamination of the human diet, with the object of minimizing some diseases considered to be related to an affluent lifestyle.

Perishable crops such as fruits and vegetables are usually not drying, because taste and aroma is lost during the drying process, therefore mostly fruits and vegetable are consumed as fresh produce. However, fruits and vegetables are dried for preservation to use in off-season or surplus production is dried and stored for minimized waste. The main fruits and vegetables are mangoes, papayas, pineapples, avocadoes, bananas, watermelons, pumpkins, tomatoes, eggfruit, bitter guard, carrots, and many more. The production of some of fruits and vegetables are highly seasonal, and during glut periods, large quantities go to waste. If the farmers could adopt a preservation technique to keep the surplus production of fruits and vegetables to be marketed during off-season periods, it would become an important income generating activity for them. Sun drying and artificial drying using dryers of the surplus production of fruits for preservation become the most feasible and cost-effective method of preservation.

14.5.1   Drying of Fruit and Vegetables

During the past decade, many advances in drying technologies have emerged with the goal of minimizing degradation of various quality attributes of food products during drying. Among an enormous number of foods that need to be dried, fruits and vegetables have received much attention as it has repeatedly been reported that these materials contain a wide array of phytochemicals, which are claimed to exert many health benefits, including antioxidant activity. In some cases where extraction of bioactive compounds cannot be performed on fresh fruits and vegetables, drying is a necessary step that needs to be conducted to keep the materials for later use. Due to their naturally heat-sensitive nature, most phytochemicals degrade significantly during drying; a careful selection of the drying technique and optimization of the drying conditions are significant.

It has been revealed that dried fruits and vegetables have also been regarded as alternative, fat-free snacks for health-conscious consumers. This implies that not only nutritional changes, but also other changes, including physical and microstructural changes, which are of importance and need to be optimized.

The drying of fruits and vegetables has been principally accomplished by convective drying. There are a number of studies that have addressed the problems associated with conventional convective drying. Some important physical properties of the products have changed such as, loss of color, change of texture, chemical changes affecting flavor and nutrients, and shrinkage. The high temperature of the drying process is an important cause for loss of quality. Lowering the process temperature has great potential for improving the quality of dried products.

14.5.2 QUALITY CHANGES DURING DRYING

There are several changes taking place in quality parameters during drying and storage. The extent of changes depends on the care taken in preparing the material before dehydration and on the process used. Major quality parameters associated with dried food products include color, visual appeal, shape of product, flavor, microbial load, retention of nutrients, porosity-bulk density, texture, rehydration properties, water activity, freedom from pests, insects, and other contaminants, preservatives, and freedom from taints and off-odors (Ratti, 2005). The state of the product (such as glassy, crystalline, or rubbery) is also important.

Quality parameters of dried fruits and vegetable can be classified into four major groups:

 i. Physical qualities
 ii. Chemical qualities
 iii. Microbial qualities
 iv. Nutritional qualities

Greater stability and quality can be achieved by maintaining the fresh or optimum conditions of the raw materials.

14.5.2.1 Physical Quality

Physical changes, such as structure, case hardening, collapse, pore formation, cracking, rehydration, caking, and stickiness can influence the quality of the final dried product. Pre-treatments given to foods before drying or optimal drying conditions are used to create a more porous structure to facilitate better mass transfer rates. Maintaining the moisture gradient levels in the solid, which is a function of drying rate, can reduce the extent of crust formation; the faster the drying rate, the thinner the crust.

Depending on the end use, hard crust and pore formation may be desirable or undesirable. The current knowledge outlines mechanisms for pore formation in foods during drying and related processes (Rahman, 2001). The glass transition theory is one of the concepts proposed to explain the process of shrinkage and collapse during drying and other related processes. According to this concept, there is a negligible collapse (more pores) in materials when it is processed below the glass transition. The higher the process temperature above the glass transition temperature (Tg), the higher the structural collapse. Methods of freeze and hot air drying can be compared based on this theory. In freeze-drying, since the temperature is below Tg′ (maximally frozen concentrated Tg), the material is in a glassy state resulting in negligible shrinkage, and therefore the final product is very porous. In hot air drying, on the other hand, since the temperature of drying is above Tg′ or Tg, the material is in a rubbery state and substantial shrinkage occurs. Hence, the food produced from hot air drying is dense and shriveled. Rahman (2001) hypothesized that since capillary force is the main factor responsible for collapse, counterbalancing this force will cause formation of pores and lower shrinkage. This counterbalancing is due to a generation of internal pressure, variation in moisture transport mechanisms, and surrounding pressure. Another factor could be the strength of the solid matrix (i.e., ice formation, case hardening, and matrix reinforcement).

Rehydration is the process of moistening dry material. In most cases, dried foods are soaked in water before cooking or consumption, thus rehydration is one of the important quality criteria. In practice, most of the changes during drying are irreversible, and rehydration cannot be considered simply as a process reversible to dehydration. In general, absorption of water is rapid at the beginning, due to surface and capillary suction. Rahman (2013) reviewed the factors affecting the rehydration process. These factors are porosity, capillaries and cavities near the surface, temperature, trapped air bubbles, the amorphous-crystalline state, soluble solids, dryness, anions, and pH of the soaking water. Porosity, capillaries, and cavities near the surface enhance the rehydration process, whereas the presence of trapped air bubbles gives a major obstacle to the invasion of fluid. When the cavities are filled with air, water penetrates the material through its solid phase. In general, the temperature strongly increases water rehydration in the early stages. There is a resistance of crystalline structures to salvation that causes the development of swelling stresses in the material, whereas amorphous regions hydrate fast. The presence of ions in water affects the volume increase during water absorption. The texture of dried products is influenced by their moisture content, composition, pH, and product maturity. The chemical changes associated with textural changes in fruits and vegetables include crystallization of cellulose, degradation of pectin, and starch gelatinization. The method of drying and process conditions also influences the texture of dried products. The quality of apples, bananas, potatoes, and carrots have changed with different drying methods such as convective, vacuum, microwave, freeze, and osmotic drying. It was found that air, vacuum, and microwave dried materials caused extensive browning in fruits and vegetables, whereas freeze drying seemed to preserve color changes, resulting in a product with improved color characteristics.

14.5.2.2 Chemical Quality

Browning, lipid oxidation, color loss, and change of flavor in foods can occur during drying and storage. Browning reactions can be categorized as enzymatic and non-enzymatic. Enzymatic browning of foods is undesirable because it develops undesirable color and produces an off flavor. The application of heat, sulfur dioxide or sulfites, and acids can help control this problem. The major disadvantage of using these treatments for food products is their adverse destructive effect on vitamin B or thiamine. Acids such as citric, malic, phosphoric, and ascorbic are also employed to lower pH, thus reducing the rate of enzymatic browning. Dipping in an osmotic solution can inhibit enzymatic browning in fruits. This treatment can also reduce the moisture content with osmotic preconcentration. Browning tends to occur primarily at the mid-point of the drying period. This may be due to the migration of soluble constituents towards the center. Browning is also more severe near the end of the drying period when the moisture level of the sample is low and less evaporative cooling is taking place. Rapid drying through a 15% to 20% moisture range can minimize the time for browning reaction.

14.5.2.3 Microbial Quality

Dried food products are considered safe with respect to microbial hazards. There is a critical water activity (airwatt) below in which no microorganisms can grow. Pathogenic bacteria cannot grow below an airwatt of 0.85 to 0.86, whereas yeast and molds are more tolerant to a reduced water activity of 0.80. Usually no growth occurs below an airwatt of about 0.62. Reducing the water activity inhibits microbial growth, but does not result in a sterile product. The heat of the drying process does reduce total microbial count, but the survival of food spoilage organisms may give rise to problems in the rehydrated product. The type of microflora present in dried products depends on the characteristics of the products, such as pH, composition, pre-treatments, types of endogenous and contaminated microflora, and methods of drying. Brining (addition of salts) in combination with drying decreases the microbial load. The dried products should be stored under appropriate conditions to protect them from infection by dust, insects, and rodents (Rahman *et al.*, 2001).

14.5.2.4 Nutritional Quality

Fruits and vegetables in their dried form are good sources of energy, minerals, and vitamins. However, during the process of dehydration, there are changes in nutritional quality. A number of vitamins such as A, C, thiamine, and polyphenols are heat sensitive and sensitive to oxidative degradation.

Sensory properties of dried foods are also important in determining quality. These include color, aroma, flavor, texture, and taste. Aroma and flavor can change due to loss of volatile organic compounds, which is the most common quality deterioration for dried products. Low temperature drying is used for foods that have high economic value, such as flavoring agents, herbs, and spices.

14.5.3 PRETREATMENTS FOR DRYING

Drying combined with some pre-treatments appears to be a cost-effective method of preservation. Several methods of pre-treatment have been widely utilized, such as immersion in chemical solutions, hot-water blanching, and physical pretreatments.

Pretreatments of some fruits (i.e., sour cherries, grapes, and apples) prior to drying have been reported to help reduce some of the undesired changes such as color and textural changes. They also reduce drying time by relaxing tissue structure and yield a good quality dried product. Chemical pretreatments include sulfating, immersion in sodium chloride, calcium chloride, or sugars, use of surfactants and impregnation with biopolymers. Among these methods sulfating is one of the most common. The effects of sulfating are well known. Immersion in solutions containing sulfites disinfects surface and reduces oxidation of liable food components facilitates drying reduces shrinkage and improved rehydration. Moreover, sulfating causes bleaching of anthocyanins, which affects color of some fruits subjected to dehydration.

Sulfites were accepted as a safe additive (generally recognized as safe) until its allergenic reaction was observed in sensitive people. If more than the permitted levels are used, it can be allergenic, causing reactions like nettle rash, angio-neurotic edema, and anaphylactic shock in ordinary individuals. Due to this, consumers tend to avoid meta-bisulfite. Furthermore, blanching has been considered to be one of the most important pretreatment technologies. The blanching pretreatment is used to inactivate the enzymes responsible for producing off flavors and unpleasant odor, maintain the freshness color, stabilize the texture and nutritional quality, expel air between the cells, and destroy the microorganisms to some extent.

Blanched fruits and vegetables have higher color attributes than non-blanched ones because of thermal inactivation of undesirable enzymes (peroxidase and lipoxygenase), and a resultant decrease in the rate of enzymatic deterioration of fruits and vegetables. Furthermore, blanching can also limit the degradation of both chlorophylls and carotenoids. However, the blanching treatment causes undesirable changes in the properties of the food, such as the loss of soluble nutrients (sugars, minerals, and vitamins). Moreover, it causes loss of aroma and negatively impacts the sensory properties associated with texture. In particular, the levels of texture softening during the blanching process can be observed by scanning electron microscopy, which has been found to result in physical changes in cell structure (such as cell separation and cell damage) of blanched carrot tissues. It was found that they appeared greater than those in untreated raw tissue. This breakdown in cellular structure has been also observed in mushrooms. Moreover, pectin is degraded and solubilized from the cell wall and the middle lamella between adjacent cell walls. This has led to a loss of adhesion between cells and turgor pressure, which ultimately destroys the membrane integrity.

Dipping or soaking a product in organic acids, such as citric acid or lactic acid, is an alternative to blanching, as these pre-treatment methods can help reduce the number of normal flora and pathogenic organisms, reduce the action of enzymes responsible for browning, and drying time. In particular, it was found that the infusion of oranges with citric acid solution (0.1, 0.25, 0.5, and 1.0% weight per volt) during the peeling process reduced the surface pH of peeled fruits (from 6.0 to less than 4.6) and extended their shelf life in comparison with fruit infused only in water. The extension of shelf life resulted primarily from the inhibition of spoilage bacteria. The effect of chemical pre-treatment on the microstructural changes has been found it rehydrated dried red bell pepper. The samples were immersed in an aqueous solution of NaCl, which was combined with $CaCl_2$ and $Na_2S_2O_5$ prior to convective drying. The results showed that the cell walls of the pretreated samples did not significantly collapse even after drying. However, a higher water holding capacity of the samples after rehydration was found.

The effect of an edible coating made by dipping the fruit in a solution containing trehalose at 0.8%, sucrose at 1.0%, and sodium chloride at 0.1%, was investigated on the cold storage of minimally processed Annurca apples. The result showed that such a coating reduced the browning phenomena, as well as resulting in a decrease in the weight loss and a reduction of organic acids, which was observed for coated samples.

14.6   DRYERS FOR DRYING OF FRUITS AND VEGETABLES

14.6.1   MULTI-CROP DRYER

The multi-crop tray type dryer which utilizes agricultural waste, such as paddy husk, for thermal energy generation does not require electricity for operation, and can be conveniently adopted at the rural level for the dehydration of fruits. This multi-crop dryer (Figure 14.11) was designed for fruits, vegetables, and chili drying to overcome the problems encountered in sun drying, such as quality deterioration, high labor requirements, a prolong drying period, and damage, especially in chili pods due to mold growth during cloudy and rainy weather conditions. The dryer has been successfully tested for dehydration of a variety of fruits and vegetables, such as carrots, leeks, bitter gourds, eggplants, jackfruit, bananas, limes; spices such as pepper and cloves; and also green leaves. The drying time in a dryer is one to two days for various crops, such as chilies, vegetables, spices, and other field crops. The average quantity dried per batch in the dryer is 100–500 kg. The dryer is a low-cost, tray-type, multi-crop dryer which can be fabricated from material available at farm level. Paddy husk, which is a byproduct of the rice milling industry, is used as the fuel for heating of drying air. Since the dryer does not require any blowers for drying air circulation or paddy husk combustion it could be used even in areas where electricity is not available.

The dryer was evaluated for its capacity, performance, optimum operational conditions, thermal efficiency, operational cost, and product quality using a variety of crops, and the results are given in Table 14.3.

14.6.2   ONION DRYING AND CURING

Before contemplating storage, it is imperative to ensure that the onion bulbs are of good quality and that they have been dried properly (with the surface moisture removed) and cured (the relatively high temperature process which allow the formation of strong, intact, outer protective skin and closure of the topped neck of the onion). Drying and curing often take place simultaneously. The optimum curing time, temperature, and RH for big onions are five to six days, 27–33°C, and 60% to 70%, respectively. The acceptable moisture loss during curing was reported to be 3% to 5% (Thompson *et al.*, 1982). If the RH falls below 60%, excessive water loss can occur resulting in a split in the outer scale (Brice *et al.*, 1997). Natural drying and curing are unable to be practiced under tropical climatic conditions due to high RH (more than 75%) during harvesting season of big onion. Initial drying is

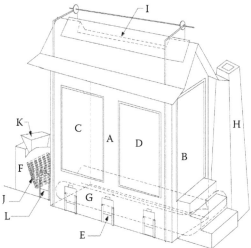

FIGURE 14.11 Multicrop dryer.

TABLE 14.3

Performance of Multicrop Dryer for Different Crops

Crop	Capacity (Kilograms)	Final Moisture (%)	Fresh/Dry Weight Ratio	Drying Cost (LKR)
Jackfruit	35	7–8	4:1	3.00–8.00
Chili	125	6–7	4:1	0.64–0.8
Carrot	20	6–7	6:1	4.25–12.00
Leeks	15	5–6	10:1	6.38–12.00
Curry leaves	10	4–5	4:1	8.50–12.00
Gotukola	15	4–5	8:1	5.67–12.00
Maize cobs	1000 (cobs)	10–12	1.5:1	0.6–0.8
Ash plantains	30	7–8	6:1	7.00–12.00
Cloves	90	10–12	3:1	0.6–0.8

used to remove surface moisture, subsequent drying to cure the outer skin and to seal the necks. This can be practically done to maintain temperature and relative humidity 25–35°C, 60% to 70%, respectively. The end of drying and curing is indicated by the necks being fully dried with no evidence of moisture when passed. The formulation of the whole outer skin and a dry neck reduces moisture losses from the bulbs during storage and reduces the risk of fungal and bacterial infections. Figure 14.12 shows the onion drying curing and storage structure developed by the Institute of Post-Harvest Technology, Sri-Lanka. This drying, curing, and storage structure is able to control inner RH through heating of ducts placed in the floor of the structure. The heating of these ducts was performed by a paddy husk–fired furnace. Accordingly, onions can be dried, cured, and maintained through storage temperature.

14.6.3   Solar Assisted Biomass Fired Dryer for Dehydration of Fruits and Vegetables

This dryer (Figure 14.13) consists of a solar collector, drying chamber, chimney, turbo ventilators, and furnace.

14.6.3.1   Solar Collector

The energy from the sun is collected here and transferred to the drying air. This air heater is covered with glass planks of 3ft × 4ft. The bed is filled with gneiss (black stones) of an average of three inches in size as a heat absorbing material. When the atmospheric air passes through the air heater, the air gets heated up before entering the drying chamber. In the absence of sun light, there is a heat exchanger by which the heat generated from the combustion of biomass could be transferred to the drying air. The dimensions of the air heater or the solar collector are 12ft × 20ft.

14.6.3.2   Drying Chamber

The expected products are dehydrated in the drying chamber. Two sides of this chamber are constructed using masonry bricks, and the other two sides are closed by a door on one side and by the air heater on the other side. The trays are made up of stainless steel wire mesh fitted with the help of slotted angles. The interior of the drying chamber is covered using an aluminum sheet. The size of the drying chamber is 12ft × 12ft × 10ft × 8ft.

14.6.3.3   Chimney

This is again constructed using masonry bricks, which are a rectangular in shape, and tapering up. The smoke from the combustion process is exhausted to the atmosphere through this chimney. Connecting the furnace and the chimney, there is a heat exchanger consisting of 7 MS pipes.

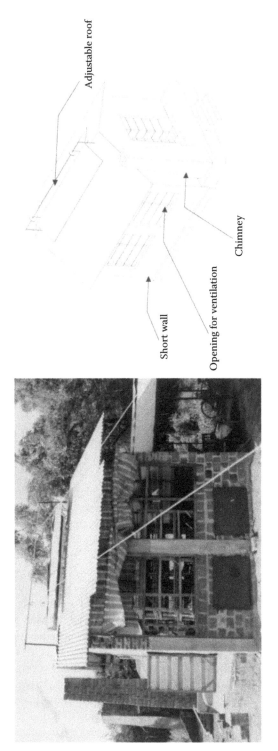

Adjustable roof

Short wall

Opening for ventilation

Chimney

FIGURE 14.12 Onion drying curing and storage structure.

FIGURE 14.13   Solar assisted biomass fired dryer.

14.6.3.4   Turbo Ventilators

There are two ventilators fixed to the top of the roof in order to facilitate the drying air flow. This air flows both horizontally and vertically through the trays on which the materials to be dried are placed. The rate of flow is controlled by an inlet flap installed at the beginning of the solar collector.

14.6.3.5   Furnace

The normal foundation is laid using brick. Special attention is paid on the preparation of cement paste during construction of the furnace. To achieve consistency in the cement paste, the following mixture proportions are used. Sand (2 portions), cow dung (1 portion), clay (1.5 portions), paddy husk ash (1 portion), and cement (0.5 portions), in paste is as follows: 2 parts of sand and 0.5 part of cement (masonry cement) should be mixed thoroughly under dry conditions. Then 1 part of paddy husk ash is added, mixed well, and combined with 1.5 parts of finely-crushed clay. Slurry was made by mixing 3 parts of water and 1 part of cow dung thoroughly and then filtering. Mixing the filtrate and the dry mixture obtained previously should be mixed to obtain the binding paste.

14.6.4   Solar Dryer for Foods at Farm Level

Solar drying relies, as does sun drying, on the sun as its source of energy. Solar drying differs from sun drying in that a structure, often of very simple construction, is used to enhance the effect of the solar insolation. Compared to sun drying, solar dryers can generate higher air temperatures and consequentially a lower relative humidity, which are both conducive to improved drying rates and lower final moisture content of the drying crops. This method has several advantages, such as less spoilage and less microbiological infestation, which leads to improved and more consistent product quality. Solar drying can also be a feasible alternative to those natural convection dryers that use wood or agricultural waste products as fuel. The saving of wood would probably be the main attraction of solar dryers. One of the common solar dryers is the solar tunnel dryer. This is popular due to the considerable reduction of drying time and significant improvement of product quality. Solar tunnel dryers have been used to dry fruits, vegetables, root crops, medicinal parts, and fish.

A passive solar tunnel dryer was designed and constructed at the Department of Mechanical Engineering, University of Moratuwa, Sri-Lanka. The tunnel dryer was tested for its performance using two chimneys, sheet metal, and Polyethylene, respectively (Figure 14.14).

14.6.5 TYPE OF DRYERS USED IN DRYING OF PERISHABLE CROPS

14.6.5.1 Tray Dryers

In tray dryers, the food is spread out, quite thinly, on trays in which the drying takes place. Heating may be by an air current sweeping across the trays, conduction from heated trays or heated shelves on which the trays lie, or radiation from heated surfaces. Most tray dryers are heated by air, which also removes moist vapors. It uses electricity, gas, and biomass as fuel for heating air.

14.6.5.2 Tunnel Dryers

These may be regarded as developments of the tray dryer, in which the trays on trolleys move through a tunnel where the heat is applied and the vapors are removed. In most cases, air is used in tunnel drying and the material can move through the dryer either parallel or counter current to the air flow. Sometimes the dryers are compartmented, and cross-flow may also be used. This dryer also uses electricity and natural gas for heating air.

FIGURE 14.14 Solar tunnel dryer at University of Moratuwa, Sri-Lanka.

14.6.5.3 Pneumatic Dryers

In a pneumatic dryer, the solid food particles are conveyed rapidly in an air stream, the velocity and turbulence of the stream maintaining the particles in suspension. Heated air accomplishes the drying and often some form of classifying device is included in the equipment. In the classifier, the dried material is separated, the dry material then passes out as product, and the moist remainder is recirculated for further drying.

14.6.5.4 Rotary Dryers

The foodstuffs are contained in a horizontal inclined cylinder through which it travels, being heated either by air flow heated by electricity or gas through the cylinder, or by conduction of heat from the cylinder walls. In some cases, the cylinder rotates, and in others the cylinder is stationary and a paddle or screw rotates within the cylinder conveying the material through.

14.6.5.5 Trough Dryers

The agriculture materials to be dried are contained in a trough-shaped conveyor belt made from mesh, and air is blown through the bed of material. The movement of the conveyor continually turns over the material, exposing fresh surfaces to the hot air.

14.6.5.6 Bin Dryers

In bin dryers, the material to be dried is contained in a bin with a perforated bottom through which warm air come from electric, gas, or biomass burning is blown vertically upward, passing through the material and drying it.

14.6.5.7 Belt Dryers

The agricultural product is spread as a thin layer on a horizontal mesh or solid belt and air passes through or over the material. In most cases the belt is moving, though in some designs, the belt is stationary and the material is transported by scrapers.

14.7 OTHER DRYING TECHNIQUES USED IN FOOD INDUSTRY

In addition to the above-explained dryers, there are many type of dryers are available for different drying processers in fruits and vegetable drying and food manufacturing. They are described below:

14.7.1 Freeze Drying

Freeze drying means freezing food at a very low temperature (-50–$80°C$) and drying. Under these conditions the water in food turns into water vapor by sublimation. This process yields the common freeze-dried coffee. It also happens to products (e.g., bread, meat, and vegetables). Freezing the food prevents taste and vitamin loss. However, freeze drying is ecologically unfriendly as freezing consumes a lot of energy. A common side effect of freeze drying is that fibers are damaged during the process, resulting in the brittle structure of freeze dried fruit. After rehydration, such fruit will not regain its original shape and texture.

14.7.2 Spray Drying

Spray drying is a method of producing a dry powder from a liquid or slurry by rapidly drying with hot gas. This is the preferred method of drying of many thermally-sensitive materials such as foods (milk) and pharmaceuticals. This spray drying technique is extensively used in industries for the transformation of wide range of products into powder form. It results in powders with good quality, low water activity, longer shelf life, and ease of transport. Spray-dried fruit juice powders have some inherent problems, such as stickiness and high hygroscopicity, due to the presence of low molecular weight sugars and acids which have low glass transition temperatures.

14.7.3 DRUM DRYING

Drum drying is a method used for drying out liquids from raw materials. In the drum drying process, pureed raw ingredients are dried at relatively low temperatures over rotating, high-capacity drums that produce sheets of drum-dried product. This product is milled to a finished flake or powder form and this operation is similar to a cloth dryer.

14.7.4 VACUUM DRYING

In vacuum drying, the air around the food is removed, causing the rapid evaporation of water inside the food. Due to the speed and the fact that it happens at room temperature guarantees that the taste, color, and nutritional value of the food are preserved. Also, the fibers are fully preserved, so after reconstitution with water, vacuum-dried fruit will reproduce the original texture of fresh fruit. Finally, the process of vacuum drying takes much less energy than freeze drying and is suitable for drying heat sensible, aromatic, agricultural products, such as spices.

14.7.5 MICROWAVE ASSISTED DRYING

Microwaves can be successfully adopted for the drying of food crops and have the advantage of less drying time. Microwaves penetrate into moist food materials, making the water molecules vibrate and generate heat. The disadvantage of microwave fixed-bed drying techniques is the results in the non-homogenous drying of materials. In this method, the penetration of microwaves inside a material strongly depends on the bed properties. The combination of fluidized bed drying methods with microwave heating for drying of moist, granular materials can help use the advantages of both drying methods. Therefore, during microwave-assisted fluidized bed drying, a more homogenous heating occurs inside the bed. Furthermore, previous investigations revealed that the combined microwave-fluidized bed drying led to marked time and energy savings, compared with conventional fluidized bed drying.

14.8 RECENT ADVANCES IN DRYING TECHNOLOGY

The emergence of new research areas in recent years has seen many drying applications not only limited to food or crop materials. The drying of advanced materials in the manufacturing of super-conducting materials or advanced ceramics, where novel spray drying technology has been developed, has obtained the desired electrical and magnetic properties. Another example is impingement jet drying of polymer solution on metal substrates to protect surfaces from corrosion. These are very well-controlled drying operations required by the market.

14.8.1 MULTI-STAGE DRYING

The concept of multi-stage drying is based on the heat and mass transfer characteristics along the drying curve of the material, which shows distinct mechanisms requiring different conditions for optimal drying rates. As a result of this, each mechanism requires different equipment or operating conditions. Multi-stage drying allows for the better control of the entire drying process resulting in higher product quality. Multistage drying systems (MSDS) make it possible to use lower temperature inlet air flows into dryers than in one dryer case. Multistage drying systems make it possible also for the decreased mass flow of drying air into the drying system compared to one dryer case, when inlet temperatures of drying air are the same than in one dryer case.

The multistage drying system enables remarkable savings in the energy costs of drying. It also offers the possibility to use secondary energy from a plant facility as the drying energy.

The first advantage makes it possible to utilize low-temperature secondary energy flows in drying. This improves the energy efficiency of power plants, etc. The second advantage makes it possible to arrange that kind of power boiler MSDS integrate, where it is possible to lead almost exit drying air flow into the combustion process as burning air flow.

14.8.2   SUPERHEATED STEAM DRYING

Replacing hot air with superheated steam in drying was an idea generated long ago and has gained interest in the present day. This is a new, revolutionary technique, which has prompted changes in operational conditions. The addition of new devices and further necessary steps for heating and cooling of products are introduced. Furthermore, either pressure or vacuum process conditions are needed to introduce materials. The basic idea of the process is to increase drying rates by improving moisture mobility in the solid material.

14.8.3   CONTACT-SORPTION DRYING

In this drying method, wet material is in contact with heated particulate medium. In the case of an inert medium, the particle-particle heat transfer induces moisture evaporation and facilitates drying. In the case of highly hygroscopic heated medium, the generated chemical potential gradient between medium and material to be dried, facilitates moisture removal. Large particles in fluidized bed drying could be dried with the introduction of another sorbent material in addition to the hot air. Sometimes this is referred to as two-phase drying. Examples of this method are used for corn drying with heated sand and large particle drying with silica gel. Also, this method was used to heat treat the soybeans for the removal of trypsin and improve digestibility (Raghavan, 1995).

14.8.4   SPOUTED BED DRYING

Spouted beds are modified fluidized beds for coarse particles belonging to Geldart's D group (larger and heavy particles). Particle movement is regular rather than random in fluidized beds. These beds have been very good for drying grains, but in recent times have been used for drying of slurries and pastes. A variation of the spouted bed emerged as the rotating jet spouted bed (RJSB). This method is recommended for drying heat sensitive particles in the falling rate period. RJSB consists of a flat-base cylindrical vessel fitted with slow rotating air distributor plates with one or several radially located spouting air nozzles.

14.8.5   HEAT PUMP DRYING

Heat pumps have been known to be energy efficient when used for drying. The advantages of the heat pump dryers are its ability to recover energy from the exhaust, and its ability to control the drying temperature and humidity. A heat pump is essentially an air conditioning system operated in reverse. The condenser is the main focus of heat exchanger. The major operations in the refrigeration cycle can be identified as follows:

a. Cooling and dehumidification of the air at the evaporator
b. In the compressor, vapor enters into the inlet and is compressed to a higher pressure. At this stage vapor is in a super-heated stage.
c. After compression, the refrigerant vapor is directed to the condenser. At the condenser, refrigerant changes from vapor to liquid. During this process, heat is ejected to the surrounding air.
d. After condensing, the refrigerant expands to reduce the pressure via an expanding device.

In the dryer, inlet drying air passes through the drying bed and picks the moisture from the drying crop material. When moist air touches the condenser, air is first cooled to its dew point. Further cooling results in water being condensed from the air. Latent heat vaporization is then absorbed by the evaporator for boiling the refrigerant. The recovered heat is pumped to the condenser. The cooled and dehumidified air then absorbs the heat at the condenser for sensible heating to the desired temperature.

A properly selected heat pump drying technology is environmental friendly. It is operated in a closed drying circuit resulting in no discharge of gas into the atmosphere. The drawback of the heat pump technology is the low moisture removal rates for atmospheric pressure freeze drying with greater residence times for stationary beds. This problem can be overcome by agitation, fluidization, and intermittent drying (Mujumdar and Alves-Filho, 2003). Any drier that uses convection as the primary mode of heat input can be fitted with a suitably designed heat pump, such as fluid bed dryers. Heat pump fluid bed drying offers better product quality, offsetting incremental increases in drying costs with a high market value of the product (Alves-Filho and Strommen, 1996).

A well-designed heat pump efficiently supplies both heating and cooling, which are required in a drying process. The heat pump evaporator can recover latent, which is heat freely available in the water vapor coming out from the wet material, and which recycle through the condenser. It is possible that it can condensate and recover valuable volatiles or to remove otherwise noxious condensable gases by controlling the surface temperature of the evaporator.

Hence, properly selected heat pump fluid bed technology is also environmentally friendly. Aside from complying with ozone depletion and global warming regulations, it operates in a closed circuit so that no gas, fumes, or fine discharge to the atmosphere and can be treated as a green drying technology.

Heat pump drying has an increasing application to the food industry for the drying of fruits, vegetables, fish, and biologically active products in many countries. The main advantage of using heat pump technology is the energy-saving potential and the ability to control drying air temperature and humidity for a wide range of drying conditions. Using a heat pump system in drying results in both latent heat and sensible heat being recovered from the exhaust air, thus improving overall thermal performance and providing effective control of air conditions at the inlet of the dryer. Heat pump system applications in drying give beneficial advantages. The evaporator recuperates sensible and latent heat from the dryer exhaust; hence, energy is recovered. Condensation of moisture occurring at the evaporator reduces the humidity of the working air, thus increasing the driving force for product drying.

14.8.6 FLUIDIZED BED DRYING

When an air stream is passed through a permeable support (distributor) on which rests the free-flowing material, the bed starts to expand when a certain velocity is reached. The superficial velocity of the air at the onset of fluidization is the minimum fluidization velocity. With a further increase in air velocity, the bed reaches a stage where the pressure across fluids in the bed drops rapidly and product is carried away by the air. The velocity at this stage is known as terminal velocity, which is an important parameter in fluidization operations. The operational velocity must remain between these two velocities.

The use of fluidization is one of the technologies commonly used in drying agro-food and other materials. It is commonly used in freezing systems. Fluidized bed drying has been recognized as a gentle, uniform drying process due to its low residual moisture content and a high degree of efficiency. In a fluidized bed, conditions are favorable for rapid heat and mass transfer, due to a thin boundary layer surrounding the food particles due to very rapid mixing. This is a very convenient method for heat-sensitive food materials, as it prevents them from overheating.

The air flow and its distribution is a primary factor which contributes to efficient fluidization. The air velocity across the bed should be even to achieve the even fluidization at every point across the bed. The proper design of a good air distributor provides this requirement. It is one of the important components in the drying system, which dominates fluidization characteristics, sanitary requirements, and removal of characteristics of the material. Mass and heat transfer is determined also by the bubble characteristics of the fluidizing medium, which is directly related to the design of the distributor.

This method is effective for spherical and small particles. For larger, irregular particles, a suitable method of fluidization has to be selected depending on the drying particle characteristics. There is a practical limit to the fluidizable height for the bed and it is also governed by the particle properties. Moisture movement through the bed of material is dependent on their physical nature, size, and shape. Product characteristics and temperature of drying also have a direct influence on the drying behavior during fluidized bed drying of grains and food materials.

14.8.7   HEAT PUMP ASSISTED FLUIDIZED BED DRYING

Figure 14.15 shows a heat pump assisted fluidized bed dryer used for experimental and research purposes at the University of Queensland. This fluidized bed dryer was a batch type plexiglass fluidizing column of 185 mm inside diameter and length of 1 m. The hot air was taken from a heat pump dehumidifier system, which was attached to the fluidizing column by flexible ducts. The usual bed height of 150 mm was used to prevent the reduction of unfluidized regions in the dryer. The hot air velocity passing through the material bed was kept at a constant value by changing the air controls

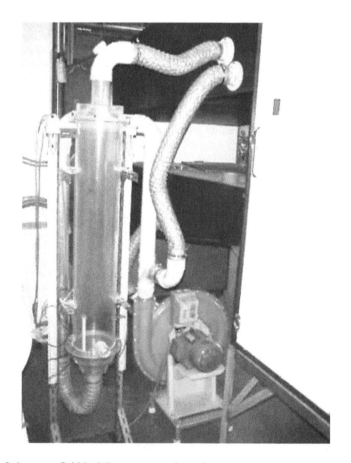

FIGURE 14.15   Laboratory fluid bed dryer connected to a heat pump.

of the fan for any drying experiment. This velocity is selected in such a way, that it has to be within the limit of fluidization and terminal velocity of materials drying and within the capability of the fan. Samples were collected from the dryer at intervals through the sample outlet.

Figure 14.16 shows the layout of the green heat pump dryer with fluidization chamber at the Norwegian University of Science and Technology (NTNU). Figure 14.17 shows the actual drying set up at NTNU.

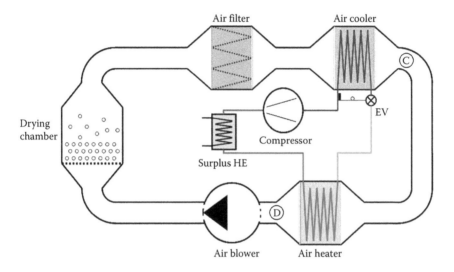

FIGURE 14.16   The layout of the green heat pump dryer with fluidization.

FIGURE 14.17   Heat pump fluid bed dryer at Norwegian University of Science and Technology.

14.9 CONCLUSION

Drying is a major operation in the agricultural and food industry. It is often used as a primary operation for preservation of food materials, or as a secondary process in some manufacturing operations. Drying, a higher energy-consuming process, is also often practiced in the industry using direct thermal energy from the sun or driers and ovens. From the viewpoint of preservation of energy, it is desirable to store food at ambient conditions. This can be achieved by drying operations which reduces moisture content and water activity to prevent spoilage under long-term storage conditions. Key benefits of drying are an increased shelf life and a reduction of product volume and weight, which facilitates easy storage and transportation.

Food drying is a complex process involving mass and heat transfer accompanied by physical and structural changes. The combined effect of higher surface area per unit volume and absence of skin shows higher drying rates in many materials. Materials with skin may experience case-hardening during drying. The expected qualities, such as texture and color at the end of drying, depend on physical changes occuring during the drying process. Drying takes place in the falling period for most of the grains and food materials. Many empirical and fundamental models are also available to describe the drying behavior more accurately. Both the increase in drying rate with drying temperature and diffusion of moisture through the material can be described by an Arrhenius relation.

The selection of a drying method depends on product nature, the situation, and provision of energy input. In terms of cost considerations, energy often becomes a key factor. As most food materials possess higher moisture levels, the energy requirement needed for moisture reduction is considerably high. Therefore, reduction of energy in the process is essential in reducing the drying cost. Moreover, reduction of the external volume or shrinkage observed during drying of some materials can cause an adverse effect on the quality of the products expected by the end-customer. Considering both minimization of energy and achieving quality is essential when selecting a drying method. Significant progress has been made in the recent years in the research and development of drying technology to improve product quality, reduce costs, and improve environmental performance. A number of innovative methods and drying technologies are also being developed.

REFERENCES

Aguilera, J. M. (2003). Drying and dried products under the microscope. *Food Sci Technol Int* 9 (3), 137–143.

Akpinar, E. K., Midilli, A., and Bicer, Y. (2006). The first law and second law analyses of thermodynamic of pumpkin drying processes. *J Food Eng* 72, 230–331.

Alves-Filho, O. and Strommen, I. (1996). Performance and improvement in Heat pump dryers. *Drying 96*. (Ed. C. Strumillo and Z. Pakowski.) Krakow, Poland, 405–15.

Brice, J. R., Currah, L., Malins, A., and Bancroft, R. (1997). *Onion storage in the tropics: A practical guide to methods of storage and their selection*. Natural Resources Institute, University of Greenwich, UK.

Brooker, D. B., Baker-Arkema, F. W., and Hall, C. W. (1992). *Drying and Storage of Grains and Oilseeds*. Van Nostrand, Reinhold, New York.

DAFF. (2005). Australian agriculture and food sector stocktake, Department of Agriculture, Fisheries and Forestry. Accessed November 8, 2016 from http://www.daff.gov.au/:data/assets/pdf_file/0005/186125/Stocktake2005.pdf.

Datta, A. K. (1990). Heat and mass transfer in microwave processing. *J Food Chem Eng Prog* 6, 47–53.

Earle, R. L. (1983). Unit Operations in Food Processing, Accessed May 5, 2016, http://www.nzifst.org.nz/unitoperations/drying1.htm.

Food and Agriculture Organization. (2011). Drying of agricultural crops. Accessed April 24, 2016. http/www.fao.org/drying/docrep.htm.

Grabowski, L. and Verma, L. R. (2003). Drying and tempering effects on thin layer drying of fruits and vegetables. *TRANS. ASAE* 29 (1), 312–9.

Jindarat, W., Rattanadecho, P., and Vongpradubchai, S. (2011). Analysis of energy consumption in microwave and convective drying process of multi-layered porous material inside a rectangular wave guide. *J Exp Therm Fluid Sci* 35, 728–37.

Momenzadeh, L., Zomorodian, A., and Mowla, D. (2010). Experimental and theoretical investigation of shelled corn drying in a microwave-assisted fluidized bed dryer using artificial neural network. *J Food Bioprod Process* 89 (1), 15–21.

Mujumdar, A. S. and Alves-Filho, O. (2003). Drying research-current state and future trends, CDROM Proceedings, Second Nordic Drying Conference (NDC'02), Copenhagen, Denmark, ISBN.

Mujumdar, A. S. and Passos, M. L. (2000). Innovation in drying technologies. *Drying Technology in Agriculture and Food Sciences*. (Ed. A. S. Mujumdar.) Enfield: Science Publishers Inc., 291–310.

Raghavan, G. S. V. (1995). Drying of agricultural products. *Handbook of Industrial Drying, Second Edition*. (Ed. A. S. Mujumdar.) New York: Marcel Dekker, Inc, 627–42.

Rahman. (2001). Change of physical and chemical properties of food during low temperature drying. *J Food Biochemistry* 19 (5), 381–9.

Rahman, M. S. (2005). Dried food properties: Challenges ahead. *Dry Technol* 23, 695–715.

Rahman, M. S. (2013). Physical and structural characteristics of dates. In: Dates: Postharvest Science. *Processing Technology and Health Benefits, First Edition*. (Ed. M. Muhammad Siddiq, S. M. Aleid and A. A. Kader.) New York: John Wiley & Sons, 157–168.

Ranjbaran, M. and Zare, D. (2013). Simulation of energetic and exergetic performance of microwave-assisted fluidized bed drying of soybeans. *Energy*, Accessed July 28 2016, http://dx.doi.org/10.1016/j.energy.2013.06.057.html.

Senadeera, Wijitha. (2009). Minimum fluidization velocity of food materials: Effect of moisture and shape. *Chem Prod Process Model* 4 (4).

Senanayake, D. P. (2007). Drying of durable and perishable crop and dryers, *Technical note iv*, Institute of Post-Harvest Technology, Anuradhapura, Sri Lanka.

Thompson, A. K. (1982). *The storage and handling of onion*. Report of Tropical Products Institute, United Kingdom G160.

Thompson, A. K., Booth, R. H., and Proctor, F. J. (1972). *Onion storage of tropics*. Tropical Science, United Kingdom 14, 19–34.

Wembly, G. and Thomsan, D. (1999). *Grain crop drying, handling, and storage rural structures in the tropics: Design and development*, The Crowood Press Ltd., 363–411.

Zare, D. and Ranjbaran, M. (2012). Simulation and validation of microwave-assisted fluidized bed drying of soybean. *Dry Technol* 30 (3), 236–347.

15 Fruit and Vegetable Packhouse

Technologies for Assessing Fruit Quantity and Quality

Kerry B. Walsh

CONTENTS

15.1 INTRODUCTION

In this chapter, an overview is presented of the operation of a fruit packhouse. Both the major sorting technologies currently in use and areas of development are considered. Thus, the chapter should provide industry context to packhouse industry personnel, students, and researchers interested in post-harvest fruit operations.

To provide some context to the operation of a single packhouse, it is useful to consider the size of the global industry. The International Society for Horticultural Science estimates the world production of fruit and vegetables to be 2.45 billion tons in 2009, of which 0.64 billion tons were fruit (Aitken and McCaffrey, 2012) (Figure 15.1). This fruit production was a 68% increase on 1990 estimates, with the greatest increase being in tropical fruits. Long distance export by road, rail, air, and sea is also increasing, as more affluent consumers in more countries are willing to pay for more

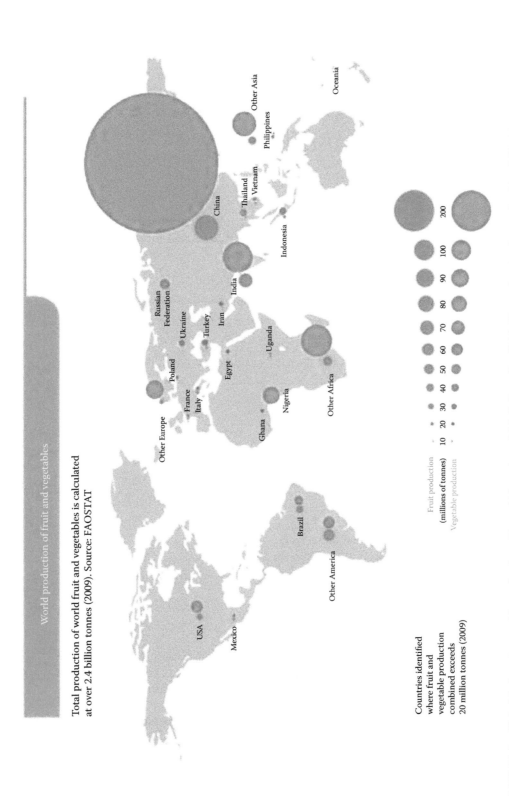

FIGURE 15.1 Global fruit and vegetable production. (From Aitken, A. and McCaffrey, D., 2012, Harvesting the Sun: A Profile of World Horticulture. http://www .harvestingthesun.org [accessed November 15, 2016].)

diversity and for wider windows of availability. About 20 large container ships are required to ship 1 million tons of produce. The "smaller picture" is also thought provoking—e.g., Australian production, almost insignificant on a global scale (not even represented on Figure 15.1), has a farm-gate value of AUS$9 billion, with export valued at $1 billion (Australian Institute of Horticulture Inc., 2016).

Production and distribution of such volumes of fresh produce requires appropriate infrastructure and labor. Indeed, horticultural production is labor-intensive compared to other agricultural production (such as grains or livestock). For example, Australia is known more for livestock than horticultural production, but over one-third of the agricultural workforce in Australia is associated with horticultural production (www.aih.org.au). Of course, the level of infrastructure varies between countries (developing cf. developed) and between the type of producer (small "local food" producer cf. corporate farm), but, with changes in China, the bulk of the world's horticultural production will now involve machinery in the harvest, processing, and transport steps of the value chain (Figure 15.2).

The impact of technology in horticulture is most obvious and most advanced in the modern packhouse. Fruit coming in from a field harvest must be washed, waxed, dried, sorted, packed, etc., with a need for rapid processing, given the perishable nature of fruit, and a need to process large volumes, given the inherently low value of fruit per unit. The packhouse thus employs many types of technology—from the mechanical engineering of bin tippers, conveyors, and box stackers, and the electrical engineering of weight and vision-based sorting systems, to the postharvest biology of cool stores and controlled atmospheres. A technical issue is that fast movements bring acceleration and deceleration, with potential to damage the fruit. Thus, there are compromises between speed and handling damage, and in how the packhouse integrates into the whole farm operation. For example, harvest of less mature, but firmer fruit may result in fruit with lower flavor, but less damage in processing.

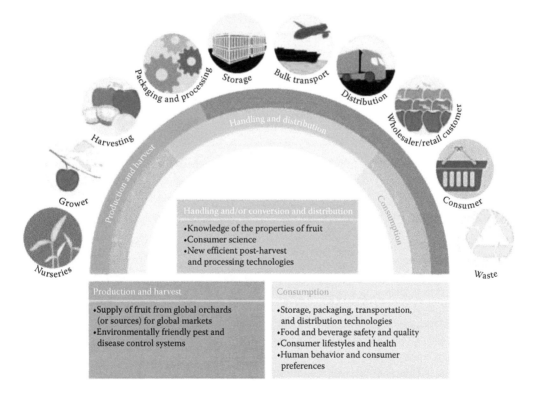

FIGURE 15.2   Value chain in horticultural production. (From Aitken, A. and McCaffrey, D., 2012, Harvesting the Sun: A Profile of World Horticulture. http://www.harvestingthesun.org [accessed November 15, 2016].)

FIGURE 15.3   Twin faces of Janus and of the packhouse manager—engineer and horticulturalist. (From https://upload.wikimedia.org/wikipedia/commons/a/a9/Janus1.JPG)

The aim of this chapter is to describe both technologies and their horticultural application, requiring a blend of the disciplines of engineering and biology. This is necessary, as problems will occur if either the electrical and mechanical engineers designing packhouse equipment fail to appreciate the physiology of the produce to be processed, or if the horticulturalist fails to appreciate the abilities and limitations of packhouse technology. Like Janus (Figure 15.3), the packhouse operator thus has two faces, and must appreciate two disciplines and speak two "languages." Definitions of terms from the two discipline areas are given in Tables 15.1 and 15.2:

The following sections address:
- Packhouse evolution: a summary of changes in packhouse technology over the last century
- A typical packhouse: a description of the operation of a "typical" packhouse, from in-feed, treatments, handling, sorting and grading, storage, and information management
- Transport and repack
- Technology providers
- Conclusions: with focus on trends in development of packhouse technology and functions

15.2   PACKHOUSE EVOLUTION

In a low technology system, fruit are sorted and packed using human labor. For example, even today (2016) the entire citrus crop of Nepal is sorted and packed by hand. Fruit are grown on steep terraces, hand harvested into dhokos (conical bamboo baskets carried on the back) or plastic crates, and then human transported to a road head to be unloaded into the tray of a utility vehicle for transport to a trans-shipping point or wholesale market, where light hand sorting to remove foreign matter or severely damaged fruit can occur (Figure 15.4). Fruit are damaged at each step with rough handling, but the intended storage life and distance of transport is low.

Automated packhouse technologies are relatively recent even for the developed world, as one-hundred years ago all postharvest fruit handling was largely as described for Nepal. Packhouse mechanization has been driven by increasing cost and decreasing availability of labor, increasing scale of operation, increasing attention to care of the fruit to extend shelf life, and an increasingly discerning consumer base with purchasing preferences made on fruit size, color, etc. Indeed, the

TABLE 15.1
Definitions of Some Horticultural Terms for the Engineer

Climacteric and non-climacteric
In climacteric fruit, there is a period of rapid change ("ripening," e.g., in skin color, flesh firmness, conversion of starch to sugar, and production of aromatic volatiles). This process is triggered by the gas ethylene. Non-climacteric fruit do not display this rapid change and are not responsive to ethylene.

Ethylene
A gas (C_2H_4) that is produced within climacteric fruit, triggering ripening, and can be used exogenously to trigger ripening.

Exocarp, mesocarp, and endocarp
In the "typical" simple fruit, the expanded ovary wall is the edible flesh, with the outer (exocarp) and inner (endocarp) layers forming a "skin" and the middle (mesocarp) layer enlarged. But in some fruit, the exocarp and endocarp may be modified.

Fruits and vegetables
Technically, a fruit is a reproductive organ, generally arising from the enlargement of the ovary walls. Thus, a tomato is a fruit, not a vegetable. A vegetable is an edible root, stem, or leaf. Fruit and leaves are generally more perishable than stems or roots.
In colloquial terms, a fruit is something eaten as a dessert, while a vegetable is eaten with a main meal (e.g., a tomato).

Maturity
Fruit are termed "physiologically" mature when they reach a certain point of development, e.g., when able to ripen for a climacteric fruit. "Harvest" maturity is achieved later, when the fruit is further developed, e.g., with higher sugar levels and thus between eating quality, with the compromise of being firm enough to handle well during harvest, packing, and transport. An analogy is reproductive maturity in humans, which is achieved before full maturity.

Quality
Fruit quality is typically assessed in terms of size, weight, skin color, flesh color, firmness, Brix (sugar content of juice), dry matter content, starch staining pattern (to indicate level of maturity in apple), and volatile level (smell).

Senescence and ripening
Following harvest, fruit and vegetables are essentially dying (sensencing) slowly. The task of the supply chain is to manage this process—to extend it. In climacteric fruit, there is a period of pronounced change (the fruit "ripens," e.g., banana) while in non-climacteric fruit, the senescence is steady (e.g., pineapple).

Simple, multiple, and aggregate fruit
A simple fruit is the product of one inflorescence (e.g., a bean or peach). A multiple fruit is the product of an inflorescence in which the fruit from multiple flowers have merged (e.g., pineapple). An aggregate fruit is the product of one flower with multiple ovaries that maintain separation (e.g., raspberry).

Berry, stonefruit, and hesperidium
Fleshy fruit are categorized by features such as the number of carpels (seed sacs) and the modification of tissues. In stonefruit (e.g., peaches), the endocarp is "stony." In citrus fruit (hesperidium), the exocarp develops oil glands and long extensions that forms juice sacs.

Storage conditions
To extend the shelf life of fruit and vegetables, temperature should be reduced (specific to fruit type) and humidity maintained (to minimize water loss). High carbon dioxide and low oxygen levels are effective in extending storage life of some fruit. In these, gas levels are actively controlled (oxygen scrubbers, carbon dioxide additions, and control circuitry), the storage is said to be a "controlled atmosphere." If the gas levels are altered through the natural respiration of the fruit and restriction to gas diffusion (e.g., fresh salad in plastic packs), the storage is said to be a "modified atmosphere."

development of packhouse technology development in the last 50 years has been spectacular. Curiously, few descriptions can be found of these systems in the horticultural literature, likely because advances represent translation of technologies from other disciplines (e.g., in camera hardware and machine vision) and have been undertaken by commercial companies rather than universities or research institutes.

TABLE 15.2

Definition of Some Engineering Terms for the Horticulturalist

ADC

Analogue to digital convertor, a device which converts an analogue signal (mV, mA) to digital information.

CCD–CMOS

Two types of camera technology. In general, CMOS is replacing CCD, but CCD maintains use in some applications (e.g., low light).

Encoder

An encoder measures the position of an object, for example, a rotary encoder can be used to measure the number of turns of a cog, or a fractional turn.

Fluorescence

Some materials will fluoresce at a longer wavelength when excited by light of a shorter wavelength. It is useful to detect some defects, e.g., oil gland damage in citrus.

Fruit conveyance

Fruit can be moved by flotation and water movement (a flume), by use of rollers, a belt or cup, or finger conveyors. Cup conveyors can be designed in various forms to suit different commodities or sensor systems.

Load cell

Used in the measurement of weight (an imposed load changes the electrical resistance of a metal object), e.g., in weight of fruit in cup conveyors, with typical accuracy to 1 g.

Machine vision

The capture of imagery by CCD or CMOS cameras, with associated algorithms to process the images to identify shape, color, defects, or other characters.

Near-infrared imaging

Visible light covers wavelengths from 400 (blue light) to 700 nm (red light). Light of wavelengths between 700 and 1100 is termed far-red or short wave near-infrared. The larger range 700–2500 nm is termed infrared. CCD and CMOS cameras are sensitive to 1000 nm. A cut-off filter can be used to exclude wavelengths greater than 700 nm to make the cameras record visible light only. Conversely, wavelengths under 700 nm can be excluded—such images are useful for bruise detection in fruit.

Near-infrared spectroscopy

Measurement of the absorption of light by fruit at specific wavelengths in the short wave near infrared is termed "spectroscopy." It can be used in the estimation of fruit attributes such as Brix or dry matter content.

Neural network

A type of machine learning algorithm, in which a machine develops a set of rules when presented with a series of inputs and desired outputs (e.g., images of fruit features and desired classifications as defects).

PLC–SCADA

A programmable logic controller (PLC) is a ruggedized computer for control of a process replacing hard wired relays and timers. A supervisory control and data acquisition system (SCADA) is a software system involving collection of data from remote locations to control a system. A SCADA may involve use of PLC units.

Solenoid

Packlines typically employ an electromagnetic device (solenoids) to tip the cup or fingers of a conveyor at the desired drop point.

Human sorting of fruit can occur on color, size, and shape by eye and on firmness by hand feel. The mechanization of fruit grading (Walsh, 2005, 2014) began with mechanical size graders, e.g., diverging belt graders or rotating drums (Figure 15.5). In a diverging belt grader, fruit travels on two belts held in a V, such that the fruit falls between the diverging belts at some point in its travel on the system, rolling to one side on a sloping side board. A human can then pack fruit based on size. In a rotating drum system, fruit enters a rotating cylinder, meeting holes of increasing size as the fruit travels the length of the cylinder. Thus, smaller fruit will drop through the cylinder earlier than larger.

FIGURE 15.4 Fruit sorting in Nepal.

(a) (b)

FIGURE 15.5 (a) Rotating drum and (b) diverging belt circa 1920. (From http://collections.museumvictoria .com.au/items/766625)

Obviously, such systems work best for spherical fruit. A later technology involved use of mechanical counterbalance through weighted tipping buckets, with the bucket carrying a piece of fruit tipping if the fruit exceeded the counterbalance weight. Such systems could be employed with non-spherical fruit.

Electronic load cells were first used in fruit packlines for the weighing of individual fruit in the 1970s. The electronic load cell (Figure 15.6a) allowed for both more accurate and precise weighing (e.g., to 1–2 g) and more rapid processing than with the counterbalance weight system. In an electronic system, the load sensor output is fed to a control system which, in turn, controls the drop of fruit through control of electric solenoids which tip conveyor cups (Figure 15.6c). Other sensors can be added to such a control system, and so it was that by the late 1970s RGB cameras were being used for grading of fruit based on size and color (Figure 15.6b).

The packline provides a structured, controlled environment for imaging of fruit. A rapid evolution of imaging systems on packlines has occurred since the 1980s, in terms of use of background (conveyor) colors to maximize contrast with the fruit and thus simplify edge detection, lighting uniformity and lamp type, camera technology, use of mirrors to view the sides of fruit or rotation of fruit passing through the field of view of the camera, etc. By the 1990s, machine learning algorithms allowed for blemish detection and categorization, e.g., distinguishing a unacceptable skin mark.

(a)   (b)

(c)

FIGURE 15.6   (a) Load cell; (b) RGB camera used on a packline; (c) solenoid for tipping a conveyor cup.

In the new century, further sensor technologies were implemented on pack lines, particularly for the detection of internal defects.

Similar to other industries, as product complexity increases, there is a trend for rationalization of the number of manufacturers, often involving a series of acquisitions. Indeed, there once was a manufacturer of simple grading equipment in almost every major growing region, with examples often to be found now only in local equipment museums. As volume of production increased and with that the need to increase rate of grading, and as market requirements for fruit of certain sizes or colors tightened, more grader capabilities were required, driving an increasingly sophisticated design. Additionally, a need developed for ancillary equipment such as bin unloaders, hydrocoolers, drying tunnels, singulators, and barcode labelers. Current generation multilane graders are sophisticated instruments, costing approximately a hundred thousand dollars per lane. The grader of tomorrow will be smarter and faster—smarter in the sense of more sensors involving new technologies, such as magnetic resonance imaging or X-ray imaging, faster in terms of handling more fruit per lane and involving more automated functions in all aspects of operations.

So, what is likely to happen to citrus sorting in Nepal? For as long as the market is local, and the return low, little change is expected. But change is occurring—a population shift from rural to urban brings the rise of a retail form supermarket in the major cities, which requires consistently sized fruit with a certain shelf life. Suddenly, there will be a need for a packhouse—a simple start involving a table to remove defective fruit and sizing by hand to satisfy a clients' specifications on fruit quality. Next, the possibility to market greater volumes of fruit at better returns, e.g., into Tibet, will develop, and with that the need to inspect all fruit, and so the need for faster grading will evolve. With increased throughput volume will come the need for more handling efficiency, bin tippers to unload fruit and brush transfers of fruit between sections of the conveyor system to avoid bruising, and then a wash to remove dirt, a wax or fungal dip, a drier, a size grader, a weight grader, a color grader, a vision defect grade, a sweetness grade, post-harvest treatments such as ethylene induced de-greening, and equipment such as baggers, fruit labellers, carton labellers, etc.

The end point of this trend in technology deployment is a "lights out factory," that is, a packhouse operating without humans. Of course, this elimination of human labor brings social change, the end of an influx of hundreds if not thousands of people into a production region for the harvest season, earning and spending money, giving low-skilled employment to youth and to unskilled labor (often with repatriation of money to developing countries). But there is nothing new in this trend—the mechanization of grain harvesting took thousands off farms in the Industrial Revolution of the 1800s. As for the Industrial Revolution, such change brings the need for workers with a different skill set, more technical capability in servicing the equipment, and more skill in use of the information and capabilities to develop new markets. Further discussion of this trend can be found at Produce Business UK (2016).

15.3   A TYPICAL PACKHOUSE

In a "typical" farm operation, fruit is harvested into field bins, which are transported with some attention-to-ride quality (e.g., road smoothness, use of trailers with spring suspension) to the packhouse. On arrival, the field bins may be consigned to immediate processing or storage (temporary or long-term). Removal of heat from the produce is key to shelf life, so short-term storage should at least involve shading of the bins, and overnight and longer storage should involve active cooling.

Typical subsequent steps include:

- Transfer from field bins, e.g., a water dump
- A transition elevator consisting of rollers to lift fruit out of the water dump
- While on rollers, produce may be subject to a high pressure wash to remove surface dirt
- Passage through a drying tunnel employing forced hot air
- Transition from rollers to a flatbed conveyor
- Human sorting on attributes not handled well by the machine grading system
- Use of a singulator to transition fruit from a flatbed to a cup conveyor
- Machine grading of fruit based on weight, color, size, etc.
- Drop of selected fruit from conveyor to a cross belt
- Packing of fruit to cartons

The elements of a packline can be configured in various ways, as can the layout of the pack-line and other functions (e.g., transport corridors, cold stores, office areas, etc.) within a packhouse (Figure 15.7). Indeed, every design will be unique to a mix of factors, including the commodities to

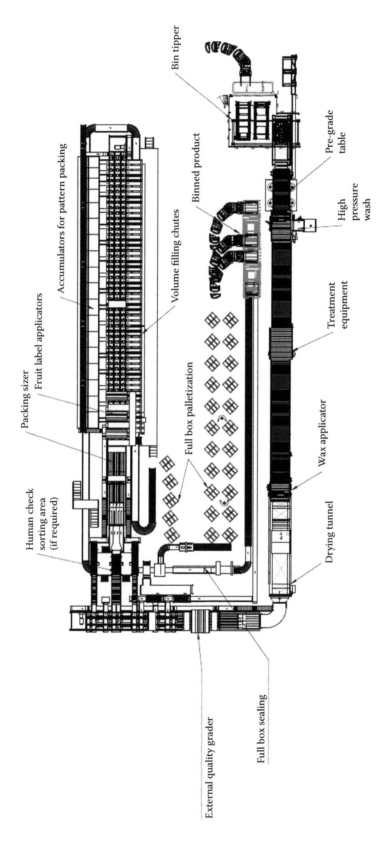

FIGURE 15.7 Schematic of a "typical" packline.

be graded; particular treatment or packing issues; constraints on location of services, such as ventilation, lighting, power supply, and transport; and consideration of corridors of movement, from field bin entry and storage, to carton supplies and to cool stores (Balls, 1986). For a discussion of these factors in context of kiwifruit grading, see Brough and Spark (1988). Optimization of layout therefore requires close collaboration between shed owner/operator and the manufacturer.

Given the cost of the investment and the impact on operations, decisions on packhouse plants must be made carefully. Following the discussion with the equipment supplier, a proposed CAD drawing of equipment layout might be followed by site visits from the equipment supplier to confirm requirements and limitations.

Some details on the typical steps employed in a packhouse include:

1. Infeed and treatment processes
2. Material handling
3. Sorting and grading
4. Labeling
5. Storage and treatments
6. Information flow

15.3.1 INFEED AND TREATMENT

Fruit and vegetables arriving to the packhouse from the field will be in a variety of container sizes, from 10 kg crates to 1 ton bins. The first challenge on beginning processing is to unload the field bins. A slow, controlled dump into a water bath is a gentle method, but alternatives include a brush roller or curtain to decrease the acceleration of the fruit, so to limit bruising forces (Figure 15.8a). In some cases, a truck may be equipped with a moving floor (conveyor) for discharge.

A chilled water bath or spray is a very effective way of removing heat from produce, given the thermal conductivity of water (hydrocooling), as well as serving in removal of some debris and in transport of the fruit (Figure 15.8b and c). A water bath can also be used in transport of fruit and as a medium for chemical treatments such as fungicides (Figure 15.8d). Typical following steps can include waxing (spraying of a wax containing solution) and drying (Figure 15.8e and f).

15.3.2 MATERIAL HANDLING

The movement of fruit through a packhouse, rapidly but without damage to the fruit, is an exercise in mechanical engineering that has attracted much commercial attention. For example, a simple Google Patent Search (October 5, 2016) on the term "fruit conveyor" yields 91,870 results. This attention has brought dramatic improvements in conveyor technology over the past decades. For example, consider conveyor belt stretch—a chain carrying cups in a cup conveyer may be 100 m in length, and with its return loop the chain carrying the cups will be 200 m. Given the thermal linear expansivity of steel (approximately 0.00001), an increase in temperature of 20°C will involve an increase in length of a 100 m steel rod by approximately 0.25 m. Such a stretch of a conveyor belt will obviously impact cup position, and thus time from the sensor system to drop point. A new approach in conveyor systems involves the use of RFID tags on cups, rather than use of encorders on cog rotation to ensure the drop of a fruit at the correct position.

Mundane, but important issues in conveyor operation include the taring of all cups in the system for weight, regular cleaning of the system with use of a good food grade oil, and the avoidance of the use of glass in all aspects of the system exposed to fruit. A primary challenge is the avoidance of forces that cause fruit bruising—with bruises often not visible for some days after the damage event. Balls instrumented with accelerometers can be passed through the entire handling chain to log the forces experienced (e.g., Bollen, 2006).

FIGURE 15.8 (a) Unloading of citrus fruit from field crates using a brush roller to slow fruit; (b) From this point the citrus fruit move on the roller bed into spray treatments, then singulation to a cup system for sorting on color, defects, size, and weight; (c) Chilled water cooling of fruit in incoming field crates; (d) fruit can be transported by a water current; (e, f) Fruit may be sprayed with waxing formulations or fungicide treatments, then dried. (Images courtesy of MAF.)

Some typical conveying systems include flat beds, roller beds, singulators, cup conveyors, roller-cups, and more exotic systems, such as those used for grape bunches (Figure 15.9). The roller-cups allow for rotation of the fruit in the cup, and thus a camera view of all of the fruit. Cups can be designed to accommodate different-sized fruit, from cherries to watermelons, and for different sensor systems (e.g., a finger design allows for detection of light passing through a fruit).

Post-sorting, fruit move to stations for packing to cartons or bags. These cartons may be packed by fruit number (Figure 15.10a and c) or weight (Figure 15.10b). Finally, the packed cartons are stacked onto pallets (Figure 15.10d). Automated systems may be used in packing and carton stacking, and in pallet movement within the packhouse.

FIGURE 15.9   (a) Transition from water bath to roller bed; (b) transition from flat bed to singulator belts; (c) transition from singulator belts to a piano key conveyor; (d) transition from roller cups to finger cups; (e) drop from finger cups to brushes and unloading to a cross belt; (f) grape bunch conveying system. (Images courtesy of MAF.)

15.3.3   Sorting and Grading

Typically, purchasers of fruit and vegetables set specifications on produce quantity (number, weight, and size) and quality (surface color, external blemishes, and internal defects). Thus, packhouses employ technologies that allow automated assessment of these attributes. Abbot (1999) produced a seminal review of noninvasive technologies for the sorting of fruit. This review has been subsequently updated by others (e.g., Ruiz-Garcia *et al.*, 2010; Kondo, 2010; Walsh, 2015). A brief description follows:

Flotation

Flotation can be used for density grading of fruit. This technique can be used for separation of fruit with spongy tissue or voids (e.g., in watermelons). The specific gravity of the solution in the flotation tank can be adjusted for better separation of defect and sound fruit, however, maintenance of solution concentration is required. The technique is not commonly used, with calculation of density more easily accomplished from fruit weight and machine vision estimated volume.

FIGURE 15.10 Vacuum-assisted lift for packing of (a) cucumbers and (d) citrus to cartons. Bag filling to weight with (b) apples. (c) Automated stacking of cartons to pallets. (Images courtesy of MAF.)

Load Cell

Fruit weight can be measured by use of load cell technology, with the cup containing the fruit skidding across the load cell and load cell output averaged over a short interval. There are several types of load cell technology—of these, the "button and washer" strain gauge is preferred for both accuracy and precision. The principle of a load cell is based on the change in the electrical resistance of a conductor when a strain is applied. The typical load cell employs a Wheatstone bridge configuration of four load cells, with output in the order of a few millivolts and accuracy of approximately 1 g, typical for most produce grading systems. Cup weight is tared while running the conveyor empty of fruit.

CCD and CMOS Cameras

A silicon charge coupled device (CCD) or complementary metal oxide semiconductor (CMOS) arrays are used to capture images through conversion of light to electrons. These sensors will respond to visible (approximately 400 to 750 nm) and also far-red (to 1100 nm) wavelengths, with cut-off filters placed within cameras to either include or exclude either visible or far-red light. CCD arrays have offered better sensitivity in the UV and far-red regions, and were preferred for low light conditions. CMOS technology holds the advantage of higher readout speeds, lower power consumption, lower fabrication costs, and better dynamic range, and with continuing improvements, CMOS will effectively replace CCD technology except for a few niche applications. The simple voltage output of each element (pixel) of a CCD or CMOS array is used to create a greyscale image. To create a color camera, pixels within the array are masked with red, green, and blue filters, typically in a Bayer pattern, with loss of some spatial resolution.

High-speed RGB and far-red CMOS cameras are commonly employed in fresh produce grading. A structured lighting environment can be created on a packline, with white LEDs typically used in combination with a reflective hood to create diffuse and uniform lighting conditions. Angled mirrors may be used to allow imaging of the sides of the fruit, and a conveyor color selected for maximal contrast to fruit color. The far-red cameras allow better discrimination of certain defects, e.g., bruises. This is presumably due to changes in free water availability in the fruit, with water absorbing at wavelengths around 840 and 960 nm (due to a third and second overtone O–H bond stretching vibration, respectively).

Machine vision algorithms allow image "segmentation," i.e., identification of fruit related pixels followed by "blobbing," i.e., grouping of pixels associated with fruit to create objects. Fruit image segmentation is typically achieved using features such as color, texture (variance between adjacent pixels), shape (perimeter features), and size (blob pixel count), to identify objects in an image (Payne and Walsh, 2014). Typical applications in fresh produce grading include estimation of size, color, shape, and surface defects. User interfaces allow for operator control of tolerances for a given grade/drop point (e.g., Figure 15.11).

Visible Near-Infrared Spectroscopy

The use of near-infrared spectroscopy (NIRS) for assessment of fruit was reviewed by Nicolai *et al.* (2007) and Herold *et al.* (2009). Fruit assessment systems typically use a partial or full transmittance geometry, with measurement of absorbance for wavelengths between 400 and 1000 nm. The technology has been used in assessment of "macro-constituents" of fruit, such as dry matter content and Brix (also known as total soluble solids, TSS, or soluble solids content, SSC) in thin-skinned fruit (apple, mango, etc.; e.g., Golic and Walsh, 2005), with a typical error of measurement of around 1%. Some claims have also been made for the measurement of fruit acidity and firmness. Firmness is a physical characteristic, related to cell turgor and cell wall characteristics, which are unlikely to be measured directly by NIRS, but an indirect correlation may exist (Subedi and Walsh, 2008). For example, immature fruit may have both higher firmness and chlorophyll levels, such that a spectroscopic assessment of firmness may be based on an assessment of chlorophyll level. Titratable acidity levels in juice of most fruit are low, around 1% weight per volume, so NIRS measurement of this attribute is also likely to be indirect (e.g., immature fruit may have lower acidity, higher chlorophyll, and lower Brix; Subedi *et al.*, 2012). Acidity can be directly measured by NIRS when acid levels are high (e.g., in lemon). Internal defect detection is also possible, e.g., of internal browning and water core of apple (Francis *et al.*, 1965; Upchurch *et al.*, 1997).

Technology uptake for sorting on "positive" quality attributes such as DM and TSS has been relatively low, a consequence of lack of market reward and losses incurred in rejecting fruit. Greater uptake success has been achieved for sorting of "negative" attributes (internal defects such as apple internal browning or water core; e.g., Khatiwada *et al.*, 2016).

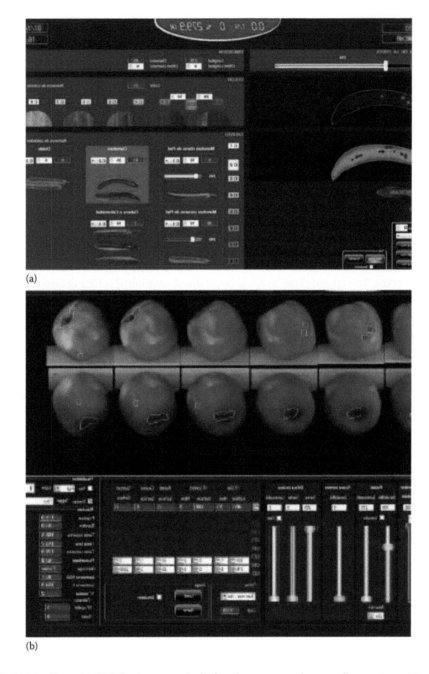

(a)

(b)

FIGURE 15.11 Example GUI for image analysis function on a produce grading system: (a) grading of cucumbers on shape (curvature), color, and defect; (b) Grading of apple on size, color, and defect. (Image courtesy of MAF.)

Typically, NIRS on-line systems employ either a partial or full transmission system (light directed to one part of the fruit and detector viewing another), rather than reflectance. In reflectance geometry, the detector views an illuminated area of the fruit. Signal level is high, but most of the signal is specular reflection (carrying no information on internal properties of the fruit), or very near surface diffuse reflections. A consequence is that any change in surface (skin) properties between batches of fruit affects the accuracy of prediction.

Commercially available equipment varies in light source (halogen, LED, or laser), optical geometry, detector figures of merit (e.g., wavelength range and resolution [full width half maximum], stray light level, and signal to noise), and type of chemometrics (e.g., partial least squares regression, multiple linear regression, support vector machine regression, and discrimination analysis) employed. The bottom line for a user, however, is a test of the technology on a new set of fruit, representative of fruit to be graded in the future.

The "typical" in-line NIRS system using an interactance geometry involves a "point" measurement, i.e., a measurement made at one point on the fruit. However, given a conveyor belt speed of 1 m.s^{-1} and a 30 ms integration time for acquisition of a spectrum, the fruit will have moved by 30 mm during the measurement, and thus a "point" measurement is effectively an average of at least part of a fruit. A full transmission geometry will effectively involve optical sampling of most of the fruit, as light also scatters in passing through the fruit.

Alternatively, imaging systems are available that capture spectra of many points on a sample (Sankaran *et al.*, 2010). Such systems are useful if the distribution of an attribute in the fruit is important. For example, Helios (2016) report use of a "chemical imaging" system for differentiation of blueberry fruit from fruit stalk. Hyperspectral imaging systems are also marketed for sorting of potato, in terms of identification of foreign objects and in identification of high sugar within cut strips (e.g., http://www.key.net and http://www.perception-park.com/sorting-potatoes-soil-stones). The reverse trend is use of only a few wavelengths, simplifying instrumentation. For example, the "IDD4" unit (MAF, France) utilizes only four wavelengths for on-line detection of fruit defects such as internal browning of apple, while the "DA-head" (Turoni, Italy) utilizes two wavelengths in assessment of fruit chlorophyll content.

X-ray Radiography and Tomography

An electron beam is used to generate X-rays. Simple transmission images using high energy or "hard" X-rays (generator voltage greater than 100 kV) allows for the detection of dense objects (e.g., stones) in fresh-cut salad packaging operations, while lower energy X-rays (20–50 kV) are applicable for the detection of voids in fresh produce. X-ray line scan systems are used in detection of voids in potato tubers (e.g., BEST, 2014), as described by Finney and Norris (1978). Claims have also been made for the detection of a number of disorders, including watercore and internal rot (e.g., Schatzki *et al.*, 1997), but commercial use is lacking. Commercial use of simple transmission X-ray imaging requires implementation of machine recognition of image features (e.g., Shahin *et al.*, 2001) and consideration of cost and the health regulations surrounding use of X-ray generators.

An application issue is that both object thickness and density affect X-ray transmission, and fruit are not of consistent thickness. X-ray computed tomography (CT) involves imaging from several perspectives to create a 3D image of a sample. Such images provide great detail on internal structure (e.g., watercore in apples; Shahin *et al.*, 1999), but the scan times are too slow and cost too high for in-line sorting. However, a recent patent proposed a fusion of camera based 3D shape analysis and line scan X-ray transmission to construct density images (Van Dael *et al.*, 2016), methodologies applicable to on-line sorting of produce.

Time or Spatial Resolved Spectroscopy

Light passing through a fruit is subject to absorption or scattering. The extent of scattering will be impacted by changes in intercellular air space distribution, and this may be related to features such as fruit firmness. Two methods have been proposed to assess scattering: time-resolved reflectance spectroscopy, in which a very short laser pulse (in picoseconds) of photons is incident onto a fruit as a point source, and the temporal profile of photons traveling through the fruit to a detector is recorded (more scattering, greater time); and spatially resolved spectroscopy, in which a continuous laser is directed onto the fruit as a point source, and the spatial profile of light re-emitted from the fruit is measured (e.g., Seifert *et al.*, 2014). This work remains experimental, with no consistent relationship to firmness yet demonstrated.

Fluorescence

In fluorescence, a material is excited by a particular wavelength of light, and a longer wavelength is emitted. In fluorescence imaging, a pulse of excitation wavelengths is followed by measurement of emitted wavelengths, typically using a CCD camera. An example in-line application is the assessment of damaged oil glands in citrus fruit. In a more sophisticated analysis, aspects of the fluorescence of chlorophyll can be used to infer information on the status of the photosynthetic system, and thus on the health of the plant tissue. Such systems are used in controlled atmosphere storage (e.g., http://www.harvestwatch.net), but have not seen use on packlines.

Magnetic Resonance and Resonance Imaging

A useful review of the application of this technology to fruit grading is provided by McCarthy and Zhang (2012). In magnetic resonance spectroscopy (NMR) and magnetic resonance imaging (MRI) the atomic nucleus is induced to spin by a radio frequency pulse, followed by detection of the radio signal emitted by the nucleus "relaxing." Information on the whole sample is produced in NMR, while 2 or 3D images are produced using MRI. Various nuclei (C, P, and H) can be excited, and several fruit quality parameters can be measured simultaneously. Most fruit MRI imaging has involved assessment of water content and mobility of the sample. An interesting "fly through" image of a citrus fruit can be found at Ellison (2016).

This technology has not seen commercial adoption due to limitations in speed and cost. However, magnet technology is advancing, with higher field strengths available at lower cost (Zhang and McCarthy, 2013). A prototype MRI unit mounted on a conveyor belt, grading fruit in batches of 10 to 12 per second has been produced by Aspect Imaging (Israel).

Firmness by Force-Deformation and Acoustic Measures

The search for a technology for on-line assessment of fruit firmness is something of a Holy Grail, as firmness is a key produce quality attribute. García-Ramos *et al.* (2005) and Khalifa *et al.* (2011) have produced useful reviews of this area.

As a consumer, you will have assessed firmness by hand feel, based on deformation under an applied load. An "air puff" on a fruit can represent an applied force, with a laser system to measure deformation. Alternatively, an accelerometer can be used to measure the deceleration force when a light impact head strikes a fruit (Chen and Ruiz-Altisent, 1996). The Sinclair "iQ" (2014) and Greefa "iFD" (intelligent Firmness Detector) were on-line versions of this technology, with the air actuated bellows of the Sinclair fruit labelling system delivering the accelerometer onto the fruit. Neither unit, however, is commercially available.

Another approach relies on acoustic response (Misrach, 2008). The dominant or resonant frequency of an object is related to its mass and density. Aweta offered equipment in which the fruit was lightly tapped, with nearby microphones collecting sound vibrations off the fruit for assessment of the resonant frequency (Walsh, 2015). Implementation into online grading will likely suffer issues related to the level of vibration noise present in such an operation.

15.3.4 LABELING

Following grading, individual fruit and cartons must be labeled. Given a retailer sells fruit as individual units, there is also requirement for individual labeling of fruit. The label typical holds a four digit PLU (price lookup number) that defines the type of produce—making for easy identification at the retailer checkout. A recent development has been the addition of a fifth digit to the PLU number to encode for GMO (genetically modified organism) or organic produce.

The task of delivering a paper label (with food grade adhesive) onto every piece of produce on a packline with fruit moving at up to 10 per second is not trivial. The dominant supplier in the produce labeler market is Sinclair (www.sinclair-intl.com), with a patented air actuated bellows technique for

(a)

(b)

(c)

ISTOCKPHOTO.COM

(d)

IMAGE COURTESY OF SUNKIST

FIGURE 15.12 Paper labelling system (a) and labelled fruit (b). (From http://www.jtechsystems.com.au /Catalogue/fruit-and-vegetable-labels) Laser etching system (c) and etched fruit (d). (From http://www .foodqualityandsafety.com/article/laser-etching-safe-for-labeling-fruit)

delivering labels (Figure 15.12a). A recently released alternative technology involves laser etching of the fruit surface (Figure 15.12b).

15.3.5 Storage and Treatment Facilities

Packhouse operation is generally inherently inefficient, in that the packhouse infrastructure services a crop with a narrow harvest window. Thus, the investment in plants is heavily utilized for a short window of time each season. For example, the typical mango packhouse might operate for six weeks in the year. Much greater efficiencies can be achieved if fruit can be stored. For example, with use of controlled atmosphere storage, apple packhouses can effectively operate throughout the year.

A packhouse typically incorporates a fan-forced cold storage facility for removal of heat from produce before shipping, and for short-term storage of fruit. A list of storage temperatures recommended for a given commodity can be found at Shipping Australia (2007).

For some commodities, other requirements include:

- Longer term storage in controlled or modified atmosphere
- Ripening or de-greening facilities involving ethylene
- Conditioning facilities involving ripening at 20°C
- Insect disinfestation treatment, involving either vapor heat treatment or gamma irradiation

These facilities need not be geographically closely associated with a packhouse. A useful review on treatments employed across South African fruit industries is provided by Dodd *et al.* (2010).

A controlled atmosphere (CA) involves the control of atmospheric composition in a storage room, typically decreasing oxygen to around 2% v/v and increasing carbon dioxide to 10% v/v (e.g., https://www.maxtend.com.au/). This gaseous composition results in a slowing of the physiological processes of the fruit, enabling longer storage. For example, apples are routinely stored in CA for nine months and longer. However, if oxygen levels fall too low, the fruit will develop defects, including off flavors and internal browning. Modified atmosphere packaging (MAP) involves allowing the respiration of the produce alter atmospheric composition within a container with a surface of known permeability to oxygen and carbon dioxide.

Climacteric fruits are fruit that ripen after harvest, such as bananas, mangoes, tomatoes, and avocados. Such fruit will ripen when exposed to ethylene, a natural ripening hormone. Other fruit, such as citrus, are not strongly climacteric, but nonetheless respond to ethylene in terms of the de-greening of skin. Ripening or de-greening facilities involve the exposure of fruit to ethylene. The ethylene may be generated from the decomposition of ethephon, by bleed of gaseous ethylene from a compressed supply, e.g., BOC Australia (2016), or from a catalytic convertor producing ethylene from ethanol. The required treatment time is dependent on the crop and the time of season, but a typical treatment involves 24 hours with an ethylene concentration of 10 μL/L.

Other treatments may be required to meet biosecurity requirements for specific destinations. Chemical treatments, such as methyl bromide to ensure consignments are insect free, have fallen out of favor. Common current treatments involve gamma irradiation, vapor heat treatment (VHT), and extended hot water dips. Such treatments, however, can cause some damage to fresh produce, reducing shelf life. Gamma irradiation involves dosing of pallets of fruit with gamma radiation, with the side benefit of delaying ripening (Figure 15.13). Produce that has been irradiated must carry the international symbol of irradiation, the "radura" (Figure 15.13a) and carry wording that the material was treated with irradiation, with either individual labeling or with a label next to the sales container. Cobalt 60 and Cesium 137 are radioactive isotopes used as gamma ray sources. Vapor heat treated (Figure 15.13) produce does not require labeling, but in general shelf life is shortened more than with irradiation. The treatment involves precise control of temperature and humidity (to 60°C and 95% RH, with control to 0.1°C and 0.1% RH) and precise timing of exposure.

15.3.6 Information Flow and Data Analytics

The most basic information flow is that from the harvest crew to the packhouse on the progress of harvesting, for the purpose of improving packhouse function, i.e., for packline scheduling. This requires information on crop maturity and speed of harvest (e.g., how many field bins are coming to the packhouse on a given day). An ideal scenario might include barcoded field bins with a load cell providing weight and wireless transmission of data from the field to the packhouse. This information flow can be automated, as seen in the sugar cane industry, in which many farms supply a single processing plant with a perishable commodity (the cut cane billets), such that harvest scheduling is critical. Groups such as AgTrix (www.agtrix.com) supply services that monitor the progress of cane harvesters in real-time. For some tree crops, dry matter content or Brix can be used as maturity or quality indicators, and decision support systems exist for field monitoring of these variables, informing the decision on when to harvest, e.g., Felix Instruments Inc. (2016).

Packhouse data is useful for inventory control and sales, e.g., associating batches to orders. The data can also be of use to feed "backward" to inform crop agronomy and "forward" to inform marketing. For example, data on packout rates or incidence of a particular defect fruit can be provided on a bin basis to inform the farm agronomist and/or for grower payment purposes (e.g., Gillgren, 2001). For example, data on the incidence rate of a defect such as internal browning in apple can be tied back to storage and pre-harvest conditions.

Of course, the packhouse is just one part of the supply chain from orchard to consumer. Efficient operation of this chain requires information flow—increasingly, information is required by value

(a)

Left: fruit stored at 20°C, 14 days, after no irradiation
Right: fruit stored at 20°C, 14 days with 640Gy

(b)

(c)

FIGURE 15.13 (a) "Radura," symbol of irradiation; (b) mango fruit 14 days after irradiation with 640Gy (from Co60)—free of fruit fly but suffering some lenticel damage; (c) a vapor heat treatment facility. (From https://commons.wikimedia.org/wiki/File:Radura-Symbol.svg; http://www.industry.mangoes.net.au/resource-collection/2016/3/6/another-piece-in-the-irradiation-puzzle; http://www.vaporheattreatment.com)

chain members for the "downstream" of the packhouse and its immediate customer. As a large packhouse acts as a produce consolidation point, gathering produce from many farms, and distributing this to many wholesalers or retailers, it is an important player in this information flow. Indeed, the fresh produce industry has been globalized—just look at product of origin labels on fresh fruit in your local retailer!

One issue with such long chains is that of ensuring product quality and safety, with greater consequences for failure given the scale of transactions. Further, social media now highlights failures, with consumers responding rapidly and widely. Indeed, consumer expectations for a response are now more immediate and adamant. Admittedly consumer expectations for consistency in fresh produce can seem unrealistic, with the consumer base now well-separated from rural life and its realities. Nonetheless, in a market led system, the "customer is always right," and thus produce supply chains must have information ready to respond to "events" and maintain consumer confidence. Governments and retailers also respond to consumer concern through stricter regulatory regimes (e.g., GlobalGAP, PTI [Produce Traceability Initiative], HACCP, ISO, British Retail Council, and FarmSafe; Walsh, 2015), so companies operating across jurisdictions need measurement technologies and IT systems that relay information (Compac Australia, 2015).

These needs drive the development of record keeping and traceability. For example, all incoming lots to a packhouse can be barcoded, allowing the produce to be followed from "paddock to plate," with managers able to view the history of movement of each unit. This allows for product recall if necessary and for detective work on the cause of a problem. An example is provided by FarmSoft software (Farm Software Review, 2016). Field information can be recorded in such a system, e.g., spray diaries, with non-compliant harvests rejected at the packhouse (e.g., if a chemical spray has been made within the regulated withholding period). An example of a current provider in these areas is MuddyBoots (Muddy Boots, 2015).

Downstream information can also be tied in. For example, produce temperature influences postharvest shelf life during transport and storage. Indeed, produce quality can be predicted from temperature history. Temperature logging devices placed within loads can communicate via wireless networks to centralize information and allow control of ripening during transport. For example, the "Xsense" system from Stepac P/L provides satellite-based location information and temperature data on produce consignments in transit, feeding a model which estimates remaining shelf life (BT9 2016). A similar technology is offered by the Web ID System (2016). Managers can use such information to redirect consignments, should transport conditions cause estimated shelf life to change markedly.

A key enabling technology is the development of wireless networks (Internet of Things, or IoT) for data access from instrumentation in orchards and in transport modules, e.g., review by Ruiz-Garcia et al. (2009). Another enabler is the developments of artificial intelligence and deep learning systems—while it is still too early to appreciate their impact, it is certainly an area to watch over the next decade, and with it the rise of commercial providers for data warehousing, sharing, and analysis. For example, Kaloxlos et al. (2014) reports on an attempt to build a "promiscuous" data management system that enables data analytic modules to be added in a system comparable to a mobile phone App store, enabling a user to customize the system to their needs.

15.4 TRANSPORT AND REPACK OPERATIONS

Only two generations ago, fresh fruit and vegetables were largely produced and consumed locally. Fresh produce chains have become truly global, with even developing countries receiving consignments of produce from growing areas "half a world" away. This is possible through the efficiency of road, rail, sea, and air transport, and the development of storage systems that maintain produce shelf life. Keys to this success are produce hygiene, temperature control, and controlled or modified atmosphere.

A useful review of transport of fresh fruit and vegetables in refrigerated containers (Figure 15.14) is provided in the Shipping Australia (2007) code. In brief, a refrigerated shipping container is an insulated box with typical external dimensions of 6.06 (length) by 2.438 (width) by 2.438 or 2.591 (height) m; or 12.192 (length) by 2.438 (width) by 2.591 or 2.895 (height) m (referred to as 6 and 12 m containers), with an associated refrigerated air supply. The refrigeration units operate at 440 V/60 Hz at sea, but at 415 V/50 Hz within Australian ports. These units can maintain produce temperature, but cannot reduce produce temperature effectively. Therefore, produce should be close to its ideal transport

FIGURE 15.14 (a) 12 m "reefer" (refrigerated container). (From http://www.cnc-line.com/services/reefer); (b) Emirates "White Container" for air freight, which uses dry ice for cooling. (From http://zooxel.com /skycargo/sometimes-it-takes-a-cool-low-tech-solution-to-get-perishables-to-market/; Courtesy of Envirotainer, http://www.envirotainer.com/en/top-Download/Photos/Container-solutions)

temperature before loading to the container. The 6 m containers have a nominal rate of heat leakage of approximately 25 W/°C temperature differential between the outside and inside of the container wall. When the external temperature is above 20°C, it is recommended that the majority of the refrigerated air should be directed over the internal container walls. If the refrigeration unit is run with doors open to ambient air, condensation of moisture will occur on evaporator coils, possibly causing ice up and malfunction after door closure. Condensation on packaging, known as "cargo sweat" can weaken fiberboard packaging. In short, long distance transport of a perishable good is risky, and requires good management.

Long distance (inter-continental) transport has also brought a need for a special case of packhouse operation, the "repack" operation. This involves repack of fruit after long distance transport or long-term storage. For example, fruit arriving in Europe from South Africa may be subject to repack, separating out damaged fruit, or sorting on stage of ripening to provide the consumer with a product that is "ripe tonight."

15.5 TECHNOLOGY PROVIDERS

The automotive industry provides a very visible example of trends in manufacturing. Before Henry Ford, the automotive manufacture was a cottage industry, often undertaken as an extension of a horse carriage fabrication business. With the arrival of the process assembly line came the advent of specialist providers, and with a fall in cost per unit, and previous providers were forced out of the market. Increasing sophistication of the product then led to a global consolidation of the market (e.g., General Motors, Ford, Toyota, etc.). However, new entries offering either lower cost of production (think of Great Wall Motors) or new technologies (think of Tesla Motors) can arrive to disrupt the market.

So it is with technology for the packhouse. A little over a century ago, such technology was new, and every significant fruit growing region would have a local provider of bespoke equipment. With the increasing sophistication of product, providers consolidated, first on a regional and then on a national level. Markets with high labor costs (e.g., Australia) were driven to adopt packhouse automation early. With increasing sophistication of product came the consolidation of manufacturers, developing economies of scale in production and servicing. Conversely, niche providers have emerged making special technology offers, and some low technology or low-cost providers have survived to supply the smaller packhouses.

The internationally dominant manufacturers (with country of headquarters) include:

- Aweta (the Netherlands)
- Compac/Taste Technology/TOMRA (New Zealand)
- Greefa (the Netherlands)
- MAF (France)
- Unitec (Italy)

15.6 A CASE STUDY

To consolidate the above discussion, consider the packline schematic shown in Figure 15.7. Fruit arrives in bins from the field. A bin tipper drops the fruit into water within a receival hopper. A creep feed elevator delivers the fruit onto a roller bed that carries the fruit past a pre-grading station. This unit may be as simple as rollers with a spacing that allows undersized fruit to drop though the gaps, moving to a cull bin. Some human sorting may also be done, e.g., to remove diseased fruit that would otherwise contaminate the line. The main line of fruit then moves through a brusher/high pressure spray treatment. This will remove scale insects and surface molds in citrus, or surface fuzz in peaches, etc. The next station involves treatment for post-harvest diseases, and would typically involve a hot water/fungicide combination. Following this treatment, fruit moves into an applicator of a food grade, water-based emulsion of wax. The wax will reduce fruit water loss and improve fruit aesthetics, and can be used as a carrier for postharvest fungicide. To dry the emulsion, fruit are typically exposed to air at 49°C for approximately 2.5 minutes, with air typically heated using a gas furnace. This completes the "pre-sizing operations."

Fruit are then singulated onto three lanes of cup conveyors, moving at a rate of up to 1 m/s, or approximately 10 fruit per second, moving through an inspection box, with white LEDs providing uniform illumination, and two cameras viewing each lane ("external quality grader"). Fruit are rotated while moving, so visible light cameras can view all sides of the fruit. The images from these color cameras can be used to estimate fruit color, presence of external blemishes, volume, and coupled with load cell weight information, density. Another set of cameras, equipped with 700 nm long pass filters provide a shortwave, near-infrared image, which allows better detection of fruit contour, and of bruises. The overlay of color and near-infrared images is also useful for distinguishing certain features, e.g., stalks and blemishes. Reject fruit (due to skin blemishes, etc.) are dropped a line that carries the fruit back, parallel to the input line, ending as "binned product." The fruit are then moved by a cross conveyor past a "human check" station. This may only be manned occasionally, when a defect become prevalent that is not recognized by the camera system. Rejects from the human check join the rejects from the external quality grader in passage to the binned product. On the main line, fruit move over a load cell to acquire weight information (packing sizer), and then labelers place a sticker on each fruit. The human operator sets criteria (color, weight, density, defects, etc.) for each fruit drop, with solenoids triggering the tipping of cups to drop fruit from the main line to either stations for pattern packing, or chutes for volume filling.

The aim in a fruit pack is to avoid compression bruising, but have fruit firmly placed, such that movement and bruising does not occur during transport. For pattern packing, uniformly sized fruit are allowed to accumulate into a single layer of the desired pattern, then a set of vacuum cups are used to transfer fruit into a carton. For volume filling, fruit flow into a carton until a desired weight is achieved, then the box is briefly vibrated to settle the fruit. Carton depth is typically at least three to four times the diameter of the fruit. Cartons will be labelled with information on the crop (locality, variety, grade, and date).

Cartons then move to stations for palletization, before movement to cool storage, awaiting transport to market.

15.7 CONCLUSION—TRENDS

15.7.1 The Last Decade

In summary, the great trend across both developed and developing nations is the declining availability of farm labor, and thus packhouse labor, or at least labor available at an affordable price given the prices achieved for produce. This trend has provided a driver for continued automation of packhouse functions, with extension into blemish detection, into carton packing and carton stacking to pallets.

An ancillary trend is the need for product traceability and for inventory control for increased efficiency in packhouse and transport operations. This trend includes use of barcoded field bins within a data management system that maintains information on packout rates (per field bin) and stock on hand (e.g., http://www.gvcustomsoftware.com.au/fruit-packing.aspx). RFID armbands have been used to record employee productivity (e.g., in carton packing). Use of mobile devices to share information from and to forklift drivers is nascent, but holds promise for improved productivity.

15.7.2 The Future Packhouse

In the near future, advanced packhouses are likely to continue investment in:

- Additional automation in carton filling and palletisation, with RFID tracking of pallets
- Automated pallet movements, e.g., https://www.transbotics.com/applications/pallet-handling
- Deployment of robots alongside human workers in fruit grading or packing (e.g., the Baxter robot)
- Improved logistics software
- Adoption of sensors for detection of internal defects such as far-red transmission systems for internal browning in apple
- Development in the machine learning space for estimation of external blemishes on a range of commodities

In the long term, expect new sensor developments for fruit internal defects, e.g., the Van Dael *et al.* (2016) system for internal density variation. Other technologies to watch are those around volatile detection, either the so-called e-noses or chemical based monitors (tags placed in fruit clamshells, which change color in the presence of a specific volatile), and the use of terahertz or magnetic resonance spectroscopy for estimation of attributes such as surface pesticides or sugar content. A Holy Grail for the fruit sorting industry is a reliable, non-destructive measure of fruit firmness, but success in this field is difficult to predict. Other trends include the movement of erstwhile packhouse functions "upstream" into the field or "downstream" into transport, as discussed in 15.7.3.

15.7.3 Moving the Packhouse to the Field

There is a saying that in prediction of future trends we tend to overestimate the amount of change in the next five years, but underestimate the amount of change that will occur in the next 20 years. For example, from a 2017 perspective it appears that driverless car technology is upon us, but adoption will likely be behind technical capacity, as social or legal issues remain to be resolved. Such issues will slow adoption for general road use, with earlier use expected in restricted environments, such as farms and packhouses.

Technology developments will allow for the replacement of human labor in field operations, as it already largely has in the packhouse. In essence, technology developments will allow a shift in functions currently carried out in the packhouse to the field. The following are areas to watch in terms of future developments:

- In-field machine vision measures of fruit number and size
- In-field measures of plant status that are related to fruit quantity or quality, such as spectral indices (e.g., The Eye, cool-farm.com)
- Automated and selective in-field harvesting
- In-field automated fruit treatment functions (hydrocooling, protective treatments, etc.)

Low-cost machine harvesting of fruit may allow the return of some artisanal practices. For example, current commercial production typically involves an attempt to produce uniform fruit, picked with non-skilled, low-cost labor in one harvest pass (optimally), followed by sorting of fruit in the packhouse. Artisanal fruit production involves selective picking of tree-ripened fruit, but these practices requires multiple harvest passes and high labor costs. Machine vision guided automated harvesting could be used for selective picking of ripe fruit, allowing a return to old practices, and removal of defective fruit, translating a packhouse function into the orchard.

15.7.4 Moving the Packhouse Downstream

Other functions are likely to move out of the packhouse, downstream into the transport chain. For example, sealed containers used in transport allow for controlled temperatures and modified atmospheres. Thus, in-transit ripening is possible.

Such a future relies on reliable monitoring of temperature and gas composition while the product is in transit. Already systems are available that provide in-transit monitoring of temperature (BT9, 2016) and gas concentrations (Felix Instruments, 2016).

ACKNOWLEDGMENT

MAF Oceania Pty. Ltd. is acknowledged for use of images.

REFERENCES

Abbot, J. A. (1999). Quality measurements of fruits and vegetables. *Postharvest Biol Technol* 15, 207–25.

Aitken, A. and McCaffrey, D. (2012). Harvesting the Sun: A Profile of World Horticulture. http://www .harvestingthesun.org (Accessed November 15, 2016).

Australian Institute of Horticulture Inc. (2016). Australian Production of Fruits. http://aih.org.au/ (Accessed October 5, 2016).

Balls, R. C. (1986). Packhouse design and operation. *Horticultural Engineering Technology*. (Ed. R.C. Balls), 1–49. http://link.springer.com/chapter/10.1007/978-1-349-07099-2_1 (Accessed November 15, 2016).

BEST. (2014). http://www.bestsorting.com/sorting-food/sorters/ixus-bulk-x-ray-sorter/ (Accessed July 12, 2014).

BOC Australia. (2016). *The Linde Group*. Retrieved from http://www.boc-limited.com.au/en/index.html (Accessed October 5, 2016).

Bollen, F. (2006). Technological innovations in sensors for assessment of postharvest mechanical handling systems. *International Journal of Postharvest Technology and Innovation*. http://www .inderscienceonline.com/doi/abs/10.1504/IJPTI.2006.009179 (Accessed October 5, 2016).

Brough, A. K. and Spark, S. A. (1988). Kiwifruit packhouses [online]. Conference on Agricultural Engineering Institute of Engineers 88 (12), Australia, 1988, 430–34. http://search.informit.com.au/documentSummary ;dn=669687027204131;res=IELENG ISBN: 0858254115 (Accessed October 5, 2016).

BT9 Intelligent Supply Chain Information. (2016). *The Xsense System*. Retrieved from http://www.bt9-tech .com/xsense-system-layout/ (Accessed February 14, 2016).

Chen, P. and Ruiz-Altisent, M. (1996). A low-mass impact sensor for high-speed firmness sensing of fruits. Proc International Conference Agricultural Engineering. Madrid, Spain. Paper 96F-003.

Compac Australia. (2015). The top 10 food trust issues impacting fruit packhouses. Retrieved from http://content.compacsort.com/blog/the-top-10-food-trust-issues-impacting-fruit-packhouses (Accessed September 5, 2015).

Dodd, M., Cronje, P., Taylor, M., Huysamer, M., Kruger, F., Lotz, E., and Merwe, K. V. D. (2010). A review of the post-harvest handling of fruits in south africa over the past twenty-five years. *South African Journal of Plant and Soil* 27 (1), 97–116.

Ellison, A. (2016). Interactive Fruits and Veggies. http://insideinsides.blogspot.com.au/p/3d-interactive-fruits -and-veggies.html (Accessed October 5, 2016).

Farm Software Review. (2016). Post harvest traceability, inventory, packing, and sales solution from farmsoft. Retrieved from https://farm-software.net/category/fruit-packing-software/ (Accessed October 5, 2016).

Felix Instruments Inc. (2016). From farm to fresh market, food quality instrumentation. https://www .felixinstruments.com/ (Accessed February 5, 2017).

Finney, E. and Norris, K. (1978). X-ray scans for detecting hollow heart in potatoes. *Am. J. Potato Res.* 55, 95–105.

Francis, F. J., Bramlage, W. J., and Lord, W. (1965). Detection of watercore and internal breakdown in delicious apples by light transmittance. *Proc. Am. Soc. Hortic. Sci.* 87, 78–84.

Garcia-Ramos F. J., Valero, C., Homer, I., Ortiz-Canavate, J., and Ruiz-Altisent, M. (2005). Non-destructive fruit firmness sensors: A review. *Spanish Journal of Agricultural Res.* 3, 61–73.

Gillgren, D. (2001). Finding the fruit: A spatial model to assess variability within a kiwifruit block. SIRC 2001. The 13th Annual Colloquium of the Spatial Information Research Centre University of Otago, Dunedin, New Zealand. http://citeseerx.ist.psu.edu/viewdoc/download?doi=10.1.1.579.9599&rep=rep1&type=pdf (Accessed March 7, 2016).

Golic, M. and Walsh, K. B. (2005). Robustness of calibration models based on near-infrared spectroscopy to the in-line grading of stonefruit for total soluble solids. *Anal. Chim. Acta.* 555, 286–91.

HELIOS. (2016). Chemical Imaging Systems. Retrieved from http://www.hyperspectral-imaging.com/products .html.

Herold, B., Kawano, S., Sumpf, B., Tillmann, P., and Walsh, K. B. (2009). Chapter 3: VIS/NIR Spectroscopy. *Optical Monitoring of Fresh and Processed Agricultural Crops*. (Ed. M. Zude), 141–249. ISBN 978-1-4200-5402-6 Boca Raton, Florida: CRC Press. http://www.sciencedirect.com/science/article/pii/S0924224409002611.

Kaloxlos, A., Groumas, A., Sarris, V., Katsikas, L., Magdalinos, P., E. Antoniou, E., Politopoulou, Z. *et al.* (2014). A cloud-based farm management system. *Computers and Electronics in Agriculture* 100, 168–79.

Khalifa, S., Komarizadeh, M. H., and Touisi, B. (2011). Usage of fruit response to both force and forced vibration applied to fruit firmness: A review. *Aust. J. Crop Sci.* 5, 516–22.

Khatiwadi, B., Subedi, P., Hayes, C., Cunha-Carlos, L. C., and Walsh, K. B. (2016). Assessment of internal flesh browning in intact apple using visible, short-wave, near-infrared spectroscopy. *Postharvest Biol. and Technol.* 120, 103–11.

Kondo, N. (2010). Automation of fruit and vegetable grading system and food traceability. *Trends in Food Science and Technology* 21, 145–52.

McCarthy, M. J., and Zhang, L. (2012). Food quality assurance and control. *eMagRes.* doi:10.1002/9780470034590.emrstm1295.

Misrach, A. (2008). Ultrasonic technology for quality evaluation of fresh fruit and vegetables in pre- and post-harvest processes. *Postharvest Biol. Technol.* 48, 315–30.

Muddy Boots. (2015). Supply chain insight, from grower to retailer. (Accessed 2016). Retrieved from http://en.muddyboots.com/about/index.

Nicolai, B. M., Beullens, K., Bobelyn, E., Peirs, A., and Saeys, W. (2007). Nondestructive measurement of fruit and vegetable quality by means of NIR spectroscopy. *Postharvest Biol. Technol.* 46, 99–118.

Payne, A. and Walsh, K. B. (2014). Machine vision in estimation of crop yield. *Plant Image Analysis: Fundamentals and Applications.* (Ed. S. Dutta-Gupta and Yasuomi Ibaraki.) CRC Press.

Produce Business UK. (2016). Compac shifts gears in race to raise the packhouse technology bar. http://www.producebusinessuk.com/services/stories/2015/12/15/compac-shifts-gears-in-race-to-raise-the-packhouse-technology-bar (Accessed October 5, 2016).

Ruiz-Garcia, L., Moreda, G. P., Lu, R., Hernandez-Sanchez, N., Correa, E. C., Diezma, B., Nicolai, and Garcia-Ramos, J. (2010). Sensors for product characterization and quality of speciality crops: A review. *Comp. Elect. in Agric.* 74, 176–94.

Sankaran, S., Mishra, A., Ehsani, R., and Davis, C. (2010). A review of advanced techniques for detecting plant diseases. *Comp. and Elect. in Agric.* 72, 1–13.

Schatzk-Broughi, T., Haff, R., Young, R., Can, I., Le, L., and Toyofuku, N. (1997). Defect detection in apples by means of X-ray imaging. *Trans ASAE* 40, 1407–15.

Seifert, B., Zude, M., Spinelli, L., and Torricelli, A. (2014). Optical properties of developing pip and stone fruit reveal underlying structural changes. *Physiol. Plant.* 153, 327–336 doi: 10.1111/ppl.12232.

Shahin, M., Tollner, E., Evans, M., and Arabnia, H. (1999). Water core features for sorting red delicious apples: A statistical approach. *Trans ASAE* 42, 1889–96.

Shahin, M., Tollner, E., and McClendon, R. (2001). AE—Automation and emerging technologies: Artificial intelligence classifiers for sorting apples based on watercore. *J. Agric. Eng. Res.* 79, 265–74.

Shipping Australia. (2007). Transport code of practice for handling fresh fruit and vegetables in refrigerated shipping containers. https://shippingaustralia.com.au/wp-content/uploads/2012/03/Code-of-Practice-Draft-2007.pdf (Accessed November 10, 2016).

Sinclair. (2014). *IQ System.* http://www.sinclair-intl.com/pages/iq_main.html (Accessed July 10, 2014).

Sonego, L., Ben-Arie, R., Raynal, J., and Pech J. (1995). Biochemical and physical evaluation of textural characteristics of nectarines exhibiting woolly breakdown: NMR imaging, X-ray computed tomography and pectin composition. *Postharvest Biol. Technol.* 5, 187–98.

Subedi, P., Walsh, K., and Owens, G. (2007). Prediction of mango eating quality at harvest using short-wave near-infrared spectrometry. *Postharvest Biol. Technol.* 43, 326–34.

Subedi, P. P. and Walsh, K. B. (2008). Noninvasive measurement of fresh fruit firmness. *Postharvest Biol. Technol.* 51, 297–304.

Subedi, P. P., Walsh, K. B., and Hopkins, D. W. (2012). Assessment of titratable acidity in fruit using short-wave near-infrared spectroscopy. Part B: intact fruit studies. *J. Near Infrared Spectro.* 20, 459–63.

Upchurch, B. L., Throop, J. A., and Aneshansley, D. J. (1997). Detecting internal breakdown in apples using interactance measurements. *Postharvest Biol. Technol.* 10, 15–19.

Van Dael, M., Lebosta, S., Herremans, E., Verboven, P., Sijbers, J., Opara, U. L., Cronje, P. J., and Nicolai, B. M. (2016). Automated quality control and selection. PCT/EP2016/055718.

Walsh, K. B., Golic, M., and Greensill, C. V. (2004). Sorting of fruit and vegetables using near-infrared spectroscopy: Application to soluble solids and dry matter content. *J. Near Infrared Spectro.* 12, 141–8.

Walsh, K. B. (2005). Commercial adoption of technologies for fruit grading, with emphasis on NIRS. FRUTIC 5, Montpellier, France. http://www.symposcience.net/exl-doc/colloque/ART-00001679.pdf.

Walsh, K. B. (2014). *Chapter 9: Postharvest Regulation and Quality Standards on Fresh Produce, in Postharvest Handling—A Systems Approach.* Third edition. (Ed. W. J. Florkowski, R. L. Shewfelt, B. Brueckner, and S.E. Prussia.) Elsevier ISBN 978-0-12-374112-7, 205–45.

Walsh, K. B. (2015). Nondestructive assessment of fruit quality. *Advances in Postharvest Fruit and Vegetable Technology*. CRC Press series on Contemporary Food Engineering. (Ed. R.B.H. Wills and J.B. Golding.) Elsevier, 40–61.

Web ID System. (2016). Wireless Sensor Monitoring Solutions. Retrieved from http://www.webidsystems.com.au/temperature_sensors.html (Accessed October 5, 2016).

Zhang, L. and McCarthy, M. J. (2013). Assessment of pomegranate post-harvest quality using nuclear magnetic resonance. *Postharvest Biol. and Technol.* 77, 59–66.

Section IV

Computer Modeling

16 Applications of Crop Modeling for Agricultural Machinery Design

Kenny Nona, Tom Leblicq, Josse De Baerdemaeker,
and Wouter Saeys

CONTENTS

16.1   INTRODUCTION

Agricultural machines have significantly changed the ways of harvesting fibrous biological materials such as wheat, barley, maize, and grass. In the nineteenth century, harvesting a hectare of wheat required large amounts of manual labor to cut, bind, transport, thresh, and clean the grain. Nowadays, this is done by large combine harvesters that have harvest capacities of almost 10 tons of grain per hour (Guinness World Records, 2014). In Europe and North America, this spectacular evolution in farming efficiency is thanks to the development of agricultural machinery. This evolution has made that a small fraction of the active farmer population is sufficient to provide food for the rest of the population. Notwithstanding, the spectacular capacities that are already realized, and the innovations in agricultural machinery currently still largely focus on further increasing the capacity and efficiency in terms of reducing the working time, harvesting losses, and required driving power. However, as the size of the machinery is approaching the limits for road transport, the focus is turning toward optimization of the internal processes to maximally exploit the available capacity.

To reach this goal, most agricultural machinery producers focus on in-field testing and tuning of existing designs. During the short field test season, these designs are evaluated in a large number of crops and conditions. At the end of the harvesting season, a decision is made with respect to the adjustments which will be introduced to the market. In the next test season, the new machine is again taken to the field where the testing and adjustment continues.

As a result, only a limited number of adaptations can be tested each year, such that changes in practice are limited to incremental adaptations with a high chance of success. However, due to the complex nature of the interaction between the crop and the machine components, it is not straight-forward to optimize the machine performance through trial and error. The suggested improvements are, therefore, limited to small changes, and the tuning is typically limited to a part of the entire machine. As a result, the combined action of the individually optimized components may be sub-optimal. Besides the high costs for building each concept and testing and tuning it in as many field conditions as possible, the quality of the performed tests will also vary between and within field test seasons. A very rainy test period, for example, results in testing the machine concepts on more wet crops only. This means that the suggested concepts have only been validated for the specific field conditions which were available at the time of testing, while the effect of other conditions on the adjusted design could not be tested.

Other industrial sectors have successfully reduced their development time and cost by replacing a large part of the prototype building and intensive testing by *in silico* development of new concepts. For example, Miller (2014) demonstrated the value of a model-based design methodology in the development of wind turbines, while similar added value was also shown for the design of pumps (Kim *et al.*, 2012). However, this concept is still hardly used in the design of agricultural machinery.

A paradigm shift from in-field testing and tuning through trial and error to simulation-based development could allow to evaluate new designs in a more time- and cost-efficient way. Moreover, the simulation-based design and optimization would allow for the evaluation of the entire process, including the total energy requirement and the building material usage. This could be established by developing a cost function that combines all these specifications. In this way, different concepts could be optimized iteratively and compared with respect to these specifications to select the most promising ones. Through sensitivity analyses, such simulations could also be used to identify the parameters in a design which has the largest influence on this cost function, leading to better machine control.

In silico design of agricultural machinery implies modeling the effect of a change in the machine settings on the crop in the machine. This requires a material model that accurately describes the behavior of the biological material inside the machines. While such material models are well-established for engineering materials, such as steel and polymers, accurate material models for fibrous biological materials are still lacking. The absence of such accurate material models is con-sidered the main reason why such simulation tools are still hardly used in agricultural machinery design.

The increasing usage of renewable energy and material sources makes fibrous biological materials, such as straw and hay, increasingly important for a farmer's income. So, these materials are increasingly harvested and transported from the fields to the farms. Balers are used to increase the density of these materials with the aim to facilitate their transport and storage. Moreover, the diversity of materials for which these machines are used is rapidly increasing. For example, nowadays *Miscanthus*, rapeseed, and even flax fibers are processed by balers. As farmers want to use the same machine to harvest these different materials, the next generation of balers should have a high and robust performance in all of these crops and conditions. This makes the current practice, which solely relies on in-field testing of new designs, even more impractical and inefficient. As a result, the demand for *in silico* tools for harvesting machinery design is rapidly increasing.

In this chapter, the state of the art in crop modeling is reviewed with a focus on fibrous biological materials (such as straw, grass silage, hay, and corn stalks) and some perspectives for the future are presented. First, some basic properties of the compression of fibrous biological material will be discussed. Then, two different approaches for modeling this behavior are discussed at the bulk level and at the particle level. Each approach will be illustrated with an application in agricultural machinery design.

16.2 COMPRESSION BEHAVIOR OF FIBROUS BIOLOGICAL MATERIAL

16.2.1 STAGES IN COMPRESSION

The compression behavior of materials is typically described in terms of the required force per area (stress) for applying a certain relative deformation (strain) to the material or vice-versa, the resulting strain from applying a certain stress. However, in the case of fibrous biological materials, this stress-deformation behavior is also time-dependent. Therefore, it is more appropriate to describe the material behavior in the stress, deformation, and time domains.

Many researchers have measured stress-strain curves for a wide range of biological materials. Most of them quantified this behavior by linking each component of the stress vector to the strain vector. A commonly returning entity is the ratio of the required stress increase $d\sigma$ to the applied relative change in volume dV/V_0. This ratio is known as the bulk modulus K (Mohsenin, 1986):

$$K = -\frac{d\sigma}{dV/V_0} \qquad (16.1)$$

In which V_0 is the initial volume of the material sample. For most fibrous biological materials, K is not constant, but changes with density (O'Dogherty and Wheeler, 1984). The reason for this behavior of the bulk can be found on a smaller scale, namely on the level of the particles which are intertwined to form this bulk. In the first stage of the compression (at low bulk densities), the compression stress mainly overcomes the inter-particle and particle-wall friction (Kaliyan and Morey, 2009). Particles thereby move across each other, and the pore space volume is strongly reduced. The high porosity also allows bending of particles, which increases their entanglement (Leblicq *et al.*, 2015). This stage is characterized by large deformations at small pressures (Hu *et al.*, 2009). In Faborode (1986), this stage is called the inertial stage due to the large particle movements. Once the particles are in close contact, it becomes very difficult to press out more air, resulting in a large increase in the compression stress and the required specific energy for compression (Adapa *et al.*, 2009). In this stage, the individual particles undergo elastic and plastic deformations, thereby further increasing the inter-particle contact (Hu *et al.*, 2009). At even higher pressures, this results in particle breakage and mechanical interlocking of the particles (Mani *et al.*, 2004). Finally, when the pores in the particles collapse, the bulk density approaches the density of the material's fibers.

O'Dogherty *et al.* (1995) reported densities for intact straw particle walls between 151 and 192 kg/m^3, dependent on the maturity, while Stelte *et al.* (2012) reported final pellet densities between 750 and 900 kg/m^3 for compression temperatures between 30 and 100°C. The definition of the bulk modulus (Equation 16.1) can be used to calculate the effort required to compress these particles. This compression effort is proportional to KAL.

The required effort in the inertial phase is calculated as the kinetic energy for moving the particles. Faborode and O'Callaghan (1986) related this kinetic energy to the density as follows:

$$E_k \sim \rho A L v^2 \tag{16.2}$$

where, ρ is the material's density, A is the area of the plunger, L is the height of the sample, and v is the applied plunger speed. Faborode and O'Callaghan (1986) defined the Cauchy number N_c as the ratio of the movement and compression efforts:

$$N_c = \frac{\rho v^2}{K} \tag{16.3}$$

In the first stage of compression, the Cauchy number will increase as the density change is dominated by movement of the particles with respect to each other. Later on, compression of the particles becomes dominant over particle movement, such that the Cauchy number decreases again. The maximal Cauchy number, therefore, indicates the start of particle deformation and hence the end of the inertial stage. The density at this point is called the critical density. Faborode and O'Callaghan (1986) found critical densities of chopped hay and barley straw around 300 kg/m^3, depending on the initial density and the maximum applied pressure. Ferrero *et al.* (1990) confirmed this by reporting the start of compressing the stalk parenchyma of wheat and barley straw to be at 310 kg/m^3.

16.2.2 TIME-DEPENDENT BEHAVIOR

The force required to compress fibrous biological material depends on the rate at which the material is compressed. Therefore, fibrous biological materials have both rate dependent and rate independent properties (Sitkei, 1986). Where the rate independency is assumed to mainly originate from the particle deformations, the rate-dependency is assumed to relate to internal friction and expulsion of air from between and within the particles. The rate-dependency is therefore more distinct at higher densities. Moreover, the applied compression speed is important for the final obtained material density. Indeed, the speed determines the time the particles have to rearrange themselves during compression: when compressing to a similar stress, a higher compression speed will typically result in a lower bulk density (Kanafojski and Karwowski, 1976).

Besides the rate-dependent increase in the stress-strain ratio during compression, Mohsenin (1986) and Sitkei (1986) also observed a time-dependent behavior of the materials. When a constant pressure is applied, the bulk density will keep increasing over time (Kanafojski and Karwowski, 1976). On the other hand, when the material is compressed until a certain density, the required force to hold this density decreases. These effects are respectively known as creep and stress relaxation. Furthermore, the stress-deformation behavior of biological materials is also influenced by the history of the compression. Haupt (1999) reported that for subsequent recompressions to a similar density, the required compressing pressure in each cycle decreases. This is known as the Mullins effect. Kutzbach (1973) described this decrease of the maximal force in subsequent compression cycles with a power law relation. In Nona *et al.* (2014), we quantified the reduction in compression effort by fitting the exponential model proposed by Faborode and O'Callaghan (1986) to stress-strain profiles acquired for different compression cycles and particle orientations. By comparing the model parameters, we found that although the total energy for recompressing the material to a similar density decreased, the compression curves became steeper, as the inter-particle contact was gradually increased.

From the above, it can be concluded that the overall bulk modulus of fibrous biological materials changes with the compression stage and duration of compression, the compression history, and the rate of the applied compression. As reported in Nona *et al.* (2014), the initial particle stacking hereby determines the quantity of the change. Therefore, only a model that accurately describes the compression behavior of fibrous biological materials as a function of stress, strain, and time will be suitable for model-based agricultural machinery design.

16.2.3 MOISTURE CONTENT AND TEMPERATURE

As the compression behavior of biological materials is influenced by their moisture content and temperature (O'Dogherty, 1989), several researchers have investigated the possibility to improve the compression behavior of biological materials by preconditioning. For example, in pellet extrusion, a high processing temperature causes the particles to sinter together into a durable pellet (Mani, Tabil, and Sokhansanj, 2004). To capture this temperature effect in their models for the hot pressing of wood, Thoemen *et al.* (2006) investigated the change of their model parameters with the processing temperature. Stelte *et al.* (2012) studied the effect of temperature on the wax layer of wheat straw particles and found that at low temperatures, the wax serves as a lubricant for inter-particle friction, while at higher temperatures no such reduced friction can be observed.

Faborode (1989) showed a significant effect of moisture content on the compression and relaxation behavior of barley straw and indicated that no durable compressed products can be produced when the moisture content of the samples is too high. He also reported that the effect of the moisture content on the material behavior is density-dependent (Faborode and O'Callaghan, 1986). Galedar *et al.* (2008) indicated that alfalfa stems with a higher moisture content required a higher shearing energy. So, high moisture contents can limit the compression of fibrous biological materials. Nona *et al.* (2014) hypothesized that the effect of moisture content on the compression behavior can be explained by differences in inter-particle friction and fluid compressibility. While initially, at low densities, an increase in moisture content would lubricate the inter-particle movement, at high densities, the lower compressibility of water compared to air creates a larger resistance against compression.

To limit the complexity of the bulk model equations, most researchers assumed that only the parameters of the bulk model change and that the model structure remains constant for a limited change in temperature or moisture content. This reasoning will also be followed here.

16.3 BULK COMPRESSION MODELING

Haupt (1999) categorized materials based on their rate-dependency and their equilibrium curve, which is the loading-unloading curve obtained after an infinite number of compression cycles. While the force-deformation relations are different for subsequent compression cycles, due to the time- and history-dependency of the materials, they eventually converge to this material's equilibrium relation. The hysteresis quantifies the difference between the loading and unloading curves and is 0 when the loading and unloading curves are the same. Haupt (1999) defined four groups of materials, according to their equilibrium loading-unloading relation and whether the material shows hysteresis or not, as illustrated in Figure 16.1.

- A rate-independent force-deformation curve and an equilibrium curve without hysteresis characterize an elastic material (Figure 16.1a).
- A rate-independent force-deformation curve and an equilibrium curve with hysteresis indicate a plastic material (Figure 16.1b).
- A rate-dependent force-deformation curve where the equilibrium curve has no hysteresis indicates visco-elasticity (Figure 16.1c).
- A rate-dependent force-deformation curve with hysteresis indicates an elasto-viscoplastic material (Figure 16.1d).

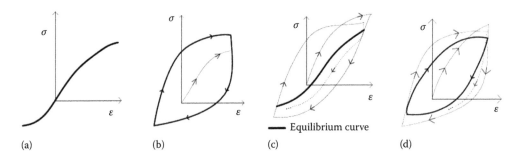

(a) (b) (c) (d)

— Equilibrium curve

FIGURE 16.1 Four theories for material behavior: (a) elastic; (b) elasto-plastic; (c) visco-elastic; (d) elasto-viscoplastic material. (With kind permission from Springer Science+Business Media: *Continuum Mechanics and Theory of Materials*, 1999, Haupt, P.)

These behaviors can be measured by applying excitations in either stress or strain, while measuring the other quantity. By varying the duration and the rate of the applied stress or strain profiles, the time dependent behavior of the materials can be quantified. For this, it can be assumed that the responses of the materials to a number of excitations are the same as the sum of the responses to each individual excitation. In material engineering, this is known as the Boltzmann superposition principle (Roylance, 2001), which leads to the following integration of a strain response (De Pascalis, Abrahams, and Parnell, 2014):

$$\sigma(t, \varepsilon) = \int_{-\infty}^{t} G(t - \tau) \frac{d\varepsilon(\tau)}{d\tau} d\tau \qquad (16.4)$$

Where $G(t)$ is the relaxation modulus, ε is the strain, and t is the time. The above equation is a general form describing the stress-strain-time behavior of a biological material and is commonly used in finite element simulations. The disadvantage of using the relation presented in Equation 16.4 is the assumption that the strain ε is the input and the stress σ is the output. Therefore, it does not allow to simulate creep. To overcome this limitation, the integral definition should be re-written as a differential equation which can be generalized in the following form (Schiessel et al., 1995):

$$a\sigma^{(n)} + b\sigma^{(n-1)} + \ldots + c\sigma = d\varepsilon^{(n)} + \ldots + e\varepsilon \qquad (16.5)$$

with $a(t)$, $b(t)$, $c(t)$, $d(t)$, $e(t)$ the material parameters and (n) indicating the nth order derivative to time. In this form, the model accepts either σ or ε as input variable.

In material engineering, the shape of $G(t)$ or the coefficients $a(t)$ to $f(t)$ are typically described by means of combinations of three fundamental mechanical elements: a spring (elasticity), a dashpot (viscosity), and a friction element (plasticity) (Mohsenin, 1986). In the following section, these fundamental properties are described mathematically.

16.3.1 Fundamental Properties

An *ideal elastic material* has a stress σ that is proportional to the relative deformation ε. For a linear spring, this is described by Hooke's law:

$$\sigma = E\varepsilon \qquad (16.6)$$

where E is the modulus of elasticity, which is also known as Young's modulus. The stress-strain relation and the symbolic representation are illustrated in Figure 16.2a.

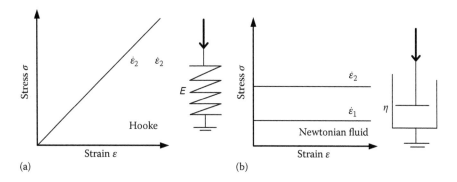

FIGURE 16.2 Ideal elastic (a) and viscous (b) behavior for $\dot{\varepsilon}_2 > \dot{\varepsilon}_1$. (Symbol reprinted from *Mechanics of Agricultural Materials*, Sitkei, G., Copyright 1986, with permission from Elsevier; and Mohsenin, N. N., 1986, *Physical Properties of Plant and Animal Materials.* Department of Agricultural Engineering, Pennsylvania State University.)

Ideal viscosity is characterized by a change in stress in response to a change in rate of deformation $\dot{\varepsilon}$ and shows similarities with a dashpot. The linear dashpot is modeled as follows, with η as the (Newtonian) viscosity:

$$\sigma = \eta\dot{\varepsilon} \tag{16.7}$$

Such a rate dependent stress-strain relation is illustrated in Figure 16.2b.

A *friction element* (or Saint Venant body) describes the ideal plastic deformation and shows close similarities to friction. The body only deforms for stresses higher than the yield stress σ_y. It is characterized by the following equations (Verhás, 1997):

$$\begin{cases} \sigma = \sigma_y & \text{if} \quad \dot{\varepsilon} > 0 \\ \sigma = -\sigma_y & \text{if} \quad \dot{\varepsilon} < 0 \\ -\sigma_y < \sigma < \sigma_y & \text{if} \quad \dot{\varepsilon} = 0 \end{cases} \tag{16.8}$$

When the applied force is taken away, the body stops moving and keeps its position. The resulting deformation is thus permanent. The corresponding stress-strain behavior and the symbol are illustrated in Figure 16.3.

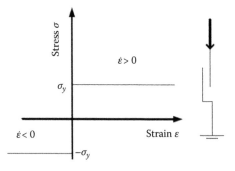

FIGURE 16.3 Ideal plastic behavior and symbol. (Redrawn from Verhás, J., 1997, *Thermodynamics and Rheology.* Dordrecht: Kluwer Academic Publishers.)

One of these ideal elements alone cannot capture the full complexity of the material behavior. Therefore, many combinations of these ideal elements have been proposed, of which the most common ones will be discussed in the following sections.

16.3.2 MAXWELL MODEL

In the simplest case, a spring and a dashpot can be placed in series. The resulting model is known as the Maxwell model. The total strain ε is calculated as the sum of the elastic and viscous strains, while the stress σ in both elements is the same (Mohsenin, 1986). By combining the equations of the basic elements (Equations 16.6 and 16.7), the following stress-strain relation is obtained:

$$\frac{1}{E}\dot{\sigma} + \frac{\sigma}{\eta} = \dot{\varepsilon} \qquad (16.9)$$

Where $\dot{\sigma}$ is the change of stress σ with time. For a constant compression rate $\dot{\varepsilon} = R$, the following solution for the time dependent stress is obtained:

$$\sigma(t) = (\sigma_{t=0} - R\ \eta)e^{-(E/\eta)t} + R\ \eta \qquad (16.10)$$

From this equation, it can be concluded that the Maxwell model gives a nonlinear compression curve with a dependency on strain rate, as shown in Figure 16.4.

During relaxation measurements, the stress decays over time. For visco-elastic fluids, the final stress value will be 0, while for visco-elastic solids the stress decays until a final value $\sigma_e \neq 0$ (Steffe, 1996). The Maxwell model, however, only describes the decay to 0. To account for this final value σ_e and to obtain a more realistic behavior, n Maxwell models can be placed in parallel with a spring to obtain the generalized Maxwell model. As this model requires the estimation of $2n+1$ parameters, this often leads to a large uncertainty in the estimated parameter values (Figure 16.5).

$$\sigma(t) = \varepsilon_0\left((E_1 - E_0)e^{-\frac{t}{T_1}} + (E_2 - E_1)e^{-\frac{t}{T_2}} + \ldots + (E_n - E_{n-1})e^{-\frac{t}{T_n}} + E_e\right) \qquad (16.11)$$

Here, T_i is the relaxation time of Maxwell element i and is given by: $T_i = \eta_i/E_i$, and E_0 is determined from the initial conditions at the start of relaxation: $E_0 = \sigma_0/\varepsilon_0$. The relaxation modulus G, as presented in Equation 16.4, is here written as a sum of exponential functions which is also known as a Prony series (Haghighi-Yazdi and Lee-Sullivan, 2011). This series is often used in combination with Boltzmann's superposition principle (Equation 16.4) to implement complex material behaviors into FEM software packages (Comsol4.3, 2012).

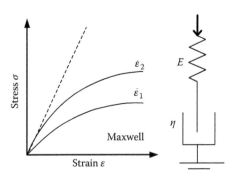

FIGURE 16.4   Stress-deformation relation and mechanical analogy of the Maxwell model with $\dot{\varepsilon}_1 < \dot{\varepsilon}_2$.

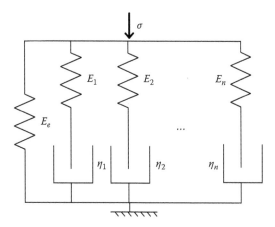

FIGURE 16.5 Mechanical analogy of the general Maxwell model.

Prony series in combination with FEM have been successfully used to describe the compression behavior of biological and non-biological materials. For example, Pierantoni *et al.* (2011) modeled the cyclic compression and fatigue of a rubber, while Chen (2000) determined the parameters of the Prony series based on loading, relaxation, and unloading data of different plastics, measured at different strain rates. The general Maxwell model with $n = 1$ is referred to as the standard linear solid. This model has been successfully used to describe the visco-elastic behavior of plastics, when the irreversible strain can be neglected (Plaseied and Fatemi, 2008). This model has also been used to describe the compression of animal feed mainly consisting of corn stalks (Munoz and Herrera, 2002). However, a piecewise fit was required to fit the loading data in each compression stage.

16.3.3 KELVIN MODEL

Besides the pure elastic and viscous movement, most biomaterials also exhibit delayed expansion. During unloading for example, the load can be removed faster than the material expands. This can be modeled with a spring in parallel with a dashpot, which is known as the Kelvin model. Here, the strains of the spring and dashpot are the same, while the stress σ is the sum of the stresses taken up by the spring and the dashpot. This results in the following equation for the stress-strain relationship:

$$\sigma = E\varepsilon + \eta\dot{\varepsilon} \tag{16.12}$$

The stress-strain relation of the Kelvin model is illustrated in Figure 16.6 for two different strain rates.

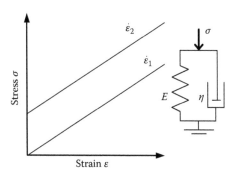

FIGURE 16.6 Stress-deformation relation and mechanical analogue of the Kelvin model with $\dot{\varepsilon}_1 < \dot{\varepsilon}_2$.

16.3.4 BURGERS MODEL

The Maxwell and Kelvin elements both have their advantages and disadvantages. While the Maxwell model is often used for describing loading and relaxation, the Kelvin model is more useful for describing delayed expansion and creep. For modeling loading, unloading, relaxation, and creep, the Kelvin and Maxwell models can be placed in series. This is known as the Burgers model, which is illustrated in symbolic form and in terms of the stress-strain relation in Figure 16.7. The combination of springs and dashpots in the Burgers model results in the following higher order differential equation:

$$\eta_k \ddot{\varepsilon} + E_k \dot{\varepsilon} = \frac{\eta_k}{E} \ddot{\sigma} + \left(1 + \frac{E_k}{E} + \frac{\eta_k}{\eta}\right) \dot{\sigma} + \frac{E_k}{\eta} \sigma \qquad (16.13)$$

where E and E_k are the elastic moduli of the springs and η and η_k are the viscosities of the dashpots in the model. The subscript k separates the parameters of the Kelvin model from those in the Maxwell model. $\ddot{\varepsilon}$ and $\ddot{\sigma}$ indicate the second order derivative of respectively ε and σ to time.

Zhao *et al.* (2009) used the Burgers model to describe the cyclic loading-unloading behavior of soft rock. The plastic strain was increased by including a damage parameter that adjusted the modulus of elasticity over time. It should, however, be noted that the Burgers model describes a decreasing stress-strain ratio, while this ratio typically increases for fibrous biological material, due to the decreasing porosity and the increasing contact between the particles (Section 16.2.1). As a consequence, many researchers had to adjust their parameters to each compression stage (Kaliyan, 2009; Thoemen, 2006).

Ren (1991) extended the Burgers model with an extra element to represent the immediate plastic strain. The resulting model is known as the Burgers-Humphrey model, which is illustrated in Figure 16.8. He was then able to simulate the behavior of wood under compression. However, he found that the elastic moduli and viscosities in his model increased exponentially with density. Thoemen *et al.* (2006) used the Burgers-Humphrey model for simulating the behavior of wood during hot pressing and had to change the model parameters during the extrusion process, which reduced their interpretability. These findings suggest that the compression of fibrous biological materials requires a nonlinearly increasing stress-strain ratio, which cannot be obtained with a combination of the linear fundamental properties. Secondly, due to the presence of the dashpot η in series, the creep strain continues to change in time, which is not feasible for the behavior of fibrous biological materials. Therefore, the use of nonlinear springs and dashpots will be proposed in the following section to obtain constitutive relations which can accurately describe the mechanical properties of fibrous biological materials.

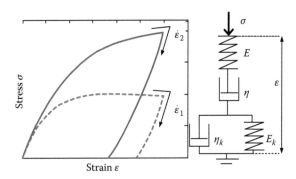

FIGURE 16.7 Typical stress-strain relation and symbolic representation of the mechanical analogy for the Burgers model with $\dot{\varepsilon}_1 < \dot{\varepsilon}_2$.

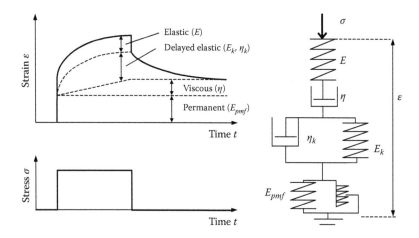

FIGURE 16.8 Symbolic representation of the Burgers-Humphrey model and its response to a step in stress. (From Thoemen, H. *et al.*, 2006, *Holz als Roh- und Werkstoff* 64, 125–33.)

16.3.5 Nonlinear Mechanical Representation of Biological Materials

In order to describe the increasing stress-strain ratio of fibrous materials under compression nonlinear springs and dashpots can be implemented. When the properties of the springs and dashpots change with strain, the parameters of the general model structure (Equation 16.5) become dependent on both strain and time: $a(\varepsilon,t)$, $b(\varepsilon,t)$, $c(\varepsilon,t)$, $d(\varepsilon,t)$, $e(\varepsilon,t)$, $f(\varepsilon,t)$. As these models involve a large number of parameters, a set of parameters should be found which can describe the full stress-strain-history behavior, including loading, unloading, relaxation, and creep.

Peleg (1983) proposed a four-element model with a hardening spring in series with a parallel connection of a softening spring, a dash-pot, and a friction element to describe visco-plastic solids such as fruits, vegetables, and paperboard packaging. The symbolic representation and the resulting stress-strain relation of this model are illustrated in Figure 16.9. However, Faborode (1989) indicated that the model was inapplicable to the first phase in compression of fibrous biological material, and proposed to expand the Peleg model with an extra dash-pot in series to represent the large permanent deformations in the first phase. The resulting model was constructed to simulate loading and creep measurements with subsequent recompressions. However, relaxation behavior was not considered. At high densities, Faborode (1989) also expected the loading curve to show a decreasing stress-strain ratio.

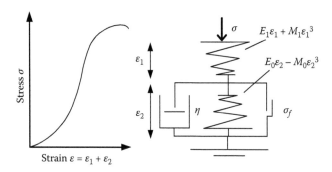

FIGURE 16.9 Symbolic representation and resulting stress-strain relation of the Peleg model. (From Peleg, K., 1983, *Journal of Rheology* 27 (5), 411–31.)

For the compaction of corn stover and switchgrass, Kaliyan (2009) only used a part of the Peleg model, but adjusted the nonlinear spring to a more general representation: $E\varepsilon + M\varepsilon^n$. The resulting model is illustrated in Figure 16.10 and has the following form:

$$\sigma = E\varepsilon + M\varepsilon^n + \eta\dot{\varepsilon} + \sigma_f \tag{16.14}$$

Kaliyan (2009) also related the elasticity (E, M) and viscosity (η) parameters to the durability of the formed briquettes. Although nonlinear elements were used, they still had to update the model parameters for different stress ranges of the compression curve. It should also be noted that they only evaluated the performance of the model for fitting the compression curve and ignored the unloading, creep, and relaxation phenomena.

Wiezlak and Gniotek (1999), as cited in Krucinska *et al.* (2004), proposed a Burgers model in series with a plastic element to describe the compression of fibers of nonwoven fabrics. The model parameters could, however, not be determined based on the loading curve alone. Therefore, Krucinska *et al.* (2004) derived a simpler model with parameters which could be estimated by fitting the model to the observed loading curves. The model scheme is illustrated in Figure 16.11. The model structure is similar to the standard linear solid (Figure 16.5, for $n = 1$), but the dashpot is defined here to be a function of both strain rate and strain. Their motivation for incorporating this extra dependency on strain was to capture the increasing difficulty to augment the permanent strain of the fabric. Unfortunately, they did not test their model for describing unloading, relaxation, or creep, such that no conclusions can be drawn with respect to its validity for describing the full stress-strain-history behavior.

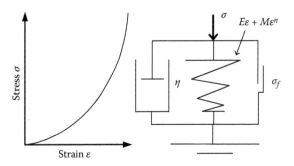

FIGURE 16.10 Symbolic representation and resulting stress-strain relation of the Kaliyan model. (From Kaliyan, N., and Morey, R. V., 2009, *Biosystems Engineering* 104, 47–63.)

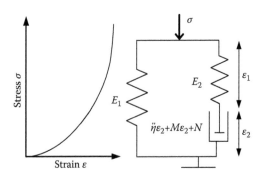

FIGURE 16.11 Symbolic representation of the Krucinska model with the resulting stress-strain relation. (From Krucinska, I. *et al.*, 2004, *Textile Research Journal* 74 (2), 127–33.)

From this review of the state of the art on material models, it can be concluded that a model structure with several nonlinear elements is required to accurately describe the full stress-strain-history behavior of fibrous biological materials.

16.4 APPLICATION OF BULK COMPRESSION MODELS IN MACHINERY DESIGN

The suggested models from the previous section consider the material as a continuum, i.e., the stress-strain-time behavior of the bulk of particles is described with a single model that relates the applied stress to the obtained strain or vice versa. The application of such models to simulate the interaction between the crop and the machine that is processing this crop, thus requires the machine to interact with the crop on a bulk level. In Section 16.2, this assumption was indicated to be valid in the density region above the critical density. This is the case in the compression chamber of an agricultural baler. Therefore, the potential of this bulk modeling approach for the design of an agricultural baler will be discussed. First, the working mechanism of a baler is described in more detail. Then, an overview of the existing models for simulating the compression process is given.

16.4.1 WORKING MECHANISM OF AN AGRICULTURAL BALER

An agricultural baler compresses fibrous biological materials such as wheat straw, grass silage, and hay, and typically does this in the field, close to the place where these crops are harvested. In the plunger-type extruders, a plunger moves back and forth in the compression chamber, thereby compressing the material in the chamber and pushing the compressed bale out. Each time the plunger retracts, new material can be thrown in the chamber, which is compressed against the material that is already present in the compression channel. The force that the plunger requires to complete one cycle is called the plunger force (Figure 16.12). Initially, the newly added material is compressed against

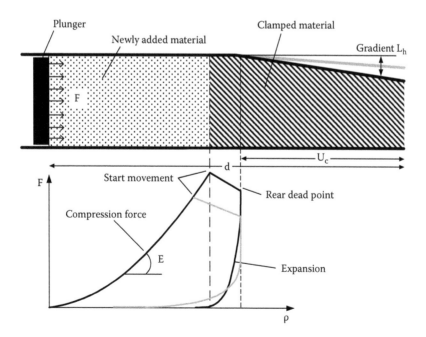

FIGURE 16.12 Schematic representation of required extrusion pressure, for two gradients of the extrusion die. (Reprinted from *Mechanics of Agricultural Materials*, Sitkei, G., copyright 1986, with permission from Elsevier.)

the clamped material, resulting in a nonlinear stress increase similar to the one that would be observed in closed chamber compression, as described in Section 16.1 (Sitkei, 1986). Klenin *et al.* (1986) also indicated a flexure point in the force curve where the newly added material approaches the density of the bale in the chamber. Since the material's bulk modulus depends on its density, the compression force only spreads out over the entire mass in the chamber once the densities in the chamber are similar. The amount of compressed material therefore changes during compression, thereby influencing the plunger force. As soon as the plunger force exceeds the static friction force between the material and the wall, the bale starts to move. Sitkei (1986) indicated that a decay in the plunger force can be observed at this point, because the dynamic friction is smaller than the static friction, as illustrated in Figure 16.12. After the movement, the plunger returns to complete its cycle, thereby allowing the compressed material to expand. The force on the plunger then decays rapidly and fully disappears when contact with the material is lost. By adjusting the gradient of the extrusion channel, the start of material movement can be regulated. A larger gradient causes the material to start moving later and thus to move less, resulting in larger compression and higher plunger forces. In Figure 16.12, the plunger force in a compression cycle for two different gradients of the die is shown. The start of movement and the point at which the plunger returns are both indicated.

In a compression chamber, the material's density is not homogeneous. Both the expansion when the plunger retracts and the expansion at the chamber's outlet cause the density to decrease. The internal stress in the material also changes with the position in the chamber. Kanafojski *et al.* (1976) measured these stresses in an agricultural baler and obtained the results presented in Figure 16.13. The solid lines indicate the stresses during material compression. The dashed lines are the remaining stresses after the expansion. Due to the friction between the walls of the compression chamber and the compressed material, the stress along the compression direction x decreases with the depth in the chamber. Similarly, for both lateral directions y and z, a decreasing trend is visible. After the expansion, the largest stresses occur where the material is held best, i.e., where the expansion is minimal.

The design of the compression chamber determines the shape of the stress and density profiles. A good design minimizes the peak stresses on the chamber walls and uniformly compresses the material to form beam-shaped bales with uniform density. The density and the stress profiles along

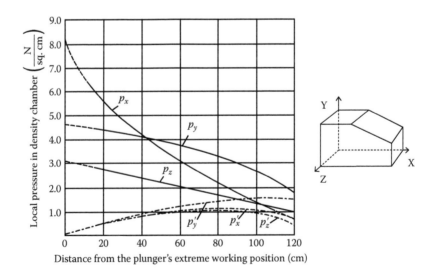

FIGURE 16.13 Pressure inside the compression chamber of an agricultural baler. The solid line is at maximal compression. The dashed line is after the rebound of the material. (From Kanafojski, C. and Karwowski, T., 1976, *Agricultural Machines Theory and Construction: Vol. 2 Crop-Harvesting Machines.* Foreign Scientific Publications Department of the National Center for Scientific, Technical and Economic Information.)

the chamber length, therefore, quantify the quality of the extrusion process and indicate the homogeneity in the compressed product. Moreover, these profiles characterize the holding and expansion zones of the material in the chamber, which can be used to identify the zones with reluctant wear due to the cyclic compression and expansion of the material in the extrusion chamber. By simulating these profiles, the effect of the chamber design on the shape of these profiles can be investigated. For this purpose, a model is required that relates the chamber design and the machine settings to the resulting stress and density profiles in the chamber. As such, the dependency of the extrusion process on the gradient and the shape of the die can be investigated. As the material in the chamber exhibits loading, unloading, relaxation, and creep, a material model describing this full stress-strain-history behavior should be used to simulate the stress and density profiles with respect to the applied machine settings and the chamber design. As the relation between the stress and the resulting deformation has been discussed in the previous section, the models in this section will relate the chamber design or machine settings to the stress profiles in the chamber.

16.4.2 Extrusion Compression Modeling

From the working mechanism of the agricultural baler, it can be observed that each plunger cycle adds a slice to the material in the chamber. This new slice is compressed against the material in the chamber, while at the chamber outlet, a compressed slice is pushed out. For each slice, Newton's second law can be applied, which results into a model for the stress profile along the chamber length (Nona *et al.*, 2013). In Figure 16.14, the partitioning into slices of the material in the compression chamber of an agricultural baler is schematically illustrated (Verhaeghe *et al.*, 2015).

In a second step, the calculated stresses can be related to the material deformations in the compression chamber by using the stress-strain relation of the material. Faborode (1986) successfully applied the following model to predict the pressure in the box for given crop parameters and plunger position:

$$P_x = \frac{K_0}{b_c} \left(e^{b_c \frac{S}{L_0 - S}} - 1 \right) e^{-\mu k \, L_R x / A_r} \tag{16.15}$$

With S, the distance the piston has moved while compressing the material; L_0, the initial height of the sample; μ, the friction coefficient between the material and the chamber; x, the position in the chamber; A_r, the cross sectional area of the compression chamber; L_R, the perimeter of this cross section, and K_0 and b_c the material parameters. Unfortunately, this relation does not consider the inclination of the side walls. As a result, this stress-deformation relation is not suitable for describing the evolution of the pressure along the length of a tapered compression chamber.

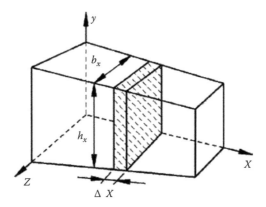

FIGURE 16.14 Schematic illustration of a slice in the compression chamber of an agricultural baler (WO Patentnr. PCT/EP2014/065,355, 2015).

Sitkei (1986) assumed the strain to increase logarithmically with the applied stress. The compression stress with respect to the position in the chamber could then be calculated with the following relation:

$$P_x = P_e e^{A(L-x)} + \frac{B}{A}\theta_t^2\left(e^{A(L-x)} - 1\right) + \frac{Bh_0^2}{\theta_t^3}\left(\frac{e^{A(L-x)}}{h_L} - \frac{1}{h_x}\right)$$
$$+ Ke^{\frac{h_x A}{\theta_t}}\left(\int_{h_0}^{h_L}\frac{e^{-\left(\frac{A}{\theta_t}\right)h_x}}{h_x}dh_x - \int_{h_0}^{h_x}\frac{e^{-\left(\frac{A}{\theta_t}\right)h_x}}{h_x}dh_x\right)$$

(16.16)

where P_e is stress at the end of the chamber, θ_t is the gradient of the top door, L and b the length and width of the compression chamber, h_x the chamber height at position x in the chamber, θ_t the inclination of the top door, and:

$$A = \frac{v}{1-v}\left(\frac{\theta_t + 2\mu}{a_m} + \frac{2\mu}{b}\right)$$

$$B = \frac{C\rho_0^2\theta_t^2}{1-v^2}\left(\frac{\theta_t + 2\mu}{a_m} + 2\frac{\mu v}{b}\right)$$

$$K = AB\frac{h_0^2}{\theta_t^4} + 2B\frac{h_0}{\theta_t^3} \quad h_x = h_0 - \theta_t x$$

$$h_L = h_0 - \theta_t L$$

where the Poisson ratio v is the ratio of the lateral expansion over the applied compression. As the Poisson ratio v cannot be easily measured, Sitkei (1986) estimated it for alfalfa to be around 0.30 for the pressure range in agricultural balers, and used this value in all calculations. This relation describes the stress decay along the chamber length, but only considers one inclined wall, which is approximated by an average value $\left(h_x \approx a_m = h_0 - \theta_t\frac{L}{2}\right)$. Solving this equation also requires the pressure P_e inside the bale, at the outlet of the extrusion chamber. This pressure, however, cannot be easily measured.

As the stress in the material at the chamber's outlet had to be measured and only the inclination of the top door had been considered, Afzalinia and Roberge (2008) derived a static mechanical relation under the assumption that the material is linear elastic and isotropic:

$$P_x = P_p e^{-Ax} + \frac{B}{A^2}\left(1 - Ax - e^{-Ax}\right)$$

(16.17)

Where:

$$A = \frac{v}{1-v}\left(\frac{\theta_t + 2\mu}{a_m}\right)$$

$$B = E\frac{b_m(\theta_t + 2\mu)(2\theta_t v h_0) + 2a_m(\theta_t + \mu)(2\theta_t h_0)}{h_0 b_0 a_m b_m(1-v^2)}$$

where a_m and b_m are the average height and width of the extrusion compression chamber.

They reported that this relation was only valid in the highest density range, which was achievable with the machine and could not be extended to the lower ranges. They also reported that the stresses on the side walls were not fitted accurately. They suggested that this could be due to the assumption of linear elasticity and isotropicity. Notwithstanding, this approach could be promising for on-line control applications due to the short computation time required for solving this equation. However, in most commercial large square balers the pressure applied to the side walls is controlled, while the position of the side walls is a result of this pressure. Therefore, the model should be rewritten as a function of the stress on the side walls instead of the side wall positions.

16.5   STEM MODELING

Bulk approaches, as discussed in the previous sections, are inadequate to model the processing of crops at densities below the critical density, because the movements of stems with respect to each other and the interactions between individual stems and between stems and machine components cannot be neglected at these low densities. As the interactions take place at the particle level, a modeling framework on this level is required. When modeling the behavior of a collection of particles, discrete element modeling (DEM) is a logical choice, as it allows to describe the behavior of each particle through its interactions with the other particles and the system elements (Tijskens *et al.*, 2003).

Discrete element modeling was originally proposed by Cundall and Strack (1979) in the field of rock mechanics. Since then, DEM has been used in a wide range of areas and applications, such as mining, mineral processing, powder metallurgy, pharmaceutical and chemical processing, and food handling. Since the beginning of the twenty-first century, DEM is also increasingly used for simulating particulate processes in agricultural machinery, such as: fertilizer spreading (Tijskens *et al.*, 2003; Van Liedekerke *et al.*, 2009), grain flow in silos (González-Montellano *et al.*, 2012), grain–straw separation (Lenaerts *et al.*, 2014), and crop compression (Leblicq *et al.*, 2016). Using DEM, the influence of particle properties and boundary conditions on the process can be assessed with a set of *in silico* experiments.

16.5.1   BASIC CONCEPTS OF DISCRETE ELEMENT MODELING

DEM is essentially a numerical technique to model the motion of an assembly of particles interacting with each other through collisions (Tijskens *et al.*, 2003). The technique sums up the forces acting on the particles and integrates Newton–Euler equations of motion to obtain velocities and positions at the next time step (Leblicq *et al.*, 2016). In doing so, the technique describes the path of every particle as time proceeds. Physically, particles in a DEM problem are approximated as rigid bodies and the contacts between them as point or line contacts. In reality, most particles are however deformable. This is accounted for by allowing the particles to overlap. As overlapping particles can only exist in computer models, this overlap is termed virtual overlap. The forces resulting from a contact between two particles are related to their virtual overlap by a contact force model (Tijskens *et al.*, 2003).

Mathematically, a DEM problem is formulated as a nonlinear system of coupled ordinary differential equations (ODE) formed by Newton's equations of motion for each individual particle:

$$m_i a_i = G_i + \sum_c F_{ci} \tag{16.18}$$

$$I_i \alpha_i = H_i + \sum_c r_{ci} \times F_{ci} \tag{16.19}$$

where a_i and α_i are the translational and rotational acceleration of the i_{th} particle, m_i and I_i designate its mass and inertia tensor, and G_i and H_i are, respectively, the body force and moment acting on the i_{th} particle. In addition, the particle experiences contact forces F_{ci} arising from contacts with neighboring particles, where the subscript c indicates that it is the c_{th} contact and r_{ci} is the position vector of that contact relative to the particle's center of mass.

The contact forces which particles exert on each other are determined by a contact model. Similar to the approaches used in bulk modeling, contact models are typically formed through combination of spring, damper, and friction elements (Junemann *et al.*, 2013).

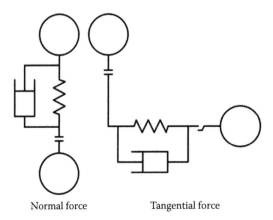

Normal force   Tangential force

FIGURE 16.15 Model of the contact force between two spherical elements. (From Nan, W. *et al.*, 2014, *Powder Technology*, 210–218.)

Contact forces are typically decomposed into a normal and a tangential component with respect to the contact surface. The normal component of the contact force acting on a particle is generally described as a function of the virtual overlap (δ) and its time derivative ($\dot{\delta}$) (Tijskens *et al.*, 2003). The simplest model assumes a linear elastic component and a linear viscous damping (Figure 16.15).

$$F_N\left(\delta, \dot{\delta}\right) = k_n\delta + c_n\dot{\delta} \tag{16.20}$$

This is the Kelvin model that has been discussed earlier at the bulk level (Section 16.3.3). For viscoelastic materials, a nonlinear model, based on Hertz contact law has been proposed (Tijskens *et al.*, 2003):

$$F_N\left(\delta, \dot{\delta}\right) = k_n\delta^{\frac{3}{2}} + c_n\delta^{\frac{1}{2}}\dot{\delta} \tag{16.21}$$

The dampers prevent the elements from oscillating, dissipate energy, and stabilize the simulation (Leblicq *et al.*, 2015).

Tangential contact forces arise when two particles take part in an oblique collision, or when contacting particles are rotating relative to each other (Tijskens *et al.*, 2003). The simplest contact model in the tangential direction describes Coulomb friction with a viscous damper, enforcing the no-slip regime. As for impact in the normal direction, realistically behaving contact models in the tangential direction also require elastic and viscous terms (Figure 16.15).

16.5.2 VIRTUAL CROP STEMS IN DEM

To obtain reliable results with DEM, the virtual particles must have realistic geometries and deform realistically during contact. Tubular particles resembling crop stems should, therefore, be created in the DEM environment. The particles should also be compressible in both longitudinal and radial direction, interact realistically, and be bendable in every direction.

16.5.2.1 Realistic Stem Geometries

Favier *et al.* (1999) developed a method for representing non-spherical, smooth-surfaced, particles in DEM using overlapping spheres of arbitrary size, whose centers are fixed in position relative to each other (Figure 16.16). This method allows to model the dynamic behavior of particles of high aspect ratio and irregular curvature. Later, also other basic geometries were used to create more complex

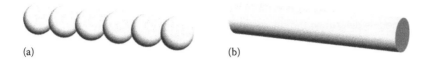

(a)   (b)

FIGURE 16.16 Representations of tubular particles. (a) glued-spheres particle; (b) cylindrical particle. (From Guo, Y. *et al.*, 2012, *Journal of Fluid Mechanics*, 1–26.)

particles, such as tubes (Lenaerts *et al.*, 2014), capsules (Leblicq *et al.*, 2016), and triangulated rounded bodies (Smeets *et al.*, 2014).

As crop stems are typically tubular, several approaches have been proposed for generating tubular particles. Kattenstroth *et al.* (2011) and Ma *et al.* (2015) developed virtual straw particles by connecting spheres, as illustrated in Figure 16.16. However, as these stems were rigid and unbendable, the deformation of the crop particles through interaction with machine components could not be simulated. To overcome this limitation, it was proposed to develop flexible filamentous particles composed of chains of rigid bodies, connected through ball and socket joints (Geng *et al.*, 2011; Guo *et al.*, 2013; and Grof and Sanek, 2013). Alternatively, Junemann *et al.* (2013) and Kajtar and Loebe (2014) proposed to create bendable crop stems by connecting spheres with cylindrical bonds. Guo *et al.* (2013) compared glued-sphere particles and true cylindrical particles and concluded that for smooth particles, the most realistic results are obtained with cylindrical particles. Therefore, Lenaerts *et al.* (2014) created segmented bendable straw stems by connecting rigid cylinders with flexible, elastic, configurable bonds to simulate grain-straw separation, and validated the obtained simulation results through comparison with experimental results (Figure 16.17). Leblicq *et al.* (2014, 2015) improved the bending model of Lenaerts *et al.* (2014) by using capsules (Figure 16.18) and demonstrated that this particle representation allows to simulate the compression of straw stems up to the critical density.

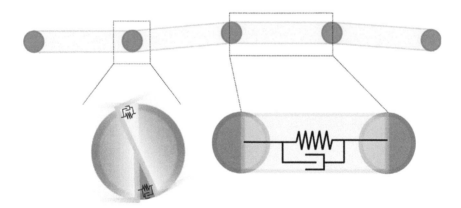

FIGURE 16.17 Bendable straw in DEM. (From Lenaerts, B. *et al.*, 2014, *Computers and Electronics in Agriculture*, 24–33.)

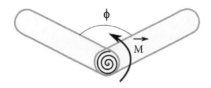

FIGURE 16.18 Two segments of a virtual stem in DEM. (From Leblicq, T. *et al.*, 2016, *Computers and Electronics in Agriculture*, 80–8.)

16.5.2.2 Traditional Contact Models and Their Limitations

As the contact models determine the interaction of particles with each other and with the machine components, they have a major impact on the simulation results. In most DEM simulations of agricultural processes, the basic contact models presented in Equations 16.20 and 16.21 have been used to model the interactions between tubular particles. While Grof *et al.* (2007), Nguyen *et al*, (2013), Junemann *et al.* (2013), Kajtar and Loebe (2014), and Leblicq *et al.* (2015a) used linear spring-damper systems; Guo *et al.* (2012b), Nan *et al.* (2014), and Lenaerts *et al.* (2014) used Hertz contact models. In all these studies, Coulomb friction was assumed between the particles. However, many researchers have shown that the deformation of crop stems is highly nonlinear (Bright and Kleis, 1964; O'Dogherty *et al.*, 1995; Annoussamy *et al.*, 2000; Yu *et al.*, 2006; Nazari Galedar *et al.*, 2008; Lenaerts *et al.*, 2014; Leblicq *et al.*, 2015). As can be seen from Figure 16.19, the assumption of linear elastic or visco-elastic behavior is only reasonable for the first part of the force-deformation curve. Therefore, most researchers have used this part of the curve to calibrate the model parameters. At larger deformations, ovalization, buckling, and cracking of the particles take place. During ovalization, the forces on the wall tend to flatten the cross section of the stem, which eventually leads to a collapse of the particle, known as buckling. This results in a highly nonlinear and plastic deformation behavior. Plastic deformation implies that stems will respond differently to recompression or further deformation, which is ignored by most DEM contact models. As the bulk behavior of a collection of stems is the sum of the behaviors of the individual stems, an incorrect description of the deformation behavior of the individual stems will most likely result in incorrect simulation results and incorrect decisions for machine optimization.

16.5.2.3 Data-Based Contact Models

Leblicq *et al.* (2015) proposed a different approach for modeling the interaction and compression of crop stems. Here, the contact spring was not assumed to be constant (as is the case in the linear elastic and the Hertz model), but depends on the overlap (δ) and the previous deformations (d) of the particle:

$$k_N = f(\delta, d). \tag{16.22}$$

The normal force was therefore defined as:

$$F_N\left(d, \delta, \dot{\delta}\right) = k_N \delta + \gamma \dot{\delta}. \tag{16.23}$$

The values for the normal spring force were determined experimentally. To obtain realistic contact forces, measured force-deformation data were used to model the interactions between particles. To

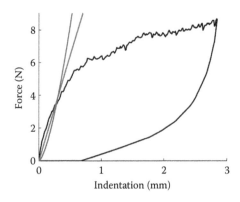

FIGURE 16.19 Measured compression profiles of wheat stems. The hysteresis is a result of the plastic deformation. Black: measured data, red: linear elastic model, blue: Hertz model.

determine the effect of plastic deformation and damage, the measurements were not performed in one run, but in a step-wise way. In every cycle, the deformation was incrementally increased with an additional 0.1 mm relative to the previous cycle, as illustrated in Figure 16.20.

For each measured data point, the maximum historical deformation (MHD) was recorded together with the force and deformation. During the first cycle, while the deformation is increasing, the MHD is also increasing. During unloading, the MHD is kept constant at the maximum deformation of the first cycle. During the second cycle, the MHD will only increase once the deformation exceeds the deformation of the first cycle. The MHD is thus always equal to or larger than the deformation. As a result, each data point consists of three values: deformation, force, and MHD. Each pair of values for the MHD and the deformation corresponds to a value of the force. These recorded relations were then used to construct lookup tables (LUTs) from the measured stem data (Figure 16.21).

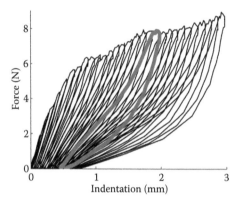

FIGURE 16.20 Stepwise deformation (experimental results). For clarity, one cycle (loading and unloading) is displayed in blue. (From Leblicq, T. *et al.*, 2016, *Computers and Electronics in Agriculture*, 80–8.)

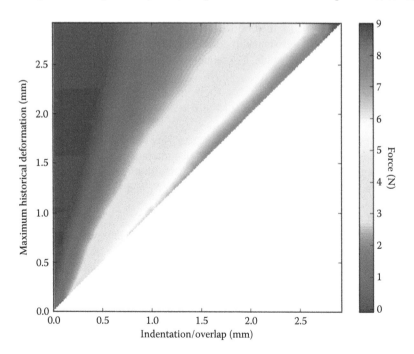

FIGURE 16.21 Look up table for the interactions between stems. (From Leblicq, T. *et al.*, 2016, *Computers and Electronics in Agriculture*, 80–8.)

When stems in DEM interact, they overlap. This virtual overlap corresponds to the measured deformation. Each segment of the virtual stems has a memory in which the maximum historical overlap/deformation (MHD) is stored. In every time step of the simulation and for each segment, the overlap with neighboring segments is calculated. From this overlap and from the stored MHD, the corresponding force can then be obtained from the LUT for that stem (Leblicq *et al.*, 2016).

16.5.2.4 Interacting Particles

Using the traditional or the databased contact models (LUTs), virtual stems were created that experience and exert (realistic) forces when compressed. This does, however, not mean that the stems will interact realistically, because one cannot assume that the contact force between interacting particles is the sum of the compression forces of both particles separately or the average of both. Also, the deformation cannot evenly be distributed over both particles. In reality, particles with less deformation resistance (e.g., stems with smaller wall thickness or stems which have already been deformed) will deform more and sustain more plastic deformation. This makes it more likely that they will be damaged more during future interactions. This process will continue until a stem is completely crushed and the stem walls make contact. At this point, the resistance to deformation is strongly increased. The extent to which each stem is deformed and damaged thus depends on the relative resistance to deformation (Leblicq *et al.*, 2016).

The virtual overlap in DEM corresponds to physical deformation. So, although many researchers assume an equal distribution of the overlap in DEM simulations, this assumption will in many cases be wrong. Stems with a higher resistance to deformation (e.g., a large wall thickness) will deform less in contact with other stems, and thus should get less overlap. The opposite is true for damaged stems, which will take up the major part of the joint deformation. When, for example, two particles with different radii (R_1 and R_2) interact, the contact forces are determined using an effective radius (R):

$$\frac{1}{R} = \frac{1}{R_1} + \frac{1}{R_2} \qquad (16.24)$$

Leblicq *et al.* (2016) developed a method to attribute the largest part of the overlap ($a\delta$) to the particle (e.g., stem) with the least resistance to deformation, while the remaining part of the overlap $(1 - a)\delta$ is attributed to the second particle. Consequently, the sum of the overlap of the two stems is equal to the total overlap.

$$a\delta + (1 - a)\delta = \delta \qquad (16.25)$$

The forces between two stems should, according to Newton's third law, always be equal in magnitude, but opposite in direction. Each time particles interact, the particular distribution of the overlap has to be determined. The determination of the distribution of the overlap corresponds to finding a value for a such that the forces of the two stems (F_1 and F_2) are equal.

16.5.2.5 Bending Models

In most studies involving flexible virtual stems, linear elastic models were used to connect two adjacent segments. One example of a linear elastic bending model for virtual crop stems was proposed by Lenaerts *et al.* (2014). The stems consist of interconnected cylinders (Figure 16.17). Two cylinders of the virtual stem are interconnected with six linear spring-dampers (Kelvin–Voigt models), positioned in an axisymmetric pattern. The stiffness of the springs determines the bending resistance and the dampers prevent oscillations by dissipating energy. The springs are not loaded if the two cylinders are in line with each other. When the virtual stem is bent, the springs on the inner side are compressed, while the springs on the outer side are stretched. When the interactions between two segments are described using linear elastic models, the entire virtual stem will also behave linear elastically.

To assess whether these linear elastic models yield realistic results, Leblicq *et al.* (2014) investigated the bending behavior of crop stems. Two consecutive phases could be distinguished during the bending of stems: ovalization and buckling. When a stem is bent, the inner side of the stem is longitudinally compressed, while the outer side is stretched. Both this compression and tension result in a resistance of the stem against the bending moment. As both have a component directed towards the center of the stem, the stresses cause a flattening of the circular cross section into an oval shape (Figure 16.22) (Leblicq *et al.*, 2015). This ovalization process is considered to be elastic. When this process continues, the flexural stiffness is reduced until the structure becomes unstable and buckles. A kink is suddenly formed, and the cross section completely flattens locally (Figure 16.23). This deformed cross section offers virtually no resistance to bending (Calladine, 1989). These processes

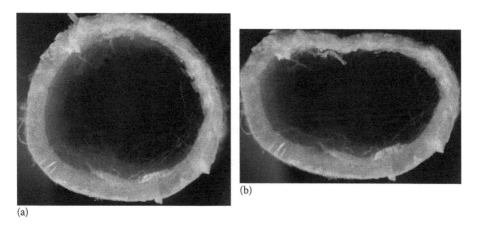

(a)   (b)

FIGURE 16.22 (a) Undeformed cross section, (b) flattened and ovalized cross section. (From Leblicq, T. *et al.*, 2016, *Computers and Electronics in Agriculture*, 80–8.)

FIGURE 16.23 Wheat stem after buckling. (From Leblicq, T. *et al.*, 2016, *Computers and Electronics in Agriculture*, 80–8.)

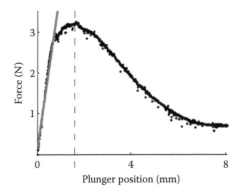

FIGURE 16.24 Typical force-deformation curve for three point bending of wheat straw. A linear elastic model was fitted to the first part of the data (red line). The dashed line indicates the end of ovalization and the start of buckling. (From Leblicq, T. *et al.*, 2016, *Computers and Electronics in Agriculture*, 80–8.)

result in highly nonlinear and plastic deformation behavior (Figure 16.24). After the initial linear elastic phase, the force increases at a slower rate and then drops rapidly for a further increase in deformation (Leblicq *et al.*, 2015).

Leblicq *et al.* (2016) proposed a different approach for modeling the bending of crop stems. A rotational spring is placed between two segments (in this case, capsules) to create the bending resistance (Figure 16.18). When the stem is bent, this angular spring (k_b) generates a moment (M) that counteracts the bending. The bending angle (θ) is chosen such that it is zero when the stem is undeformed, and thus the angle increases with increasing deformation. An extra damper (c_b) is responsible for stabilizing the simulation by dissipating energy.

$$M = k_b\theta + c_b\dot{\theta} \qquad (16.26)$$

When the angular spring (k_b) has a constant value, the virtual stem also behaves linear elastic when bent. To create stems with realistic bending behavior, the value of k_b should be made variable and function of the angle and the previous deformation (the maximum historical angle, *MHA*) (Leblicq *et al.*, 2016):

$$k_b = f(\theta, MHA) \qquad (16.27)$$

To determine realistic values for k_b, a data-based approach was used similar to the one used for stem compression (Section 16.5.2.4). In this approach, LUTs are constructed based on measurement data where the *MHA* is stored for each segment of the virtual stems. In every time step of the simulation and for each pair of neighboring segments, the angle between those segments is determined. For this angle and for the stored *MHA*, the corresponding rotational spring stiffness can be obtained from the LUT. When this LUT is used in the simulations, the resistance to bending is changed depending on the bending angle and the degree of historical deformation (*MHA*). It was shown that this method is also useful for stems consisting of different numbers of segments, for stems having different lengths and for different support distances (Leblicq *et al.*, 2016).

16.6   CROP FLOW SIMULATIONS IN AGRICULTURAL MACHINERY

The virtual stems presented in Section 16.5 can be used to simulate the processing of crop stems. Two examples will be discussed in this section: In the first example, traditional contact and bending models are used, while in the second example, the data-based models developed by Leblicq *et al.* (2016) are used.

16.6.1 EXAMPLE 1: GRAIN-STRAW SEPARATION

In this first example, the separation of grain and straw in a combine harvester will be simulated. Grain and straw have a different shape and density. Grain can, therefore, be separated from straw by accelerating the mixture. In the separation section of a combine harvester, the grain kernels, which have been released from the ears through threshing, have to be expelled from the straw layer. As the combine harvester is a continuous-flow machine, the residence time of the grain–straw mixture in the threshing and separation unit limits the time available for the grain to migrate through the straw layer (Lenaerts *et al.*, 2014).

In this context, an idealized experiment representative for the separation on straw walkers was described by Beck (1992). The setup consists of a straw layer contained in a vertically, sinusoidally, oscillating box. After the straw has been agitated for a certain time, a layer of grain kernels is released at once on top of the oscillating straw layer. Grain passage through the grating at the bottom of the box is recorded as a function of time. The grain sinks through the straw layer, while it disperses laterally. This results in a sigmoidal curve of the separated fraction as a function of time. The lower the area density of the straw layer, the faster the kernels can sink and the steeper the separation curve will be (Beck, 1992).

This experiment was repeated in simulation by Lenaerts *et al.* (2014). The bendable virtual stems are discussed in Section 16.5.2.2 and displayed in Figure 16.17. The grain kernels were approximated by spheres. It should be noted here, that this is a rough approximation of real grain kernels, rather having an ellipsoid shape. More realistic grain kernels could be created by using composite particles made up of several overlapping spheres, as was done by Ma *et al.* (2015) for rice grain kernels. In the simulations, a (virtual) box was used which oscillated at the same frequency and amplitude as the box in the experiments (Figure 16.25). After releasing the grain layer above the straw mass (Figure 16.26), the kernels fall into the box, resisted by aerodynamic drag forces. The straw restrains the kernels' movement. As the straw is randomized, the particle trajectory lengths differ as well as their velocity on the trajectories. When the kernels approach the floor, they pass through freely, as there is no contact detection between the kernels and the floor of the box (Figure 16.27). Both the experiments and the simulations were performed at different grain and straw

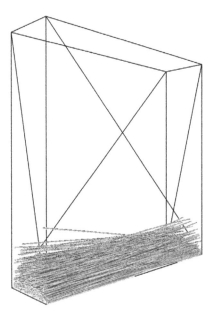

FIGURE 16.25 Oscillating box and moving straw particles. (From Lenaerts, B. *et al.*, 2014, *Computers and Electronics in Agriculture*, 24–33.)

FIGURE 16.26 Grain initialization. (From Lenaerts, B. *et al.*, 2014, *Computers and Electronics in Agriculture*, 24–33.)

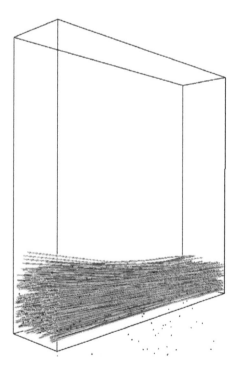

FIGURE 16.27 Separation. (From Lenaerts, B. *et al.*, 2014, *Computers and Electronics in Agriculture*, 24–33.)

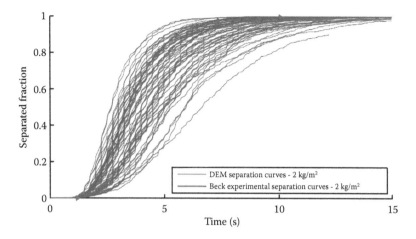

FIGURE 16.28 Separation curves (both measurement and simulations) resulting from astraw area density of 2 kg/m². (From Lenaerts, B. *et al.*, 2014, *Computers and Electronics in Agriculture*, 24–33.)

densities. In post-processing, the cumulative separated fraction was plotted as a function of time and compared with the measured, separated fraction (Figure 16.28).

This simulation approach was successfully validated against the experiment data (Lenaerts *et al.*, 2014). However, it should be noted that the fact that the Hertz contact model and the linear elastic bending model yielded sufficiently accurate results, which most likely can be explained by the limited deformation of the individual stems in this case. As this would not be the case when the stems deform further, this case will be investigated in the next example.

16.6.2 EXAMPLE 2: STRAW COMPRESSION

In this second example, the compression of straw stems will be discussed, as this happens in the first parts of a baler (feeding and pre-compression), a combine harvester (elevator, threshing unit, and an axial separation unit) or a forage harvester (feeding and cutting). Since the purpose of a baler is to deform (compress) straw stems to high densities, the traditional contact and bending models are not suitable to simulate these processes.

The crop feeding in the square baler is illustrated in Figure 16.29. The crop has been gathered in the field in lines and is picked-up by the pick-up in the front part of the baler. This pick-up brings the crop to the rotor which brings it to the pre-compression chamber. Before feeding the crop into the machine it can be cut (chopped). When the pre-compression chamber is full, the crop pushes the filling plates or filling sensor away, which triggers the feeding cycle. The stuffer will feed the crop into the bale chamber, where it will be compressed with very high pressure by the plunger against the narrowing bale chamber and the resistance of the previous bale (as discussed in Section 16.4.1). In all these processes (with the exception of compression to higher densities in the bale chamber), the density of the crop is below the critical density and a DEM approach is considered the most appropriate.

To simulate the compression of crop stems in these processes, Leblicq *et al.*, (2016) both measured and simulated the compression of stems in a compression box (Figures 16.30 and 16.31). The stems were placed parallel to each other on the bottom of a compression box. Before compression, the box was shaken to randomize the orientation of the stems. This initialization method was chosen, because it can both be simulated and reproduced in experiments. After the randomization, the stems were compressed and the force on the plunger was recorded.

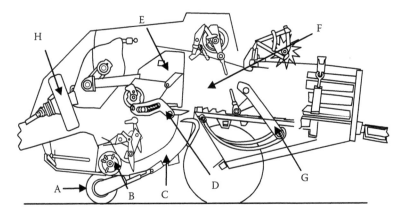

FIGURE 16.29 Cut-away of square baler: (A) pick-up; (B) rotor to throw the crop to the pre-compression chamber; (C) pre-compression chamber; (D) stuffer; (E) plunger driven by a crank shaft mechanism; (F) bale chamber; (G) knotter needles which are activated when the bale has reached a given length; (H) flywheel driven by the tractor PTO (power take-off).

FIGURE 16.30 Bulk compression measurements. (a) stem orientations after initialization; (b) bulk compression. (From Leblicq, T. *et al.*, 2016, *Computers and Electronics in Agriculture*, 80–8.)

The LUTs (required for the deformation models) were created based on stem measurements carried out on stems sampled from the same field as the stems that were used in the bulk compression experiments. Each stem was randomly given two LUTs from the stem database (one for the stem-stem interactions and one for the stem-box interactions). As no significant differences could be observed between the measured and simulated stress-strain profiles (Figure 16.32), it was concluded that the DEM simulations and the corresponding data based models are able to correctly predict the stem processing.

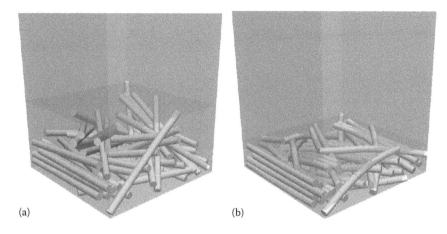

(a) (b)

FIGURE 16.31 Bulk compression simulations. (a) stem orientations after initialization; (b) bulk compression. (From Leblicq, T. *et al.*, 2016, *Computers and Electronics in Agriculture*, 80–8.)

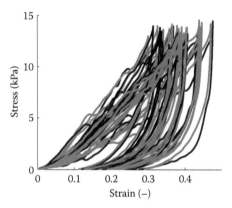

FIGURE 16.32 Bulk compression (black: measurements, blue: simulations with data-based deformation models). (From Leblicq, T. *et al.*, 2016, *Computers and Electronics in Agriculture*, 80–8.)

16.7 CONCLUDING REMARKS AND FUTURE PERSPECTIVES

Thanks to the rapid increase in computational power, the implementation of complex material models in numerical simulation methods such as DEM and FEM is becoming feasible. It was shown that above the critical density, the relative movements of the particles are limited, such that the material can be modeled as a bulk by means of FEM simulations. However, when the density is still below the critical value, the particle movements cannot be neglected, and DEM is considered the best modeling approach. When these material models are able to describe the machine-crop interactions with sufficient accuracy, they can form the basis for *in silico* design of the next generation of agricultural machinery.

The future of agricultural machinery development therefore lies in improving the accuracy of the material and machine models. When successful, this will allow to estimate the effect of a change in the machine design with sufficient accuracy to decide on the most promising designs. Therefore, the aim of field testing will change from the current practice of optimizing designs through trial and error, to validation of the simulations and in-field evaluation of the designs which have been optimized in simulation. As such, these simulation tools are expected to speed up the innovation process and to reduce the costs for field testing, as is currently done in other industrial sectors.

REFERENCES

Adapa, P., Tabil, L., and Shoenau, G. (2009). Compaction characteristics of barley, canola, oats and wheat straw. *Biosystems Engineering* 104, 335–44.

Afzalinia, S. and Roberge, M. (2008). Modeling of pressure distribution inside the compression chamber of a large, square baler. *Transactions of the ASABE, Annual meeting of the American Society of Agricultural and Biological Engineers*, 1143–52.

Afzalinia, S. and Roberge, M. (2013). Modeling of the pressure-density relationship in a large cubic baler. *Journal of Agricultural Science and Technology* 15, 35–44.

Annoussamy, M., Richard, G., Recous, S., and Guérif, J. (2000). Change in the mechanical properties of wheat straw due to decompostion and moisture. *Applied Engineering in Agriculture*, 657–64.

Bright, R. and Kleis, R. (1964). Mass shear strength of haylage. *Transactions of the ASAE*, 100–1.

Calladine, C. R. (1989). *Theory of Shell Structures*. Cambridge University Press.

Chen, T. (2000). Determining the Prony series for a viscoelastic material from time-varying strain data. *NASA/TM-2000-210123, NAS 1.15:210123, ARL-TR-2206, L-17978*.

Comsol4.3. (2012). *Structural Mechanics Module User's Guide*. Retrieved from \url{http://plato.fab.hs-rm.de /web-mathematik/mathematik/semester8/mechanik/StructuralMechanicsModuleUsersGuide.pdf}.

Cundall, A. and Strack, D. (1979). A discrete numerical model for granular assemblies. *Geotechnique*, 47–65.

De Pascalis, R., Abrahams, I. D., and Parnell, W. J. (2014). On nonlinear viscoelastic deformations: A reappraisal of Fung's quasi-linear viscoelastic model. *Proceedings of the Royal Society A* 470 (2166), 1–18.

Faborode, M. (1989). Moisture effects in the compaction of fibrous agricultural residues. *Biological Wastes*, 61–71.

Faborode, M. O. (1986). *The Compression and Relaxation Behavior of Fibrous Agricultural Materials*. Newcastle upon Tyne.

Faborode, M. and O'Callaghan, J. O. (1986). Theoretical analysis of the compression of fibrous agricultural materials. *Journal of Agricultural Engineering Research* 35, 175–91.

Faborode, M. and Callaghan, J. O. (1989). A rheological model for the compaction of fibrous agricultural materials. *Journal of Agricultural Engineering Research* 42 (3), 165–78.

Favier, J., Abbaspour-Fard, M., Kremmer, M., and Raji, A. (1999). Shape representation of axisymmetrical, nonspherical particles in discrete element simulation using multi-element model particles. *Engineering Computations*, 467–80.

Ferrero, A., Horabik, J., and Molenda, M. (1990). Density-pressure relationships in compaction of straw. *Canadian Agricultural Engineering* 33 (1), 107–11.

Galedar, M., Jafari, A., Mohtasebi, S., Tabatabaeefar, A., Sharifi, A., O'Dogherty, M. *et al.* (2008). Effects of moisture content and lavel in the crop on the engineering properties of Alfalfa stems. *Biosystems Engineering* 101, 199–208.

Geng, F., Li, Y., Wang, X., Yuan, Z., Yan, Y., and Luo, D. (2011). Simulation of dynamic processes on the transverse section of a rotary dryer and its comparison with video-imaging experiments. *Powder Technology*, 175–82.

González-Montellano, C., Gallego, E., Ramírez-Gómez, Á., and Ayuga, F. (2012). Three-dimensional discrete element models for simulating the filling and emptying of silos: Analysis of numerical results. *Biosystems Engineering*, 22–32.

Grof, Z., Kohout, M., and Stepanek, F. (2007). Multi-scale simulation of needle-shaped particle breakage under uniaxial compaction. *Chemical Engineering Science*, 1418–29.

Guo, Y., Curtis, J., Wassgren, C., Ketterhagen, W., and Hancock, B. (2013). Granular shear flows of flexible, rodlike particles. *Powders and Grains*, 491–4.

Guo, Y., Wassgren, C., Hancock, B. Ketterhagen, W., and Curtis, J. (2013). Validation and time step determination of discrete element modeling of flexible fibers. *Powder Technology*, 386–95.

Haghighi-Yazdi, M., and Lee-Sullivan, P. (2011). Modeling linear viscoelasticity in glassy polymers using standard rheological models.

Haupt, P. (1999). *Continuum Mechanics and Theory of Materials*. Springer.

Hu, J.-J., Lei, T. Z., Xu, G.-Y., Shen, S.-Q., & Liu J. W. (2009). Experimantal study of stress relaxation in the process of cold molding with straw. *BioResources.com*, 4 (3), 1158–67.

Junemann, D., Kemper, S., and Frerichs, L. (2013). Simulation of stalks in agricultural processes—Applications of the discrete element method. *Landtechnik*, 164–7.

Kajtar, P. and Loebe, S. (2014). Diskrete element simulation von Halmgut. *Land Technik AgEng 2014*.

Kaliyan, N., and Morey, R. V. (2009). Constitutive model for densification of corn stover and switchgrass. *Biosystems Engineering* 104, 47–63.

Kanafojski, C. and Karwowski, T. (1976). *Agricultural machines theory and construction: Vol. 2 Crop-Harvesting Machines*. Foreign Scientific Publications Department of the National Center for Scientific, Technical and Economic Information.

Kattenstroth, R., Harms, H., and Lang, H. (2011). Systematic alignment of straw to optimize the cutting process in a combines straw chopper. *Proc. of Land. Technik AgEng*.

Kim, J., Oh, K., Pyun, K., Kim, C., Choi, Y., and Yoon, J. (2012). Design optimization of a centrifugal pump impeller and volute using computational fluid dynamics. *Earth and Environmental Science*, 15 (032025).

Klenin, N., Popov, I., and Sakun, V. (1986). *Agricultural machines*. (Ed. D. V. Kothekar.) A.A. Balkema.

Krucinska, I., Jalmuzna, I., and Zurek, W. (2004). Modified rheological model for analysis of compression on nonwoven fabrics. *Textile Research Journal* 74 (2), 127–33.

Kutzbach, H. D. (1973). Die grundlagen der halmgutverdichtung. *Grundlagen der Landtechnik*, 23 (1), 23–4.

Leblicq, T., Smeets, B., Ramon, H., and Saeys, W. (2016). A discrete element approach for modeling the compression of crop stems. *Computers and Electronics in Agriculture*, 80–8.

Leblicq, T., Vanmaercke, S., Ramon, H., and Saeys, W. (2015). Mechanical analysis of the bending behavior of plant stems. *Biosystems Engineering* 129, 87–99.

Lenaerts, B., Aertsen, T., Tijskens, E., De Ketelaere, B., Ramon, H., De Baerdemaeker, J., and Saeys, W. (2014). Simulation of grain-straw separation by discrete element modeling with bendable straw particles. *Computers and Electronics in Agriculture*, 24–33.

Mani, S., Tabil, L., and Sokhansanj, S. (2003). An overview of compaction of biomass grinds. *Powder Handling & Processing*, 160–8.

Mani, S., Tabil, L., and Sokhansanj, S. (2004). Evaluation of compaction equations applied to four biomass species. *Canadian Biosystems Engineering* 46, 3.55–3.61.

Miller, S. (2014). Developing complex turbines with model-based design. *Developing Complex Turbines with Model-Based Design*.

Mohsenin, N. N. (1986). *Physical Properties of Plant and Animal Materials*. Department of Agricultural Engineering, Pennsylvania State University.

Munoz, G. and Herrera, P. (2002). Multidimensional modeling of agricultural fibrous materials in densification: Compression stage.

Nazari, M., Jafari, A., Mohtasebi, S. S., Tabatabaeefar, A., Sharifi, A., O'Dogherty, M. *et al.* (2008). Effects of moisture content and level in the crop on the engineering properties of alfalfa stems. *Biosystems Engineering*, 199–208.

Nguyen, D., Kang, N., and Park, J. (2013). Validation of partially flexible rod model based on discrete element method using beam deflection and vibration. *Powder Technology*, 147–52.

Nona, K. D., Kayacan, E., and Saeys, W. (2013). Compression modeling in the large square baler. *VDI-Berichte* 2193, 387–92.

Nona, K. D., Lenaerts, B., Kayacan, E., and Saeys, W. (2014). Bulk compression characteristics of straw and hay. *Biosystems Engineering* 118, 194–202.

O'Dogherty, M. and Wheeler, J. (1984). Compression of straw to high densities in closed cylindrical dies. *Journal of Agricultural Engineering Research* 29 (1), 61–72.

O'Dogherty, M. J. (1989). A review of the mechanical behavior of straw when compressed to high densities. *Journal of Agricultural Engineering Research* 44, 241–65.

O'Dogherty, M. J., Huber, J. A., Dyson, J., and Marshall, C. J. (1995). A study of the physical and mechanical properties of wheat straw. *Journal of Agricultural Engineering Research* 62, 133–42.

Peleg, K. (1983). A rheological model of nonlinear viscoplastic solids. *Journal of Rheology* 27 (5), 411–31.

Pierantoni, M., Monte, M. D., Papathanassiou, D., Rossi, N. D., and Quaresimin, M. (2011). Viscoelastic material behaviour of PBT-GF30 under thermo-mechanical cyclic loading. *Procedia Engineering* 10 (0), 2141–6.

Plaseied, A. and Fatemi, A. (2008). Deformation response and constitutive modeling of vinyl ester polymer including strain rate and temperature effects. *Journal of Materials Science* 43, 1191–9.

R7.1, L. D. (2014). *LS-DYNA keyword user's manual: Volume I*. Livermore Software Technology Corporation.

Ren, S. (1991). *Thermo-hygro-rheological behavior of materials used in the manufacture of wood-based composites*. College of Forestry.

Roylance, D. (2001). *Engineering viscoelasticity*. Course notes, Cambridge, MA: MIT Press.

Schiessel, H., Metzler, R., Blumen, A., and Nonnenmacher, T. (1995). Generalized viscoelastic models: Their fractional equations with solutions. *Journal of Physics A: Mathematical and General* 28, 6567–84.

Sitkei, G. (1986). *Mechanics of Agricultural Materials*. Amsterdam: Elsevier.

Smeets, B., Odenthal, T., Keresztes, J., Vanmaercke, S. V., Tijskens, E., Saeys, W. *et al.* (2014). Modeling contact interactions between triangulated rounded bodies for the discrete element method. *Comput. Methods Appl. Mech. Engrg.*, 219–38.

Steffe, J. F. (1996). *Rheological Methods in Food Process Engineering*. Freeman Press.

Stelte, W., Clemons, C., Holm, J. K., Ahrenfeldt, J., Henriksen, U. B., and Sanadi, A. R. (2012). Fuel pellets from wheat straw: The effect of lignin glass transition and surface waxes on pelletizing properties. *Bioenergy Research* 5(2), 450–8.

Thoemen, H., Haselein, C., and Humphrey, P. (2006). Modeling the physical processes relevant during hot pressing of wood-based composites. Part II. Rheology. *Holz als Roh- und Werkstoff* 64, 125–33.

Tijskens, E., Ramon, H., and De Baerdemaeker, J. (2003). Discrete element modeling for process simulation in agriculture. *Journal of Sound and Vibration*, 493–514.

Van Liedekerke, P., Tijskens, E., and Ramon, H. (2009). Discrete element simulations of the influence of fertilizer physical properties on the spread pattern from spinning disc spreaders. *Biosystems engineering*, 392–405.

Verhaeghe, D., Coen, T., and Nona, K. (2015). WO Patent No. PCT/EP2014/065, 355.

Verhás, J. (1997). *Thermodynamics and Rheology*. Dordrecht: Kluwer Academic Publishers.

Wiezlak, W. and Gniotek, K. (1999). Indices characterizing rheological properties of geo-nonwovens during compression. *Fibres and Textiles in Eastern Europe* 7 (3), 40–4.

Yu, M., Womac, A., Igathinathane, C., Ayers, P., and Buschermohle, M. (2006). Switchgrass ultimate stresses at typical biomass conditions available for processing. *Biomass and Bioenergy*, 214–9.

Zdenek, G. and Frantisek, S. (2013). Distribution of breakage events in random packings of rodlike particles. *Physical Review*, 1–5.

Zhao, Y., Cao, P., Wang, W., and Wan, W. (2009). Viscoelasto-plastic-rheological experiment under circular increment step load and unload and nonlinear creep model of soft rocks. *Journal Cent. South Univ. Technol.* 16, 488–94.

Zheng, M., Yaoming, L., and Lizhang, X. (2015). Discrete-element method simulation of agricultural particles' motion in variable-amplitude screen box. *Computers and Electronics in Agriculture*, 92–9.

17 Three-Dimensional (3D) Numerical Modeling of Morphogenesis in Dehydrated Fruits and Vegetables

C. M. Rathnayaka, H. C. P. Karunasena,
Wiji Senadeera, Lisa Guan, and Y. T. Gu

CONTENTS

17.1   INTRODUCTION

Plant-sourced food items, such as fruits and vegetables, are an integral part of the human diet. Fruits and vegetables, by their nature, contain up to 90% water (Jangam, 2011), which induces internal microbial activities resulting in rapid spoilage. Therefore, the removal of water from food matter makes it more resistant to spoilage (Chen and Mujumdar, 2009) since microorganisms cannot thrive in dry environments (Delong, 2006). Drying is the oldest method of economically removing moisture from food materials, resulting in traditional (as well as innovative) dried food products (Jangam, 2011). Recently, there has been a significant increase in the consumption of dehydrated food in the market (De la Fuente-Blanco *et al.*, 2006). This necessitates to efficiently produce high quality dried food products, where a close control of moisture content of the product has to be maintained, as shown in the drying curve in Figure 17.1. It is clear that the moisture tends to remove rapidly in a given bulk food sample at the beginning of a drying process, followed by a reducing trend, due to the collapse of the food structure, as elaborated in Figure 17.2a and 17.2b. Here, a bulk scale deforming and shrinking behavior of a fresh apple sample is presented, before and after drying. As evidenced by Figures 17.1 and 17.2, the bulk food material undergoes gradual physical alterations leading to morphological changes with the removal of moisture. Depending on these variations in the bulk scale, it is evident that the cellular structure of the food material should similarly undergo consequent deformations during the process.

Figure 17.3 presents microscopic images of the apple cellular structure before and after drying, obtained through digital light microscopy, and the corresponding microscopic quantitative geometric

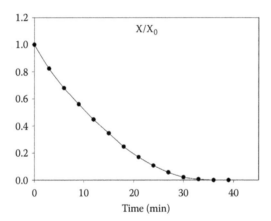

FIGURE 17.1   Typical drying curve (variation of normalized moisture content with drying time). This curve is for apple under 70°C.

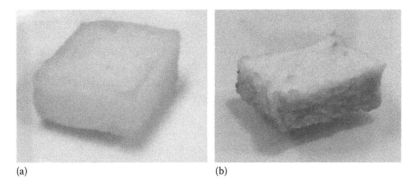

(a)   (b)

FIGURE 17.2   Bulk scale morphological behavior of apple samples. (a) Fresh sample; (b) dried sample.

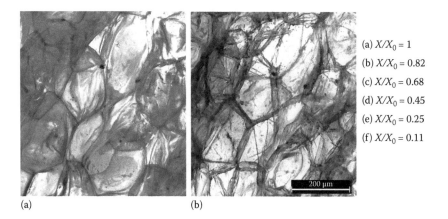

(a) $X/X_0 = 1$

(b) $X/X_0 = 0.82$

(c) $X/X_0 = 0.68$

(d) $X/X_0 = 0.45$

(e) $X/X_0 = 0.25$

(f) $X/X_0 = 0.11$

(a)   (b)

FIGURE 17.3 Apple cellular structure (a) before and (b) after drying at 70°C, observed through the digital light microscope (the same set of cells were imaged).

parameters measured are given in Table 17.1. Since the significant fall in the cell area, diameter, and perimeter have occurred during drying, it is evident that there is a substantial morphological alteration occurring in the food microstructure as well. This alteration can further be classified as an irregular deformation, since the secondary or derived parameters like roundness, elongation, and compactness have varied due to drying. This fundamentally agrees with the phenomena of bulk level morphological behavior having a cumulative effect of microscopic level deformations (Taiz and Zeiger, 2002). On the other hand, these cellular and tissue level structural changes have a strong impact on the dried product quality and the market value. Hence, advanced studies, such as the numerical modeling of such morphological changes, are critical for the field of food engineering.

Within this context, two of the key parameters will be highlighted: moisture content (Mayor, Silva, and Sereno, 2005; Ramos *et al.*, 2004; Hills and Remigereau, 1997; Lewicki and Pawlak, 2003) and the drying temperature (Ratti, 2001). Here, the moisture content is strongly correlated with the cell turgor pressure (Bartlett, Scoffoni, and Sack, 2012) while the temperature is related to the relative humidity. To better understand the effects exerted by these driving forces on the plant morphogenesis, various theoretical (Crapiste, Whitaker, and Rotstein, 1988) and empirical (Mayor, Silva, and Sereno, 2005; Ramos *et al.*, 2004) models have been developed. In this regard, numerical modeling has proven to be a better option over the experimental procedures due to its efficiency in terms of time, cost, and resources (Jangam, 2011). It is highly versatile, and produces detailed results supporting enhanced fundamental understanding on different aspects associated with morphogenesis

TABLE 17.1

Measured Microscopic Geometric Parameters of the Apple Sample Subjected to Drying Experiments at 70°C (Karunasena, Hesami et al., 2014)

Cellular Geometric Parameter	Fresh State ($X/X_0 = 1.000$)	Dried State ($X/X_0 = 0.014$)
Area (μm^2)	21341	8749
Diameter (μm)	164	105
Perimeter (μm)	601	374
Roundness	0.727	0.769
Elongation	1.459	1.600
Compactness	0.839	0.811

of food structures during drying. For decades, numerical modeling has investigated morphological behavior of various types of materials. However, until recently it had not been used for studying the micro-structural morphogenesis of plant food matter during drying. Lately though, it has attracted much attention as a viable tool to serve this purpose (Karunasena et al., 2014).

In this background, one of the main intentions of this chapter is to assess various attempts made to numerically model plant cells and tissues during drying, along with subsequent phenomena. The chapter evaluates the accuracy, versatility, and the potential influence toward the enhancement of the drying operation performance both in the current context and future.

17.2 VARIOUS NUMERICAL MODELING TECHNIQUES USED FOR SIMULATING PLANT FOOD DRYING

17.2.1 MODELS BASED ON THE FINITE ELEMENT METHOD (FEM) AND THE FINITE DIFFERENCE METHOD (FDM)

Plant cell numerical models which employ the Finite Element Method (FEM) or the Finite Difference Methods (FDM) mainly focus on correlating the micro-mechanical features of the plant cellular structure to the macroscopic characteristics of the bulk plant structure (Zhu and Melrose, 2003; Lardner and Pujara, 1980; Gao and Pitt, 1991). In addition, these models have investigated the response of the plant cellular structure under external mechanical loading with the use of the elasticity theory for large deformations on a lattice of regular ideal 3D hexagonal cells (Zhu and Melrose, 2003). Numerical prediction of the cell wall stiffness has been realized by modeling the force-deformation behavior (Wang, Wang, and Thomas, 2004; Rathnayaka et al., 2017a).

FEM– and FDM–based plant cell or tissue models could accurately capture the deformational properties of solid components, such as the cell wall or tissue boundaries. However, both FEM and FDM exhibit many limitations, such as the lack of ability to handle multiphase phenomena and large deformations, particularly in discrete domains such as the plant cellular structure. Additional complications arise when modeling the internal cell fluid. As a result, there are some FEM models which try to model the combined effect of the cell fluid and the cell wall by virtually exerting a distributed force applied on the internal surface to account for the effect of the turgor pressure (Wu and Pitts, 1999). The existing FEM– and FDM–based models are mostly restricted to fresh plant cells and tissues (Smith, Moxham, and A.P.J., 1998; Wu and Pitts, 1999; Pitt, 1982; Pathmanathan et al., 2009; Loodts et al., 2006; Wang, Wang, and Thomas, 2004; Gao and Pitt, 1991; Nilsson, Hertz, and Falk, 1958). Therefore, such models have limitations in representing shrinkage, porosity development, or surface wrinkling, which are unique phenomena observed in the food cellular structure during drying. In this background, some of these conventional models have oversimplified the cell walls to be impermeable (Zhu and Melrose, 2003) or have imposed artificial restrictions to the moisture transport through the cell wall (Smith, Moxham, and A.P.J., 1998). The applicability of such FEM and FDM methods to model drying characteristics of plant cellular structure would be limited.

17.2.2 MODELS DEVELOPED BY COMBINING FEM AND OTHER TECHNIQUES

17.2.2.1 Basic Vertex Cell Models

Vertex models possess the ability to approximate the fresh plant cell behavior, mimicking biological growth and division of true cells (Honda, Yamanaka, and Dan-Sohkawa, 1984) as well as cell migration and pattern formation (Fleischer and Barr, 1994). There are considerable drawbacks in this approach, especially when modeling drying related characteristics such as shrinkage, porosity development, and surface wrinkling. The recent 3D vertex cell model, which has been developed to study the response of cell aggregates under mechanical loading, is confined to limited deformations (Honda, Tanemura, and Nagai, 2004). Hybrid vertex models have been developed to overcome the limitations of these basic vertex models (Ishimoto and Morishita, 2014). However, these hybrid models have critical limitations, such as confining to linear cell walls and oversimplified turgor pressure representation, causing them to be not directly applicable for modeling cells or tissues undergoing drying.

17.2.2.2   Vertex Tissue Drying Models

There have been several recent attempts to numerically model and simulate dried plant materials using an improved form of vertex modeling that involve polygon-shaped cells following some other vertex models (Abera *et al.*, 2013) and a virtual tissue generator which follows cell growth concepts (Fanta *et al.*, 2014). This vertex tissue model investigates the relationship between water transport characteristics and the structural deformations observed in plant cellular structure under slow dehydration, which usually happens during the general storage conditions. This approach has a limited ability to describe high moisture content reductions and the subsequent large deformations observed during real drying operations. For instance, the model can represent about 30% of overall moisture reduction, which is not sufficient to model a true drying scenario, where the overall moisture removal can be as high as 95%. At the same time, polygon-shaped cells with linear cell wall segments tend to remain straight, and do not accurately represent the deformed or wrinkled cell walls detected during real drying operations. Further, this numerical approach is not straightforward, as it requires a number of routines or subroutines and auxiliary software (Fanta *et al.*, 2014). These limitations arise mainly due to the inherent shortcomings encountered in grid-based modeling approaches when dealing with excessive deformations.

17.3   USE OF NOVEL, MESHFREE-BASED APPROACHES TO OVERCOME LIMITATIONS OF GRID-BASED TECHNIQUES

Meshfree modeling methods are able to overcome the limitations of conventional grid-based techniques. They do not operate on an underlying fixed mesh to discretize the system domain as it is in classical FEM or FDM methods. Alternatively, a scheme of particles are used to represent the problem domain, where the particles are not connected permanently to each other or they do not have predetermined neighbors (Liu and Liu, 2010). This makes it possible for the meshfree techniques to produce more stable solutions for complex physical problems, such as large deformations and multiphase phenomena (Liu and Liu, 2003b).

Meshfree approaches have proven to be stable as well as versatile with high accuracy and efficiency applicable to a wider range of problems (Belytschko *et al.*, 1996). At present, there is a wide range of meshfree methods developed to suit different applications (Liu, 2010; Liu and Liu, 2003b; Liu and Gu, 2005; Li and Liu, 2002; Belytschko *et al.*, 1996; Idelsohn and Onate, 2006; Nguyen *et al.*, 2008; Frank and Perré, 2010). Among them, the most popular techniques are: Smoothed-Particle Hydrodynamics (SPH) (Liu and Liu, 2010), Element-Free Galerkin (EFG) method (Belytschko, Lu, and Gu, 1994), Point Interpolation Method (PIM) (Liu and Gu, 2002), Meshless Local Petrov-Galerkin (MPLG) method (Atluri and Zhu, 1998), Local Radial Point Interpolation Method (LRPIM) (Liu and Gu, 2001), and the boundary point interpolation method (BPIM) (Liu and Gu, 2004).

17.3.1   Smoothed Particle Hydrodynamics (SPH)

Smoothed Particle Hydrodynamics (SPH) is the most developed and versatile meshfree particle method (Liu and Liu, 2010). By nature, it is a Lagrangian method and was originally developed for astrophysical studies (Gingold and Monaghan, 1977; Lucy, 1977). In SPH, a given system is characterized by a set of particles that are not connected through a fixed mesh. The particles characterize the properties of the system and interact with neighboring particles inside an influence domain defined by a smoothing kernel or weight function (Liu and Liu, 2003b). The governing equations (e.g., Navier-Stokes equations) are discretized with the use of the particles, and the field properties such as density, mass, pressure, position, velocity, and acceleration are computed based on the influence domain of each particle (Liu and Liu, 2010; Perré, 2011). Every SPH particle represents a small portion of the entire system depending on the number of particles used to represent a given domain (Frank and Perré, 2010). The pressure of the system is evolved through an Equation of State (EOS) and the viscosity of the fluid is also incorporated into computations. As time elapses, particle properties get updated, and more importantly, particles are displaced according to the force fields developed on them, highlighting the Lagrangian nature of the computational method.

Compared to conventional grid-based approaches, SPH has been proven to perform well in advanced modeling applications, such as large deformations related to multiphase flows, free surfaces, deformable boundaries, and moving interfaces (Liu and Liu, 2010; Morris, Fox, and Zhu, 1997). For instance, when simulating explosions, large deformations exist, and the system is essentially composed of discretized domain segments moving relative to each other, which has been successfully modeled with SPH (Zhang, 1976; Liu and Liu, 2010). Likewise, high velocity impact (HVI) systems and subsequent shock wave propagation problems (Zukas, 1990) with discontinuities and phase fragmentations can effectively and efficiently be simulated with SPH (Liu and Liu, 2003b; Frank and Perré, 2010). The material boundaries, moving boundaries, and free surface flows could be tracked in SPH simulations, even in complicated systems whereas for grid-based methods such tasks would be comparatively difficult. The versatility of SPH could allow one to couple SPH and Molecular Dynamics (MD) or SPH and Dissipative Particle Dynamics (DPD) for examining multiscale problems in biomechanics.

Other SPH applications include multi-phase fluid modeling (Colagrossi and Landrini, 2003), coastal modeling (Monaghan, 1994), environmental and geophysical flows (Tartakovsky and Meakin, 2005), heat and mass transfer (Chen, Beraun, and Carney, 1999), ice and cohesive grains (Gutfraind and Savage, 1998), microfluidics (Liu and Liu, 2005), elastic and plastic flows (Randles and Libersky, 1996), fracture of solids (Benz and Asphaug, 1995), metal forming (Bonet and Kulasegaram, 2000), magneto-hydrodynamics (Ala *et al.*, 2007), red blood cell deformation within blood flow (Polwaththe-Gallage *et al.*, 2015), and traffic flow applications (Rosswog and Wagner, 2002).

The main drawbacks of SPH are associated with the computational aspects. For example, the time step for the computations reduces significantly in the presence of spatial discontinuities (Bhojwani and Engin, 2007). SPH is more computationally expensive than other meshfree methods, as it approximates each particle's field properties through its neighboring particles while an increased number of non-interconnected particles are required to achieve a desirable accuracy during the computations. Another notable limitation is the deficiencies in handling shock wave problems. Finally, the error estimation in SPH is not always accurate.

17.3.2 Coarse-Grained Approaches

In coarse-grained (CG) modeling approaches, the number of degrees of freedom of a system are brought down by using simpler approximations (Tozzini, 2005). A network of characteristic particles is employed to represent the whole system while the properties are concentrated into these representative particles. The relations among CG particles could be made more sophisticated (if necessary) to replicate the real physical description. The CG particle network behaviors are very similar to their counterparts in SPH. CG methods have been extensively used to model and simulate various categories of biopolymers, as well as the ones that make up the plant cell walls (Petridis *et al.*, 2014; Fan and Maranas, 2015; Lu, Maranas, and Milner, 2016).

17.3.3 SPH in Plant Cell Modeling

There have been several efforts to numerically model the macro- and micro-mechanics of plant materials using a coupled SPH and DEM approach (Van Liedekerke, Ghysels *et al.*, 2010;

Van Liedekerke *et al.*, 2011; Van Liedekerke, Tijskens *et al.*, 2010). This approach has been used to simulate both individual plant cells and their aggregates under external loading (Van Liedekerke, Ghysels *et al.*, 2010). In these "first generation" SPH-DEM coupled models, a given cell is defined using two model segments, namely the cell fluid and cell wall. Cell fluid represents the protoplasm of actual cells, which is the interior liquid volume and is modelled with SPH. Cell wall is the outer boundary of the cell, which is essentially a solid material with viscoelastic characteristics and modeled with DEM (Van Liedekerke *et al.*, 2011). The cell wall hydraulic conductivity has also been integrated into the model description via a constitutive relationship (Van Liedekerke, Ghysels *et al.*, 2010; Rathnayaka *et al.*, 2017a).

A "second generation" of models have been developed based on the fundamentals of the above "first-generation" models by incorporating novel force fields and model features in order to represent the cellular drying mechanisms. Accordingly, two-dimensional (2D) individual plant cells (Karunasena, Senadeera, Gu *et al.*, 2014; Karunasena, Senadeera, Brown *et al.*, 2014b) and cell aggregates (Karunasena, Gu *et al.*, 2015a; Karunasena, Brown *et al.*, 2015; Karunasena, Gu *et al.*, 2015b; Karunasena, Senadeera, Brown *et al.*, 2014a; Karunasena, Senadeera, Brown *et al.*, 2014c) have been modeled to study various drying phenomena such as shrinkage, porosity development, and case hardening. The model outcomes have compared with experimental findings on real plant cells and tissues such as apple, grape, potato, and carrot. A summary of strengths and limitations of various numerical modeling techniques employed for the simulation of plant food cells or tissues is provided in Table 17.2 (Rathnayaka *et al.*, 2017a).

DEM was initially developed to solve problems in soil mechanics (Mishra and Rajamani, 1992). The DEM particles can have different geometry and physical properties including the specific shape, dimension, and material properties. The interaction between DEM particles is usually indirect. The plant cell wall can be treated as a continuum thin membrane consisting of various biopolymers. DEM is conceptually inappropriate to model the plant cell wall. The literature suggests that conceptually, a coarse-grained (CG) approach could be more suited for modeling plant cells or tissues (Petridis *et al.*, 2014; Fan and Maranas, 2015; Lu, Maranas, and Milner, 2016). Furthermore, there is a stronger conceptual and fundamental matching in an SPH-CG coupling than in an SPH-DEM coupling.

Compared to the above 2D meshfree models, 3D meshfree models (Rathnayaka *et al.*, 2016) are being developed in order to investigate plant food drying phenomena in a more realistic scale, which is detailed out in this chapter. As such, a novel SPH-CG approach is employed considering its fundamental capability to better simulate key cellular drying characteristics such as: large deformations, complex physics, multiphase phenomena, as well as the conceptual agreement of the coupled numerical frameworks. This work assumes that cells are the elementary building blocks that build up bulk plant biological structures, where different physiological and biochemical functions are implemented by different types of cells in a given plant. Out of the main tissue varieties in plants, ground tissue systems are responsible for the largest part, in which the parenchyma cells are the majority (Taiz and Zeiger, 2002). This study focuses on parenchyma cells, particularly on individual cells where model formulation and computational implementation can conveniently be understood. The following sections present details of the steps involved in developing the model, its numerical implementation, validation, and scope for future work.

17.4   THE FUNDAMENTALS OF THE COUPLED SPH-CG METHODOLOGY USED IN THE 3D SINGLE-CELL MODEL

17.4.1   3D PARTICLE ARRANGEMENT WITHIN A CELL

Following the original approach of Van Liedekerke *et al.* (Van Liedekerke, Ghysels *et al.*, 2010; Van Liedekerke *et al.*, 2011; Van Liedekerke, Tijskens *et al.*, 2010) and Karunasena *et al.* (Karunasena, Senadeera, Gu *et al.*, 2014; Karunasena, Senadeera, Brown *et al.*, 2014b; Karunasena, Hesami *et al.*, 2014; Karunasena, Gu *et al.*, 2015a; Karunasena, Brown *et al.*, 2015; Karunasena, Gu *et al.*, 2015b;

TABLE 17.2

Strengths and Limitations of Various Numerical Modeling Approaches Utilized to Simulate Plant Food Cellular Structures (Rathnayaka *et al.*, 2017a)

Numerical Modeling Technique	Strengths	Limitations
FEM/FDM based methods (grid-based) (Abera *et al.*, 2013; Fanta *et al.*, 2014; Liu *et al.*, 2010)	Well suited for modeling solid materials Relatively more stable More user-friendly software and consequent Graphical User Interface (GUI) Have been in use for a long time	Deficiencies in modeling multiphase phenomena Deficiencies in dealing with large deformations and complicated physics Lack of ability to efficiently simulate the combined effect of heat and mass transfer
Vertex Models (grid-based) (Ishimoto and Morishita, 2014; Rudge and Haseloff, 2005)	Specialized in investigating cell growth aspects Strong biological foundation Well suited to study cell division	Inability to effectively deal with large deformations and moisture contents Neglecting the curved nature of the cell walls Inefficient simulations procedure
"First generation" of meshfree methods for fresh plant cells (meshfree) (Van Liedekerke, Ghysels *et al.*, 2010)	Ability of demonstrating large deformations and multiphase phenomena Well suited to study plant tissue deformations under different mechanical loading schemes	Not considering drying phenomena Increased computational cost when dealing with large scale tissues
"Second generation" of meshfree models for both fresh and dry cells/tissues (meshfree) (Karunasena, Senadeera, Brown *et al.*, 2014a)	Drying phenomena of plant cellular structures could be studied in detail Efficient incorporation of multiphase interactions and complicated physics Versatile modeling and simulation scheme	The analysis is only limited to 2D scale The heterogeneity in plant cellular structure has not been addressed Cells/tissues take the ideal shapes (irregularities have been omitted)

Karunasena, Senadeera, Brown *et al.*, 2014c; Karunasena, Senadeera, Brown *et al.*, 2014a), a cell is modeled as composed of two main components: cell fluid and cell wall. It also follows the fundamental assembly of a plant parenchyma cell, where the solid flexible wall enclosing the mass of cell fluid (Lewicki and Pawlak, 2003; Wang, Wang, and Thomas, 2004; Pitt, 1982; Taiz and Zeiger, 2002; Smith, Moxham, and A.P.J., 1998; Wu *et al.*, 1985; Jarvis, 1998). The cell fluid is taken as an incompressible, high viscous, homogeneous Newtonian liquid and modeled with Smoothed-Particle Hydrodynamics (SPH). The cell wall is considered to be a stiff, semipermeable membrane with viscoelastic characteristics similar to an incompressible neo-Hookean solid material. A Coarse-Grained (CG) method is used for the numerical implementation of the cell wall model.

A physical equilibrium has been assumed between the fluid turgor pressure and cell wall hoop stress. In other words, the stretchable cell wall holds the fluid mass inside via balancing the force exerted by fluid and creating tension within the cell wall. Accordingly, the basic geometry of a single cell has been considered to be spherical (Nilsson, Hertz, and Falk, 1958) and the cell fluid has been considered as a solid spherical object, while the cell wall has been considered as a hollow, 3D spherical shell (see Figure 17.4). Thermodynamically, the whole cell has been regarded as an isothermal system (Karunasena, Senadeera, Gu *et al.*, 2014).

The cell fluid and cell wall are discretized using a particle scheme where the whole cell is represented as a unit composed of non-interconnected set of particles, as presented in Figure 17.4, following the fundamentals of meshfree particle based methods (Liu and Liu, 2003b;

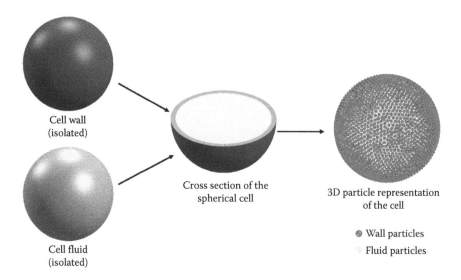

FIGURE 17.4 3D particle placement of the single cell model consisting of cell fluid particles and cell wall particles.

Cleary and Monaghan, 1999; Monaghan, 2005). The relationship among the particles has been described using a set of governing equations, which are presented in Sections 17.4.2 and 17.4.3. Due to the fundamental adaptability of this particle based modeling, the aggregation of single cells in order to develop a tissue is straightforward (Van Liedekerke, Tijskens *et al.*, 2010). Furthermore, the adopted modeling framework is highly flexible to investigate subcellular mechanisms during drying, which is a key factor to be incorporated to understand the complex mechanisms driving cellular and bulk morphological changes.

17.4.2 Modeling of the Cell Fluid

Since the water content of the cell fluid (protoplasm) is generally very high, the cell fluid is approximated to be an incompressible homogeneous Newtonian fluid similar to water. Navier-Stokes equations are used to model its behavior. The fluid viscosity is taken to be higher than water and low Reynolds number flow characteristics have been hypothesized (Karunasena, Senadeera, Brown *et al.*, 2014a; Morris, Fox, and Zhu, 1997). The model accounts for four types of force fields to fully define the interaction of the cell fluid: pressure forces (F^p), viscous forces (F^v), wall-fluid attraction forces (F^a), and wall-fluid repulsion forces (F^{rw}). The resultant total force F_i on any fluid particle i is found given in Equation 17.1 of Table 17.3 (see Figure 17.5).

The pressure force ($F^p_{ii'}$) and the viscous force ($F^v_{ii'}$) for any fluid particle i is defined as the summed influence of the respective forces exerted by its surrounding fluid particles i'. This is derived through the standard SPH momentum equation for weakly compressible and low Reynold's number flows (Morris, Fox, and Zhu, 1997). The relevant formulations are shown in Equation 17.2 and Equation 17.3, in Table 17.3. As the smoothing kernel, the cubic spline has been employed. It is one of the most widely applied kernel functions in SPH studies because it is well known as a stable and computationally efficient kernel (see Equation 17.4) (Liu and Liu, 2003b). From Equation 17.2, Equation 17.3, and Equation 17.4, it could be seen that the smoothing kernel (W) and the s value is significant in determining pressure and viscous forces ($F^p_{ii'}$ and $F^v_{ii'}$) in the fluid domain. The smoothing length (h) should vary according to the SPH standard protocol to maintain an ideal number of particles inside each and every influence domain. To execute this, a variable smoothing length procedure is implemented through a simple geometrical relationship (see Equation 17.5).

TABLE 17.3
Governing Equations and Corresponding Formulation for the Fluid Particle Scheme

No.	Equation	Formulation	Description
17.1	Total force on a fluid particle	$F_i = F_{ii'}^p + F_{ii'}^v + F_{ik}^{rw} + F_{ik}^a$	i': neighboring fluid particles k: interacting wall particles
17.2	Pressure force on a fluid particle	$F_{ij}^p = -m_i \sum_{i'} m_j \left(\dfrac{P_i}{\rho_i^2} + \dfrac{P_{i'}}{\rho_{i'}^2} \right) \nabla_i W_{ii'}$	m: fluid particle mass P: cell turgor pressure ρ: particle density μ: dynamic viscosity W: smoothing kernel
17.3	Viscous force on a fluid particle	$F_{ij}^v = m_i \sum_{i'} m_{i'} \left(\dfrac{\mu_i + \mu_{i'}}{\rho_i \rho_{i'}} \right) \mathbf{v}_{ii'} \dfrac{1}{r_{ii'}} \dfrac{\partial W_{ii'}}{\partial r_{ii'}}$	h: smoothing length s: the ratio of $r_{ii'}/h$
17.4	Smoothing kernel function	$W_{ii'}(s,h) = \dfrac{2}{3\pi h^3} \begin{cases} \dfrac{2}{3} - s^2 + \dfrac{s^3}{2} & 0 < s < 1 \\ \dfrac{(2-s)^3}{6} & 1 < s < 2 \\ 0 & s > 2 \end{cases}$	$r_{ii'}$: distance between particle i and i'
17.5	Variable smoothing length	$h = \left(\dfrac{D}{D_0} \right) h_0$	D: current cell diameter D_0: initial cell diameter h_0: initial smoothing length
17.6	Equation of state (EOS)	$P_i = P_T + K \left[\left(\dfrac{\rho_i}{\rho_0} \right)^7 - 1 \right]$	P_T: initial turgor pressure K: fluid compression modulus ρ_i: current density ρ_0: initial density
17.7	Density–general definition	$\rho_i = \dfrac{m_i}{v_i}$	m: fluid particle mass v_i: corresponding fluid volume
17.8	Density–time derivation	$\dfrac{d\rho_i}{dt} = m_i \dfrac{d}{dt}\left(\dfrac{1}{v_i}\right) + \dfrac{1}{v_i}\dfrac{dm_i}{dt} = \dfrac{d\rho_i^*}{dt} + \dfrac{\rho_i}{m_i}\dfrac{dm_i}{dt}$	ρ_i^*: density (assuming constant particle mass)
17.9	Density–SPH approximation of continuity	$\dfrac{d\rho_i^*}{dt} = \sum_{i'} m_{i'} V_{ii'} \cdot \nabla_i W_{ii'} = m_i \sum_{i'} V_{ii'} \cdot \nabla_i W_{ii'}$	
17.10	Mass transfer from/to the cell	$\dfrac{dm_i}{dt} = -\dfrac{A_c L_p \rho_i}{n_f}(P_i + \Pi),$	L_P: cell wall permeability n_f: number of fluid particles A_C: cell wall surface area
17.11	LJ repulsion forces	$F_{ik}^{rw} = \sum_k f_{ik}^{rw} x_{ik}$	f_{ik}^{rw}: magnitude of the force x_{ik}: position vector
17.12	LJ attraction forces	$F_{ik}^a = \sum_k f_{ik}^a x_{ik}$	f_{ik}^a: magnitude of the force x_{ik}: position vector

An Equation of State (Equation 17.6) is used to maintain the relationship between the density and the pressure. Fluid compression modulus (K) represents the compressibility of the cell fluid in order to ensure that the fluid is sufficiently incompressible (Van Liedekerke, Ghysels *et al.*, 2010). The general definition for density is stated by Equation 17.7. And the differentiation of that relationship with respect to time leads to Equation 17.8. A modified version of the SPH continuity equation is used to compute the fluid density during simulations as given in Equation 17.9. In Equation 17.8, the first term in the right-hand side denotes the density variation as affected by cell volume variation, while the second term represents the change in water content due to the cell wall permeability which is defined in Equation 17.10. The difference between the cell turgor pressure and the osmotic potential $\Pi (\Pi < 0)$ creates a net rate of mass transfer occurring across the semi-permeable cell wall. This fluid mass transfer rate can be represented as a constitutive equation as given by Equation 17.10 while accounting for the cell wall hydraulic conductivity (L_p) (Taiz and Zeiger, 2002).

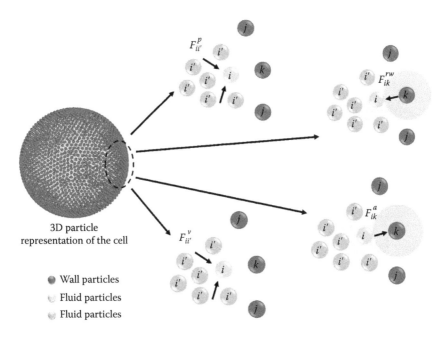

FIGURE 17.5 Force fields on the fluid particle domain: pressure forces (F^p), viscous forces (F^v), wall-fluid repulsion forces (F^{rw}), and wall-fluid attraction forces (F^a).

If there is a net moisture transfer into the cell (inflation), ideally the density increases, leading to pressure increments. To physically balance this high pressure of the cell fluid, it tends to push the cell wall radially outward. As a result, the cell wall stretches longitudinally and develops corresponding stresses within cell wall. After a number of cycles, the forces become balanced in the system and the cell wall ceases to deform further (Van Liedekerke, Tijskens et al., 2010). The boundary conditions have been set through the repulsion forces F_{ik}^{rw} and attraction forces F_{ik}^a among the wall and fluid particles of the cell. They have been defined as Lennard-Jones (LJ) type of forces, which will be further discussed in the next section. They are defined in Equations 17.11 and 17.12.

17.4.3 MODELING OF THE CELL WALL

Since the cell wall exhibits both elastic and plastic characteristics, it is difficult to explain the behavior of the plant cell wall membrane using a simple linear elasticity theory. In addition, the irreversible energy dissipation of the cell wall has to be attributed to structural and viscous damping, which strongly depends on the time scale of the numerical investigation (Veytsman and Cosgrove, 1998; Van Liedekerke, Tijskens et al., 2010). During short time scales, it could be considered that the composition of the cell wall membrane material stays constant. Under such circumstances, cell wall can be considered as a thin walled structure, represented by an elasto-visco-plastic or a hyperelastic constitutive law. In this investigation, a coarse-grained (CG) approach was used where the particles are evenly dispersed on the surface of a 3D spherical shell. It has been assumed that these wall particles are locally connected to each other and that they react with each other through discrete forces. Each particle possesses physical properties corresponding to the respective cell wall segment.

There are four types of force fields that define the physical behavior of the cell wall particle scheme: stiff forces (F^e), damping forces (F^d), wall-fluid attraction forces (F^a), and wall-fluid repulsion forces (F^{rf}) as presented in Figure 17.6. The total force (F_k) on any wall particle k is defined as given in Equation 17.13 in Table 17.4. A spring network model is used to determine the stiff forces using elastic energy functions, as shown in Figure 17.6 and Equation 17.14 of Table 17.4. A linear

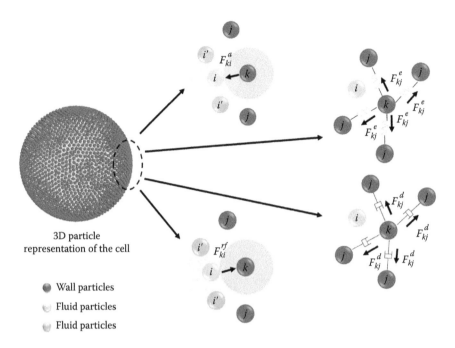

FIGURE 17.6 Force fields on the wall particle domain: stiff forces (F^e), damping forces (F^d), wall-fluid attraction forces (F^a), and wall-fluid repulsion forces (F^{rf}).

dashpot model has been used as given in Equation 17.15 in order to describe the damping forces (F^d) which account for the viscous properties of the cell wall.

The boundary conditions have been defined in this model using Lennard-Jones (LJ) type wall-fluid attraction forces (F^a) and wall-fluid repulsion forces (F^{rf}). Wall-fluid repulsion forces ensure that all fluid particles are appropriately repulsed and retained within the system boundary (i.e., wall). It is necessary to have a sufficiently strong LJ repulsive force field to prevent slip conditions (i.e., fluid particles penetrating the wall). The force F^{rf}_{ki} on a wall particle k is given by Equations 17.16 and 17.17 (Van Liedekerke, Ghysels *et al.*, 2010). In order to prevent the unrealistic separations between the wall and the fluid, wall-fluid attraction forces (F^a_{ki}) are used. These forces play a pivotal role in determining the morphological characteristics of the single cells system and are effective at extremely dried states. The attraction forces F^a_{ki} on a wall particle are defined similar to F^{rf}_{ki} using a LJ type of force field with an LJ contact strength of f^a_0.

17.4.4 Execution Methodology for the Numerical Scheme

The model described in the preceding section was used to model an apple cell with the physical properties, as given in Table 17.5. The model performance was then compared and validated with published literature on both experimental and numerical aspects (Mayor, Silva, and Sereno, 2005; Karunasena, Hesami *et al.*, 2014; Wu and Pitts, 1999; Rathnayaka *et al.*, 2017b). In order to generate the initial 3D particle schemes for both the fluid and the wall, COMSOL Multiphysics was used. The Leapfrog method (Liu and Liu, 2003a) was employed for the time integration while Courant-Friedrichs-Lewy (CFL) stability criterion was utilized to determine the size of the time step so that the model is stable (Liu and Liu, 2003a; Colagrossi *et al.*, 2012).

During the simulations, the associated turgor pressure variations of the cell fluid radially push or pull the cell wall causing either shrinkage or inflation of the cell. Cellular properties like cell volume, equivalent diameter, and surface area tend to change as the cell model evolves with time. When the turgor pressure changes, the rate of mass transfer also changes as given by the mass transfer equation

TABLE 17.4

Governing Equations and Corresponding Formulation for the Wall Particle Scheme

No.	Equation	Formulation	Description
17.13	Total force on a wall particle	$F_k = F_{kj}^e + F_{kj}^d + F_{ki}^{rf} + F_{ki}^a$	k: wall particle i: surrounding fluid particles j: bonded wall particles
17.14	Stiff force on a wall particle	$F_{kj}^e = \dfrac{Gl_0 t_0}{\sqrt{3}}\left(\lambda - \dfrac{1}{\lambda^5}\right)n$	G : shear modulus ($\approx E/3$) E: Young's modulus $\lambda = l/l_0$: current stretch ratio l: current length l_0: initial length t_0: cell wall thickness
17.15	Damping force on a wall particle	$F_{kj}^d = -\gamma v_{kj}$	γ: damping constant v_{kj}: relative velocity
17.16	LJ repulsion force	$F_{ki}^{rf} = f_{ki}^{rf} x_{ki}$	f_{ki}^{rf}: magnitude of the force x_{ki}: position vector
17.17	LJ force magnitude	$f_{ki}^{rf} = \begin{cases} f_0^{rf}\left[\left(\dfrac{r_0}{r_{ki}}\right)^8 - \left(\dfrac{r_0}{r_{ki}}\right)^4\right]\left(\dfrac{1}{r_{ki}^2}\right)\left(\dfrac{r_0}{r_{ki}}\right) & \left(\dfrac{r_0}{r_{ki}}\right) \geq 1 \\ \\ 0 & \left(\dfrac{r_0}{r_{ki}}\right) < 1 \end{cases}$	r_0: initial gap r_{ki}: current gap f_0^{rf}: LJ contact strength

(Equation 17.10) (Karunasena, Senadeera, Brown *et al.*, 2014a). This repeats in a cyclic manner until the system achieves steady state. In the developed numerical model, unique kind of boundary conditions exist where a set of distributed wall particles repulse the fluid through Lennard-Jones (LJ) forces. In order to further intensify the repulsion force field at the wall-fluid interface, a set of massless virtual particles have been introduced to the system (Van Liedekerke, Ghysels *et al.*, 2010; Karunasena, Senadeera, Brown *et al.*, 2014a).

Considering the computational efficiency of the entire numerical scheme, an approach based on the moisture difference has been used instead of a time domain–based approach (Karunasena, Senadeera, Brown *et al.*, 2014a; Van Liedekerke, Ghysels *et al.*, 2010). For the simulations, cell turgor pressure was varied from 200 kPa (fresh cell) to 180 kPa, 140 kPa, 100 kPa, 60 kPa, and 20 kPa in order to model different dryness states such as: $X/X_0 = 0.9$, $X/X_0 = 0.7$, $X/X_0 = 0.5$, $X/X_0 = 0.3$, and $X/X_0 = 0.1$, respectively. During the model evolution, cell mass changes were achieved by keeping the particle number constant while allowing individual mass values to evolve. This approach was adopted to minimize the computational cost and increase the stability of the numerical scheme (Van Liedekerke *et al.*, 2011; Karunasena, Senadeera, Gu *et al.*, 2014).

In order to quantify the morphological variations in the model predictions, a set of geometrical parameters were used such as: cell area (A), diameter* (D), perimeter (P), roundness[†] (R), elongation[‡] (EL), and compactness[§] (C). The variation of these parameters were analyzed with the

---

[*] $\sqrt{4A/\pi}$
[†] $4\pi A/P^2$
[‡] $\sqrt{4A/\pi}/(Dmajor)$
[§] D_{major}/D_{minor}

TABLE 17.5

Set of Physical Properties Used for Modeling an Apple Cell Using the Proposed SPH-CG 3D Model

Parameter	Value	Reference
Initial cell radius (r)	75 μm	(Hills and Remigereau, 1997)
Cell wall shear modulus (G)	18 MPa	(Wu and Pitts, 1999)
Initial thickness of the cell wall (t)	126 nm	(Wang, Wang, and Thomas, 2004)
Initial cell fluid mass (m_i)	1.767×10^{-9} kg	Set (Karunasena, Senadeera, Gu et al., 2014)
Initial cell wall mass (m_j)	1.767×10^{-10} kg	Set (Karunasena, Senadeera, Gu et al., 2014)
Cell wall damping ratio (γ)	5×10^{-6} Ns/m	Set (Karunasena, Senadeera, Gu et al., 2014)
Cell fluid viscosity (μ)	0.1 Pas	Set (Van Liedekerke, Ghysels et al., 2010; Van Liedekerke et al., 2011)
SPH smoothing length (h)	9.3 μm	Set (Liu and Liu, 2003b; Karunasena, Senadeera, Gu et al., 2014; Van Liedekerke, Ghysels et al., 2010)
Turgor pressure of fresh cell (P_T)	200 kPa	(Wang, Wang, and Thomas, 2004)
Osmotic potential of fresh cell ($-\pi$)	−200 kPa	Equal to - P_T (Van Liedekerke, Ghysels et al., 2010; Karunasena, Senadeera Gu et al., 2014)
Cell wall permeability (L_P)	2.5×10^{-6} m³/Ns	(Taiz and Zeiger, 2002)
Cell fluid compression modulus (K)	20 MPa	Set (Karunasena, Senadeera, Gu et al., 2014)
Number of fluid particles (n_f)	3082	Set
Number of wall particles (n_w)	2067	Set
LJ contact strength for attraction forces (f_0^a)	1×10^{-11} Nm⁻¹	Set
LJ contact strength for repulsion forces (f_0^r)	5×10^{-12} Nm⁻¹	Set

dry-basis moisture content X ($mass_{water}/mass_{dry\ solid}$). Normalized parameters (A/A_0, D/D_0, P/P_0, R/R_0, EL/EL_0, and C/C_0) were used and compared with experimental findings of apple cellular structure for the sake of a better comparison (Karunasena, Hesami et al., 2014; Mayor, Silva, and Sereno, 2005).

17.4.5 DEVELOPMENT OF THE COMPUTER SOURCE CODE OF THE NUMERICAL MODEL

The above mentioned governing equations and related physical properties were coded into a C++ computer program. The fundamental structure and flow of this source code is depicted as a schematic diagram in Figure 17.7. The source code consists of three main sections: definition and initialization of parameters and arrays; SPH-CG calculations running in an iterative manner; and processing of outputs of the model. In the development of this source code, an existing FORTRAN SPH source code (Liu and Liu, 2003b) and a C++ based SPH-DEM source code (Karunasena, Senadeera, Brown et al., 2014a) were referred. The source code was executed on a multi-core high performance computer as a parallel code via OpenMP (Open Multi-Processing) (Chandra, 2001). For visualizations of the code outputs such as particle positions, Open Visualization Tool (OVITO version 2.6.1) (Alexander, 2010) was used and geometric calculations on those outputs were done using the Image J software (Version 1.48v) (Abràmoff, Magalhães, and Ram, 2004; Schneider, Rasband, and Eliceiri, 2012).

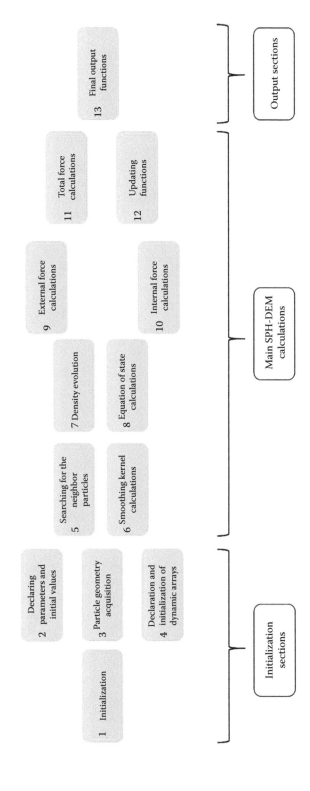

FIGURE 17.7 Schematic diagram in representation of the SPH-CG based C++ source code used for the numerical model.

17.5 MORPHOLOGICAL PREDICTIONS AND VALIDATION OF THE 3D SPH-CG SINGLE CELL MODEL

17.5.1 SIMULATION OF CELLULAR DRYING AND ASSOCIATED DEFORMATIONS

Different cell dryness levels were simulated starting from a fresh (fully turgid) state (P_T = 200 kPa and X/X_0 = 1.0) to an extremely dried state (P_T = 20 kPa and X/X_0 = 1.0). The recent grid-based models that study fruit tissue dehydration (Fanta *et al.*, 2014) can only predict down to 30% of moisture reduction from the fresh state. A remarkable achievement of the SPH-CG approach is its ability to simulate an extreme drying state of X/X_0 = 0.1, in which 90% of the initial moisture content is removed. This emphasizes the ability of the proposed meshfree-based SPH-CG approach to effectively model multiphase phenomena and large deformations more effectively even at 3D. The model outcomes are presented as 3D images in Figure 17.8 and as 2D, sectional side views as shown in Figure 17.9 in order to make the results more clarified. Furthermore, in order to validate model predictions, results from a digital light microscopy-based experimental investigation on the apple cellular structure during drying are presented in Figure 17.10. According to Figure 17.10, there is a gradual and obvious shrinkage in the entire cellular structure with the increasing degree of dryness, as clearly predicted by the 3D SPH-CG single cell model. In terms of shrinkage and surface wrinkling phenomena, there exists a close agreement, where the 3D model has reproduced those effects, as shown in Figures 17.9 and 17.10.

17.5.2 QUANTIFICATION OF SHRINKAGE THROUGH NORMALIZED GEOMETRICAL PARAMETERS

The cell area (A), Feret diameter (D), perimeter (P), roundness (R), elongation (EL), and compactness (C) have been used in their normalized forms to record morphological behavior of the developed, meshfree, 3D single cell drying model. Two separate sources of experimental findings (Karunasena, Hesami *et al.*, 2014; Mayor, Silva, and Sereno, 2005) have been compared with the quantified model predictions to test the validity of the model as presented in Figure 17.11. Simultaneously, model predictions of Karunasena *et al.* (Karunasena, Senadeera, Brown *et al.*, 2014b) in their 2D meshfree-based (SPH-DEM) single-cell drying models have also been used here for a better comparison.

Model predictions of the normalized cell area (A/A_0) are in agreement with the experimental findings (see Figure 17.11a). Nevertheless, the degree of agreement decreases toward the extremely dried states (X/X_0 ≤ 0.3). The shrinkage closely agrees with the 2D model predictions of Karunasena *et al.*, except for minor deviations towards the very low moisture levels (X/X_0 ≤ 0.3). It follows the same trend for the normalized cell diameter (D/D_0) and perimeter (P/P_0), as given in Figure 17.11b and c.

When roundness (R/R_0) is considered, the experimental results exhibit a constant value close to 1 throughout the entire moisture domain where the model predictions of this study agree with that of the experimental findings (see Figure 17.11d). However, there are minor discrepancies around the low moisture content region (X/X_0 ≤ 0.5). Further, 2D meshfree model prediction shows a discrepancy of similar or higher degree compared to the experimental findings. Next, the model predictions for the cell elongation (EL/EL_0) also are in close agreement with the results of the experimental investigations and follows a similar trend as of 2D meshfree SPH-DEM model predictions (see Figure 17.11e). The magnitude of the parameter stays around 1 during drying. This indicates that the morphological behavior of the cell is rather symmetrical.

However, there are irregularities in morphological trends at extremely dried conditions (X/X_0 ≤ 0.2), due to the highly wrinkled nature of the cell wall after extended drying. These models are not yet capable of replicating this sharp growth of surface wrinkling behavior at very low moisture content values. The recently developed 2D SPH-DEM single cell drying models (Karunasena, Senadeera, Brown *et al.*, 2014b) show a similar deficiency. This is a key limitation of

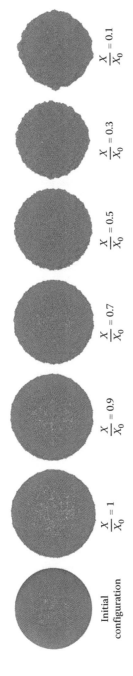

FIGURE 17.8 3D single cell-based simulations on different states of dryness. Wall particles (brown), fluid particles (yellow).

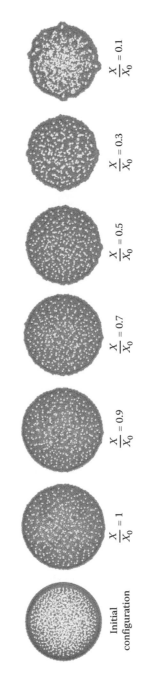

FIGURE 17.9 Section side views of 3D cells presented in Figure 17.8. Wall particles (brown), fluid particles (yellow).

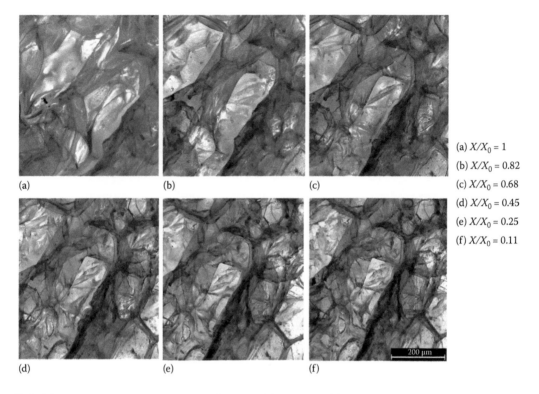

(a) $X/X_0 = 1$
(b) $X/X_0 = 0.82$
(c) $X/X_0 = 0.68$
(d) $X/X_0 = 0.45$
(e) $X/X_0 = 0.25$
(f) $X/X_0 = 0.11$

(a)   (b)   (c)

(d)   (e)   (f)

FIGURE 17.10 Digital light microscopic images of the apple cellular structure at different dryness stages: (a) $X/X_0 = 1$; (b) $X/X_0 = 0.82$; (c) $X/X_0 = 0.68$; (d) $X/X_0 = 0.45$; (e) $X/X_0 = 0.25$; (f) $X/X_0 = 0.11$.

the current single cell models arising due to absence of interactions among cells. This emphasizes the importance of developing tissue (or multiple cell) models accounting for cell-cell interactions.

The model predictions of this study conform well to both the experimental studies, as well as the predictions of Karunasena *et al.*,'s 2D meshfree model when it comes to normalized compactness (C/C_0), as given in Figure 17.11f. Trends can be described similar to that of normalized roundness and elongation discussed above. The absence of intercellular interactions affects the derived geometric parameters such as R/R_0, EL/EL_0, and C/C_0 more severely, than the basic parameters such as A/A_0, D/D_0, and P/P_0.

17.6 CLOSING REMARKS

Food drying is one of the key techniques of preserving fruits and vegetables. Considering the wide application of food drying and the amount of investment on operation, food engineers and technologists continuously search for methods to optimize the dried food quality, processing efficiency, and market value. In this regard, numerical modeling can save time, money, and other resources spent particularly on experimenting with real-life food samples. Grid-based methods and meshfree (or grid-free) methods are the two most viable techniques applicable for numerical modeling of food structures during drying. However, the conventional grid-based methods are not well suited for this purpose due to their deficiencies in modeling large deformations, multiphase phenomena, and complex physics, which are the key characteristics to be addressed when modeling plant cells or tissues during drying.

Recently, researchers have investigated the applicability of meshfree methods in simulating drying phenomena in cellular scale leading to successful results. This chapter initially discussed the information about different numerical modeling techniques and a recent application of meshfree methods for modeling plant cellular structures. Next, a detailed discussion on the development of a

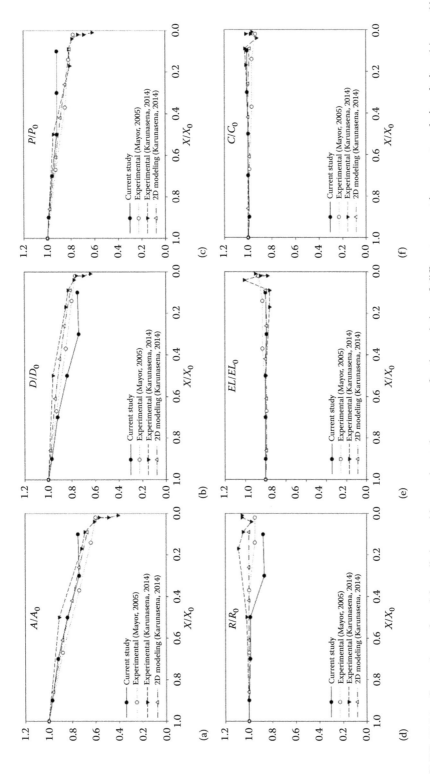

FIGURE 17.11 Comparison of model predictions with experimental outcomes for the variation of different geometric parameters during drying: (a) area (A/A_0); (b) diameter (D/D_0); (c) perimeter (P/P_0); (d) roundness (R/R_0); (e) elongation (EL/EL_0); and (f) compactness (C/C_0).

novel 3D single plant cell drying model using a SPH-CG numerical approach was presented. The 3D model predictions have been validated with the outcomes of experimental findings on apple cells qualitatively and quantitatively. The comparisons show that the developed 3D single cell model has a fairly good ability to approximate the behavior of real plant cells during drying. Based on this single cell model, it is possible to develop more sophisticated tissue models, as the adopted numerical framework supports expansion.

To develop a pragmatic approach, it demands a deeper analysis and research to extend 3D single cell modeling scheme toward higher and complicated levels of structure. Firstly, the multiple cell/ tissue level could be considered and secondly, the bulk material level. There are some qualities derived due to the hierarchical structure of the food matter; thus, it would be difficult to directly extrapolate the energy consumption (or efficiency) of the drying operations of the bulk food matter using cellular or tissue scale models. In order to develop such bulk scale models, multiscale modeling approaches will have to be employed. In doing so, CG modeling methods are a viable candidate.

ACKNOWLEDGMENTS

The authors acknowledge funding from the ARC Discovery Project (DP150100828) and ARC Linkage Project (LP150100737) from the Australian Research Council. The authors of this study kindly acknowledge the support given by high performance computing facilities provided by Queensland University of Technology (QUT), Brisbane, Queensland, Australia; the financial support provided by the Chemistry, Physics, and Mechanical Engineering (CPME) scholarship provided by the Science and Engineering Faculty (SEF), QUT; and the first author specifically appreciates the sincere support provided by the University of Moratuwa, Sri Lanka.

REFERENCES

Abera, Metadel, Solomon, K., Fanta, Workneh, Verboven, Pieter, Ho, Quang T., Carmeliet, Jan, and Nicolai, Bart M. (2013). Virtual fruit tissue generation based on cell growth modeling. *Food and Bioprocess Technology* 6 (4), 859–69. doi: 10.1007/s11947-011-0775-4.

Abràmoff, Michael D., Magalhães, Paulo J., and Ram, Sunanda J. (2004). Image processing with image. *J. Biophotonics International* 11 (7), 36–42.

Ala, Guido, Francomano, E., Tortorici, Adele, Toscano, E., and Viola, Fabio. (2007). Corrective meshless particle formulations for time domain maxwell's equations. *Journal of Computational and Applied Mathematics* 210 (1), 34–46.

Alexander, Stukowski. (2010). Visualization and analysis of atomistic simulation data with OVITO—the open visualization tool. *Modeling and Simulation in Materials Science and Engineering* 18 (1), 015012.

Atluri, S. N. and T. Zhu. (1998). A new, Meshless Local Petrov-Galerkin (MLPG) approach in computational mechanics. *Computational Mechanics* 22 (2), 117–27. doi: 10.1007/s004660050346.

Bartlett, Megan K., Scoffoni, Christine, and Sack, Lawren. (2012). The determinants of leaf turgor loss point and prediction of drought tolerance of species and biomes: A global meta-analysis. *Ecology Letters* 15 (5), 393–405. doi: 10.1111/j.1461-0248.2012.01751.x.

Belytschko, T., Krongauz, Y., Organ, D., Fleming, M., and Krysl, P. (1996). Meshless methods: An overview and recent developments. *Computer Methods in Applied Mechanics and Engineering* 139 (1–4), 3–47. doi: http://dx.doi.org/10.1016/S0045-7825(96)01078-X.

Belytschko, T., Lu, Y. Y., and Gu, L. (1994). Element-free Galerkin methods. *International Journal for Numerical Methods in Engineering* 37 (2), 229–56. doi: 10.1002/nme.1620370205.

Benz, W. and Asphaug, E. (1995). Simulations of brittle solids using smooth-particle hydrodynamics. *Computer Physics Communications* 87 (1–2), 253–65. doi: http://dx.doi.org/10.1016/0010-4655(94)00176-3.

Bhojwani, S., and The University of Texas at El Paso. (2007). Smoothed-particle hydrodynamics modeling of the friction stir welding process. *Mechanical Engin.* University of Texas at El Paso.

Bonet, J. and Kulasegaram, S. (2000). Correction and stabilization of smooth particle hydrodynamics methods with applications in metal forming simulations. *International Journal for Numerical Methods in Engineering* 47 (6), 1189–214.

Chandra, Rohit. (2001). Parallel programming in OpenMP: Morgan Kaufmann.

Chen, J. K., Beraun, J. E., and Carney, T. C. (1999). A corrective smoothed particle method for boundary value problems in heat conduction. *International Journal for Numerical Methods in Engineering* 46 (2), 231–52.

Chen, Xiao Dong and Mujumdar, Arun S. (2009). Drying technologies in food processing: John Wiley & Sons.

Cleary, Paul W. and Monaghan, Joseph J. (1999). Conduction modeling using smoothed particle hydrodynamics. *Journal of Computational Physics* 148 (1), 227–64. doi: http://dx.doi.org/10.1006/jcph.1998.6118.

Colagrossi, Andrea, Bouscasse, B., Antuono, M., and Marrone, S. (2012). Particle packing algorithm for SPH schemes. *Computer Physics Communications* 183 (8), 1641–53.

Colagrossi, Andrea and Landrini, Maurizio. (2003). Numerical simulation of interfacial flows by smoothed particle hydrodynamics. *Journal of Computational Physics* 191 (2), 448–75.

Guillermo, H. Crapiste, Stephen, Whitaker, and Enrique, Rotstein (1988). Drying of cellular material—I. A mass transfer theory. *Chemical Engineering Science* 43 (11), 2919–28. doi: http://dx.doi.org/10.1016 /0009-2509(88)80045-9.

De la Fuente-Blanco, S., Riera-Franco De Sarabia, E., Acosta-Aparicio, V. M., Blanco-Blanco, A., and Gallego-Juárez, J. A. (2006). Food drying process by power ultrasound. *Ultrasonics* 44:e523-e527.

Delong,, D. (2006). *How to Dry Foods*. Penguin Publishing Group.

Fan, Bingxin and Maranas, Janna K. (2015). Coarse-grained simulation of cellulose Iβ with application to long fibrils. *Cellulose* 22 (1), 31–44.

Fanta, Solomon Workneh, Abera, Metadel K., Aregawi, Wondwosen A., Ho, Quang Tri, Verboven, Pieter, Carmeliet, Jan, and Nicolai, Bart M. (2014). Microscale modeling of coupled water transport and mechanical deformation of fruit tissue during dehydration. *Journal of Food Engineering* 124, 86–96.

Fleischer, Kurt and Barr, Alan H. (1994). A simulation testbed for the study of multicellular development: The multiple mechanisms of morphogenesis. *Santa Fe Institute Studies in the Sciences of Complexity*.

Frank, Xavier and Perré, Patrick. (2010). The potential of meshless methods to address physical and mechanical phenomena involved during drying at the pore level. *Drying Technology* 28 (8), 932–43. doi: 10.1080 /07373937.2010.497077.

Gao, Q. and Pitt, R. E. (1991). Mechanics of parenchyma tissue based on cell orientation and microstructure. *Transactions of the ASAE*, 34.

Gingold, Robert A., and Monaghan, Joseph J. (1977). Smoothed particle hydrodynamics—Theory and application to non-spherical stars. *Monthly Notices of the Royal Astronomical Society* 181, 375–89.

Gutfraind, Ricardo and Savage, Stuart B. (1998). Flow of fractured ice through wedge-shaped channels: Smoothed particle hydrodynamics and discrete-element simulations. *Mechanics of Materials* 29 (1), 1–17.

Hills, Brian P. and Remigereau, Benoit. (1997). NMR studies of changes in subcellular water compartmentation in parenchyma apple tissue during drying and freezing. *International Journal of Food Science & Technology* 32 (1), 51–61.

Honda, H., Yamanaka, H., and Dan-Sohkawa, M. (1984). A computer simulation of geometrical configurations during cell division. *Journal of Theoretical Biology* 106 (3), 423–35. doi: http://dx.doi.org/10.1016/0022 -5193(84)90039-0.

Honda, Hisao, Tanemura, Masaharu, and Nagai, Tatsuzo. (2004). A three-dimensional vertex dynamics cell model of space-filling polyhedra simulating cell behavior in a cell aggregate. *Journal of Theoretical Biology* 226 (4), 439–53. doi: http://dx.doi.org/10.1016/j.jtbi.2003.10.001.

Idelsohn, Sergio R. and Onate, Eugenio. (2006). To mesh or not to mesh. That is the question… *Computer Methods in Applied Mechanics and Engineering* 195 (37), 4681–96.

Ishimoto, Yukitaka and Morishita, Yoshihiro. (2014). Bubbly vertex dynamics: A dynamical and geometrical model for epithelial tissues with curved cell shapes. *Physical Review E* 90 (5), 052711.

Jangam, Sachin V. (2011). An overview of recent developments and some R&D challenges related to drying of foods. *Drying Technology* 29 (12), 1343–57.

Jarvis, M. C. (1998). Intercellular separation forces generated by intracellular pressure. *Plant, Cell, & Environment* 21(12), 1307–10. doi: 10.1046/j.1365-3040.1998.00363.x.

Karunasena, H. C. P., Brown, R. J., Gu, Y. T., and Senadeera, W. (2015). Application of meshfree methods to numerically simulate microscale deformations of different plant food materials during drying. *Journal of Food Engineering* 146 (0), 209–26. doi: http://dx.doi.org/10.1016/j.jfoodeng.2014.09.011.

Karunasena, H. C. P., Gu, Y. T., Brown, R. J., and Senadeera, W. (2015a). Numerical investigation of plant tissue porosity and its influence on cellular level shrinkage during drying. *Biosystems Engineering* 132 (0), 71–87. doi: http://dx.doi.org/10.1016/j.biosystemseng.2015.02.002.

Karunasena, H. C. P., Hesami, P., Senadeera, W. Gu, Y. T., Brown, R. J., and Oloyede, A. (2014). Scanning electron microscopic study of microstructure of gala apples during hot air drying. *Drying Technology* 32 (4), 455–68. doi: 10.1080/07373937.2013.837479.

Karunasena, H. C. P., Senadeera, W., Brown, R. J., and Gu, Y. T. (2014a). A particle-based model to simulate microscale morphological changes of plant tissues during drying. *Soft Matter* 10 (29), 5249–68. doi: 10 .1039/C4SM00526K.

Karunasena, H. C. P., Senadeera, W., Brown, R. J., and Gu, Y. T. (2014b). Simulation of plant cell shrinkage during drying—A SPH–DEM approach. *Engineering Analysis with Boundary Elements* 44 (0), 1–18. doi: http://dx.doi.org/10.1016/j.enganabound.2014.04.004.

Karunasena, H. C. P., Senadeera, W., Gu, Y. T., and Brown, R. J. (2014). A coupled SPH-DEM model for micro-scale structural deformations of plant cells during drying. *Applied Mathematical Modelling* 38 (15–16), 3781–801. doi: http://dx.doi.org/10.1016/j.apm.2013.12.004.

Karunasena, H. C. P., Gu, Y. T., Brown, R. J., and Senadeera, W. (2015b). Numerical investigation of case hardening of plant tissue during drying and its influence on the cellular-level shrinkage. *Drying Technology* 33 (6), 713–34.

Karunasena, H. C. P., Senadeera, Wijitha, Brown, Richard J., and Gu, YuanTong. (2014c). A meshfree model for plant tissue deformations during drying. *ANZIAM Journal* 55, C110–C137.

Lardner, T. J. and Pujara, P. (1980). Compression of spherical cells. *Mechanics Today* 5, 161–76.

Lewicki, Piotr P., and Pawlak, Grzegorz. (2003). Effect of drying on microstructure of plant tissue. *Drying Technology* 21 (4), 657–83. doi: 10.1081/DRT-120019057.

Li, Shaofan, and Liu, Wing Kam. (2002). Meshfree and particle methods and their applications. *Applied Mechanics Reviews* 55 (1), 1–34.

Liu, G. R. and Gu, Y. T. (2001). A local radial point interpolation method (lrpim) for free vibration analyses of 2D solids. *Journal of Sound and Vibration* 246 (1), 29–46. doi: http://dx.doi.org/10.1006/jsvi.2000.3626.

Liu, G. R. and Gu, Y. T. (2002). Comparisons of two meshfree local point interpolation methods for structural analyses. *Computational Mechanics* 29 (2), 107–21. doi: 10.1007/s00466-002-0320-4.

Liu, G. R. and Gu, Y. T. (2004). Boundary meshfree methods based on the boundary point interpolation methods. *Engineering Analysis with Boundary Elements* 28 (5), 475–87. doi: http://dx.doi.org/10.1016 /S0955-7997(03)00101-2.

Liu, G. R. and Liu, M. B. (2003a). *Smoothed Particle Hydrodynamics: A Meshfree Particle Method.* Singapore: World Scientific Publishing Co.

Liu, Gui-Rong. (2010). *Meshfree Methods: Moving Beyond the Finite Element Method.* CRC Press.

Liu, Gui-Rong and Gu, Yuan-Tong. (2005). *An Introduction to Meshfree Methods and Their Programming.* Springer Science & Business Media.

Liu, Gui-Rong and Liu, M. B. (2003b). *Smoothed Particle Hydrodynamics: A Meshfree Particle Method.* World Scientific.

Liu, M. B. and Liu, G. R. (2005). Meshfree particle simulation of micro-channel flows with surface tension. *Computational Mechanics* 35 (5), 332–41.

Liu, M. B. and Liu, G. R. (2010). Smoothed particle hydrodynamics (SPH): An overview and recent developments. *Archives of Computational Methods in Engineering* 17 (1), 25–76.

Liu, Zishun, Hong, Wei, Suo, Zhigang, Swaddiwudhipong, Somsak, and Zhang, Yongwei. (2010). Modeling and simulation of buckling of polymeric membrane thin film gel. *Computational Materials Science* 49 (1), S60–S64.

Loodts, Jimmy, Tijskens, Engelbert, Wei, Chunfang, Vanstreels, E. L. S., Nicolaï, Bart, and Ramon, Herman. (2006). Micromechanics: Simulating the elastic behavior of onion epidermis tissue. *Journal of Texture Studies* 37 (1), 16–34. doi: 10.1111/j.1745-4603.2006.00036.x.

Lu, Keran, Maranas, Janna K., and Milner, Scott T. (2016). Ion-mediated charge transport in ionomeric electrolytes. *Soft Matter* 12 (17), 3943–54.

Lucy, Leon B. (1977). A numerical approach to the testing of the fission hypothesis. *Astronomical Journal* 82, 1013–24.

Mayor, L., Silva, M. A., and Sereno, A. M. (2005). Microstructural changes during drying of apple slices. *Drying Technology* 23 (9–11), 2261–76.

Mishra, B. K. and Rajamani, Raj K. (1992). The discrete element method for the simulation of ball mills. *Applied Mathematical Modeling* 16 (11), 598–604. doi: http://dx.doi.org/10.1016/0307-904X(92) 90035-2.

Monaghan, J. J. (1994). Simulating free surface flows with SPH. *Journal of Computational Physics* 110 (2), 399–406. doi: http://dx.doi.org/10.1006/jcph.1994.1034.

Monaghan, J. J. (2005). Smoothed particle hydrodynamics. *Reports on Progress in Physics* 68 (8), 1703.

Morris, Joseph P., Fox, Patrick J., and Zhu, Yi. (1997). Modeling low Reynolds number incompressible flows using SPH. *Journal of Computational Physics* 136 (1), 214–26. doi: http://dx.doi.org/10.1006/jcph.1997.5776.

Nguyen, Vinh Phu, Rabczuk, Timon, Bordas, Stéphane, and Duflot, Marc. (2008). Meshless methods: A review and computer implementation aspects. *Mathematics and Computers in Simulation* 79 (3), 763–813.

Nilsson, S. Bertil, Hellmuth Hertz, C., and Falk, Stig. (1958). On the relation between Turgor Pressure and Tissue Rigidity, II. *Physiologia Plantarum* 11 (4), 818–37. doi: 10.1111/j.1399-3054.1958.tb08275.x.

Pathmanathan, P., Cooper, J., Fletcher, A., Mirams, G., Murray, P., Osborne, J., Pitt-Francis, J., Walter, A., and Chapman, S. J. (2009). A computational study of discrete mechanical tissue models. *Physical Biology* 6 (3), 036001.

Perré, Patrick. (2011). A review of modern computational and experimental tools relevant to the field of drying. *Drying Technology* 29 (13), 1529–41.

Petridis, Loukas, O'Neill, Hugh M., Johnsen, Mariah, Fan, Bingxin, Schulz, Roland, Mamontov, Eugene, Maranas, Janna, Langan, Paul, and Smith, Jeremy C. (2014). Hydration control of the mechanical and dynamical properties of cellulose. *Biomacromolecules* 15 (11), 4152–9. doi: 10.1021/bm5011849.

Pitt, R. E. (1982). Models for the rheology and statistical strength of uniformly stressed vegetative tissue. *Transactions of the ASAE* 25 (6), 1776–84.

Polwaththe-Gallage, Hasitha-Nayanajith, Saha, Suvash C., Sauret, Emilie, Flower, Robert, and Gu, Yuantong. (2015). A coupled SPH-DEM approach to model the interactions between multiple red blood cells in motion in capillaries. *International Journal of Mechanics and Materials in Design*, 1–18. doi: 10.1007/s10999-015-9328-8.

Ramos, Inês N., Cristina, Silva, L. M., Sereno, Alberto M., and Aguilera, José M. (2004). Quantification of microstructural changes during first stage air drying of grape tissue. *Journal of Food Engineering* 62 (2), 159–64. doi: http://dx.doi.org/10.1016/S0260-8774(03)00227-9.

Randles, P. W. and Libersky, L. D. (1996). Smoothed particle hydrodynamics: Some recent improvements and applications. *Computer Methods in Applied Mechanics and Engineering* 139 (1), 375–408.

Rathnayaka, C. M., Karunasena, H. C. P., Gu, Y. T., Guan, L., and Senadeera, W. (2016). A 3-D meshfree numerical model to analyze cellular scale shrinkage of different categories of fruits and vegetables during drying. *Proceedings of the 7th International Conference on Computational Methods*.

Rathnayaka, C. M., Karunasena, H. C. P., and Gu, Y. T. (2017a). Application of 3D imaging and analysis techniques for the study of food plant cellular deformations during drying. *Drying Technology*, 1–14. doi: https://doi.org/10.1080/07373937.2017.1341417.

Rathnayaka, C. M., Karunasena, H. C. P., Gu, Y. T., Guan, L., and Senadeera, W. (2017b). Novel trends in numerical modeling of plant food tissues and their morphological changes during drying—A review. *Journal of Food Engineering* 194, 24–39. doi: http://dx.doi.org/10.1016/j.jfoodeng.2016.09.002.

Ratti, Cristina. (2001). Hot air and freeze-drying of high-value foods: A review. *Journal of Food Engineering* 49 (4), 311–9.

Rosswog, Stephan, and Wagner, Peter. (2002). Towards a macroscopic modeling of the complexity in traffic flow. *Physical Review E* 65 (3), 036106.

Rudge, Tim and Haseloff, Jim. (2005). A computational model of cellular morphogenesis in plants. *Advances in Artificial Life*, 78–87. Springer.

Schneider, Caroline A., Rasband, Wayne S., and Eliceiri, Kevin W. (2012). NIH image to image J: 25 years of image analysis. *Nat Methods* 9 (7), 671–5.

Smith, A. E., Moxham, K. E., and A. P. J. (1998). On uniquely determining cell-wall material properties with the compression experiment. *Chemical Engineering Science* 53 (23), 3913–22. doi: http://dx.doi.org/10.1016/S0009-2509(98)00198-5.

Taiz, Lincoln and Zeiger, Eduardo. (2002). *Plant Physiology*. New York: Sinauer.

Tartakovsky, Alexandre M. and Meakin, Paul. (2005). Simulation of unsaturated flow in complex fractures using smoothed particle hydrodynamics. *Vadose Zone Journal* 4 (3), 848–55.

Tozzini, Valentina. (2005). Coarse-grained models for proteins. *Current Opinion in Structural Biology* 15 (2), 144–50. doi: http://dx.doi.org/10.1016/j.sbi.2005.02.005.

Van Liedekerke, Paul, Ghysels, Pieter, Tijskens, Engelbert, Samaey, Giovanni, Roose, Dirk, and Ramon, Herman. (2011). Mechanisms of soft cellular tissue bruising. A particle-based simulation approach. *Soft Matter* 7 (7), 3580–91.

Van Liedekerke, Paul, Ghysels, Pieter, Tijskens, Engelbert, Samaey, Giovanni, Smeedts, B., Roose, D., and Ramon, H. (2010). A particle-based model to simulate the micromechanics of single-plant parenchyma cells and aggregates. *Physical Biology* 7 (2), 026006.

Van Liedekerke, Paul, Tijskens, E., Ramon, H., Ghysels, P., Samaey, G., and Roose, D. (2010). Particle-based model to simulate the micromechanics of biological cells. *Physical Review E* 81 (6), 061906.

Veytsman, Boris A. and Cosgrove, Daniel J. (1998). A model of cell wall expansion based on thermodynamics of polymer networks. *Biophysical Journal* 75 (5), 2240–50.

Wang, C. X., Wang, L., and Thomas, C. R. (2004). Modeling the mechanical properties of single suspension-cultured tomato cells. *Annals of Botany* 93 (4), 443–53.

Wu, Hsin- I., Spence, Richard D., Sharpe, Peter J. H., and Goeschl, John D. (1985). Cell wall elasticity: I. A critique of the bulk elastic modulus approach and an analysis using polymer elastic principles. *Plant, Cell, & Environment* 8 (8), 563–70. doi: 10.1111/j.1365-3040.1985.tb01694.x.

Wu, Naiqiang and Pitts, Marvin J. (1999). Development and validation of a finite element model of an apple fruit cell. *Postharvest Biology and Technology* 16 (1), 1–8. doi: http://dx.doi.org/10.1016/S0925-5214 (98)00095-7.

Zhang, S. Z. (1976). *Detonation and its Applications*. Press of National Defense Industry, Beijing.

Zhu, H. X. and Melrose, J. R. (2003). A mechanics model for the compression of plant and vegetative tissues. *Journal of Theoretical Biology* 221 (1), 89–101. doi: http://dx.doi.org/10.1006/jtbi.2003.3173.

Zukas, Jonas A. (1990). *High Velocity Impact Dynamics*. Wiley-Interscience.

Index

Page numbers followed by f and t indicate figures and tables, respectively.